Making Things Talk

Third Edition

Tom Igoe

Making Things Talk

Third Edition

Tom Igoe

Making Things Talk

by Tom Igoe

Published by Maker Media, Inc.
1700 Montgomery Street, Suite 240, San Francisco, CA 94111

Maker Media books may be purchased for educational, business, or sales promotional use.

You can send comments and questions to us by email at books@makermedia.com..

Print History
September 2007
First Edition

September 2011
Second Edition

August 2017
Third Edition

Editor: Patrick Di Justo
Proofreader: Elizabeth Welch
Cover Designer: Tom Igoe
Production Editor: Maureen Forys, Happenstance Type-O-Rama
Book Production: Jeff Lytle, Happenstance Type-O-Rama
Indexer: Valerie Perry, Happenstance Type-O-Rama
Cover Photograph: Tom Igoe

ISBN: 978-1-68045-215-0
[TI]

2017-07-21 First Release

to Frank Igoe

who showed me the potential of computers

and to Red Burns

who showed me their potential to benefit humans

Contents

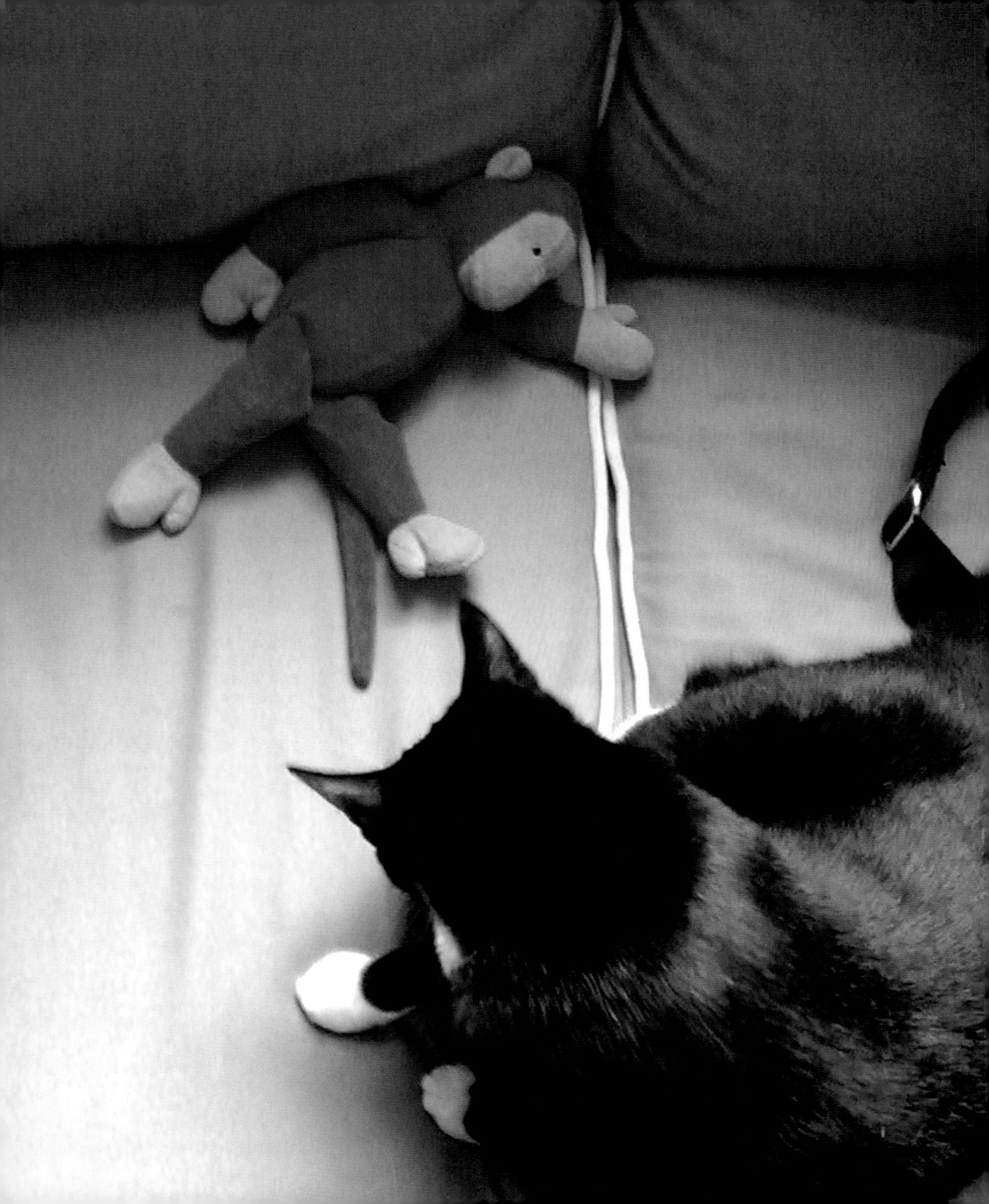

Making Things Talk

MAKE: PROJECTS

Preface

A few years ago, Neil Gershenfeld wrote a smart book called *When Things Start to Think*. In it, he discussed a world in which everyday objects and devices are endowed with computational power: in other words, today. He talked about the implications of devices that exchange information about our identities, abilities, and actions. It's a good read, but I think he got the title wrong. I would have called it *When Things Start to Gossip,* because—let's face it—even the most exciting thoughts are worthwhile only once you start to talk to someone else about them. *Making Things Talk* teaches you how to make things that have computational power talk to each other, and about giving people the ability to use those things to communicate.

For a couple of decades now, computer scientists have used the term *object-oriented programming* to refer to a style of software development in which programs and subprograms are thought of as objects. Like physical objects, they have properties and behaviors. They inherit these properties from the *prototypes* from which they descend. The canonical form of any object in software is the code that describes its type. Software objects make it easy to recombine objects in novel ways. You can reuse a software object if you know its *interface*—the collection of properties and methods to which its creator allows you access (as well as the documents so that you know how to use them). It doesn't matter how a software object does what it does, as long as it does it consistently. Software objects are most effective when they're easy to understand and when they work well with other objects.

In the physical world, we're surrounded by all kinds of electronic objects: clock radios, toasters, mobile phones, music players, children's toys, and more. It can take a lot of work and a significant amount of knowledge to make a useful electronic gadget—it can take almost as much knowledge to make those gadgets talk to each other in useful ways. But that doesn't have to be the case. Electronic devices can be—and often are—built up from simple modules. As long as you understand the interfaces, you can make anything from them. Think of it as *object-oriented hardware*. Understanding the ways in which things talk to each other is central to making this work, regardless of whether the object is a toaster, an email program on your laptop, or a networked database. All of these objects can be connected if you can figure out how they communicate. This book is a guide to some of the tools for making those connections.

X

Who This Book Is For

This book is written for people who want to make things talk to other things. Maybe you're a science teacher who wants to show your students how to monitor weather conditions at several locations around your school district simultaneously, or a sculptor who wants to make a whole room of choreographed mechanical sculptures. You might be an industrial designer who needs to be able to build quick mockups of new products, modeling both their forms and their functions. Maybe you're a cat owner, and you'd like to be able to play with your cat while you're away from home. This book is a primer for people with little technical training and a lot of interest. This book is for people who want to get projects done.

The main tools in this book are personal computers, web servers, and microcontrollers, the tiny computers inside everyday appliances. Over the past decade, microcontrollers and their programming tools have gone from being arcane items to common, easy-to-use tools. Elementary school students are using the tools that baffled graduate students only a decade ago. During that time, my colleagues and I have taught people from diverse backgrounds (few of them computer programmers) how to use these tools to increase the range of physical actions that computers can respond to, sense, and interpret.

In recent years, there's been a rising interest among people using microcontrollers to make their devices not only sense and control the physical world, but also talk to other things about what they're sensing and controlling. If you've built something with a Basic Stamp or a Lego Mindstorms kit, and want to make that thing communicate with things you or others have built, this book is for you. It is also useful for software programmers familiar with networking and web services who want an introduction to embedded network programming.

If you're the type of person who likes to get down to the very core of a technology, you may not find what you're looking for in this book. There aren't detailed code samples for Bluetooth or TCP/IP stacks, nor are there circuit diagrams for Ethernet controller chips. The components used here strike a balance between simplicity, flexibility, and cost. They use object-oriented hardware, requiring relatively little wiring or code. They're designed to get you to the end goal of making things talk to each other as quickly as possible.

X

What You Need to Know

In order to get the most from this book, you should have a basic knowledge of electronics and programming microcontrollers, some familiarity with the internet, and access to both.

Many people whose programming experience begins with microcontrollers can do wonderful things with some sensors and a couple of servomotors, but they may not have done much to enable communication between the microcontroller and other programs on a personal computer. Similarly, many experienced network and multimedia programmers have never experimented with hardware. If you're either of these people, this book is for you. Because the audience of this book is diverse, you may find some of the introductory material a bit simple, depending on your background. If so, feel free to skip past the stuff you know to get to the meatier parts.

If you've never used a microcontroller, you'll need a little background before starting this book. I recommend you read my previous book, **Physical Computing: Sensing and Controlling the Physical World with Computers** (Thomson), co-authored with Dan O'Sullivan, which introduces the fundamentals of electronics, microcontrollers, and physical interaction design. Some of the hardware in that book is dated, but the concepts still stand. Massimo Banzi's **Getting Started With Arduino** (Make) is an indispensable companion as well.

You should also have a basic understanding of computer programming before reading much further. If you've never done any programming, check out the Processing programming environment at www.processing.org. Processing is designed to teach nonprogrammers how to program, yet it's powerful enough to do a number of advanced tasks.

This book includes code examples in a few different programming languages. They're all fairly simple examples, so if you don't want to work in the languages provided, you can use the comments in these examples to rewrite them in your favorite language.

X

Contents of This Book

This book explains the concepts that underlie networked objects and then provides recipes to illustrate each set of concepts. Each chapter contains instructions for building working projects that make use of the new ideas introduced in that chapter.

In Chapter 1, you'll encounter the major programming tools in the book and get to "Hello World!" on each of them.

Chapter 2 introduces the most basic concepts needed to make things talk to each other. It covers the characteristics that need to be agreed upon in advance, and how keeping those things separate in your mind helps troubleshooting. You'll build a simple project that features one-to-one serial communication between a microcontroller and a personal computer using Bluetooth radios as an example of modem communication. You'll learn about data protocols, modem devices, and address schemes.

Chapter 3 introduces a more complex network: the internet. It discusses the basic devices that hold it together, as well as the basic relationships among those devices. You'll see the messages that underlie some of the most common tasks you do on the internet every day, and learn how to send those messages. You'll write your first set of programs to send data across the internet based on a physical activity in your home.

In Chapter 4, you'll build your first standalone network-connected device. You'll get more experience with command-line connections to the network, and you'll connect a microcontroller to a web server without using a desktop or laptop computer as an intermediary.

Chapter 5 takes the network connection a step further by explaining socket connections, which allow for longer interaction. You'll learn how to write your own server program that you can use to connect devices together through the network. You'll connect to this server program from a browser and from a microcontroller, so that you can understand how different types of devices can connect to each other through the same server.

Chapter 6 introduces wireless communication. You'll learn some of the characteristics of wireless, along with its possibilities and limitations. Several short examples in this chapter enable you to say "Hello World!" over the air in a number of ways using infrared and radio.

Chapter 7 offers a contrast to the socket connections of Chapter 5 by introducing message-based protocols like UDP on the internet, and 802.15.4 for radio networks. Instead of using the client-server model from earlier chapters, here you'll learn how to design conversations where each object in a network is equal to the others, exchanging information one message at a time.

Chapter 8 is about location. It introduces a few tools to help you locate things in physical space, and it offers some thoughts on the relationship between physical location and network relationships.

Chapter 9 deals with identification in physical space and network space. You'll learn a few techniques for generating unique network identities based on physical characteristics. You'll also learn a bit about how to determine a networked device's characteristics.

Chapter 10 introduces mobile telephony networks, covering many of the things that you can now do with phones and phone networks.

Throughout, you'll learn troubleshooting techniques, security techniques, and details of the structure of the internet and other networks.

X

On Buying Parts

You'll need a lot of parts for all of the projects in this book. As a result, you'll learn about a lot of vendors. Because there are no large electronics parts retailers in my city, I buy parts online all the time. If you're lucky enough to live in an area where you can buy from a brick-and-mortar store, good for you! If not, get to know some of these online vendors.

Jameco (jameco.com), Digi-Key (www.digikey.com), and Farnell (www.farnell.com) are general electronics parts retailers, and they sell many of the same things. Others, like SparkFun (www.sparkfun.com), and Adafruit (www.adafruit.com) and Seeed Studio (www.seeedstudio.com) carry specialty components, kits, and bundles that make it easy to do popular projects. A full list of suppliers is included in the Appendix. Feel free to substitute parts for things with which you are familiar.

Because it's easy to order goods online, you might be tempted to communicate with vendors entirely through their websites. Don't be afraid to pick up the phone as well. Particularly when you're new to this type of project, it helps to talk to someone about what you're ordering and to ask questions. You're likely to find helpful people at the end of the phone line for most of the retailers listed here.

X

Using Code Examples

This book is here to help you get your job done. In general, you may use the code in this book in your programs and documentation. You do not need to contact us for permission unless you're reproducing a significant portion of the code.

For example, writing a program that uses several chunks of code from this book does not require permission. Selling or distributing a CD-ROM of examples from Maker Media books does require permission. Answering a question by citing this book and quoting example code does not require permission. Incorporating a significant amount of example code from this book into your product's documentation does require permission.

We appreciate attribution. An attribution usually includes the title, author, publisher, and ISBN. For example: "*Making Things Talk: Practical Methods for Connecting Physical Objects*, by Tom Igoe. Copyright 2011 Maker Media, 978-1-4493-9243-7." If you feel that your use of code examples falls outside fair use or the permission given above, feel free to contact us at permissions@oreilly.com.

X

Using Circuit Examples

In building the projects in this book, you're going to break things and void warranties. If you're averse to this, put this book down and walk away. This is not a book for those who are squeamish about taking things apart without knowing whether they'll go back together again.

Even though we want you to be adventurous, we also want you to be safe. Please don't take any unnecessary risks when building this book's projects. Every set of instructions is written with safety in mind; ignore the safety instructions at your own peril. Be sure you have the appropriate level of knowledge and experience to get the job done in a safe manner.

Please keep in mind that the projects and circuits shown in this book are for instructional purposes only. Details like power conditioning, automatic resets, RF shielding, and other things that make an electronic product certifiably ready for market are not included here. If you're designing real products to be used by people other than yourself, please do not rely on this information alone.

X

Note on the Second Edition

Two general changes prompted the first rewriting of this book: the emergence of an open source hardware movement, and the growth of participatory culture, particularly around making interactive things. The community surrounding Arduino, and the open source hardware movement more generally, has grown quickly. The effects of this are still being realized, but one thing is clear: object-oriented hardware and physical computing are becoming an everyday reality. Many more people are making things with electronics now than I could have imagined in 2005.

Before any technology is adopted in general use, there has to be a place for it in the popular imagination. People with no knowledge of the technology must have some idea what it is and for what it can be used. Prior to 2005, I spent a lot of time explaining to people what physical computing was and what I meant by "networked objects." Nowadays, everyone knows the Wii controller or the Kinect as an example of a device that expands the range of human physical expression available to computers. These days, it's difficult to find an electronic device that isn't networked.

While it's been great to see these ideas gain a general understanding, what's even more exciting is seeing them gain in use. People aren't just using their Kinects for gaming, they're building them into assistive interfaces for physically challenged clients. They're not just playing with the Wii, they're using it as a musical instrument controller. People have become accustomed to the idea that they can modify the use of their electronics—and they're doing it.

When I joined the project, my hope for Arduino was that it might fill a need for something more customizable than consumer electronic devices were at the time, yet be less difficult to learn than microcontroller systems. I thought the open source approach was a good way to go because it meant that hopefully the ideals of the platform would spread beyond the models we made. That hope has been realized in the scores of derivative boards, shields, spinoff products, and accessories that have popped up in the last several years. It's wonderful to see so many people not just making electronics for others to build on, but doing it in a way that doesn't demand professional expertise to get started.

The growth of Arduino shields and libraries has been big enough that I almost could have written the second edition so that you wouldn't have to do any programming or circuit building. There's a shield or a library to do almost every project in this book. However, you can only learn so much by fitting premade pieces together, so I've tried to show some of the principles underlying electronic communications and physical interfaces. Where there is a simple hardware solution, I've indicated it but shown the circuit it encloses as well. The best code libraries and circuit designs practice what I think of as "glass-box enclosure"—they enclose the gory details and give you a convenient interface, but they let you look inside and see what's going on if you're interested. Furthermore, they're well-constructed so that the gory details don't seem that gory when you look closely at them. Hopefully, this edition will work in much the same way.

X

Note on the Third Edition

Two hopeful changes prompted the previous rewriting of this book: the emergence of an open source hardware movement, and the growth of participatory culture, particularly around making interactive things. The changes that prompted this rewrite are more cautionary: the spread of consumer devices that collect data about all the physical activities of our lives, and the gradual obfuscation of where that data goes. These come wrapped in products and services that offer us simplicity and convenience. We need to consider these trends and their consequences carefully. An understanding of the technologies that underlie them will help us to make informed choices about the role they play in our lives.

Despite what the title (chosen more than ten years ago) might suggest, this book is not about the "Internet of Things." The ideals behind that term run counter to those I hope that this book promotes. Kevin Ashton, director of MIT's Auto-ID Center, reportedly coined the term in 1999. In his essay "That 'Internet of Things' Thing" (www.rfidjournal.com/articles/view?4986), he noted that most of the data collected on the internet at that time had been collected and entered by people. He went on to argue that people are unreliable collectors of data, and that leads to unreliable data. He stated instead that "We need to empower computers with their own means of gathering information, so they can see, hear and smell the world for themselves, in all its random glory. RFID and sensor technology enable computers to observe, identify and understand the world—without the limitations of human-entered data."

While I agree with him that computers can serve our needs better when they have the capability to sense a wider range of physical activity, I think that a world where data is collected automatically, without human oversight and judgment, is not one in which I would like to live.

Though I don't think Ashton wants to live in a world of total surveillance any more than I do, his vision is being interpreted in such a way as to make that a reality. Consumer-connected devices gather information about our activities at home, in the gym, in the car, and everywhere else we go, in order to learn our patterns and deliver us ever more convenient services. That's great as long as we understand who the custodians of that data are, what they are collecting, and what the terms of our relationship with those custodians include. Unfortunately, that level of transparency has not yet been realized in the devices and services we're enthusiastically inviting into our lives.

My Nest knows my temperature preferences and auto-adjusts, my Hue brings up a warm glow as I approach the house, and Alexa knows just what I we want to hear on the stereo. I can relax, I'm told, because the data is "in the cloud." But there is no cloud. There routers and switches and servers, and they're owned by companies with whom we have relationships. I believe we can maintain the upper hand in those relationships when we have a better understanding of the operation of these devices and the networks through which they communicate.

I want you to know how these devices convert your actions into data, how they transmit that data to servers, and where they send those readings. For that reason I haven't used many of the cloud-based data services for connected devices in this book. The internet and the Worldwide Web are built on a number of open and collaboratively derived standards like the Internet Protocol (IP) and the Hypertext Transport Protocol (HTTP), and there is value in knowing those standards before you start using cloud-based services that rely on them.

The internet has become a less innocent place in the years since this book was written. The events of 2016, from the Mirai botnet attacks to the network activities surrounding global political change, have made it clear that we can no longer see the internet as a place to escape from the physical world. Along with its potential, it poses similar dangers to our privacy, our individual agency, and our well-being. It is now necessary that anyone using the internet must have a basic understanding of the security tools that make it a safer place to conduct our activities. Although this is not a book on network security, I've attempted to include a few basic tips on that in the pages that follow.

What makes the internet great is not its ability to store information about us, but its ability to allow us to interact with each other in new and unexpected ways. Sensors and creative design thinking expand the range of physical expression that our networked devices might sense and interpret. Although this capability comes with the potential for abuse, I believe we can manage that danger if we understand how things work. Ashton says, "In the real world, things matter more than ideas." That's true only if we keep in mind that in the best world, people matter more than things.

Software Reference

There have been a number of large changes made to the software platforms used in this book since the first edition. Some of the tools from the first and second editions are no longer in the book, replaced by new tools. Others, like Arduino and Processing, are still here, but very different than they were originally. Whenever you encounter a software tool, check the tool's website for the latest and greatest.

The code for this book can be found online on my GitHub repository at github.com/tigoe/MakingThingsTalk2/ in the 3rd edition directory, and I'll write about any changes on the blog, www.makingthingstalk.com.

Hardware Reference

To keep the focus on communications between physical devices, I've chosen to use the Arduino 101 and MKR1000 as the reference microcontroller designs for this edition. Most of the code in this book will still work on an Arduino Uno, though some of it may strain the Uno's limits a bit. Many of the Arduino-compatible boards, particularly the newer ones based on the ARM Cortex M0 or the ESP8266, will work well for these projects too.

You'll encounter some embedded Linux projects in this book as well. The Raspberry Pi is my reference platform for these projects, though all of the embedded projects should work on a BeagleBone too, with a little tweaking.

X

Acknowledgments

This book is the product of many conversations and collaborations. It would not have been possible without the support and encouragement of my own network, which continues to grow over time.

The Interactive Telecommunications Program in the Tisch School of the Arts at New York University has been my home for more than the past decade. It is a lively and warm place to work, crowded with many talented people. This book originally grew out of a class, Networked Objects, that I taught there for several years. I hope that the ideas herein represent the spirit of the place and give you a sense of my enjoyment in working there.

Red Burns, the department's founder, supported me since I first entered this field. She indulged my many flights of fancy and brought me firmly down to earth when needed. On every project, she challenged me to make sure that I use technology not for its own sake, but always so it empowers people.

Dan O'Sullivan, my colleague and now chair of the program, introduced me to physical computing and then generously allowed me to share in teaching it and shaping its role at ITP. He is a great advisor and collaborator, and offered constant feedback as I worked. Most of the chapters started with a rambling conversation with Dan. His fingerprints are all over this book, and it's a better book for it.

My faculty and staff colleagues past and present have been offered valued support and advice over the years and the editions. Marianne Petit, Clay Shirky, Daniel Rozin, Dan Shiffman, Shawn Van Every, Benedetta Piantella, Mimi Yin, Allison Parrish, Gabriel Barcia-Colombo, Luke

Dubois, Katherine Dillon, Nancy Hechinger, Marina Zurkow, Jean-Marc Gauthier, George Agudow, Edward Gordon, Midori Yasuda, Eric Rosenthal, Megan Demarest, Nancy Lewis, Robert Ryan, John Duane, Ben Light, Marlon Evans, Tony Tseng, Brian Kim, Gloria Sed, Matthew Berger, Yen-An Chen, Ahmad Arshad, Dante DelGiacco, and Anna Gallagher tolerated all kinds of insanity in the name of physical computing and networked objects, and made things possible for the other faculty and me, as well as the students.

Many other faculty, staff, research residents, and students have left their mark on this book: Jamie Allen, Mustafa Bağdatlı, Matthew Belanger, Ithai Benjamin, David Boyhan, Caroline Brown, Christian Cerrito, Dennis Crowley, Jody Culkin, John Dimatos, Patrick Dwyer, Zach Eveland, Robert Faludi, Doria Fan, John Farrell, Xiaoyang Feng, Jeff Feddersen, Scott Fitzgerald, Thomas Gerhardt, Gabriela Gutiérrez, Meredith Hasson, Patrick Hebron, Liesje Hodgson, Todd Holoubek, Jeremiah Johnson, Craig Kapp, Peter Kerlin, Kacie Kinzer, Raffi Krikorian, Jihyun Lee, Rune Madsen, Adi Marom, Zannah Marsh, Surya Mattu, Carlyn Maw, Corey Menscher, Ariel Nevarez, David Nolen, Rory Nugent, Michael Olson, Matt Parker, Dustyn Roberts, Matt Richardson, Paul Rothman, Maria Paula Saba, John Schimmel, Michael Schneider, Greg Shakar, Yining Shi, Andy Sigler, Peiqi Su, Deqing Sun, James Tu, Tymm Twillman, Karl Ward, Antonius Wiriadjaja, Rosalie Yu, and Jingwen Zhu, to name but a few.

I have relied on the software and hardware tools of many people here. Casey Reas and Ben Fry made the software side of this book possible by creating Processing and the Processing foundation (processingfoundation.org). Lauren McCarthy's development p5.js is pivotal to this edition and to my current teaching. The originators of Arduino and Wiring made the hardware side of this book possible: Massimo Banzi, Gianluca Martino, David Cuartielles, and David Mellis on Arduino; Hernando Barragán on Wiring; and Nicholas Zambetti bridging the two. I have been lucky to work with them, and it has been a pleasure and an honor to work with the talented staff that Arduino has added over the years.

I've tried to use and cite many hardware vendors in this book, but a few have been valued advisors as well. Nathan Seidle at Spark Fun, Limor Fried and Philip Torrone at Adafruit, Windell Oskay and Lenore Edman at Evil Mad Science Labs, Eric Pan at Seeeed Studio.

The third edition has been heavily influenced by a few colleagues: Don Coleman, Sandeep Mistry, and Alasdair Allan. Their perspective on networks and devices has heavily influenced mine, and their software brilliance has made most of the new material possible.

This book looks better graphically thanks to Fritzing, an open source circuit drawing tool available at www.fritzing.org: Reto Wettach, André Knörig, and Jonathan Cohen Thanks also to Giorgio Olivero and Jody Culkin for additional drawings.

Thanks to the faculty and students I've worked with at the Royal College of Art's Interaction Design program, UCLA's Digital Media | Arts program, the Interaction Design program at the Oslo School of Architecture and Design, Interaction Design Institute Ivrea, and the Copenhagen Institute of Interaction Design. Thanks to Mark Hansen and Michael Krisch at the Brown Center for Media Innovation at Columbia University for giving me a great place to write during this edition.

Many networked object projects inspired this writing. Thanks to those whose work illustrates the chapters: Tuan Anh T. Nguyen, Joo Youn Paek, Doria Fan, Mauricio Melo, and Jason Kaufman; Tarikh Korula and Josh Rooke-Ley of Uncommon Projects; Jin-Yo Mok, Alex Beim, Andrew Schneider, Gilad Lotan and Angela Pablo; Mouna Andraos and Sonali Sridhar; Tim Dye of Sonoma Technologies; Gabriel Barcia-Colombo; Kate Hartman, Kati London, and Sai Sriskandarajah; Frank Lantz and Kevin Slavin of Area/Code; Sarah Johansson; Benedetta Piantella and Justin Downs of Groundlab; Meredith Hasson, Ariel Nevarez, and Nahana Schelling, Timo Arnall, Elnar Sneve Martinussen, Jørn Knutsen, Mustafa Bağdatlı Frances Gilbert and Jake, Pedro Oliveira and Xuedi Chen. Apologies to Anton Chekhov. Thanks to Tali Padan for the comedic inspiration. Thanks to Giana Gonzalez, Younghui Kim, Jennifer Magnolfi, Jin-Yo Mok, Matt Parker, Andrew Schneider, Gilad Lotan, Angela Pablo, James Barnett, Morgan Noel, Aaron Parsekian, Noodles, and Monski for modeling projects in the chapters.

Thanks to numerous contributors to the platforms used here: Michael Shiloh, Mikal Hart, Michael Margolis, Paul Stoffregen, Adrian McEwen, Alexander Brevig, Ryan Mulligan, Keith Casey, Bonifaz Kaufmann, Andreas Göransson, Helena Bisby, Michael Adams, and Joël Gähwiler. Rob Faludi remains my source on all things XBee- and Digi-related. Thanks to the support team at Lantronix: Garry Morris, Gary Marrs, and Jenny Eisenhauer.

Geoff Smith, Chris Heathcote, Durrell Bishop, Mike Kuniavsky, Tod Kurt, Bjoern Hartmann, Eric Paulos, Wendy Ju, Lars Erik Holmquist, and the folks at the "Sketching in Hardware" workshops helped me see this work as part of a larger community, and introduced me to a lot of new tools.

At Make, Dale Dougherty encouraged of all of my ideas, dealt patiently with my delays, and indulged me when I wanted to try new things. Brian Jepson and Patrick Di Justo have been critical to this book's success as its editors. Scott Fitzgerald has not only been a valued technical editor, but also a close colleague and sounding board across all three editions. Nancy Kotary, Katie Wilson, and Tim Lillis made it look great. Sherry Huss and the Maker Faire team made it possible to bring this book to a wider audience. Thanks to all of the MAKE team. Thanks to my agents Studio B as well.

Thanks for those who helped in last-minute production of examples, photos, projects, and more, this time and the last two. It would not be done without Denise Hand, Ben Light, Clive Thompson, Joe Hobaica, Max Whitney, and Jennifer Magnolfi.

Thanks to my family and friends who listened to me rant enthusiastically or complain bitterly as this book progressed. Much love to you all. Thanks finally to Denise for going down this road with me again, and making it more enjoyable. I hope there are many others that we travel together.

X

We'd Like to Hear from You

Please address comments and questions concerning this book to the publisher:

Maker Media, Inc.
1700 Montgomery Street, Suite 240
San Francisco, CA 94111

We have a website for this book, where we list errata, examples, and any additional information. You can access this page at: http://shop.oreilly.com/product/0636920031369.do

To comment or ask technical questions about this book, send email to: bookquestions@oreilly.com

Maker Media, Inc. is devoted entirely to the growing community of resourceful people who believe that if you can imagine it, you can make it. Maker Media encourages the Do-It-Yourself mentality by providing creative inspiration and instruction.

For more information about Maker Media, Inc., visit us online: http://makermedia.com

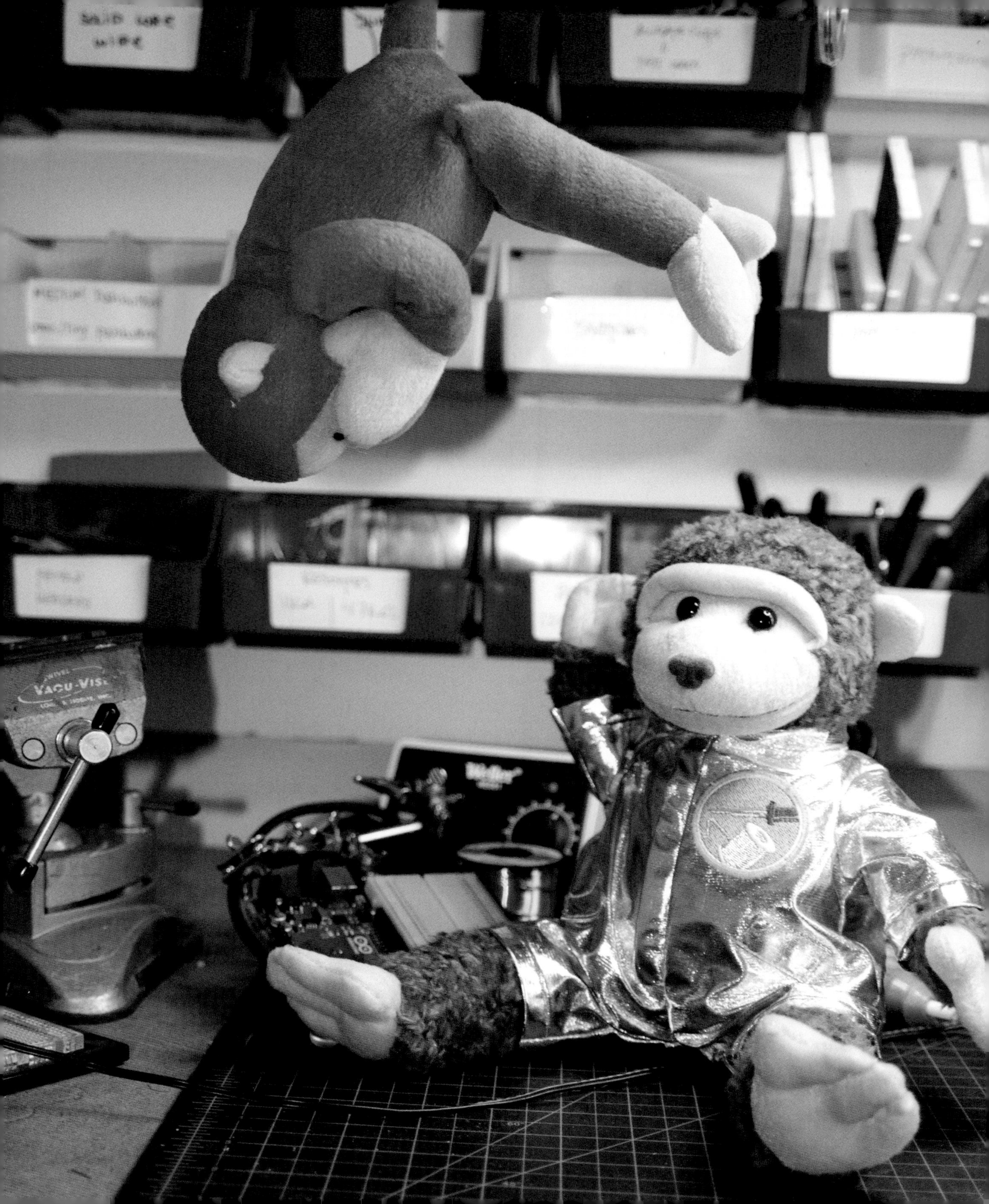

1

MAKE: PROJECTS 

The Tools

This book is a cookbook of sorts, and this chapter covers the key ingredients. The concepts and tools you'll use in the rest of the book are introduced here. There's enough information on each tool to get you to the point where you can make it say **"Hello World!"** Chances are you've used some of the tools in this chapter before—or ones just like them. Skip past the things you know and jump into learning the tools that are new to you. Take note of where to go for more information on each tool. You may want to explore some of the less-familiar tools on your own to get a sense of what they can do. The projects in the following chapters only scratch the surface of what's possible for most of these tools. The book also provides references for further investigation.

◂◂ Happy Feedback Machine by Tuan Anh T. Nguyen
The main pleasure of interacting with this piece comes from the feel of flipping the switches and turning the knobs. The lights and sounds produced as a result are secondary, and most people who play with it remember how it feels rather than its behavior.

It Starts with the Stuff You Touch

All of the objects that you'll encounter in this book—tangible or intangible—will have certain behaviors. Software objects will send and receive messages, store data, or both. Physical objects will move, light up, or make noise. The first question to ask about any object is: what does it do? The second is: how do I make it do what it's supposed to do? Or, more simply, what is its interface?

An object's interface is made up of three elements. First, there's the *physical interface*. This is the stuff you touch—knobs, switches, keys, and other sensors—that react to your actions. The connectors that join objects—the wires, sockets, and plugs— are also part of the physical interface. Every network of objects begins and ends with a physical interface. People construct mental models of how a system works based on the physical interface. A computer is much more than the keyboard, mouse, and screen, but that's what we think of it as, because that's what we see and touch. You can build all kinds of wonderful functions into your system, but if those functions aren't apparent in the things people see, hear, and touch, they will never be used. If the physical interface isn't good, the rest of the system suffers.

Second, there's the *software interface*—the commands that you send to the object to make it respond. In some projects, you'll invent your own software interface; in others, you'll rely on existing interfaces to do the work for you. The best software interfaces have simple, consistent functions that result in predictable outputs. Unfortunately, not all software interfaces are as simple as you'd like them to be, so be prepared to experiment a little to get some software objects to do what you think they should do. When you're learning a new software interface, it helps to approach it mentally in the same way you approach a physical interface. Don't try to use all the functions at once; first, learn what each function does on its own. You don't learn to play the piano by starting with a Bach fugue—you start one note at a time. Likewise, you don't learn a software interface by writing a full application with it—you learn it one function at a time. There are many projects in this book; if you find any of their software functions confusing, write a simple program that demonstrates just that function, then return to the project.

Finally, there's the *electrical interface*—the pulses of electrical energy sent from one device to another to be interpreted as information. Unless you're designing new objects or the connections between them, you never have to deal with this interface. When you are designing new objects or the networks that connect them, however, you have to understand a few things about this interface, so that you know how to match up objects that might have differences in their electrical interfaces.

X

It's About Pulses

In order to communicate with each other, objects use *communications protocols*.A protocol is a series of mutually agreed-upon standards for communication between two or more objects.

Serial protocols like USB, MIDI, and the Human Interface Device (HID) specification connect computers to printers, hard drives, keyboards, mice, and other peripheral devices. Network protocols like Ethernet and TCP/IP connect multiple computers through network hubs, routers, and switches.

A communications protocol usually defines the rate at which messages are exchanged, the arrangement of data in the messages, and the grammar of the exchange. If it's a protocol for physical objects, it will also specify the electrical characteristics, and sometimes even the physical shape of the connectors.

The commands to make an object do something rely on protocols in the same way that clear instructions rely on good grammar—you can't give useful instructions if you can't form a good sentence. Protocols don't specify what happens between objects, however. That's up to the designer and the user.

One thing that all communications protocols have in common—from the simplest chip-to-chip message to the most complex network architecture—is this: it's all about pulses of energy. Digital devices exchange information by sending timed pulses of energy across a shared connection. The USB connection from your mouse to your computer uses two wires for transmission and reception, sending timed pulses of electrical energy across those wires. Likewise, wired network connections like Ethernet are made up of timed pulses of electrical energy sent down the wires. For longer distances and higher bandwidth, the electrical wires may be replaced with fiber-optic cables , which carry timed pulses of light. In cases where a physical connection is inconvenient or impossible, the transmission can be sent using pulses of radio energy between radio transceivers (a *transceiver* is two-way radio, capable of transmitting and receiving). The meaning of data pulses is independent of the medium that's carrying them. You can use the same sequence of pulses whether you're sending them across wires, fiber-optic cables, or radios. If you keep in mind that all of the communication you're dealing with starts with a series of pulses—and that somewhere there's a guide explaining the sequence of those pulses—you can work with any communication system you come across.

X

Computers of All Shapes and Sizes

You'll encounter a few different types of computers in this book, grouped according to their physical interfaces. The most familiar of these is the personal computer. Whether it's a desktop or a laptop, it's got a keyboard, screen, and mouse, and you probably use it just about every working day. These three elements—the keyboard, the screen, and the mouse or trackpad—make up its physical interface.

The second group are mobile devices that you use every day: mobile phones and tablet computers. Although the line between high-end tablets and personal computers is rapidly blurring, there is one big difference: currently, mobile platforms tend to be designed more for media consumption than for software development. Although this is changing, it's still more common and more convenient to develop software for all kinds of computers on a laptop or desktop than it is on a tablet.

The third type of computer you'll encounter in this book, the *microcontroller*, has no physical interface that humans can interact with directly. It's just an electronic chip with input and output pins that can send or receive electrical pulses. Using a microcontroller is a three-step process:

1. You connect sensors to the inputs to convert physical energy like motion, heat, and sound into electrical energy.
2. You attach motors, speakers, and other devices to the outputs to convert electrical energy into physical action.
3. You write a program to determine how the input changes affect the outputs.

Microcontrollers generally don't have an operating system; they just run the program you write for them. In other words, the microcontroller's software and physical interface are whatever you make of them. The Uno and many of the other Arduinos and derivatives you'll encounter in this book do not have an operating system.

There's a related type of computer called the *embedded microprocessor* or *single-board computer* that you'll encounter as well, such as the Raspberry Pi. These are similar in appearance and price to microcontrollers, but unlike the former, they're powerful enough to run an operating system.

You'll interact with embedded devices similarly to how you interact with the fourth type of computer in this book, the *network server*. A server is basically the same as a desktop computer—it may even have a keyboard, screen, and mouse. Even though it can do all the things you expect of a personal computer, its primary function is to send and receive data over a network. Any computer can act as a server, if it's programmed to respond to client requests.

Most people don't think of servers as physical things because they only interact with them over a network, using their local computers as physical interfaces to the server. A server's most important interface for most users' purposes is its software interface.

The fifth group of computers is a mixed bag: music synthesizers, networked cameras, and motor controllers, to name a few. Some of them will have fully developed physical interfaces, some will have minimal physical interfaces but detailed software interfaces, and most will have a little of both. Even though you don't normally think of these devices as computers, they are. When you think of them as programmable objects with interfaces that you can manipulate, it's easier to figure out how they can all communicate, regardless of their end function.
X

Good Habits

Connecting objects is a bit like love: you never know what your partner is thinking, you only know what he or she said. Since you don't know the receiver's inner workings, you never really know how they're interpreting your message; you only know how they respond. There are a thousand ways for your message to get lost or garbled in transmission.

In the same way, you may know exactly what message your local computer is sending, how it's sending it, and what all the bits mean, but the remote computer only has the bits it receives. There are four principles for good communications (in love or networking):

- Listen more than you speak.
- Never assume that what you said is what they heard.
- Agree on how you're going to say things in advance.
- Ask politely for clarification when messages aren't clear.

Listen More than You Speak

The best way to maintain a good relationship is to be a good listener. Listening is more difficult than speaking. You can speak anytime you want, but since you never know when the other person is going to say something, you have to listen all the time. In networking terms, this means you should write your programs such that they're listening for new messages most of the time, and sending messages only when necessary. It's often easier to send out messages all the time rather than figure out when it's appropriate, but that often causes problems. It usually doesn't take a lot of work to limit your sending, and the benefits far outweigh the costs.

Never Assume

What you say is not always what the other person hears. Perhaps they misinterpreted you, or perhaps they just didn't hear you clearly. If you assume that the message got through and continue on obliviously, you're in for a world of hurt. Likewise, you may be inclined to first work out all the logic of your system—and all the steps of your messages before you start to connect things—then build it, and finally test it all at once. Avoid that temptation.

It's good to plan the whole system out in advance, but build it and test it in small steps. Most of the errors that occur in these projects happen in the communication between objects. Always send a quick "Hello World!" message from one object to the others, and make sure that the message got there intact before you proceed to the more complex details. Keep that "Hello World!" example on hand for testing when communication fails.

Getting the message wrong isn't the only misstep you can make. The projects in this book involve building the physical, software, and electrical elements of the interface. A common mistake people make when developing projects like these is to assume that the problems are all in one place. I've often sweated over a bug in the software of a message, only to find out later that the receiving device wasn't even connected. Don't assume that communication errors are in the element of the system with which you're most familiar. They're most often in the element with which you're least familiar, and therefore avoiding. When you can't get a message through, think about every link in the chain from sender to receiver, and check every one. Then check the links you overlooked.

Agree on How You Say Things

In good relationships, you develop a shared language based on shared experience. You learn the best ways to say things so that your partner will be most receptive, and you develop shorthand for expressing things that you repeat all the time. Good data communications also rely on shared ways of saying things, or *protocols*. Sometimes you can design a protocol for all the objects in your system, and other times you have to rely on existing ones. If you're working with a previously established protocol, make sure you understand all the parts before you start trying to use it. If you have the luxury of making up your own protocol, make sure you've considered the needs of both the sender and receiver. For example, you might decide to use a protocol that's easy to program on your web server but that turns out to be impossible to handle on your microcontroller. A little thought to the strengths and weaknesses on both sides of the transmission will make things flow much more smoothly.

Ask Politely for Clarification

Messages get garbled in countless ways. Perhaps you hear something that may not make much sense, but you act on it, only to find out that your partner said something entirely different from what you thought. It's always best to ask nicely for clarification to avoid making a stupid mistake. Likewise, in network communications, it's wise to check that any messages you receive make sense. When they don't, ask for a repeat transmission. It's also wise to check, rather than assume, that a message was sent. Saying nothing can be worse than saying something wrong. Minor problems can become major when no one speaks up to acknowledge that there's an issue. The same thing can occur in network communications. One device may wait forever for a message from the other side, not knowing, for example, that the remote device is unplugged. When you don't receive a response, send another message. Don't resend it too often, and give the other party time to reply. Acknowledging messages may seem like a luxury, but it can save a whole lot of time and energy when you're building a complex system.

X

Tools

As you'll be working with the physical, software, and electrical interfaces of objects, you'll need physical tools, software, and (computer) hardware.

Physical Tools

If you've worked with electronics or microcontrollers before, chances are you have your own hand tools already. Figure 1-1 shows the ones used most frequently in this book. They're common tools that can be obtained from many vendors. A few are listed in Table 1-1.

In addition to hand tools, there are some common electronic components that you'll use all the time. They're listed as well, with part numbers from the retailers featured most frequently in this book. Not all retailers will carry all parts; you might have to shop around.

Software Tools

You'll use a number of different software tools and programming languages in this book, to program and configure the different types of computers mentioned earlier.

Processing

The multimedia programming environment used in this book is called Processing. It's a free, open source tool available at www.processing.org. Based on Java, it's made for designers, artists, and others who want to get something

Figure 1-1. See the list below for number references.

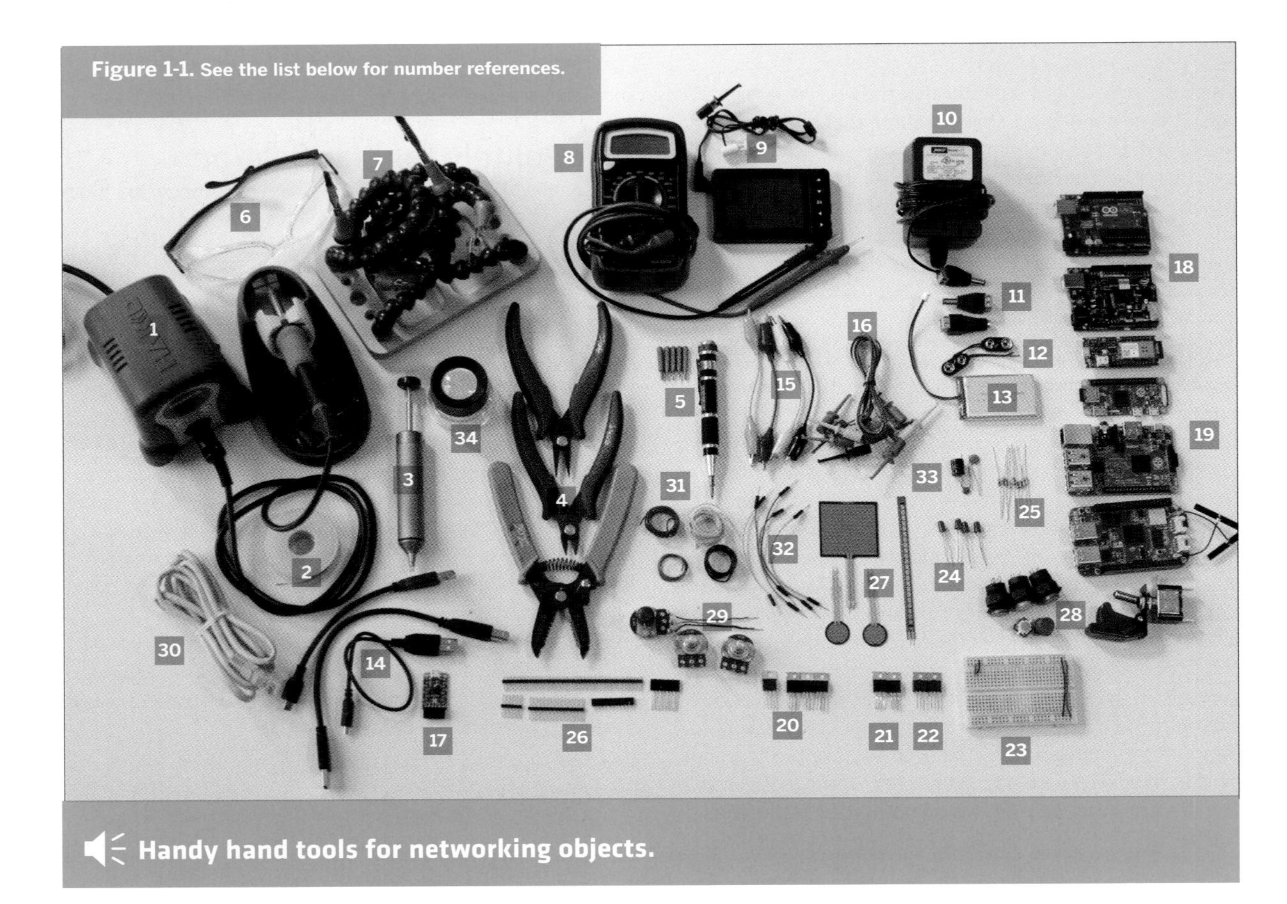

Handy hand tools for networking objects.

1 Soldering iron Middle-of-the-line is best here. Cheap soldering irons die fast, but mid-range irons like the Hakko FX-888D or the Weller WLC-100 or SP series work great for small electronic work. Avoid the Cold Solder irons. They solder by creating a spark, and that spark can damage static-sensitive parts like microcontrollers. Digi-key (digikey.com): 1691-1083-ND; Adafruit (www.adafruit.com): 1204; Farnell (www.farnell.com): 2320747

2 Solder 21-23 AWG solder is best. Get lead-free solder if you can; it's less harmful for you. Jameco (www.jameco.com): 2210116; Farnell: 419266

3 Desoldering pump This helps when you mess up while soldering. Jameco: 2214889; Adafruit: 148; Farnell: 1792724

4 Wire stripper, diagonal cutter, needlenose pliers Avoid the 3-in-1 versions of these tools. They'll only make you grumpy. These three tools are essential for working with wire, and you don't need expensive ones to have good ones. Wire stripper: Jameco: 159291; Farnell: 609195; SparkFun (www.sparkfun.com): TOL-12630
Diagonal cutter: Jameco: 161411; Farnell: 3125397; SparkFun: TOL-10447
Needlenose pliers: Jameco: 217891; Farnell: 3127199; SparkFun: TOL-08793

5 Mini-screwdriver Get one with both Phillips and slotted heads. You'll use it all the time. Jameco: 127271; Farnell: 4431212

6 Safety goggles Always a good idea when soldering, drilling, or other tasks. SparkFun: SWG-11046; Farnell: 1696193

7 Helping hands These make soldering much easier. Jameco: 681002; Farnell: 1367049; SparkFun TOL-11784

8 Multimeter You don't need an expensive one. As long as it measures voltage, resistance, amperage, and continuity, it'll do the job. Jameco: 220759; Farnell: 7430566; SparkFun: TOL-12966

9 Oscilloscope Professional oscilloscopes are expensive, but the DSO Nano is only about $100 and a valuable aid when working on electronics. SparkFun: TOL-11702 (v2); Seeed Studio (www.seeedstudio.com): 109990013; Adafruit: 468

10 9–12V DC power supply You'll use this all the time, and you've probably got a spare from some dead electronic device. Make sure you know the polarity of the plug so you don't reverse polarity on a component and blow it up! Most of the devices shown in this book have a DC power jack that accepts a 2.1mm inner diameter/5.5mm outer diameter plug, so look for an adapter with the same dimensions. Jameco: 170245 (12V, 1000mA); Farnell: 1176248 (12V, 1000mA); SparkFun: TOL-00298

11 Power connector, 2.1mm inside diameter/5.5mm outside diameter You'll need this to connect your microcontroller module or breadboard to a DC power supply. This size connector is the most common for the power supplies that will work with the circuits you'll be building here. There are both plug and socket connectors. It's wise to get one or two of both. SparkFun PRT-10288; Jameco:

159611; Digi-Key: CP-024A-ND; Farnell: 3648102

12 **9V Battery snap adapter** When you want to run a project off battery power, these adapters are a handy way to do it. SparkFun: PRT-09518; Adafruit: 80; Digi-Key: 1568-1237-ND; Jameco: 2207056; Farnell: 1650675 and 1737256

13 **Lithium-polymer (LiPo) battery** Many devices in this book can be powered from rechargeable LiPo batteries. They offer a compact and rechargeable alternative. They come in a variety of sizes, but for general network projects, 3.7V and 800-2000mA is good. SparkFun: PRT-13813 and PRT-08483; Adafruit: 258 and 2011

14 **USB cables** You'll need both USB A-to-B (the most common USB cables) and USB A-to-micro-B (the kind that's common with phones) for the projects in this book. SparkFun: CAB-00512, CAB-13244; Farnell: 1838798, 2444222

15 **Alligator clip test leads** It's often hard to juggle the five or six things you have to hold when metering a circuit. Clip leads make this much easier. Jameco: 10444; RS (www.rs-online.com): 161-6511; SparkFun: PRT-12978.

16 **IC hooks** are also handy for tightly spaced connections when alligator clips won't work. Jameco: 135299; SparkFun: CAB-00501

17 **Serial-to-USB converter** This converter lets you speak TTL serial from a USB port. You'll use these with many devices. SparkFun: DEV-14050 and DEV-12935; Adafruit: 3309 and 284; SeeedStudio: 317990026

18 **Microcontroller module** The microcontrollers shown here are an Arduino Uno, an Arduino 101, and an Arduino MKR1000. SparkFun: DEV-13787; Digi-Key 1660-1003-ND and 1659-1005-ND; RS: 913-9999 and 124-0657. See also store.arduino.cc

19 **Embedded processor** The Raspberry Pi Zero W, Pi 3, and BeagleBone Green are shown here. SparkFun DEV-13825; Adafruit: 3055 and 3400; Seeed Studio: 102010048 and 114990584 RS: 896-8660; Farnell: 2525225. See www.raspberrypi.org for more.

20 **Voltage regulator** Voltage regulators take a variable input voltage and output a constant (lower) voltage. The two most common you'll need for these projects are 5V and 3.3V. Be careful when using a regulator that you've never used before. Check the data sheet to make sure you have the pin connections correct. 3.3V: Digi-Key: 497-1491-5-ND; Jameco: 242115; Farnell: 1703357; RS: 438-4885 5V: Digi-Key: LM7805CT-ND; Jameco: 51262; Farnell: 9756078; RS: 918-1971

21 **TIP120 Transistor** Transistors act as digital switches, allowing you to control a circuit with high current or voltage from one with lower current and voltage. There are many types of transistors; the TIP120 is one used in a few projects in this book. Note that the TIP120 looks just like the voltage regulator next to it. Sometimes electronic components with different functions come in the same physical packages, so you need to check the part number written on the part. Digi-Key: TIP120-ND; Jameco: 32993; Farnell: 9804005

22 **IRF520 MOSFET Transistor** The IRF520 is a useful substitute for the TIP120 in some projects, as it requires almost no current to turn on. Jameco: 209226 Digi-Key: IRF520IR-ND Farnell: 9103031

23 **Solderless breadboard** Having a few around can be handy. I like the ones with two long rows on either side so that you can run power and ground on both sides. Jameco: 20723 (2 bus rows per side); Farnell: 4692810; Digi-Key: 438-1045-ND; SparkFun: PRT-12615 or PRT-12002

24 **Spare LEDs for tracing signals** LEDs are to the hardware developer what print statements are to the software developer. They let you see quickly whether there's voltage between two points, or whether a signal is going through. Keep spares on hand. Jameco: 333973; Farnell: 1208851; Digi-Key: 160-1144-ND

25 **Resistors** You'll need resistors of various values for your projects. Common values are listed in Table 1-1.

26 **Header pins** You'll use these all the time. It's handy to have both male and female ones around. The extra-long ones are handy too. Jameco: 103377; Digi-Key: A26509-20-ND; Farnell: 1593411

27 **Analog sensors (variable resistors)** There are countless varieties of variable resistors to measure all kinds of physical properties. They're the simplest of analog sensors, and they're very easy to build into test circuits. Flex sensors and force-sensing resistors are handy for testing a circuit or a program. **Flex sensors:** Jameco: 150551; Adafruit: 182 SparkFun: SEN-08606 **Force-sensing resistors:** SparkFun: SEN-09375; Adafruit: 166; IDigi-Key: 1027-1001-ND

28 **Pushbuttons** There are two types you'll find handy: the PCB-mount type, like the ones you find on many microcontroller boards, used here mostly as reset buttons for breadboard projects; and panel-mount types used for interface controls for end users. But you can use just about any type you want. **PCB-mount type:** Digi-Key: SW400-ND; Jameco: 119011; SparkFun: COM-00097 **Panel-mount type:** Digi-Key: GH1344-ND; SparkFun COM-09181

29 **Potentiometers** You'll need potentiometers to let people adjust settings in your project. Jameco: 29082; SparkFun: COM-09939; RS: 91A1A-B28-B15L; Farnell: 350072, RS 249-9294

30 **Ethernet cables** A couple of these will come in handy, even though WiFi is ubiquitous these days. Digi-Key: N002-004-BK-ND; Farnell 1526202

31 **Black, red, blue, yellow wire** 22 AWG solid-core hook-up wire is best for making solderless breadboard connections. Get at least three colors, and always use red for voltage and black for ground. A little organization of your wires can go a long way.
Black: Jameco: 36792
Blue: Jameco: 36768
Green: Jameco: 36822
Red: Jameco: 36856
Yellow: Jameco: 36920
Mixed: Jameco: 2153705, SparkFun PRT-11375

32 **Hookup wires** An alternative to solid core wires, for testing purposes only. You'll reuse them, so one or two sets is enough. Jameco: 20723, SparkFun: PRT-12796; RS: 791-6463; Adafruit: 759

33 **Capacitors** You'll need capacitors of various values for your projects. Common values are listed in Table 1-1.

34 **Magnifying loupe** As components get smaller, a magnifier becomes handy to read the part numbers on them. SparkFun: TOL-09316

You'll find a number of component suppliers in this book. I buy from different vendors depending on who's got the best and the least expensive version of each part. Feel free to substitute your favorite vendors. Watch out for poor-quality copies, though. One good way to check a vendor's legitimacy is to check to see if the original manufacturer lists the vendor on their distributors page. If not, be wary.

You'll find a list of parts for all the projects in each chapter at the opening of each chapter. Where appropriate, alternate components and vendors will be listed there as well. A full list of parts and vendors used for the production of this book can be found in the Appendix.

Table 1-1. Common components for electronic and microcontroller work.

D Digi-Key (www.digikey.com)
J Jameco (www.jameco.com)
RS RS (www.rs-online.com)
F Farnell (www.farnell.com)
SF SparkFun (www.sparkfun.com)

RESISTORS	
100Ω	**D** 100QBK-ND, **J** 690620, **F** 9337660, **RS** 755-0707
220Ω	**D** 220QBK-ND, **J** 690700, **F** 9339299, **RS** 707-7612
470Ω	**D** 470QBK-ND, **J** 690785, **F** 9339531, **RS** 707-8659
1K	**D** 1.0KQBK-ND, **J** 690865, **F** 9339051, **RS** 707-8669
10K	**D** 10KQBK-ND, **J** 691104, **F** 9339060, **RS** 707-7745
22K	**D** 22KQBK-ND, **J** 691180, **F** 9337814, **RS** 739-7140
100K	**D** 100KQBK-ND, **J** 691340, **F** 9337695, **RS** 707-8940
1M	**D** 1.0MQBK-ND, **J** 691585, **F** 9337709, **RS** 131-700
CAPACITORS	
0.1μF ceramic	**D** 399-4151-ND, **J** 15270, **F** 3322166, **RS** 716-7135
1μF electrolytic	**D** P10312-ND, **J** 94161, **F** 8126933, **RS** 475-9009
10μF electrolytic	**D** P11212-ND, **J** 29891, **F** 1144605, **RS** 762-1736
100μF electrolytic	**D** P10269-ND, **J** 158394, **F** 1144642, **RS** 762-1746
VOLTAGE REGULATORS	
3.3V	**D** 497-1491-5-ND, **J** 242115, **F** 1703357, **RS** 438-4885
5V	**D** LM7805CT-ND, **J** 51262, **F** 9756078, **RS** 918-1971
ANALOG SENSORS	
Flex sensors	**D** 905-1000-ND, **J** 150551, **RS** 708-1277
FSRs	**D** 1027-1000-ND, **J** 2128260
LED	
5mm, Green clear	**D** 160-1144-ND, **J** 34761, **F** 1855510, **RS** 228-5944
5mm, Red, clear	**D** 160-1665-ND, **J** 94511, **F** 1855570, **RS** 848-6480
TRANSISTORS	
2N2222A	**D** P2N2222AGOS-ND, **J** 38236, **F** 1611371, **RS** 295-028
TIP120	**D** TIP120-ND, **J** 32993, **F** 9804005, **RS** 774-3653
IRF520	**D** IRF520IR-ND, **J** 209226, **F** 9103031, **RS** 541-1180
DIODES	
1N4004-R	**D** 1N4004-E3/54GICT-ND, **J** 35991, **F** 9556109, **RS** 628-9029
3.3V zener (1N5226)	**D** 1N5226B-TPCT-ND, **J** 743488, **F** 1700785 **RS** 805-0110
PUSHBUTTONS	
PCB	**D** SW400-ND, **J** 119011, **F** 1555981
Panel Mount	**D** GH1344-ND, **J** 2231822, **F** 1634684, **RS** 718-2213
SOLDERLESS BREADBOARDS	
various	**D** 438-1045-ND, **J** 20723, 20601, **F** 4692810
HOOKUP WIRE	
red	**D** C2117R-100-ND, **J** 36856, **F** 1662031
black	**D** C2117B-100-ND, **J** 36792, **F** 1662027
blue	**J** 36768, **F** 1662034
yellow	**J** 36920, **F** 1662032
JUMPER WIRE SET	
Various	**J** 20723, **SF** PRT-12796, **F** 2396146, **RS** 791-6463, **AF** 759
POTENTIOMETER	
10K	**D** 987-1649-ND **J** 29081 **F** 1760793
HEADER PINS	
straight	**D** A26509-20-ND, **J** 103377, **F** 1056427, **SF** PRT-00116
right angle	**D** S1121E-36-ND, **F** 1056429, **SF** PRT-00553
HEADERS	
female	**D** ED7102-ND, **F** 1122344, **SF** PRT-00115
BATTERY SNAP	
9V	**D** 1568-1237-ND, **J** 2207056, **F** 1650675, **SF** PRT-09518

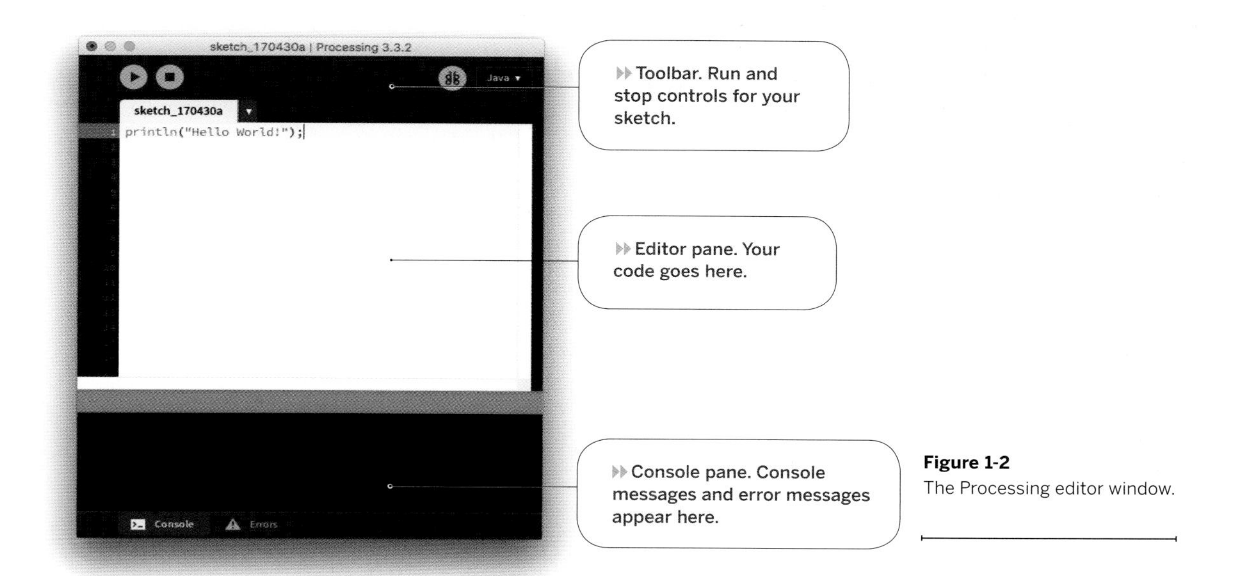
sketch_170430a | Processing 3.3.2
Java
sketch_170430a
println("Hello World!");
Console
Errors
Toolbar. Run and stop controls for your sketch.
Editor pane. Your code goes here.
Console pane. Console messages and error messages appear here.

Figure 1-2
The Processing editor window.

done without having to know all the gory details of programming. It's a useful tool for explaining programming ideas because it takes relatively little Processing code to make big things happen, such as opening a network connection, connecting to an external device through a serial port, or controlling a camera. Because it's based on Java, you can include Java classes and methods in your Processing programs. It runs on macOS, Windows, and Linux. There's also Processing for Android phones. If you don't like working in Processing, you should be able to use this book's code samples and comments as pseudocode for whatever multimedia environment you prefer. Once you've downloaded and installed Processing on your computer, open the application. You'll get a screen that looks like Figure 1-2.

Try It Here's your first Processing program. Type this into the editor window, and then press the Run button on the top-left side of the toolbar.

```
println("Hello World!");
```

It's not too flashy a program, but it's a classic. It should print `Hello World!` in the message box at the bottom of the editor window. It's that easy.

Programs in Processing are called *sketches*, and all the data for a sketch is saved in a folder with the sketch's name. The editor is very basic, without a lot of clutter to get in your way. The toolbar has buttons to run and stop a sketch, create a new file, open an existing sketch, save the current sketch, or export to a Java applet. You can also export your sketch as a standalone application from the File menu. Files are normally stored in a subdirectory of your **Documents** folder called **Processing** (in version 3 of Processing, it's called **Processing3**), but you can save them wherever you like.

Try It Here's a second program that's a bit more exciting. It illustrates some of the main programming structures in Processing.

» NOTE: **All code examples in this book will have comments indicating the context in which they're to be used: Processing, Arduino, node.js, and so forth.**

```
/*
  Triangle drawing program
  Context: Processing

  Draws a triangle whenever the mouse button is not pressed.
  Erases when the mouse button is pressed.
*/

// declare your variables:
float redValue = 0;    // variable to hold the red color
float greenValue = 0;  // variable to hold the green color
float blueValue = 0;   // variable to hold the blue color

// the setup() method runs once at the beginning of the program:

void setup() {
  size(320, 240);      // sets the size of the applet window
  background(0);       // sets the background of the window to black
  fill(0);             // sets the color to fill shapes with (0 = black)
  smooth();            // draw with antialiased edges
}

// the draw() method runs repeatedly, as long as the applet window
// is open.  It refreshes the window, and anything else you program
// it to do:

void draw() {
  // Pick random colors for red, green, and blue:
  redValue = random(255);
  greenValue = random(255);
  blueValue = random(255);

  // set the line color:
  stroke(redValue, greenValue, blueValue);

  // draw when the mouse is up (to hell with conventions):
  if (mousePressed == false) {
    // draw a triangle:
    triangle(mouseX, mouseY, width/2, height/2,pmouseX, pmouseY);
  }
  // erase when the mouse is down:
  else {
    background(0);
    fill(0);
  }
}
```

Every Processing program has two main functions, `setup()` and `draw()`. `setup()` runs once at the beginning of the program. It's where you set all your initial conditions, like the size of the applet window, initial states for variables, and so forth. `draw()` is the main loop of the program. It repeats continuously until you close the applet window.

In order to use variables in Processing, you have to declare the variable's data type. In the preceding program, the variables `redValue`, `greenValue`, and `blueValue` are all `float` types, meaning that they're floating decimal-point numbers. Other common data types you'll use are `int`s (integers), `char`s (one-byte ASCII character values) `boolean`s (true or false values), `String`s of text, and `byte`s.

Java, and therefore Processing, is a *strongly typed* language, meaning that you have to declare variables before you use them, and that the language will set aside a specific amount of space for each variable you declare. For example, a byte takes one byte of memory, an int takes four bytes, and so forth. Each language you'll see in this book will handle variables slightly differently from the others, so pay attention to how you declare and use variables in each language.

Like C, Java, and JavaScript, Processing uses C-style syntax. All functions have a data type, just like variables (and many of them are the `void` type, meaning that they don't return any values). All lines end with a semicolon, and all blocks of code are wrapped in curly braces. Conditional statements (if-then statements), for-next loops, and comments all use the C syntax as well. The preceding code illustrates all of these except the for-next loop.

Here's a typical for-next loop. Try this in a sketch of its own (to start a new sketch, select New from Processing's File menu).

```
for (int myCounter = 0; myCounter <=10; myCounter++) {
  println(myCounter);
}
```

BASIC and Python users: If you've never used a C-style for-next loop, it can seem forbidding. What this bit of code does is establish a variable called `myCounter`. As long as a number is less than or equal to 10, it executes the instructions in the curly braces. `myCounter++` tells the program to add one to `myCounter` each time through the loop. The equivalent BASIC code is:

```
for myCounter = 0 to 10
  Print myCounter
next
```

This same code in Python would look like so:

```
for myCounter in range (0, 10):
  print myCounter
```

Processing is a fun language to play with because you can make interactive graphics very quickly. It's also a simple introduction to Java for beginning programmers. If you're a Java programmer already, you can include Java directly in your Processing programs. Processing is expandable through code libraries.

Since Processing is intended as a tool for exploration of visual design, it doesn't feature many of the standard user interface widgets, like buttons and option menus. That frees you up as a designer, but means you have to be more adventurous as a programmer to create your own UI elements from scratch in Processing.

For more on the syntax of Processing, see the language reference guide at www.processing.org. To learn more about programming in Processing, check out **Processing: A Programming Handbook for Visual Designers and Artists**, by Casey Reas and Ben Fry (MIT Press), the creators of Processing, or their **Getting Started with Processing** (Maker Media). Daniel Shiffman's excellent introduction, **Learning Processing** (Morgan Kaufmann), or his more advanced **The Nature of Code** (natureofcode.com) are also great resources.

X

Command-Line Interfaces and Remote Servers

One of the most effective programming tools you'll use when making the projects in this book is a terminal program, which gives you access to the *command-line interface (CLI)* of a computer. If you've never used a command-line interface before, you'll find it's a mental shift from using a graphical user interface (GUI), but it's a powerful complement. You'll use the command-line interface to control your own computer, and to access both web servers and embedded computers.

Before graphical user interfaces, command-line interfaces were the main way to use a computer. For many networking professionals, they're still the main way, as the command line is fast and efficient once you learn it. Command-line interfaces take much less computational power than graphical user interfaces, so they're useful for interacting with low-power embedded processors and microcontrollers as well.

The most common version of the command-line interface is found on operating systems that are based on Unix, including BSD, Linux, Apple's macOS, and others. These operating systems that share a common command-line style are called *POSIX (Portable Operating System Interface)-style operating systems*. The command-line instructions you'll see in this book all assume you're using a POSIX command-line interface.

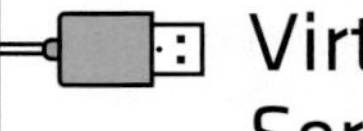

Virtual Private Servers

Many of the programs you'll write in this book are web server programs, and you'll find it convenient to have a web host on which to run them. Nowadays, web services are very simple to set up and run. If you don't have a web host, consider looking for one that offers virtual private servers (VPs). This service creates a virtual server that only you can access. It allows you to figure some things out, and if you make a horrible mess of it, you can destroy the whole server with a click on a web panel. There are several hosting services will even give you low-cost or even free VPS service, as long as you're not getting a lot of web traffic to your site. Digital Ocean (www.digitalocean.com), Amazon Web Services (aws.amazon.com), BlueHost (www.bluehost.com), and DreamHost (www.dreamhost.com) all offer affordable VPS services. If you're an educator or student, check for special deals as well.

One reason to consider a low-cost web host as opposed to a free one is that many hosts also offer one-click installation tools on your server. From the host's web panel, you choose the operating system and any programming environments or tools you want installed, and voilà! You've got a custom-configured server. Depending on how restrictive your web hosting service is, you may or may not get command-line access, so check for that as well.

Connecting to the Command Line on Your Computer

To connect to the command line on your own computer, you'll need to launch a terminal emulation program, which gives you access to your command line.

macOS and Linux

On macOS, the program is called Terminal, and you can find it in the **Utilities** subdirectory of the **Applications** directory. On Linux, look for a program called Terminal, xterm, rxvt, or Konsole.

Windows

If you're using Windows, the native CLI is based on DOS, and is different from POSIX CLIs. You can get to it by by typing `cmd` in the Start Menu search field. As of this writing, Microsoft has released a POSIX CLI on Windows 10 (search for "bash on Windows 10"), which is currently enabled only in developer mode. It's still in development but looks promising. Meanwhile, you may want to install Cygwin (www.cygwin.com), which offers a POSIX-like command-line interface for Windows. Where possible, the command-line exercises in this book were tested using Cygwin. If you install Cygwin, make sure to include the Net package when you run the installer (it's included in the Packages screen of the install). This will install many of the useful POSIX network command-line tools.

Connecting to a Remote Web Host

Although you can use the command line on your own computer, you'll use it often in this book to access a remote computer as well, either a web host's computer or an embedded microprocessor. Most web hosting providers are based on Linux, BSD, Solaris, or some other POSIX operating system.

On Windows computers, there are a few remote access programs available, but the one that you'll use here is called PuTTY, from www.puttyssh.org. Download the Windows-style installer and run it. On macOS and Linux, you can use OpenSSH, which is included with both operating systems, and can be run in the Terminal program with the command `ssh`.

macOS and Linux

Open your terminal program. You'll get a plain-text window with a greeting like this:

```
Last login: Wed Feb 22 07:20:34 on ttyp1
ComputerName:~ username$
```

The $ is called the *command prompt* in POSIX systems. When you see it followed by a cursor, you know the system's waiting for you to type something. Anytime you see it in this book, you should type what follows it on the command line.

Type ssh username@myhost.com at the command line to connect to your web host. Replace username and myhost.com with your username and host address. Enter your password when prompted, and you'll be connected.

Windows

On Windows, you'll need to start up PuTTY (see Figure 1-3). To get started, type myhost.com (replace with your web host's name) in the Host Name field, choose the SSH protocol, and then click Open. Log in with your username and password.

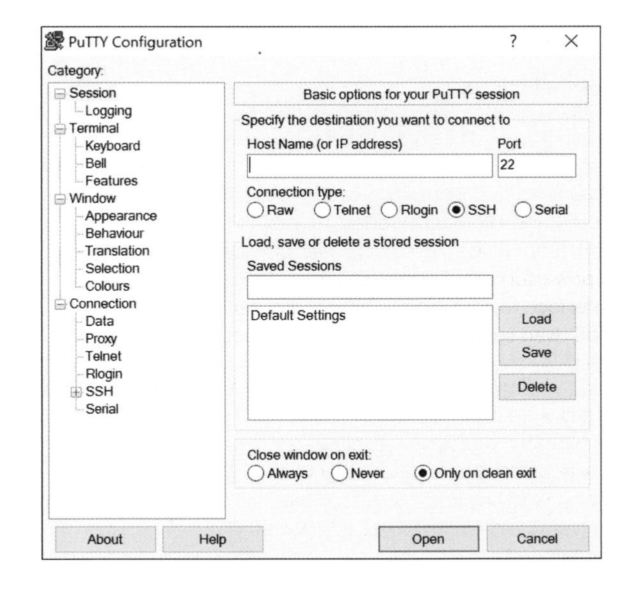

Figure 1-3
The main PuTTY window.

When you connect, you'll be asked to accept an ssh key. This is a unique identifier that allows your computer to identify the host in future logins. Accept it when prompted.

Using the Command Line

Once you've connected to the remote web server, you should see something like this:

```
Last login: Wed Feb 22 08:50:04 2016 from 216.157.45.215
[userid@myhost ~]$
```

Now you're at the command prompt of your web host's computer, and any command you give will be executed on that computer. Start off by learning what directory you're in. To do this, type:

```
$ pwd
```

which stands for "print working directory." It asks the computer to list the name and pathname of the directory in which you're currently working. (You'll see that many POSIX commands are very terse, so you have to type less. The downside of this is that it makes them harder to remember.) The server will respond with a directory path, such as:

```
/home/igoe
```

This is the *home directory* for your account. To find out what files are in a given directory, use the list (`ls`) command, like so:

```
$ ls -l .
```

NOTE: The dot is shorthand for "the current working directory." Similarly, a double dot is shorthand for the directory (the parent directory) that contains the current directory.

The `-l` means "list long." You'll get a response like this:

```
total 44
drwxr-xr-x  13 igoe users 4096 Apr 14 11:42 public_html
drwxr-xr-x   3 igoe users 4096 Nov 25  2005 share
```

This is a list of all the files and subdirectories of the current working directories, as well as their attributes. The first column lists who's got permissions to do what (read, modify, or execute/run a file). The second lists how many links there are to that file elsewhere on the system. The third column tells you who owns it, and the fourth tells you the group (a collection of users) to which the file belongs. The fifth lists its size, and the sixth lists the date it was last modified. The final column lists the filename. In a POSIX environment, all files whose names begin with a dot are invisible. Some files, like password files and other configuration files, need to be invisible. You can get a list of all the files, including the invisible ones, using the `-a` modifier for `ls`, this way:

```
$ ls -la
```

To make a new directory, type:

```
$ mkdir directoryname
```

This command will make a new directory in the current working directory. If you then use `ls -l` to see a list of files in the working directory, you'll see a new line with the new directory. If you then type cd directoryname to switch to the new directory and ls -la to see all of its contents, you'll see only two listings:

```
drwxr-xr-x  2 tqi6023 users 4096 Feb 17 10:19 .
drwxr-xr-x  4 tqi6023 users 4096 Feb 17 10:19 ..
```

The first file, **.**, is a reference to this directory itself. The second, **..**, is a reference to the directory that contains it. Those two references will exist as long as the directory exists. You can't change them.

To remove a directory, type:

```
$ rmdir directoryname
```

`rmdir` can remove only empty directories, so make sure that you've deleted all the files in a directory before you remove it. `rmdir` won't ask you if you're sure before it deletes your directory, so be careful. Don't remove any directories or files that you didn't make yourself.

Many web hosts will automatically create a subdirectory of your home directory called **html** or **public_html** as a place for your public HTML files to live. If there's no such directory, create one using `mkdir` like so:

```
$ mkdir html
```

To move around from one directory to another, there's a "change directory" command, cd. To get into the **html** directory, for example, type:

```
$ cd html
```

To go back up one level in the directory structure, type:

```
$ cd ..
```

To return to your home directory, use the ~ symbol, which is shorthand for your home directory:

```
$ cd ~
```

If you type `cd` on a line by itself, it also takes you to your home directory.

If you want to go into a subdirectory of a directory, for example the **html** directory inside your home directory, you'd type cd ~/html. You can type the *absolute path* from the main directory of the server (called the *root*) by placing a **/** at the beginning of the file's pathname. Any other file pathname is called a *relative path*.

Controlling Access to Files

Type `ls -l` to get a list of files in your current directory and to take a closer look at the permissions on the files. For example, a file marked `drwx------` means that it's a directory, and that it's readable, writable, and executable by the system user who created the directory (also known as the owner of the file). Or, consider a file marked `-rw-rw-rw`. The `-` at the beginning means it's a regular file (not a directory) and that the owner, the group of users to which the file belongs (usually, the owner is a member of this group), and everyone else who accesses the system can read and write to this file. The first `rw-` refers to the owner, the second refers to the group, and the third refers to the rest of the world. If you're the owner of a file, you can change its permissions using the `chmod` (***ch**ange **mod**e*) command:

```
$ chmod go-w filename
```

The options following `chmod` refer to which users you want to affect. In the preceding example, you're removing write permission (`-w`) for the group (`g`) that the file belongs

Figure 1-4
The nano text editor. The commands along the bottom are all invoked using the control key. For example, control-K will cut the selected line of text.

to, and for all others (o) besides the owner of the file. To restore write permissions for the group and others, and to also give them execute permission, you'd type:

```
$ chmod go+wx filename
```

A combination of u for user, g for group, and o for others, and a combination of + and - and r for read, w for write, and x for execute gives you the capability to change permissions on your files for anyone on the system. Be careful not to accidentally remove permissions from yourself (the user). Also, get in the habit of not leaving files accessible to the group and other users unless it's necessary.

Creating, Viewing, and Deleting Files

Two other command-line programs you'll find useful are nano and less. nano is a text editor. It's very bare-bones, so you may prefer to edit your files using your favorite text editor on your own computer and then upload them to your server. But for quick changes right on the server, nano is great. To make a new file, type:

```
$ nano filename.txt
```

The nano editor will open up. Figure 1-4 shows how it looks like after I typed in some text.

All the commands to work in nano are keyboard commands you type using the control (Ctrl) key. For example, to exit the program, type Ctrl-X. The editor will then ask whether you want to save, and prompt you for a filename. The most common commands are listed along the bottom of the screen.

Once you've created a file, you can delete it using the rm command, like this:

```
$ rm filename
```

Like rmdir, rm won't ask whether you're sure before it deletes your file, so use it carefully.

While nano is for creating and editing files, less is for reading them. less takes any file and displays it to the screen one screenful at a time. To see the file you just created in nano, for example, type:

```
$ less filename.txt
```

You'll get a list of the file's contents, with a colon (:) prompt at the bottom of the screen. Press the spacebar for the next screenful. When you've read enough, type q to quit. There's not much to less, but it's a handy way to read long files.

There are three other useful commands for viewing files: `cat`, `head`, and `tail`.

`cat` dumps the contents of a file to the screen with no page breaks and no way to control it:

```
$ cat filename.txt
```

`head` and `tail` let you view either the beginning or the end lines of a long file. You specify how many lines with an additional parameter, like so:

```
$ head -5 filename.txt
```

This shows the first five lines of **filename.txt**. Likewise,

```
$ tail -10 filename.txt
```

shows you the last ten lines of **filename.txt**.

`cat`, `head` and, `tail` come in handy when you start to combine programs together, as explained below.

Combining Programs Together

Unix was built on a philosophy of "small pieces loosely joined." In other words, each command-line program (often called *processes*, though one program can sometimes start multiple processes) should be designed to do one thing well, and they should allow you to combine them together and automate them flexibly. Since the output of all command-line programs is text, it's easy to use the results of one program as the input of another. For example, if you're in a directory with dozens of files and you list files using `ls`, the first files will scroll past your screen, but you can take the output of `ls` and send it to `less`, and it will show up paginated. You do this like so:

```
$ ls -la . | less
```

The vertical slash (|) is called a *pipe* in command-line environments, and what you're doing is piping the output of one program into another for further processing.

You can also redirect the output of a program from the screen to somewhere else, like a file. For example, if you wanted to save the list of files in a directory to a file, you could do this:

```
$ ls -la . > fileList.txt
```

This example would either create `fileList.txt` if it doesn't exist, or it would replace whatever's in it with the output of the `ls` command. If you wanted to append the list output to the file rather than replacing it, you'd do this:

```
$ ls -la . >> fileList.txt
```

One reason this works is that POSIX systems view many things as *data streams*, and therefore interchangeable. For example, the text you see on the screen is called the *standard out*, or *stdout*, data stream. The bytes coming from the keyboard are the *standard in*, or *stdin*, data stream. A file is also a data stream, which is why you can send bytes to and from files, the stdin, and the stdout using pipes and redirection.

You can think of a data stream like a tube full of bytes. The first byte that goes in is the first one that comes out. This is referred to as *first-in, first-out*, or *FIFO*. You can get pretty fancy when you combine this with some of the commands and concepts you've seen already. You'll see some uses of these later in the chapter and the book.

There are many other commands available in the command shell. For more information, type help at the command prompt to get a list of commonly used commands. For any command, you can get its user manual by typing man commandname.

When you're ready to close the connection to your server, or a terminal session on your local computer, type `logout`. Some variants of Linux use `exit` instead.

If you're using a POSIX-style personal computer, whether it's a macOS machine, a Linux machine, or a single-board computer, all of these commands should work just as well in the terminal program of your desktop as they do via an ssh connection to a remote server. Most of them will work in Cygwin on a Windows machine as well. You'll also be able to use them on single-board computers, as you'll see later in this chapter.

For more on getting around Unix and Linux systems using the command line, see **Linux for Makers** by Aaron Newcomb (Maker Media).

X

Node.js

The web server programs in this book are written mostly in node.js, a framework for using the JavaScript programming language within an operating system. When JavaScript was originally released, it was used only in web browsers, to enable sophisticated user interaction that HTML, the HyperText Markup Language, didn't provide. It's become the standard programming language for browser-based applications. In 2009, Ryan Dahl and other programmers at Joyent wanted to use JavaScript for server-side programming as well, so they developed node.js. It's become a very popular tool for server-side programming and has found many other applications as well.

One of the great things about node.js is that you can run it on your personal computer or on an embedded computer just as well as you can on a web server. So you can test your programs locally before you upload them to your web host. You can also can make client-server applications that run only on your home network, without ever needing the internet. For most of the examples here, you'll use your personal computer as the server to simplify things.

Unlike Processing, there is no default GUI for node.js. You use it from the command line. Download the installer for it from nodejs.org and install it on your personal computer. Then open a terminal window like you did in the last section and you're ready to use node.js. Start by typing:

```
$ node -v
```

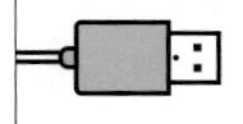

Installing Your Toolchain

Node.js relies on a few other programming tools that you may not have if you're a new programmer. The combination of compilers, text processors, and debuggers is called a *toolchain*, and there are some standard ones for every operating system. On Windows, most of these come with Visual Studio, and on macOS, they come with XCode.

Windows users can install Microsoft Visual Studio Community Edition to address this. It's free from www.visualstudio.com/downloads/. Install it before you install node.js and forget about it.

macOS users can install XCode from the Mac App Store. Once XCode is installed, you'll also need to install the XCode command-line tools. Do this from the command line like so:

```
$ xcode-select --install
```

Once you've got your toolchain in place, you're ready to install and use node.js.

You'll get a reply like this:

```
v6.9.5
```

This tells what version of node.js is installed on your computer. The code in this book was written using node.js version 6.9.5. If you're running that version or later, you'll be fine.

Try It Here's your first node.js program. Open your favorite text editor, type in the code shown here, and save it on your computer with the name **hello.js** in your home directory.

```
console.log("Hello world!");
```

Now, open a terminal program and navigate to the directory in which you saved this program. Type the following to see the results:

```
$ node hello.js
```

You should get the following response:

```
Hello world!
```

About as exciting as your first Processing sketch, isn't it? X

A Node.js Web Server

Node.js was originally designed for writing web server programs, and that's the first thing most people do with it.

A *server* is a program that provides a service to other programs on the network, called *clients*. More specifically, a *web server* responds to requests made using the *HyperText Transport Protocol (HTTP)*. Its responses are usually in the form of HTML files, images, sound files, and other elements of a website. Your browser, a client, makes a connection to the server to request a page. The server program accepts the connection and delivers a response, and the connection is closed.

You'll learn more about HTTP, clients, and servers in Chapter 3, but for now, let's write a simple server in node.js.

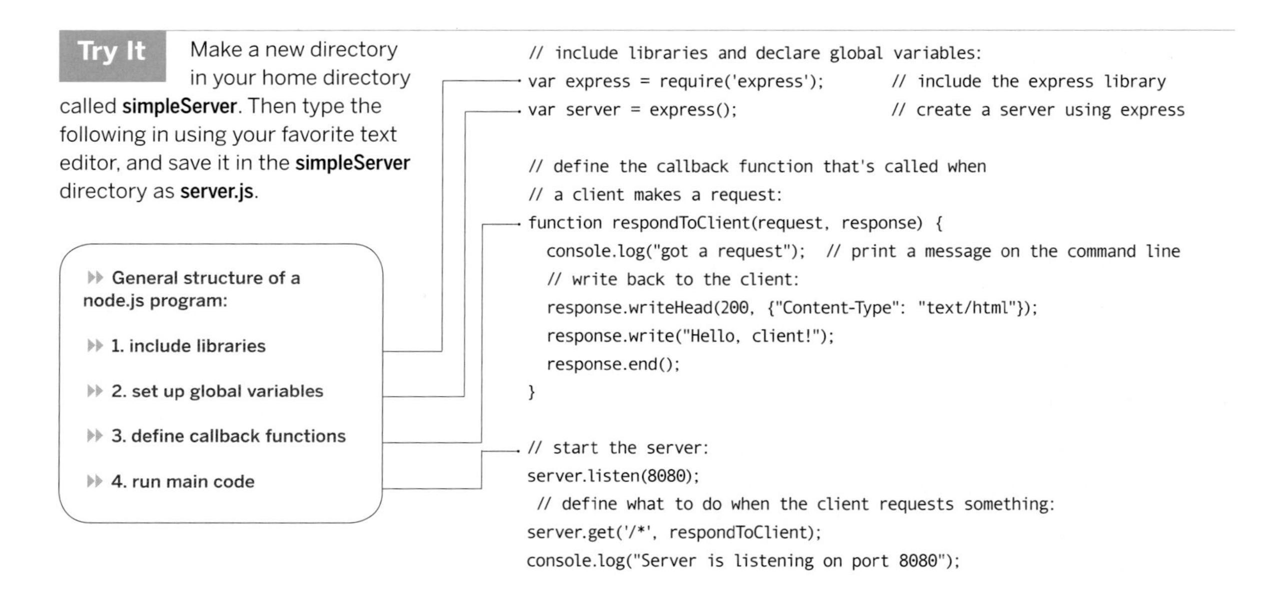

Try It Make a new directory in your home directory called **simpleServer**. Then type the following in using your favorite text editor, and save it in the **simpleServer** directory as **server.js**.

General structure of a node.js program:

- **1. include libraries**
- **2. set up global variables**
- **3. define callback functions**
- **4. run main code**

```
// include libraries and declare global variables:
var express = require('express');        // include the express library
var server = express();                  // create a server using express

// define the callback function that's called when
// a client makes a request:
function respondToClient(request, response) {
  console.log("got a request");  // print a message on the command line
  // write back to the client:
  response.writeHead(200, {"Content-Type": "text/html"});
  response.write("Hello, client!");
  response.end();
}

// start the server:
server.listen(8080);
 // define what to do when the client requests something:
server.get('/*', respondToClient);
console.log("Server is listening on port 8080");
```

Make sure you're connected to the internet, then navigate to the **simpleServer** directory on the command line, and type:

```
$ npm install express
```

The *node package manager (npm)* will install a library called express.js that you'll use in this server. You'll see a progress bar, then a few warnings, and the command prompt will return. Then type:

```
$ node server.js
```

You should see the response

```
Server is listening on port 8080
```

You won't see a command prompt after it, because your program is still running, unlike the hello.js program, which quit after it ran the one line. Open a browser window and enter the following in the address bar:

```
http://localhost:8080
```

You should get the page seen in Figure 1-5. On the command line, you should see a response as well, like so:

```
Got a request!
```

Every time your web browser makes a request to the server program, you'll see the request response. Sometimes the browser will make two requests when you enter the address in the address bar. One of them is a request for the *favicon*, which is the little icon next to the page's address in the address bar.

To quit the server program, type control-C on the command line.

Figure 1-5
The results of your first node.js server, in a browser.

Elements of Node.js

Node.js programs (sometimes called *scripts*) are structured differently than Processing sketches. Let's go through some of the differences.

First, JavaScript (and therefore node.js) is a *weakly typed* language. You don't have to declare the data types of your variables; you just need the keyword `var` to declare them. JavaScript will try to figure out the data type depending on how you use them.

Second, JavaScript is a *functional language*. This means that you can put functions into variables, not just values. This will make more sense as you see it in practice. You can see it in the previous program. The first line makes a copy (called an *instance*) of node.js's express library in the variable `express`. Then you can call on all the functions of that library from the `http` variable. In the second line, you call the `express()` function that makes an instance of the whole express library, which you're putting into the variable called `server`. Down at the bottom of the program, you're calling a function from that server object, `listen()`. This function listens for new requests from clients.

Third, JavaScript is *asynchronous*. This means that each command starts immediately after the previous one, even if the previous command is not finished. When you call a function, you often give it a *callback function* to run when it's ready to deliver some results. You can see this in the **server.js** program's second line. The `server.get()` function has a callback function, `respondToClient()`, that runs whenever a new client makes a request. You can respond to several requests at once because JavaScript is asynchronous. The server doesn't have to finish one response in order to start another response.

JavaScript uses a C-style syntax like Java and C. It uses semicolons at the ends of each statement, encloses code blocks in curly braces, and uses the C-style syntax for if-then statements and for-loops.

Node.js server scripts are generally structured like this:

1. Include any libraries you need to include using `require();`
2. Declare any variables that will be used throughout the program (*global variables*)
3. Declare any callback functions that will be used by main code
4. Run the main code

You can see this structure in the **server.js** program.

For more on the syntax of node.js, see the language reference guide at www.nodejs.org. To learn more about programming in node.js, check out **Learning Node: Moving to the Server-Side**, by Shelley Powers (O'Reilly), For a good introduction to JavaScript, see **Learning JavaScript** by Ethan Brown (O'Reilly).

X

HTML5 and Web Applications

HTML5, the most recent version of the HyperText Markup Language standard, offers a lot of useful tools for interactive applications. It supports a greater amount of interaction and control over how web pages look and behave. In fact, what makes HTML5 interesting is not just the markup language itself, but the possibilities it offers when combined with JavaScript and CSS3, the Cascading Style Sheets standard.

So, how do HTML5, JavaScript, and CSS3 work together? Roughly, you could say that HTML gives you nouns, CSS gives you adjectives and adverbs, and JavaScript gives you the verbs to put them into action. HTML describes the basic page structure and the elements within it: forms, input elements, blocks of text, and so forth. HTML organizes the things you see, like text, buttons, option menus, and other items, into HTML documents, and describes them using the *Document Object Model (DOM)*. CSS primarily describes the characteristics of the visual elements: the colors, fonts, spacing, etc. The scripting language JavaScript allows you to make connections between the elements of a page, between pages in a browser, and between the browser and remote servers. The grammar analogy isn't perfect, but the point is that the three tools give you a wide range of means to present information, listen to user input, generate interactive responses, and send and retrieve data to locations other than the browser—whether the data is on the user's hard drive or a remote server.

Previously, applications running within a web browser had very limited access to the hardware of the computer on which they were running. Operating system manufacturers felt that it was unsafe to allow an application that you downloaded from the internet to access your hard drive, camera, microphone, or other computer hardware. Of course, now that nearly every application is downloaded from the internet, that thinking seems dated. You could still download a program that does malicious things—but, by now, there are tools for managing the trustworthiness of an online source. At the same time, more of the data we need to access lives online today, whether on a social media site or a web data storage service like Google Docs or Dropbox. There's no need for a browser to access your hard drive if your files are online.

The blurring of the distinction between browser security and general security is good news if you like to build physical interfaces. HTML5 and JavaScript include methods to access some of the hardware of your computer or phone. You'll see camera applications, GPS, and more in the chapters that follow. You can also read the accelerometer and gyrometer, if there is one on your device.

The bad news is that not all of these new methods are universally agreed upon. Not every browser gives you access to all of these features, and not every browser implements them the same way. So, while it's easy to access devices external to your phone through the browser, getting access to the ones on your phone may take a bit more work.

p5.js

Since you're writing web servers in node.js, you'll need to write web pages as well to provide user interfaces. You can add interactive elements to any HTML page in a web browser using JavaScript. There are thousands of JavaScript libraries and frameworks available for working in a browser, depending on your tastes and your application. If you're familiar with Processing, then p5.js is a good one to use because makes your browser-based JavaScript look a lot like a Processing sketch. You'll use it throughout this book.

p5.js is open source and free to use. You can learn all about it and find its source code and reference documentation at p5js.org. For further reading, see **Getting Started with p5.js**, by Lauren McCarthy (Maker Media).

p5.js projects can be written in any text editor, and that's how you'll handle them for this book. To get started, go to p5js.org/download/ and download the complete library. There's a directory called **empty-example** that contains all the required files for a p5.js project, in the right arrangement. Copy that directory and rename it and you'll have all of the files you need in the right directories: an HTML page called **index.html** at the root of your project, and in a subdirectory called **libraries**, the **p5.js** library and its addons, **p5.dom.js** (for manipulating the DOM of the HTML page), and **p5.sound.js**. Also in the root of the project is **sketch.js**, the file in which you'll write all of your code.

To add the p5.js libraries (or any JavaScript library) to your HTML page, add a `<script>` tag in the head of your HTML with the path to the script. You can either provide a URL path to a remote copy of the script, or a file path to a file that's local to your server. To add any addon library, you just add another `<script>` tag. If you look at the default **index.html** page in the p5.js example project, you'll see the script tags for all three of the libraries mentioned, and one for your **sketch.js** file. The head of the HTML page will look like this:

```
<!DOCTYPE html>
<html>
  <head>
   <meta charset="utf-8">
   <script type="text/javascript" src="p5.min.js"></script>
   <script type="text/javascript" src="libraries/p5.dom.js">
   <script type="text/javascript" src="sketch.js">
   <title></title>
  </head>
```

For most projects, you can use this default page exactly as is. When you need extra libraries for projects later in the book, they'll be mentioned.

If you prefer to create your p5.js projects on the command line, you can install p5-manager using the node package manager like so:

```
$ sudo npm install -g p5-manager
```

Then you can create all the files for a new p5.js project with one line, like so (substitute your project name):

```
$ p5 generate --bundle myProject
```

or

```
$ p5 g -b myProject
```

Although p5.js is inspired by Processing, and attempts to be similar to it in style, there are some differences. p5.js scripts have a `setup()` and a `draw()` function like Processing sketches, but because p5.js is JavaScript, there are a few differences. The syntax is JavaScript, not Java. That means variables are declared with `var`, functions with `function myFunction()`, and so forth.

A minimal p5.js script looks like similar to a Processing sketch:

```
function setup() {
  // runs once when the page loads
}

function draw() {
  // runs repeatedly while the page is open
}
```

Since much of the interaction in web pages is driven by the user, you need the `draw()` function only if you're running an animation or something else that's independent of the user's actions. Many p5.js scripts do not have a `draw()` function as a result.

Try It Make a new p5.js project, open the **sketch.js** file, and enter the code to the right. Save the file, then open the project's **index.html** file in a browser.

p5.js is good for making responsive web interfaces like this. When you click the button, it will move, and update the text div with its position. As you can see, this sketch has no `draw()` function, since everything is driven by the mouse or the touch on the screen (if you're running it on a mobile device).

For more on the p5.js syntax, see the reference page at p5js.org/reference.

You'll see more of p5.js in the chapters that follow.

```
/*
  Sneaky button
  context: p5.js
  Moves a button when you click it.
*/
var myButton, responseDiv;     // DOM elements

function setup() {
  createCanvas(windowWidth, windowHeight);  // create the canvas
  myButton = createButton('click me');      // create the button
  myButton.touchEnded(changeButton);        // set button's listener function
  myButton.position(10, 10);               // position the button
  responseDiv = createDiv('catch me');     // create a text div
  responseDiv.position(10, 40);            // position it
}

// runs when you release the mouse or stop touching the screen:
function changeButton() {
  var x = random(windowWidth) - myButton.width;    // a new x position
  var y = random(windowHeight) - myButton.height; // a new y position
  myButton.position(x, y);                        // move the button
  responseDiv.html(x + ',' + y);                  // update the responseDiv
}
```

Serial Communication Tools

The remote-access programs in the earlier section were *terminal emulation programs* that gave you access to remote computers through the internet, or to the command line of your own computer, but that's not all a terminal emulation program can do. They can also connect to a computer's serial ports. Before most people had a home internet service provider, connectivity was handled through modems attached to the serial ports of computers. Back then, many users connected to bulletin boards (BBS) and used menu-based systems to post messages on discussion boards, download files, and send mail to other users of the same BBS. Nowadays, serial ports are mostly software drivers associated with various USB and other peripheral devices. In microcontroller programming, they're used to exchange data between the computer and the microcontroller. For the projects in this book, you'll find that using a terminal program to connect to your serial ports is indispensable. There are several freeware and shareware terminal programs available. CoolTerm is an excellent piece of freeware by Roger Meier available from freeware.the-meiers.org. It works on macOS, Windows, and Linux, and it's my personal favorite these days. If you use it, do the right thing and make a donation because it's developed in the programmer's spare time. For Windows users, PuTTY (Figure 1-6) can also open serial ports. Alternatively, you can keep it simple and stick with a classic: the GNU `screen` program running in a terminal window. macOS users can use `screen` as well, though it's less full-featured than CoolTerm.

Serial Communication Using CoolTerm

To get started with CoolTerm, open it and click the Options icon. In the Options tab, you'll see a pulldown menu for the port. In macOS, the port names are similar to this: /dev/tty.usbmodem1441. In Windows, they're `COM1`, `COM2`, `COM3`, and so forth. To find your port for sure, check the list when your serial device is unplugged, then plug it in and click Re-scan Serial Ports in the Options tab. The new port listed is your device's serial connection. To open the serial port, click the Connect button in the main menu. To disconnect, click Disconnect.

Serial Communication Using GNU Screen

Linux and macOS users can also use CoolTerm (Figure 1-7), or you can use the GNU `screen` program. On Ubuntu 15, you will need to install `screen` from the command line like so:

```
$ sudo apt-get install screen
```

Who's Got the Port?

Serial ports aren't easily shared between applications. In fact, only one application can have control of a serial port at a time. If PuTTY, CoolTerm, or the screen program has the serial port open to an Arduino module, for example, the Arduino IDE can't download new code to the module. When an application tries to open a serial port, it requests exclusive control of it either by writing to a special file called a lock file, or by asking the operating system to lock the file on its behalf. When it closes the serial port, it releases the lock on the serial port. Sometimes when an application crashes while it's got a serial port open, it won't release the port, so no other application can open the port. When this happens, restart the operating system, which clears all the locks. Alternatively, you could wait for the operating system to figure out that the lock should be released, but that's less reliable. To avoid this problem, make sure that you close the serial port whenever you switch from one application to another. Otherwise, you may find yourself restarting your machine a lot.

To get started with `screen` on Linux (or macOS), open a terminal window and type:

```
$ ls /dev/tty.*     # Mac OS X
$ ls /dev/tty*      # Linux
```

This command will give you a list of available serial ports. The names of the serial ports in macOS and Linux are more unique, but they're more cryptic than the COM1, COM2, and so on that Windows uses. Pick your serial port and type:

```
$ screen portname datarate
```

For example, to open the serial port on an Arduino board (discussed shortly) at 9600 bits per second, you might type screen /dev/tty.usbmodem1441 9600 on macOS. On Linux, the command might be screen /dev/ttyUSB0 9600. The screen will be cleared, and any characters you type will be sent out the serial port you opened. They won't show up on the screen, however. Any bytes received in the serial port will be displayed in the window as characters. To close the serial port, press Ctrl-A followed by Ctrl-\.

In the next section, you'll use a serial communications program to communicate with a microcontroller.

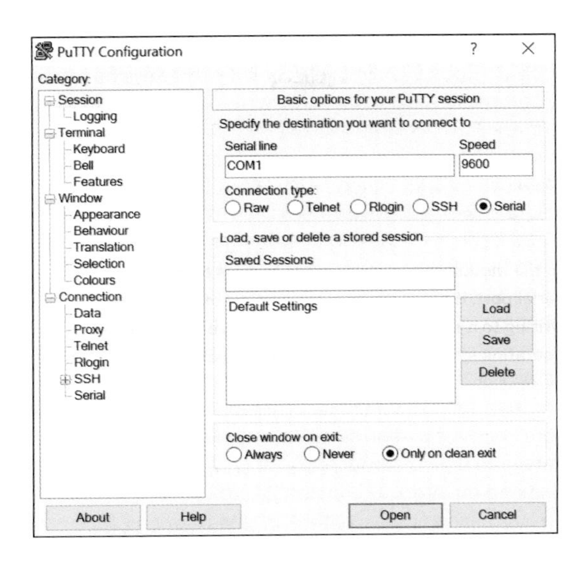

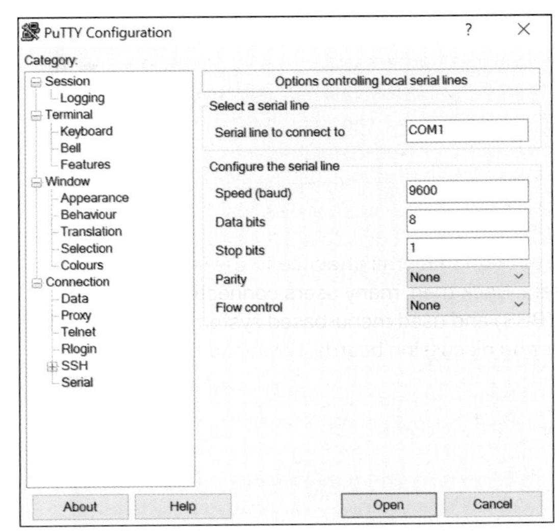

Figure 1-6
Configuring a serial connection in PuTTY.

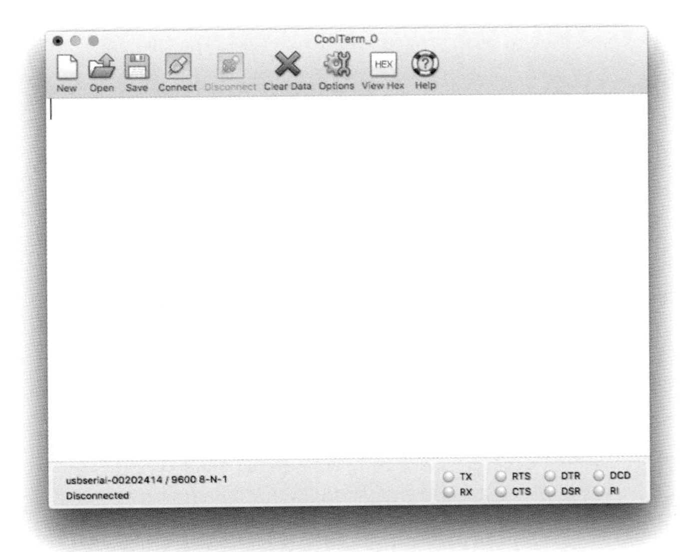

Figure 1-7
The CoolTerm serial terminal program.

Hardware

All of the hardware you'll use in the projects that follow can be broken down into a few categories: diagnostic tools and hand tools, like the ones shown in Figure 1-1; electronic components, sensor, and actuator modules that translate between the physical world and the computational world; consumer devices that have programmable interfaces, such as networked cameras, computers, and mobile devices; and microcontrollers and microprocessors.

Microcontrollers

Look at the things around you. If they use electricity, and have any behavioral logic, there's probably a microcontroller inside. Your electric toothbrush, the brakes in your car, your thermostat, even motion-activated light switches in your house are all probably powered by a microcontroller. They are the tiny computers that interface with the physical world through motors, switches, sensors, lights, and other things you see and touch.

All microcontrollers share a few properties. They communicate with the world by turning on or off the voltage on their output pins, or reading voltage changes on their input pins. These are called *general-purpose input and output* pins (usually abbreviated to *GPIO*, or just I/O). Many microcontrollers can read variable voltage changes on a subset of GPIO pins, called *analog input* pins, using *analog-to-digital converters (ADCs)*. All microcontrollers can communicate with other computers using one or more forms of digital *serial communication*. A microcontroller contains everything you need to run a computer program for a single application: the processor, memory (RAM and ROM), and any peripheral features, like timers, voltage comparators, and communications hardware.

The cheapest and most ubiquitous microcontrollers on the market today are 8-bit controllers, meaning that they can process data and instructions in 8-bit chunks. For many physical tasks, like reading sensors or controlling motors, you don't need a lot of computational power. Nowadays, 32-bit devices, which can process instructions in 32-bit chunks, are getting cheap enough to replace the 8-bit devices. Your personal computer is likely using a 64-bit processor, and your mobile phone is likely using a 32-bit processor.

There are dozens of microcontroller brands on the market. Platforms like Microchip's PIC controllers (www.microchip.com) and Atmel's AVR controllers (www.atmel.com), Texas Instruments' MSP430 (www.ti.com) are excellent microcontrollers, and are used in industrial and consumer applications worldwide. ARM (www.arm.com) has created a 32-bit architecture that's very popular today. They have licensed their designs to many companies, including some of the ones listed earlier. Of course industry giants like Intel (www.intel.com) make microcontrollers as well, and you'll see one of theirs in this book, along with various ARM, Atmel AVR microcontrollers, and perhaps a few others.

Although the microcontrollers themselves are cheap (between $1 and $10 apiece), there can be a pretty significant time investment in getting set up, as the tools for programming these controllers from scratch assume a level of technical knowledge—both in software and hardware—greater than what's needed for other tools listed here. This is where hobbyist platforms come in handy. Platforms like Parallax's Basic Stamp 2 (BS-2), PICAXE, Wiring, and Arduino and the Arduino-compatible derivatives you'll use here are all based on the controllers listed earlier. Each offers different features and capabilities.

Arduino, Wiring, and Derivatives

The main microcontroller used in this book is the Arduino module. The Arduino (www.arduino.cc) and Wiring (www.wiring.co) microcontroller modules both came out of the Institute for Interaction Design in Ivrea, Italy, in 2005. They were originally based on the same microcontroller family, Atmel's ATmega series (www.atmel.com), and they're both programmed in C/C++. The "dialect" they speak is based on Processing, as are the software *integrated development environments* (*IDE*s) they use. You'll see that some Processing commands have made their way into Arduino and Wiring, such as the `setup()` and `loop()` methods (Processing's `draw()` method was originally called `loop()`), the `map()` function, and more.

At the time of the first edition of this book, there was one Wiring board, a handful of variants of Arduino, and almost no derivatives. Now, there are several Arduino models that cover multiple microcontroller architectures. Wiring has seen several versions, culminating in the current Wiring S board, and there are scores of Arduino-compatible derivatives from other companies. Figure 1-8 shows some current and past variations. Though you'll find some differences, most code written for an Arduino board should work on most derivatives. Where there is difference, it will be mentioned in the book. Other platforms have adopted this API, so it's increasingly common to program a microcontroller using the commands from Wiring and Arduino, even if the controller you're using isn't either one of those. If you're using a third-party board, check with the manufacturer of your individual board to see how compatible it is.

In this edition, I've chosen to use 32-bit controllers and controllers with built-in radio communications as the default boards. The Arduno Uno, on which the previous edition was based, is an 8-bit board, as is the current Wiring S board, so some code shown here may not run on them. Generally, the principles in this edition will apply to any microcontroller, however. Each project will list the recommended microcontroller for it, and the features used. All the code will work with the Arduino IDE version 1.8.2 or later, available from www.arduino.cc.

The software and hardware for Arduino, Wiring, and many of the derivatives is open source, and can be downloaded from their respective websites. The boards can be purchased from various online retailers, listed on the sites earlier and in the appendix. Or, if you're a hardcore hardware geek and like to make your own printed circuit boards, you can download the plans to do so. If you're just getting started, I recommend purchasing premade boards first. Once you know your way around the hardware, you can consider making your own boards. There's a project in Chapter 2 that introduces you to making your own on a breadboard.

One of the best things about Wiring and Arduino is that they are cross-platform; they work well on macOS, Windows, and Linux. This is a rarity in microcontroller development environments. Another good thing is that, like Processing, they can be extended. Just as you can include Java classes and methods in your Processing programs, you can include C/C++ code in your Wiring and Arduino programs. For more on how to do this, visit their respective websites.

Figure 1-8. Arduino and compatibles past & present, Wiring, and others: **1.** Arduino Uno **2.** ATtiny85 **3.** LilyPad Arduino **4.** Wiring **5.** Arduino Due **6.** Arduino Pro **7.** Arduino c.2005 **8.** Arduino c.2006 **9.** Arduino Fio **10.** Arduino Micro **11.** Arduino MKR1000 **12.** Adafruit Feather Huzzah! 8266 **13.** SparkFun 8266 Thing **14.** Arduino Leonardo **15.** Red Bear BLE Nano **16.** Arduino 101.

For an excellent introduction to Arduino, see Massimo Banzi's book **Getting Started with Arduino** (Make).

Other Microcontroller Platforms

The number of hobbyist microcontroller platforms has exploded in the past few years. Two trends particularly exciting to watch: higher-level languages on microcontrollers, and the ubiquitous connectivity afforded by inexpensive radios.

Microcontrollers have traditionally been programmed with lower-level languages like C and assembly languages to optimize for speed and efficiency. However, as processors have gotten faster, it's becoming possible to support higher-level scripting languages like Python and JavaScript, which are easier for beginning programmers. 32-bit controllers like the ARM family make this more feasible. The same processor family can also be found in platforms like the MicroPython board, the BBC Micro:bit, Espruino, and NodeMCU. All run on similar processors. If you're a Python or JavaScript fan, check these out.

Processor speed and processing power isn't the only thing that's changed. Radios are getting smaller and cheaper, and it's increasingly common that microcontroller boards come with a communications radio on board. Both the MKR1000 and the 101 have built-in radios, wireless Ethernet (WiFi) and Bluetooth LE, respectively. Many others do as well, like Espressif's ESP8266 boards, which are WiFi native, and RedBear Labs' BLE Micro and BLE Nano, which are Bluetooth LE native. If the trend continues, it will soon be the default that microcontrollers come with a built-in radio.

Microntroller Features

Microcontrollers are just small computers. Like every computer, they have inputs, outputs, a power supply, and communications ports to connect to other devices. Figure 1-9 shows the functional parts of the 101 and the MKR1000. Both have connections for power and ground as well as input and output (GPIO) pins. Some pins that also function as analog inputs. Other pins afford various forms of serial communication as well: *asynchronous serial communication* (which you'll learn about in a few pages) through the *Universal Asynchronous Receiver-Transmitter (UART)*, and two forms of *synchronous serial communication*: *Serial-Peripheral Interface (SPI)* and *Inter-Integrated Circuit communications (I2C)*. You'll see more on both of these in Chapter 2 and later chapters. Some microcontrollers are able to communicate via USB natively as well. Others, like the Arduino Uno, connect to USB through an external USB-to-serial converter chip.

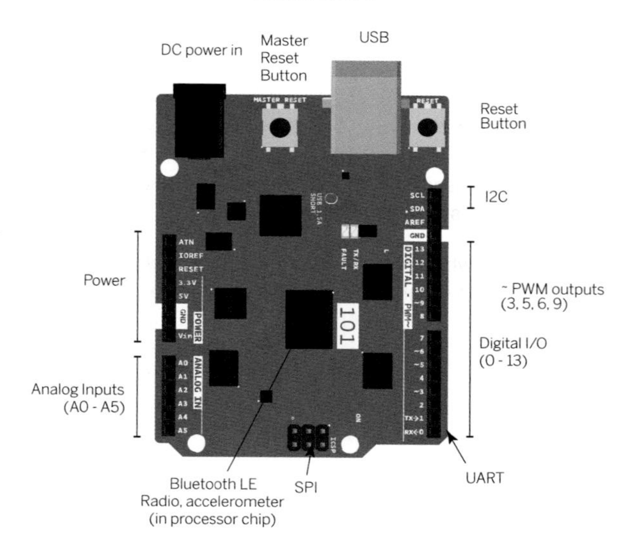

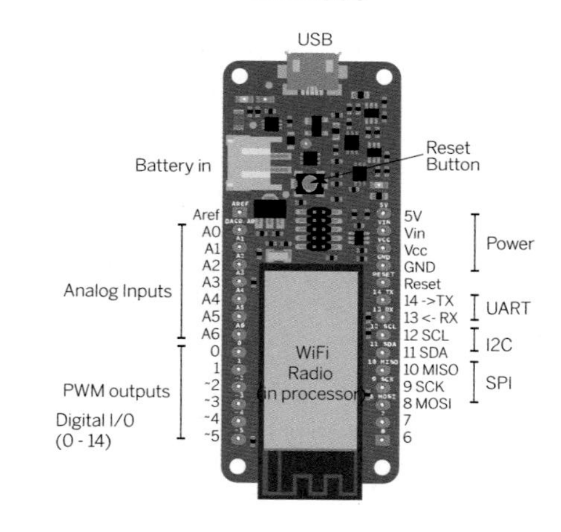

Figure 1-9
Functional parts of a microcontroller. Note that the pin's function numbers (e.g., Analog 2, digital 2, etc.) do not correspond to the physical pin numbers. Technical documentation usually refers to the functional pin number, not the physical pin.

Most microcontroller pins can be used for more than one purpose, depending on what you connect to them and how you program the microcontroller. For example, some of the digital I/O pins are also marked with a tilde (~). These pins are connected to a circuit internal to the chip that can pulse those pins on and off with precise timing, and can modulate that pulse in order to produce a varying average voltage. This function, called *pulse width modulation (PWM)*, is used to dim light sources and to control the speed of motors, among other things.

Regardless of what microcontroller you're working with, you'll find some combination of these features:

GPIO (general-purpose input and output)
analog input
PWM (pseudo-analog output)
serial communications:
  UART (asynchronous)
  SPI (synchronous)
  I2C (synchronous)

Some will have more features, and some less, depending on size and price. When you're deciding what controller to use for your project, consider whether it's got the necessary combination of these features.

Which Microcontroller Board Should I Get?

In this edition, I've chosen to work with on a small subset of the available Arduino boards and derivatives, representing three different microcontrollers. There are many other options, and much of the code contained herein may work on other boards, but these are the ones I think do the best job, and use most frequently myself.

The first is the Arduino 101 board, based on the Intel Curie controller. The Curie actually has two 32-bit processors in it: an x86 processor and an ARC processor working together. It's got the same pin layout as the Arduino Uno, and has a built-in gyroscope, accelerometer, real-time clock, and Bluetooth LE radio.

Previous editions of this book used Arduino boards that operated on 5 volts and were compatible with 5-volt external components and modules, but the ones used in this edition are all 3.3 volt modules. Some of them, like the 101, are 5-volt tolerant, but make sure to check that any external components or circuits you use can operate at 3.3 volts before connecting them to these boards.

The second is the MKR1000 board, based on Atmel's SAMD21 Cortex-M0+ controller. This board has a smaller layout and a battery connector and charging circuit, so it's better for mobile device projects. It's also a built-in real-time clock and a WiFi radio on board.

The third is Espressif's ESP8266 microcontroller. The ESP8266 is very inexpensive, yet has a 32-bit processor and built-in WiFi radio. There are several companies that make boards with this processor, and it can be a bit more complex to work with, depending on whose you're using. It doesn't have as many I/O pins as the other two boards either. However, it works well for applications that don't require a lot of I/O, or as a dedicated WiFi controller for another microcontroller.

As of this writing, there's no official ESP8266 board from Arduino.cc, but there are many variants made by other companies. Of the ESP8266 boards I've tested, the Adafruit's Huzzah! ESP8266 boards and SparkFun's ESP8266 Thing boards have been the simplest with which to work. Both Adafruit and SparkFun have confusing names for their products in this area, however: Adafruit carries the Feather Huzzah! ESP8266 and the slightly cheaper Huzzah! ESP8266 breakout board. SparkFun carries the ESP8266 Thing and the ESP8266 Thing Dev Board for the same price. The Feather Huzzah! ESP8266 and the Thing Dev Board are the simpler boards to use: you program them directly from their USB ports. The Huzzah! ESP8266 breakout board and the ESP8266 Thing require you to use an external USB-to-serial adapter. When in doubt, use the Feather Huzzah! 8266 or the Thing Dev Board.

Other specialty boards will show up in some projects as well, but these will be the defaults.

X

Your First Microcontroller Program

Download the latest Arduino software from www.arduino.cc and then follow these instructions.

Setup on macOS

Double-click the downloaded file to unpack it, and you'll get a disk image that contains the Arduino application. Drag the application to your **Applications** directory. Once you're installed, open the Arduino application and you're ready to go.

Setup on Windows 10

Download the Windows installer, unzip it, and run it. You'll be asked for permission to install a few times, for the compiler and for the USB drivers. When the drivers are installed, you're ready to go.

Shields and Modules

One of the features that makes Arduino easy to work with are the add-on modules called shields, which allow you to add preassembled circuits to the main module. The layout of pins was designed to be standard to support these add-ons, and the Arduino Uno rev3 layout has been adopted by some other microcontroller platforms as well. For most applications you can think of, there's a third-party company or individual making a shield to do it. The growth of shields was a major factor in the spread of Arduino, and the well-designed and documented ones make it possible to build many projects with no electronic experience whatsoever. The shield footprint, like the board designs, is available online at www.arduino.cc.

Since the last edition of this book, new microcontroller modules in a variety of form factors have come to market. As a result, breadboard-compatible versions of many shields have become popular as well. Figure 1-10 shows various shields and modules. Most of the projects in this book will use the breadboard-friendly modules, since they can work with either of the two default boards, the MKR1000 and the 101.

Beware! Not every shield or module is compatible with every board. Voltage differences are one common source of problems, as many platforms are transitioning from 5 volts to 3.3 volts. Before you buy, check the technical specifications of your board and of any shield or module you want to use to be sure it they are compatible with each other.

X

Figure 1-10
Shields and external modules enable a wide range of applications.

Setup on Linux

Arduino for Linux depends on the flavor of Linux you're using. For Ubuntu 14 and later, download the package from www.arduino.cc, unzip the package, open a terminal window, and do the following:

```
$ sudo mv arduino-1.8.2 /opt
$ cd /opt/arduino-1.8.2/
$ chmod +x install.sh
$ ./install.sh
```

Once the install script is run, you'll have an Arduino icon on the Desktop to launch the application. You'll also need to add yourself to the `dialout` group so that you can open serial ports:

```
$ sudo usermod -a -G dialout $USER
```

Now you're ready to launch Arduino. Connect the module to your USB port and double-click the Arduino icon to launch the software. The editor looks like Figure 1-11.

The environment is based on the Processing IDE and has New, Open, and Save buttons on the main toolbar. In Arduino, the Run function is called Verify, and there is an Upload button as well. Verify compiles your program to check for any errors, and Upload both compiles and uploads your code to the microcontroller module. There's an additional button, Serial Monitor, that you can use to receive serial data from the module while you're debugging. X

Updates to the Arduino IDE software occur frequently. The notes in this book refer to Arduino version 1.8.2. By the time you read this, the specifics may be slightly different, so check www.arduino.cc for the latest details.

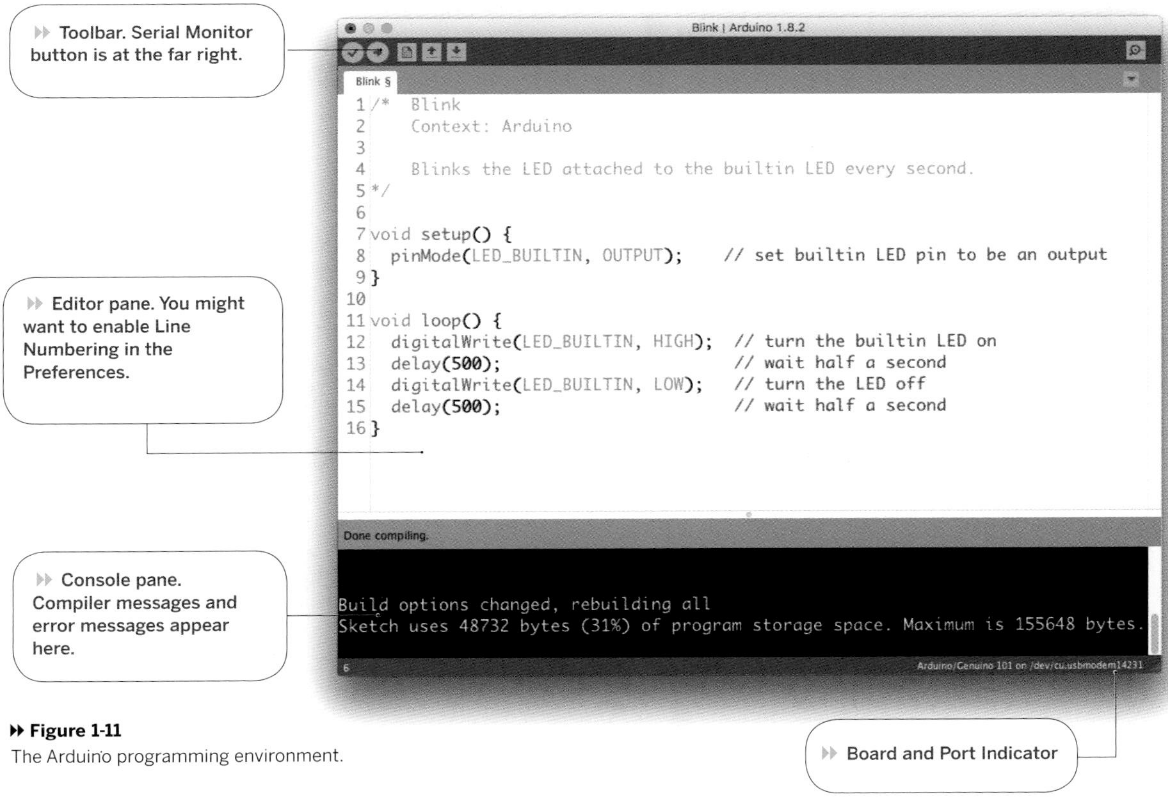

Toolbar. Serial Monitor button is at the far right.
Blink | Arduino 1.8.2
Blink §
1 /* Blink
2   Context: Arduino
3
4   Blinks the LED attached to the builtin LED every second.
5 */
6
7 void setup() {
8   pinMode(LED_BUILTIN, OUTPUT);   // set builtin LED pin to be an output
9 }
10
11 void loop() {
12   digitalWrite(LED_BUILTIN, HIGH); // turn the builtin LED on
13   delay(500);   // wait half a second
14   digitalWrite(LED_BUILTIN, LOW);   // turn the LED off
15   delay(500);   // wait half a second
16 }
Editor pane. You might want to enable Line Numbering in the Preferences.
Done compiling.
Build options changed, rebuilding all
Sketch uses 48732 bytes (31%) of program storage space. Maximum is 155648 bytes.
Console pane. Compiler messages and error messages appear here.
Arduino/Genuino 101 on /dev/cu.usbmodem14231
Board and Port Indicator

Figure 1-11
The Arduino programming environment.

Try It Like Processing programs, Arduino programs are often referred to as *sketches*. Here's your first sketch.

```
/*  Blink
    Context: Arduino

    Blinks the LED attached to the board every second.
  */

void setup() {
  pinMode(LED_BUILTIN, OUTPUT);    // set built-in LED pin to be an output
}

void loop() {
  digitalWrite(LED_BUILTIN, HIGH);  // turn the built-in LED on
  delay(500);                        // wait half a second
  digitalWrite(LED_BUILTIN, LOW);   // turn the LED off
  delay(500);                        // wait half a second
}
```

Type the code into the editor. Click Tools→Board to choose your Arduino model. If your board's not there, add it using the Boards Manager (see sidebar). Then click Tools→Serial Port to choose the serial port of your board. On the Mac or Linux, the serial port will have a name like this: /dev/tty.usbmodem1421 (Arduino 101). On Windows, it should be COMx, where x is some number (for example, COM5).

NOTE: On Windows, COM1–COM4 are generally reserved for built-in serial ports, regardless of whether your computer has them.

Once you've selected the port and model, click Verify to compile your code. When it's compiled, you'll get a message at the bottom of the window saying `Done compiling`. Then click Upload. This will take a few seconds. Once it's done, you'll get a message saying `Done uploading`, and a confirmation message in the serial monitor window that says:

```
Sketch uses xx bytes (yy%) of program storage space. Maximum
is zz bytes.
```

The values for the sketch size and the maximum sketch size will vary depending on which board you're using. Once the sketch is uploaded, you'll see the LED marked L (for "LED") blink on and off once per second. A blinking LED is the microcontroller equivalent of "Hello World!"

The Boards Manager

Some of boards used in this book may not yet included with the download of the IDE as of this writing. In order to use them, you'll need to use the IDE's Boards Manager to install the definition files for these boards. In Arduino 1.6.8 and later, click Tools→Boards→Boards Manager to open the Boards Manager.

To install a new board, search the list for the board's name. If the board is already installed, it will say "Installed" in the board listing. If not, choose the latest version and click Install.

If you're installing a third-party board, there's an extra step. You'll need to add the URL of the board's software repository to the IDE. For example, to install the ESP8266 (any variant), click on Preferences and look for the Additional Boards Manager URLs box. Enter the board's URL in the box. For the ESP8266, it's: http://arduino.esp8266.com/stable/package_esp8266com_index.json. Then click OK and restart the IDE. When you restart, you'll find the boards you're using in the Tools→Boards menu. If you're working with other third-party boards, follow this same procedure.

▸▸ NOTE: **If it doesn't work, you might want to seek out some external help. The Arduino Learning section has many tutorials (www.arduino.cc/en/Tutorial). The Arduino forums (www.arduino.cc/forum) are full of helpful people who love to answer these sort of things.**

Although the I/O pins all have numbers that you normally use to refer to them in code, this sketch uses `LED_BUILTIN` for the built-in LED's pin number. The built-in LED is on different pins for each board. For the 101, and all the Uno-layout boards, the built-in LED is attached to pin 13, but on the MKR1000, it's attached to pin 6. The keyword `LED_BUILTIN` is reserved for the built-in LED on every board, and is assigned to the proper number in the board's definition file. You'll never have to worry about the definition files; just know that they exist and that they describe all the features of each different microcontroller board.

X

Try It This next Arduino program listens for incoming serial data. It adds one to whatever serial value it receives, and then sends the result back out. It also blinks an LED on the board regularly—on the same pin as the last example—to let you know that it's still working.

```
/*
    Simple Serial
    Context: Arduino
    Listens for an incoming serial byte, adds one to the byte
    and sends the result back out serially.
    Also blinks the built-in LED every second.
 */

int inByte = 0;               // variable to hold incoming serial data
long blinkTimer = 0;          // keeps track of how long since the LED
                              // was last turned off
int blinkInterval = 1000;     // a full second from on to off to on again

void setup() {
  pinMode(LED_BUILTIN, OUTPUT);     // set pin 13 to be an output
  Serial.begin(9600);             // configure the serial port for 9600 bps
}

void loop() {
  // if there are any incoming serial bytes available to read:
  if (Serial.available() > 0) {
        inByte = Serial.read();  // then read the first available byte,
        Serial.write(inByte+1);  // add one to it and send the result out
  }

  // Meanwhile, keep blinking the LED.
  // after a half a second, turn the LED on:
  if (millis() - blinkTimer >= blinkInterval / 2) {
    digitalWrite(LED_BUILTIN, HIGH);       // turn the LED on
  }
  // after a half a second, turn the LED off and reset the timer:
  if (millis() - blinkTimer >= blinkInterval) {
    digitalWrite(LED_BUILTIN, LOW);  // turn the LED off
    blinkTimer = millis();            // reset the timer
  }
}
```

Serial Communication with a Microcontroller

One of the most frequent tasks you'll use a microcontroller for in this book is to communicate serially with another device, either to send sensor readings over a network or to receive commands to control motors, lights, or other outputs from the microcontroller. The sketch on the previous page shows how that works. First, you'll configure the serial connection for the right data rate. Then, you'll read bytes in, write bytes out, or both, depending on what device you're talking to and how the conversation is structured.

The Serial Monitor in the Arduino IDE provides you with a way to send and receive serial messages to your Arduino module.To send bytes from the computer to the microcontroller module, first compile and upload this program. Then click the Serial Monitor icon (the rightmost icon on the toolbar). The screen will change to look like Figure 1-12. The serial data rate should be set to 9600 bits per second (bps or baud) by default.

Type any letter in the text entry box and press Enter or click Send. The module will respond with the next letter in sequence. For every character you type, the module adds one to that character's ASCII value, and sends back the result.

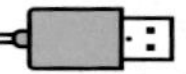

Where's My Serial Port?

The USB serial port that's associated with the Arduino module is actually a software driver that loads every time you plug in the module. When you unplug, the serial driver deactivates and the serial port will disappear from the list of available ports. You might also notice that the port name changes when you unplug and plug in the module. On Windows machines, you may get a new COM number. On Macs, you'll get a different alphanumeric code at the end of the port name.

It may take a few seconds after you plug your board in before the port shows up in the Port menu. That's because the board's bootloader needs to finish before your computer can identify it. Some bootloaders, like the 101's, take longer than others.

Never unplug a USB serial device when you've got its serial port open; you should close the Serial Monitor before you unplug anything. Otherwise, you're likely to crash the application, and possibly the whole operating system.

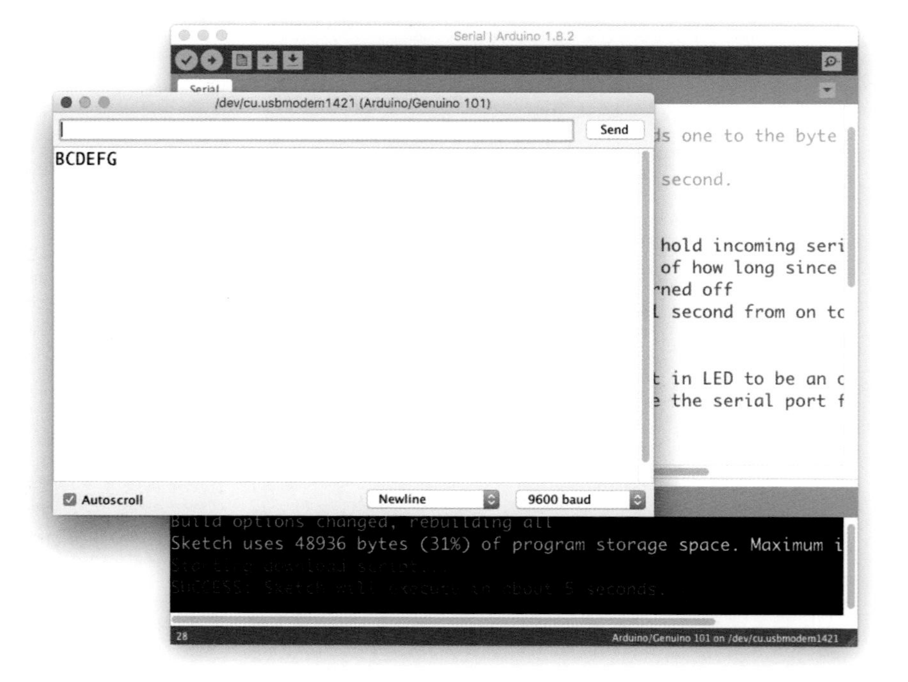

Figure 1-12
The Serial Monitor in Arduino, running the previous sketch. The user typed ABCDEF.

Connecting Components to the Module

Microcontroller boards don't have many sockets for connections other than the I/O pins, so you'll need to keep a solderless breadboard handy to build subcircuits for your sensors and actuators (output devices). Figure 1-13 shows a standard setup for connections between the two.

Basic Input-Output Circuits

There are two basic I/O circuits that you'll use a lot in this book: digital input and analog input. If you're familiar with microcontroller development, you're already familiar with them. Any time you need to read a sensor value, you can start with one of these. Even if you're using a custom sensor in your final object, you can use these circuits as placeholders, just to see any changing sensor values.

Digital Input

A digital input to a microcontroller is nothing more than a switch. The switch is connected to voltage and to a digital input pin of the microcontroller. A high-value resistor (10 kilohms is good) connects the input pin to ground. This is called a *pulldown resistor*. Other electronics tutorials may connect the switch to ground and the resistor to voltage. In that case, you'd call the resistor a *pullup resistor*. Pullup and pulldown resistors provide a reference to power (pullup) or ground (pulldown) for digital input pins. When a switch is wired as shown in Figure 1-14, closing the switch sets the input pin high. Wired the other way, with the switch connected to ground and a pullup resistor to power, closing the switch would set the input pin low.

Analog Input

The circuit at the top of Figure 1-15 is called a *voltage divider*. The variable resistor and the fixed resistor divide the voltage between them. The ratio of the resistors' values determines the voltage at this connection. For example, two equal resistors in series will halve the voltage. A 1-kilohm resistor in series with a 2-kilohm resistor (with the former connected to the input voltage) will give you an output voltage that's two-thirds of the input voltage. If you connect the analog-to-digital converter of a microcontroller to this point, you'll see a changing voltage as the variable resistor changes. You can use any kind of variable resistor: photocells, thermistors, force-sensing resistors, flex-sensing resistors, and more.

The *potentiometer*, also shown in Figure 1-15, is a special type of variable resistor. It's a fixed resistor with a wiper that slides along its conductive surface. The resistance changes between the wiper and both ends of the resistor as you move the wiper. Basically, a potentiometer (*pot* for short) is two variable resistors in one package. If you connect the ends to voltage and ground, you can read a changing voltage at the wiper.

There are many other circuits you'll learn in the projects that follow, but these are the staples of all the projects in this book.

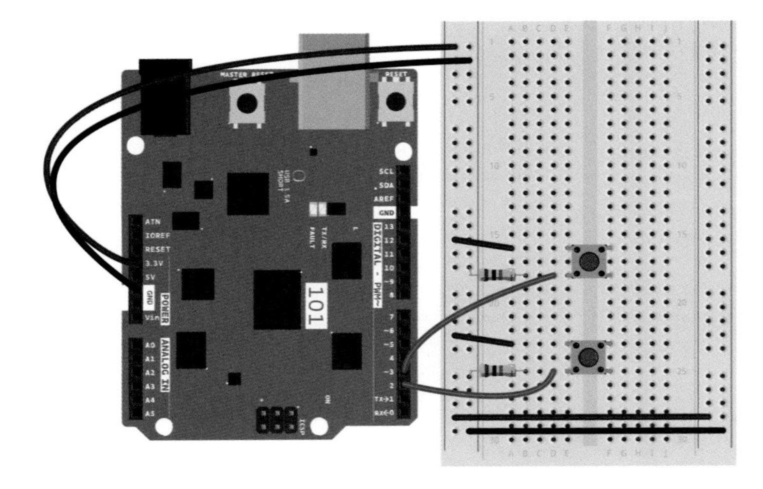

Figure 1-13
Arduino connected to a breadboard. +3.3V and ground run from the module to the long rows of the board. This way, all sensors and actuators can share the +3.3V and ground connections of the board. *Control* or *signal* connections from each sensor or actuator run to the appropriate I/O pins. In this example, two pushbuttons are attached to digital pins 2 and 3 as digital inputs.

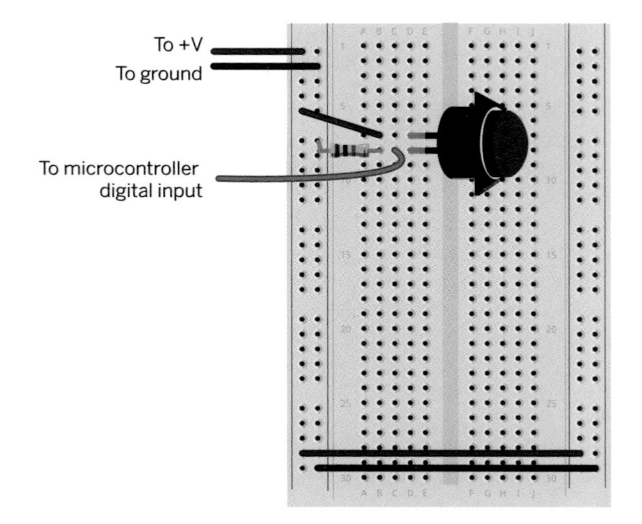

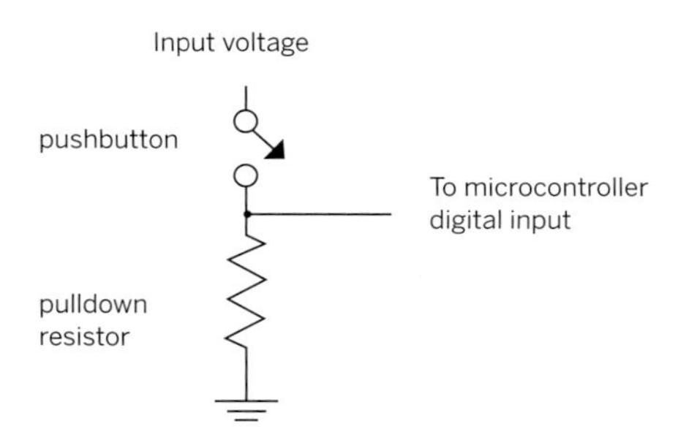

Figure 1-14
Digital input to a microcontroller.
Left: breadboard view.
Right: schematic view.

Document What You Make

You'll see a lot of circuit diagrams in this book, as well as flowcharts of programs, system diagrams, and more. The projects you'll make with this book are systems with many parts, and you'll find it helps to keep diagrams of what's involved, which parts talk to which, and what protocols they use to communicate. I used a few drawing tools heavily in this book, all of which I recommend for documenting your work:

Adobe Illustrator (www.adobe.com/products/illustrator.html). You really can't beat it for drawing things, even though it's expensive and takes time to learn well. There are many libraries of electronic schematic symbols freely available on the web.

Affinity Designer (affinity.serif.com/en-us/designer/) Affinity is a newcomer on the block, and Designer is still in beta for Windows as of this writing. It's simpler, faster, and less expensive than Illustrator for many things, and handles Scalable Vector Graphics (SVG) files better than Illustrator.

Fritzing (www.fritzing.org). Fritzing is an open source tool for documenting, sharing, teaching, and designing interactive electronic projects. It's a good tool for learning how to read schematics, because you can draw circuits as they physically look, and then have Fritzing generate a schematic of what you drew. Fritzing also has a good library of vector graphic electronics parts that can be used in other vector programs. This makes it easy to move from one program to another in order to take advantage of all three.

The figures here were cobbled together from all three tools, combining the work of many designers, including Jody Culkin and Giorgio Olivero, with a few details from André Knörig and Jonathan Cohen's Fritzing drawings.

It's a good idea to keep notes on what you do as well, and share them publicly so others can learn from them. I rely on a combination of three note-taking tools: blogs powered by Wordpress (www.wordpress.org) at www.makingthingstalk.com, tigoe.net/blog, and tigoe.net/pcomp/code; a github repository (github.com/tigoe); and a stack of Maker's Notebooks (www.makershed.com, part no. 9781449358976).

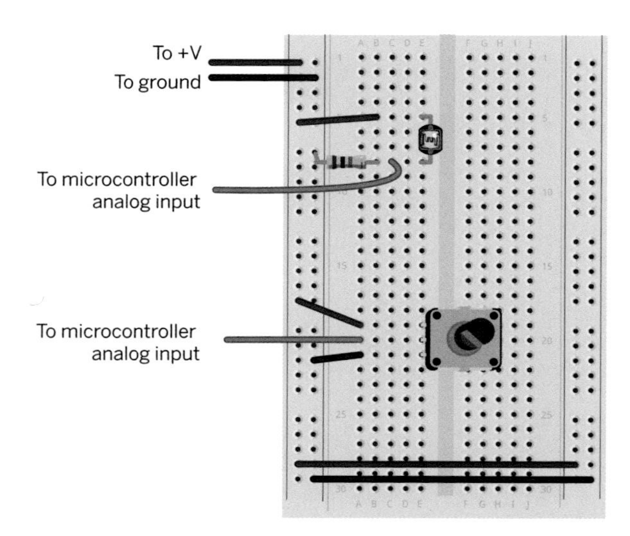

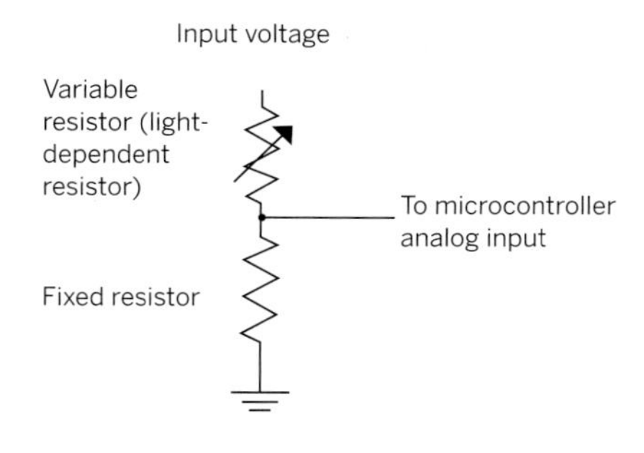

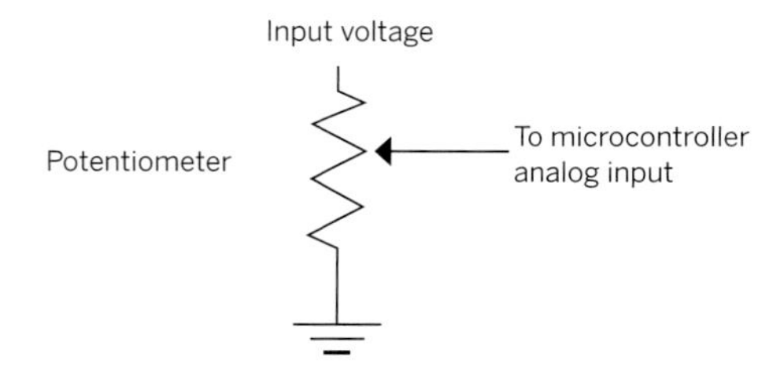

Figure 1-15
Analog input to a microcontroller. This circuit shows two different analog inputs: a light-dependent resistor in a voltage divider, and a potentiometer. These would connect to two different analog inputs.
Left: breadboard view.
Right: schematic view.

Specialty Circuits and Modules

You'll see a number of specialty circuits and modules throughout this book. You'll also build a few of your own circuits for specific projects. All of the circuits will be shown on a breadboard like those in Figures 1-13 through 1-15, but you can build them any way you like. If you're familiar with working on printed circuit boards and prefer to build your circuits that way, feel free to do so.

Each breadboard figure will also have an accompanying schematic diagram, which is the standard way to show electrical relationships between components in a circuit. Schematics don't show the physical layout of the circuit; they just show how the components are connected to each other.

Since the projects can run on different Arduino and compatible variants, the schematic view will typically just show the pins to which you're connecting, like these figures do. X

You will encounter variations on many of the modules and components used in this book. Be sure to check the data sheet on whatever component or module you're using, as your version may vary from what is shown here.

Microprocessors, or Single-Board Computers

Typically, microprocessors, unlike microcontrollers, do not have as many peripheral features on the chip: memory, analog I/O, and video co-processors have to be added separately. This is changing, however, and the line between microcontrollers and microprocessors is blurring. In the past five years, microprocessor modules, commonly known as *single-board computers*, have gained more peripheral features, and have gotten cheap and simple enough for hobbyists to use. These boards are more complex than microcontrollers. They can run a full operating system, including filesystems, multitasking, and all that you'd expect. Most of them run some variation of Linux, although Windows 10 has a version optimized for single-board computers as well. Most of these have several digital GPIO pins, but few of them have analog pins. Not all of them have GPIO that's appropriate for use at 3.3 volts or higher, however, so check your board's specifications before you buy.

In the projects that follow, you'll see the Raspberry Pi, a popular single-board computer at the moment. There are many other embedded Linux microprocessors in the market, including Texas Instruments' BeagleBone, an open source hardware board with performance very similar to the Raspberry Pi. Much of what you'll see will work on boards like the BeagleBone as well, with some minor changes.

The Raspberry Pi 3 model B is running an ARM Cortex A7 processor from Broadcom, with 1GB of RAM, running at 1.2GHz. It also has a separate video processor, HDMI output, WiFi, and Bluetooth. It runs a version of the Linux operating system called Raspbian, which you can access from the command-line interface or from its desktop GUI if you connect a monitor to it.

When choosing an embedded microprocessor board, you have to consider not only memory, processor speed, and physical features, but also whether there are easily available drivers for peripherals like WiFi, cameras, audio, and so forth. These are the features that make embedded boards really handy, and if they're not available for a given board, your life gets difficult. This is one area where the Raspberry Pi does well.

Raspberry Pi

The Raspberry Pi first came out in February 2012. Announced as a $35 computer, it made a big splash in the hobbyist community, and has remained very popular as a platform to learn about programming and networks. They've released a few variations since then. Of them all, the Raspberry Pi 3 is the fastest, nearly as powerful as a low-end laptop processor. At the other end, the Pi Zero W is a stripped-down low-power board that sells for $10, and it also has built-in WiFi.

To call the Pi a $35 computer is a bit misleading. To use it like a personal computer, you need to add a monitor, a keyboard, and a hard drive, and by then you've spent a few hundred dollars. Even then, you might be disappointed with its performance. The graphic interface can feel slow compared to a current laptop or desktop computer. It's more effective to think of them as small, cheap network

Figure 1-16
Embedded processors: Raspberry Pi 3, Raspberry Pi Zero W, BeagleBone Green, Arduino Yún.

servers and clients, and they're more responsive when you interact with them through the command-line interface or through a web browser. When combined with a microcontroller, an embedded microprocessor makes it easy to build a browser-based interface for sensors, lights, and motors. When used this way, the price tag is more accurate, and the performance is excellent for these kinds of applications.

To use the Pi for any applications in this book, you'll need four essential components:

- A 5-volt power supply with USB Micro connector. Get one that can supply at least 2 amps of current. Many USB charging adapters for mobile phones will do the job well.
- A microSD card. Get one that's at least 8GB. 16GB is better. Get several.
- A USB-to-serial cable. You'll use this to connect to the Pi's command-line interface before you get it connected to your network. There are a few workable models, like Adafruit's USB-to-TTL Serial Cable for Raspberry Pi **AF:** 954 or FTDI Friend **AF:** 284
- A WiFi USB dongle, unless you're using the Pi 3 or the Zero W. You can connect the Pi to your network using an Ethernet cable if your router is within cable range, but most of the time WiFi is more convenient. Raspberry Pi has an official Pi WiFi adapter (**AF**: 2638), but you can also use other, slightly less expensive models, like the Miniature WiFi (802.11b/g/n) Module from Adafruit (**AF**: 814), which uses Realtek's WiFi radio. Before buying a third-party adapter, check the Raspberry Pi forums to make sure that the one you want has drivers for the operating system you're using on your board.

Getting Started with the Raspberry Pi

To get started, download the latest version of Raspbian, the Raspberry Pi Linux distribution, from www.raspberrypi.org/downloads/raspbian/. Download Raspbian Jessie, version 4.4 or later. The Lite version will work fine for this book, and is easier to make secure. Follow the instructions on that site to install the operating system on a microSD card. When it's ready, insert the card into your Raspberry Pi. Connect your USB-to-serial adapter as shown in Figure 1-17; then plug it into your personal computer. Install any needed adapters for your USB-to-serial adapter and you should then see it show up as a serial port in CoolTerm or whatever serial terminal application you're using. Open a serial connection at 115200 bps, and press the spacebar a few times. You'll get a login prompt like so:

```
Raspbian GNU/Linux 8 raspberrypi ttyAMA0
raspberrypi login:
```

The default login for the Raspbian OS is `pi`, and the default password is `raspberry`. Log in!

Once you're on the command-line interface of the Pi, you're running a Linux computer. All the commands you

> **For versions of Raspbian from April 1, 2017, and after, you'll need to edit the configuration file to enable the serial connection. Before putting your Raspbian SD card in the Pi the first time, add the following to the end of the config.txt file that's on the card, on a line by itself:**
>
> ```
> [code style]
> enable_uart=1
> ```

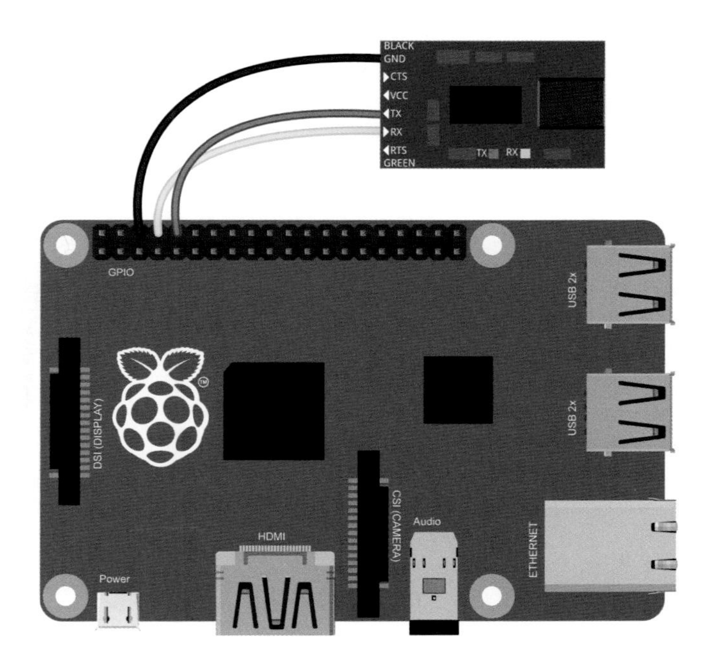

Figure 1-17
Raspberry Pi 2/3 Model B with USB-to-Serial adapter attached. The serial pins are the same as this on all Pi models.

learned in the "Using the Command Line" section are applicable here.

First, use the `raspi-config` command to make two configuration changes. At the command prompt, type:

```
$ sudo raspi-config
```

You'll see a menu that you can scroll through using the arrow keys and select items from using the Enter key. Choose option 1, Expand Filesystem. When it's done, choose option 2, Change Password, and set your own password.

You're connected serially to the Pi, but you're not yet connected to the network. To do that, you either need an Ethernet cable that you can connect to your router, or a WiFi adapter like the ones mentioned earlier.

Connecting via Ethernet

To connect using an Ethernet cable, plug one end of your cable into one of the Ethernet jacks on your router and the other end into your Raspberry Pi. If your router is set up to use *Dynamic Host Configuration Protocol (DHCP)*, like most routers are by default these days, then it will automatically assign your Pi a network address. You can check by typing:

```
$ ifconfig eth0
```

You should get a response like this:

```
eth0      Link encap:Ethernet  HWaddr b8:27:eb:63:37:4a
          inet addr:192.168.0.17  Bcast:192.168.0.255
           Mask:255.255.255.0
          inet6 addr: fe80::ffc5:e696:3b93:5b47/64
           Scope:Link
          UP BROADCAST RUNNING MULTICAST  MTU:1500
           Metric:1
          RX packets:110 errors:0 dropped:0 overruns:0
           frame:0
          TX packets:103 errors:0 dropped:0 overruns:0
           carrier:0
          collisions:0 txqueuelen:1000
          RX bytes:38147 (37.2 KiB)  TX bytes:12819
           (12.5 KiB)
```

This tells you that you're connected to the network on the wired Ethernet port (called eth0), and that its network address is 192.168.0.17. If you don't have an address, you're not connected to the network. You'll learn more about the structure of network addresses in Chapter 3.

Connecting via WiFi

If you're using one of the WiFi adapters mentioned earlier, then the drivers for it are included in Raspbian Jessie by default. Plug it into the USB port. To check, type:

```
$ iwconfig wlan0
```

If you get a response that begins like this:

```
wlan0     IEEE 802.11bgn  Nickname:"<WIFI@REALTEK>"
```

then you know the radio is there and working. Try this command to see nearby WiFi hotspots:

```
$ sudo iwlist wlan0 scan
```

You will get a list of WiFi hotspots nearby and their credentials. If it's a long list, try piping the output through `less` to paginate it:

```
$ sudo iwlist wlan0 scan | less
```

Then you can see the listings page by page using the spacebar, and close the list with `q`, as you saw earlier.

The commands you just used, `ifconfig` and `iwconfig`, are tools for managing your network interfaces (hence ifconfig, for "interface configuration"). The arguments you gave them, `eth0` and `wlan0`, refer to two network interfaces on your computer, the Ethernet port and the wireless LAN (local area network) port. You'll learn more about the details of network interfaces later, but now let's configure the WiFi port. There's a command-line tool that comes with the Raspbian distribution called `wpa_cli` that does this well.

Configuring WiFi Using wpa_cli

The `wpa_cli` program lets you discover, set, and save network configurations from the command line. First, you need to create a new network record using `wpa_cli`. Then you'll set the network to which you want to connect. Then you'll set the password, enable the network, and save the configuration. This program has an interactive mode, so you can start it, enter several commands in a row, then quit it. Start it by typing:

```
$ sudo wpa_cli
```

You'll see the `wpa_cli` intro banner next, ending with the following lines:

```
Selected interface 'wlan0'
Interactive mode
>
```

The `>` is your prompt to enter commands. Type:

```
add_network
```

> The sudo command in POSIX systems gives an ordinary user permissions to do things only a system administrator (called a superuser) can do. Normally when you run a command using sudo, you're asked for the password before the command executes. In this case, wpa_cli needs to be run as a superuser in order to save the network configuration. On the Raspberry Pi, you'll run most commands using sudo.

If there are no configured networks, you'll get the response:

```
0
>
```

This is the number of the network you're going to configure. Now type:

```
> set_network 0 ssid "Your_SSID"
> set_network 0 psk "password"
```

Replace Your_SSID and password with the name and password for your WiFi network. If you're connecting to a network with no password, add

```
> set_network 0 key_mgmt NONE
```

Next, type:

```
> enable_network 0
> save_config
```

You've enabled your Raspberry Pi to access the network configuration you set up, and you've saved the configuration to disk. Type `quit` to exit the `wpa_cli` program. You can check that you're connected using `ifconfig wlan0` and `iwconfig wlan0` as you did before. This time, you should see a network address returned by `ifconfig` and a network name from `iwconfig`.

Testing Your Network Connection with curl

Whether you connected via Ethernet or WiFi, you should now have a good connection to the network. You can test your network connection by making a web request. The command-line program curl is the best way to do this. Curl is a command-line web browser, of sorts.Type:

```
$ curl http://www.example.com
```

ssh Connection to a Single-Board Computer

Once your single-board computer is connected to your local network, you can access it via ssh. You'll need to know its network address, which you obtained from the Raspberry Pi when you ran `ifconfig` from the serial terminal connection. Open a terminal window and type `ssh username@xx.xx.xx.xx`. Replace `username` with the name appropriate to your device. For the Raspberry Pi, it's `pi`. For many others, it's `root`. Replace `xx.xx.xx.xx` with the network address of your board. For example, my Pi's address is 192.168.0.23, so I would type `ssh pi@192.168.0.23`. When the remote board asks for your password, enter it, and then you'll be logged in just like any other remote server.

You should get a response that includes a lot of HTML, ending with the following:

```
<body>
<div>
<h1>Example Domain</h1>
nts. You may use this
    domain in examples without prior coordination or asking
for permission.</p>
    <p><a href="http://www.iana.org/domains/example">More
information...</a></p>
</div>
</body>
</html>
```

If you got this result, do a little dance of celebration. Your Raspberry Pi is now connected to the internet! The `curl` command won't display the web page you requested; it will just give you the raw HTML page for which you asked. To see what the same page looks like, go to the same URL in a browser.

Set Up Housekeeping

Once you're set up and on a network, you might want to update your system and install a current version of node.js. The following are the steps I run with any new Pi.

Update the software package list using `apt`, the Raspbian software package manager:

```
$ sudo apt-get update
```

Remove old versions of node.js:

```
$ sudo apt-get remove --purge node* npm*
```

Install node version manager (nvm) using `curl`:

```
$ sudo curl -o- https://raw.githubusercontent.com/creationix/
nvm/v0.33.1/install.sh | bash
```

Log out and log back in:

```
$ sudo logout
raspberrypi login: pi
```

Then install node.js version 6.9.5:

```
$ nvm install v6.9.5
```

Make shortcut links so you can call node and the node package manager (npm) easily from the command line:

```
$ sudo ln -s /home/pi/.nvm/versions/node/v6.9.5/bin/node /usr/
bin/node
$ sudo ln -s /home/pi/.nvm/versions/node/v6.9.5/bin/npm /usr/
bin/npm
```

Finally, install Processing:

```
$ curl https://processing.org/download/install-arm.sh | sudo sh
```

Now that your Pi is working and on the network, you're ready to move on. To keep going, reboot and get started! To shut it down, type:

```
$ sudo poweroff
```

When it responds that it's powered down, close your serial connection and disconnect the board.
X

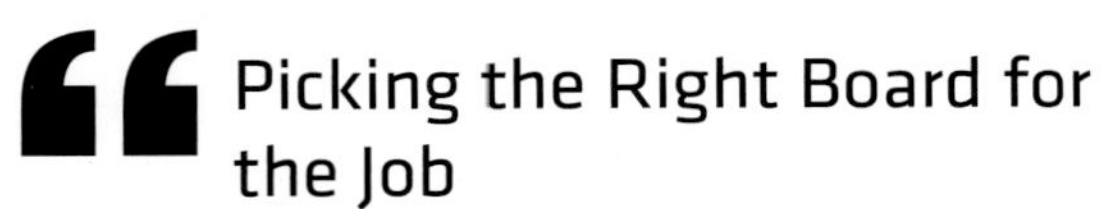

Picking the Right Board for the Job

When you have to choose a microcontroller or embedded processor for a project, there are a few factors you should consider.

First, who's in the conversation? Every device interacts with a person or other devices or both. When you're communicating with humans, your device needs to respond in human time, using language that humans can understand. When it's communicating with other devices, it needs to respond in network time, using language those devices can understand.

Human reaction time is somewhere between 10 and 30 milliseconds; that's a long time in computer time. A device that controls the automatic braking system on a car must have much better reaction time than a human, or it's not much use. When a device has a real-world constraint like this that determines its minimum reaction time, it's referred to as "real time." Real time can mean seconds or nanoseconds, depending on the other things with which you're interacting.

Operating Systems and "Real Time"

All computers run programs, but in a microcontroller, there's only one program running. That means that 100 percent of the processor's work is dedicated to your program. You have near absolute control over when you read input pins or change output pins, because no other program can interrupt yours.

Single-board computers, because they have more memory and faster processors, can run *operating systems*. An operating system is a program that manages a processor's time and memory. A typical operating system features a *scheduler* that manages which other programs have access to the processor; a *filesystem* for sharing external storage like disks or SD cards; and a *Basic Input-Output System (BIOS)* that initializes the various peripheral devices like keyboard, drives, and screen. The whole system is started by a *bootloader*, which is a small program that lives in the first few blocks of the processor's program memory and runs first when you start the computer.

Because an operating system is trying to share the processor between many programs, most operating systems don't guarantee a minimum response time to physical I/O. They're not "real-time systems."

If you can't read the processor's physical I/O pins faster than human reaction time, you might miss the moment when someone pushed the button attached to that input. You have to constantly read the pin, keep track of whether it's changed, and hope that your polling is frequent enough. If you have to have faster response, you typically use a microcontroller with no operating system, or use a simpler "real-time operating system," or RTOS.

Real-time operating systems are operating systems that guarantee a minimum time between I/O operations but leave many core functions out. They usually handle I/O and basic scheduling and not much else. In a processor that has multiple processing cores, an RTOS (pronounced ARE-toss) is useful for splitting your program across the cores. The Arduino 101 is a multicore processor, and it runs a real-time operating system. RTOSs are also useful if you want to run multiple functions at the same time, as they schedule the time that each task gets on the processor or processing cores.

Most RTOSs are not very user-friendly for beginning programmers. The 101's RTOS acts more like a bootloader in that it starts up the cores, the Bluetooth radio, and the internal motion sensor, then hands over all control to your sketch. You don't interact with it directly, but you do get access to the functions and I/O devices that it is managing.

Network Time

Microcontrollers and RTOSs are useful for real-time applications like human interaction, vehicle control, or other high-speed applications, but they can be inconvenient for applications that involve more complex operations like network transactions, file management, media control, and all the things operating systems are designed to handle well. These tasks can sometimes take longer than the minimum time between operations that an RTOS is expected to provide. This is where single-board computers come in handy. They provide all the convenience of a full operating system. Why write a custom program to do a task that's handled millions of times a day by an existing command-line tool?

Even the most common network interactions can involve dozens of devices, and if you've ever waited for a web page to load, you know that they don't guarantee real-time response. When you need to combine network communication with human interaction, consider using a combination of devices. You already do this every day. Your keyboard and mouse are probably run by microcontrollers running a single program, communicating via USB to your laptop's CPU, which is running an operating system. In many of the projects that follow, you'll see microcontrollers and single-board computers working together.

Security

Do you know every program that's running on your computer? What automated tasks are running in the background to check for upgrades, notify the manufacturer when you have a crash, update your files on the network, and so forth? Most of us only see the tip of the iceberg of what our computer is doing most of the time.

Computers connected to the public internet are vulnerable to attacks. You read about this in the news regularly. Programs that you may not be using, but that are available on your computer to do routine housekeeping, can be exploited by someone who has knowledge of what's bundled with your operating system and can figure out how to get access to your machine. A little searching on the internet will yield stories of digital phones, sound systems, and other household appliances that run embedded operating systems that have been compromised by network attacks. Vulnerability is an everyday fact of network participation.

One of the advantages that devices without an operating system offer is that they generally have less software for outside parties to exploit. You can't exploit a program that's not available on the device you're attacking. If you know all the code that your device is running, you're more likely to understand its vulnerabilities.

As object-oriented hardware becomes more common, it's unfortunately less common that a programmer might know every single line of code running on her device. You'll see WiFi radios in this book, for example, that run firmware you don't have to write yourself, and you'll use libraries that take care of some technical details for you. The trade-off for that convenience is that you're less aware of what that firmware can do. Even so, the firmware on a WiFi radio for a microcontroller is likely to contain far fewer potential vulnerabilities than the programs and utilities bundled with an operating system for an embedded processor. This is an important reason why you might choose to use a microcontroller with no operating system or an RTOS rather than a full operating system. If the task at hand is simple, keep the software choices simple as well.

X

Using an Oscilloscope

Most of what you'll be building in this book involves computer circuits that read a changing voltage over time. Whether your microcontroller is reading a digital or analog input, controlling the speed of a motor, or sending data to a personal computer, it's either reading a voltage or generating a voltage that changes over time. The time intervals it works in are much faster than yours. For example, the serial communication you just saw involved an electrical pulse changing at about 10,000 times per second. You can't see anything that fast on a multimeter. This is when an oscilloscope is useful.

An oscilloscope is a tool for viewing the changes in an electrical signal over time. You can change the sensitivity of its voltage reading (in volts per division of the screen) and of the time interval (in seconds, milliseconds, or microseconds per division) at which it reads. You can also change how it displays the signal. You can show it in real time, starting or stopping it as you need, or you can capture it when a particular voltage threshold (called a *trigger*) is crossed.

Oscilloscopes were once beyond the budget of most hobbyists, but lately, a number of inexpensive ones have come on the market. The DSO Nano from Seeed Studio, shown in Figure 1-18, is a good example. At about $100, it's a really good value if you're a dedicated electronic hobbyist. It doesn't have all the features that a full professional scope has, but it does give you the ability to change the volts per division and seconds per division, and to set a voltage trigger for taking a snapshot. It can sample up to 1 million times a second, which is more than enough to measure most serial applications. The image you see in Figure 1-18 shows part of the output of an Arduino sending the message "Hello World!" Each block represents one bit of data. The vertical axis is the voltage measurement, and the horizontal measurement is time. The Nano was sampling at 200 microseconds per division in this image, and 1 volt per division vertically. The scope's leads are attached to the ground pin of the Arduino and to the serial transmit pin.

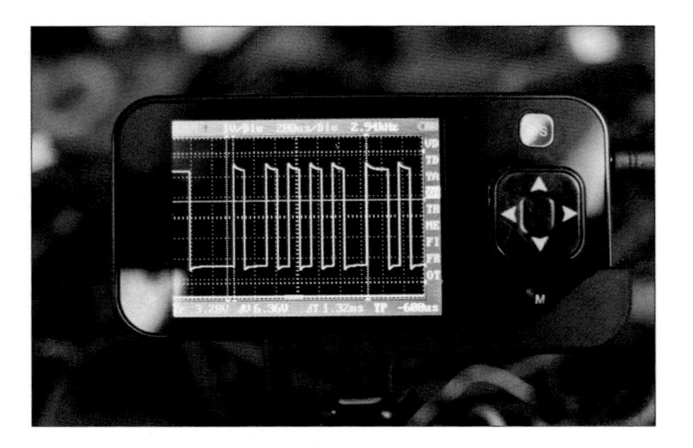

Figure 1-18
DSO Nano oscilloscope reading a serial data stream.

Similar to an oscilloscope, a logic probe like Saleae's Logic line of probes (www.saleae.com) lets you sample signals from multiple lines at once. For example, with a logic probe you could sample both the transmit and receive lines of a serial connection at the same time. Unlike an oscilloscope, a logic probe generally only shows the signals on your computer, and doesn't show it to you in real time. The least expensive of Saleae's Logic line, the Logic 4, is about the same price as a DSO Nano, and is a useful addition to a digital hardware toolkit.

Besides inexpensive hardware scopes, many software scopes are also available, both as freeware and as paid software. These typically use the audio input of your computer to sample the incoming voltage. The danger, of course, is that if you send in too much voltage you can damage your computer. For this reason, I prefer a hardware scope. But if you're interested in software scopes, a web search on `software oscilloscope` and your operating system will yield plenty of useful results.

X

It Ends with the Stuff You Touch

Though most of this book is about the fascinating world of making things talk to each other, it's important to remember that you're most likely building your project for the enjoyment of someone who doesn't care about the technical details under the hood.

Even if you're building it only for yourself, you don't want to have to fix it all the time. All that matters to the person using your system are the parts that she can see, hear, and touch. All the inner details are irrelevant if the physical interface doesn't work. So don't spend all of your time focusing on the communication between devices and leave out the communication with people. In fact, it's best to think about the specifics of what the person does and sees first.

There are a number of details that are easy to overlook but are very important to humans. For example, many network communications can take several seconds or more. In a screen-based operating system, progress bars acknowledge a person's input and keep him informed as to the task's progress. Physical objects don't have progress bars, but they should incorporate some indicator as to what they're doing—perhaps as simple as playing a tune or pulsing an LED gently while the network transfer's happening.

Find your own solution, but make sure you give some physical indication as to the invisible activities of your objects.

Don't forget the basic elements, either. Build in a power switch or a reset button. Include a power indicator. Design the shape of the object so that it's clear which end is up. Make your physical controls clearly visible and easy to operate. Plan the sequence of actions you expect a person to take, and lay out the physical affordances for those actions sensibly. You can't tell people what to think about your object—you can only show them how to interact with it through its physical form. There may be times when you violate convention in the way you design your controls—perhaps in order to create a challenging game or to make the object seem more "magical"—but make sure you're doing it intentionally. Always think about the participant's expectations first.

By including the person's behavior in your system planning, you solve some problems that are computationally difficult but easy for human intelligence. Ultimately, the best reason to make things talk to each other is to give people more reasons to talk to each other.

X

2

MAKE: PROJECTS

The Simplest Network

The most basic network is a one-to-one connection between two objects. This chapter covers the details of that two-way communication, beginning with the characteristics that have to be agreed upon in advance. You'll learn about some of the logistical elements of network communications: data protocols, flow control, and addressing. You'll practice all of this by building two examples using one-to-one serial communication between a microcontroller and a personal computer. You'll also learn about how you can replace the cable connecting the two with Bluetooth radios. Finally, you'll learn how to program very low-end microcontrollers, so you can start to distribute the computational needs of your projects among different processors.

◂◂ Joo Youn Paek's Zipper Orchestra (2006)
This is a musical installation that lets you control video and music using zippers. The zippers are wired to a microcontroller using conductive thread, and the microcontroller communicates serially with a multimedia computer that drives the playback of the zipper movies and sounds as you zip.
Photo courtesy of Joo Youn Paek

Supplies for Chapter 2

DISTRIBUTOR KEY
A Arduino Store (store.arduino.cc)
AF Adafruit (adafruit.com)
D Digi-Key (www.digikey.com)
F Farnell (www.farnell.com)
J Jameco (jameco.com)
RS RS (www.rs-online.com)
SF SparkFun (www.sparkfun.com)
SS Seeed Studio (www.seeedstudio.com)

PROJECT 1: Type Brighter RGB LED Serial Control

» **1 microcontroller module** This project will run on just about any Arduino-compatible board. The MKR1000 and the Arduino 101 are shown, but the Uno will work as well.
MKR1000: **AF** 3156, **RS** 124-0657, **A** ABX00004, GBX00011 (Africa and EU), **D** 1659-1005-ND
Arduino 101: **D** 1660-1003-ND, **J** 2239331, **SF** DEV-13787, **AF** 3033, **F** 2520713, **RS** 913-9999, **SS** 114990575, **A** ABX00005, GBX00005 (Africa and EU)
Uno: **D** 1050-1024-ND, **J** 2151486, **SF** DEV-11021, **A** A000099, **AF** 50, **F** 1848687, **RS** 715-4081, **SS** ARD132D2P

» **1 RGB LED, common cathode** It's like three LEDs in one! This emits red, green, and blue on separate channels.
D 754-1492-ND, **J** 2125181, **SF** COM-00105, **F** 2290374, **RS** 861-4290

» **1 220 ohm resistor** **D** 220QBK-ND, **J** 690700, **F** 9339299, **R** 707-7612

» **1 solderless breadboard** **D** 438-1045-ND, **J** 20723 or 20601, **SF** PRT-12615 or PRT-12002, **F** 4692810, **AF** 64, **SS** 319030002 or 319030001

» **1 personal computer**

» **All necessary converters to communicate serially from microcontroller to computer** For the microcontroller modules, all you'll need is a USB cable. Get the right one for your module.

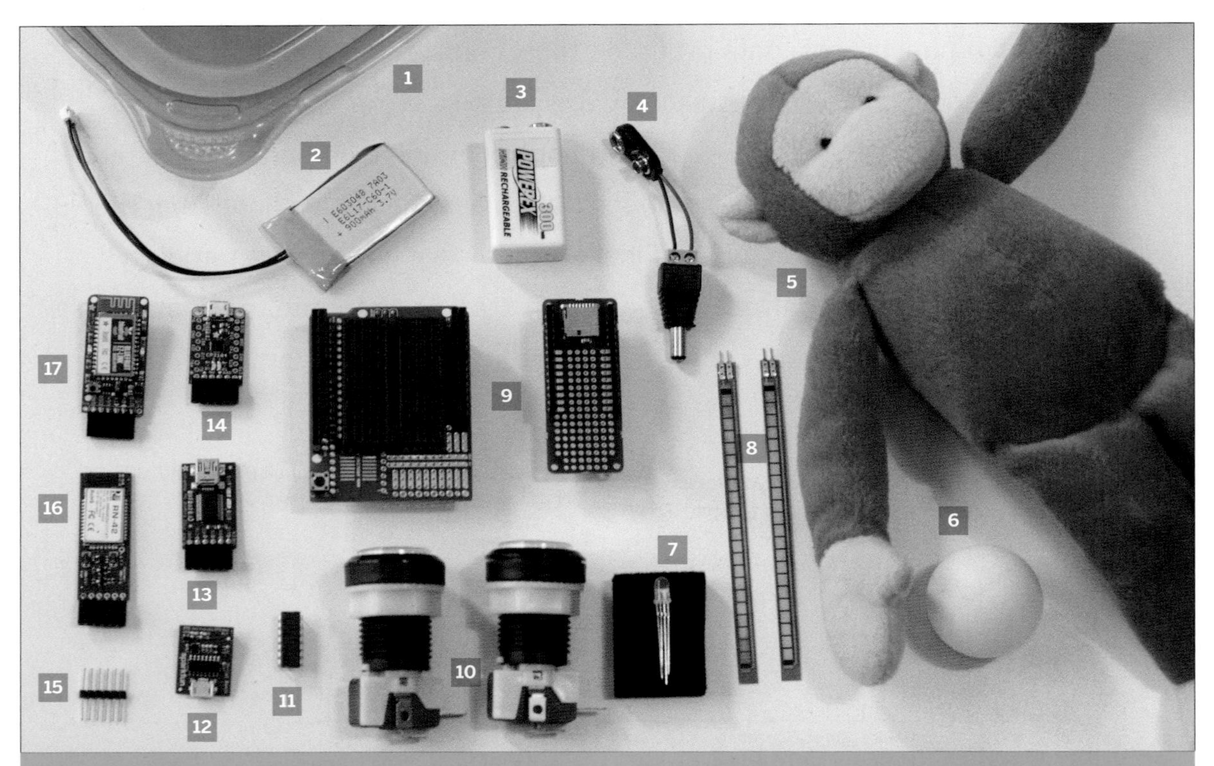

Figure 2-1. New parts for this chapter: **1.** Project box **2.** Lithium polymer battery **3.** 9V battery **4.** 9V battery clip and power jack **5.** Pink monkey **6.** Ping-pong ball **7.** RGB LED **8.** Flex sensors **9.** Prototyping shields **10.** Pushbuttons **11.** ATtiny85 microcontroller **12.** Serial Breakout CH340G **13.** Serial Breakout FTDI Friend **14.** Serial Breakout CP2104 Friend **15.** Long header pins **16.** Bluetooth Mate **17.** Bluefruit EZ-Link Serial

- 1 ping-pong ball Get a white one from your local sports store.

PROJECT 2: Monski Pong

- 1 microcontroller module This project will run on just about any Arduino-compatible board. The MKR1000 and the Arduino 101 are shown, but the Uno will work as well.
  MKR1000: **AF** 3156, **RS** 124-0657, **A** ABX00004, GBX00011 (Africa and EU), **D** 1659-1005-ND
  Arduino 101: **D** 1660-1003-ND, **J** 2239331, **SF** DEV-13787, **AF** 3033, **F** 2520713, **RS** 913-9999, **SS** 114990575, **A** ABX00005, GBX00005 (Africa and EU)
  Uno: **D** 1050-1024-ND, **J** 2151486, **SF** DEV-11021, **A** A000099, **AF** 50, **F** 1848687, **RS** 715-4081, **SS** ARD132D2P
- 2 flex sensor resistors **D** 905-1000-ND, **J** 150551, **SF** SEN-10264, **AF** 182, **RS** 708-1277
- 2 momentary switches Available from any electronics retailer. Pick the one that makes you the happiest. **D** GH1344-ND or SW400-ND, **J** 2231822 or 119011, **SF** COM-09337, **F** 1634684, **RS** 718-2213
- 4 10-kilohm resistors **J** 29082, **SF** COM-09939, **F** 350072, **RS** 249-9294
- 1 solderless breadboard **D** 438-1045-ND, **J** 20723 or 20601, **SF** PRT-12615 or PRT-12002, **F** 4692810, **AF** 64, **SS** 319030002 or 319030001
- 1 personal computer
- All necessary converters to communicate serially from microcontroller to computer Just like the previous project.
- 1 small pink monkey aka Monski. You may want a second one for a two-player game.

PROJECT 3: Wireless Monski Pong

- 1 completed Monski pong project from project 2.
- 1 9V battery and snap connector **D** 1568-1237-ND, **J** 2207056, **SF** PRT-09518, **A** 80, **F** 1650675
- DC power plug, 2.1mm ID, 5.5mm OD If you got any but the Farnell battery connector, you don't need this part. **D** CP3-1000-ND , **J** 28760, **SF** PRT-10287, **A** 369 **F** 1737256
  If you're using the MKR1000, you'll need a Lithium Polymer battery instead: **SF** PRT-13813 or PRT-08483; **A** 258 or 2011
- 1 Bluetooth Serial module **AF** 1588 **SF** WRL-12580 or WRL-12576
- 1 project box to house the microcontroller, battery, and radio board.

PROJECT 4: Making Your Own Arduino-Compatible Board

- 1 Arduino-compatible microcontroller The Uno, MKR1000, and the Arduino 101 work best for this
  MKR1000: **AF** 3156, **RS** 124-0657, **A** ABX00004, GBX00011 (Africa and EU), **D** 1659-1005-ND
  Arduino 101: **D** 1660-1003-ND, **J** 2239331, **SF** DEV-13787, **AF** 3033, **F** 2520713, **RS** 913-9999, **SS** 114990575, **A** ABX00005, GBX00005 (Africa and EU)
  Uno: **D** 1050-1024-ND, **J** 2151486, **SF** DEV-11021, **A** A000099, **AF** 50, **F** 1848687, **RS** 715-4081, **SS** ARD132D2P
- 1 ATtiny84 microcontroller You could also use the ATtiny85 if you want an even smaller board. **D** ATtiny84A-PU-ND, **SF** COM-11232, **F** 1455160, **RS** 738-0684
- 1 RGB LED, common cathode Same one as in Project 1 **D** 754-1492-ND, **J** 2125181, **SF** COM-00105, **F** 2290374, **RS** 861-4290
- 1 220-ohm resistor **D** 220QBK-ND, **J** 690700, **F** 9339299, **R** 707-7612
- 1 solderless breadboard **D** 438-1045-ND, **J** 20723 or 20601, **SF** PRT-12615 or PRT-12002, **F** 4692810, **AF** 64, **SS** 319030002 or 319030001
- Jumper wires **J** 20723, **SF** PRT-12796, **F** 2396146, **RS** 791-6463, **AF** 759

ALL PROJECTS

- USB-to-TTL serial adapter These come in handy for connecting to all sorts of serial devices. They require an additional USB-A-to-Mini-B or Micro-B cable. Get one. You'll use it all the time. **D** 36-84-4-ND, **J** 216452, **SF** DEV-09716 or DEV-14050, **A** 3309 or 284, **SS** 317990026.

Figure 2-2
USB-to-serial adapters, L to R: Adafruit FTDI Friend, CP2104 Friend; SparkFun Serial Basic Breakout CH340G

Layers of Agreement

Before you can get things to talk to each other, you have to lay some ground rules for the communication between them. These agreements can be broken down into a few layers, each of which builds on the previous ones:

- **Physical**
How are the physical inputs and outputs of each device connected to the others? How many connections between the two devices do you need to enable communication between them?

- **Electrical**
What voltage levels will you send to represent the bits of your data? 5 volt? 3.3 volt? Some other voltage?

- **Logical**
What does a voltage change represent? When high voltage represents 1 and low voltage represents 0, it's called *true logic*. When it's reversed—high voltage represents 0 and low voltage represents 1—it's called *inverted logic*.

- **Data**
What's the timing of the bits? Are the bits read in groups of 8, 9, or 10? More? Are there bits at the beginning or end of each group to punctuate the groups?

- **Application**
How are the groups of bits arranged into messages? What is the order in which messages have to be exchanged for something to get done?

This is a simplified version of a common model for thinking about networking, called the *Open Systems Interconnect (OSI)* model. Networking issues are never really this neatly separated, but if you keep these layers distinct in your mind, troubleshooting any connection will be much easier. Thinking this way gives you somewhere to start looking for the problem, and a way to eliminate parts of the system that are not the problem.

No matter how complex the network, the communication between electronic devices is all about changes in energy. Devices that use *serial communication* share an electrical connection. The sender changes the voltage of the connection at an agreed-upon time interval. Each interval represents one bit of information. The sender changes the voltage to send a value of 0 or 1 for the bit in question, and the receiver reads whether the voltage is high or low.

There are two methods (see Figure 2-4) that the sender and receiver can use to agree on the rate at which they send bits. In *asynchronous serial communication*, the rate is agreed upon mutually and clocked independently by sender and receiver. In *synchronous serial communication*, the rate is controlled by the sender, who pulses a separate connection high and low at a steady rate. Synchronous serial communication is used mostly for communication with simpler devices, such as the communication between a computer processor and its memory chips. This chapter concentrates on asynchronous serial communication, because its principles underlie the networks in the rest of the book, from Ethernet connections to radio connections.

X

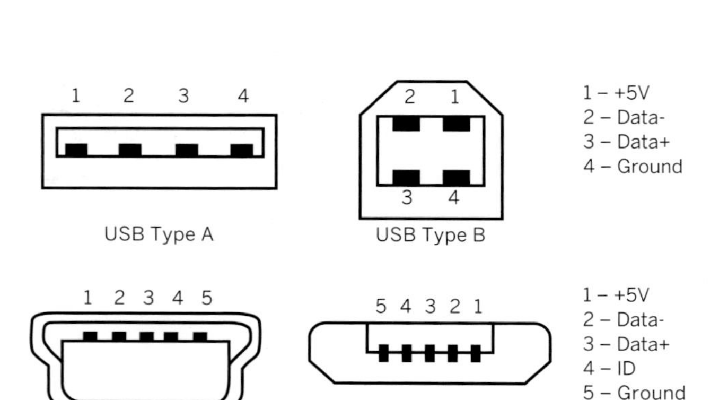

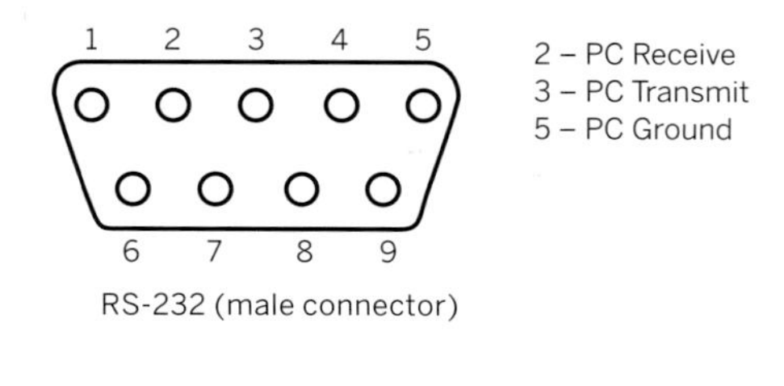

Figure 2-3
Physical connections: USB, RS-232 serial.

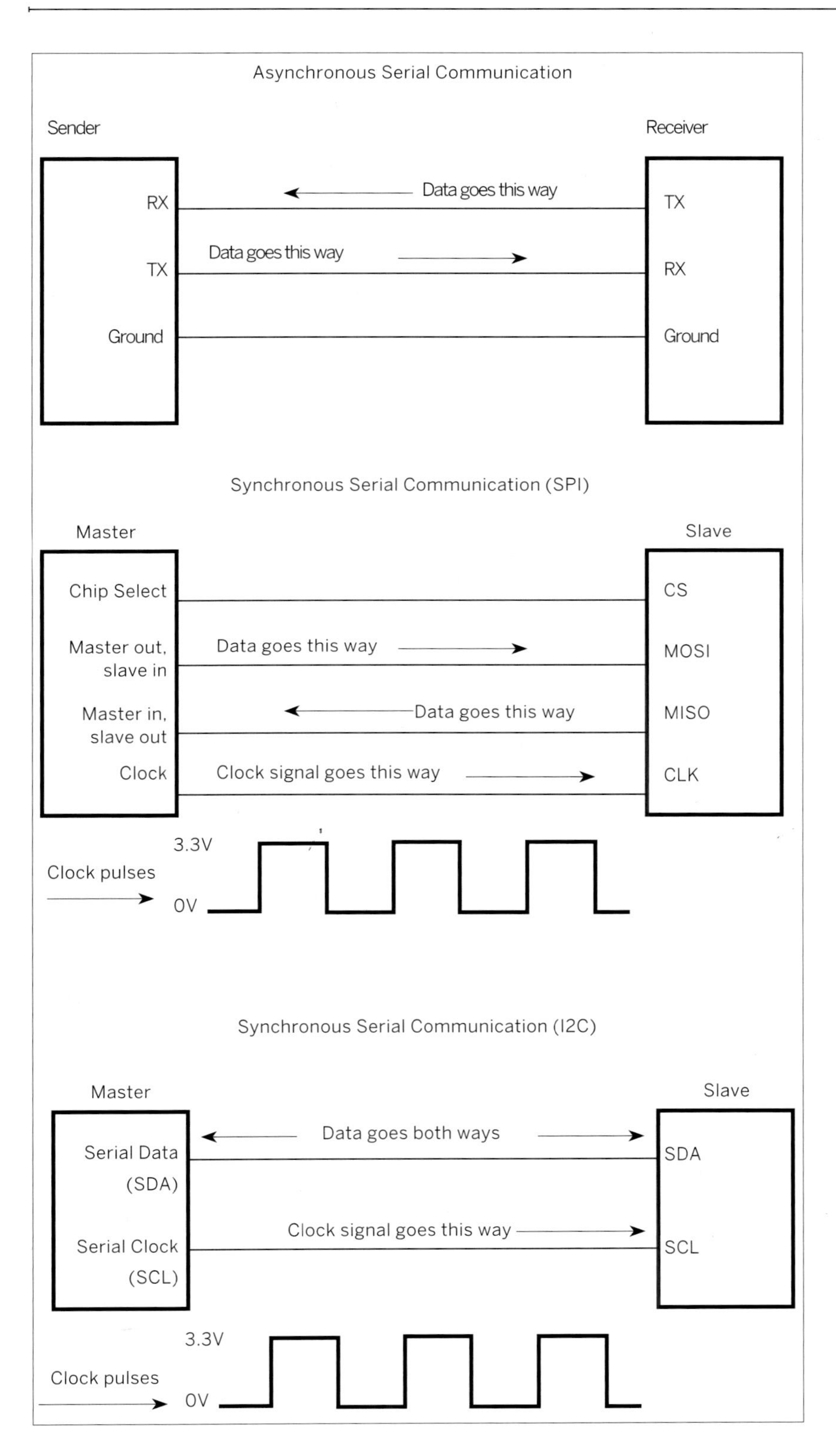

Figure 2-4
Types of serial communication

Asynchronous serial:
Two devices each have their own clock. Data is exchanged at an agreed-upon rate.

Synchronous serial: Serial-peripheral interface (SPI)
Master device sends a clock signal to slave device. Master signals to the slave using Chip Select pin. Data is exchanged when the clock changes its voltage.

Synchronous serial: Inter-integerated Circuit (I2C)
Master sends a clock signal to slave. Master signals to the slave using slave's address. Data is exchanged when the clock changes its voltage.

Making the Connection: The Lower Layers

You're already familiar with one form of serial communication, between a microcontroller and a personal computer. In the last chapter, you connected a microcontroller module to a personal computer through the USB port. That connection is an example of *asynchronous serial communication* involving two serial protocols: TTL Serial and USB.

First, there's the protocol that the microcontroller speaks, called *TTL serial* (short for *Transistor-Transistor Logic*):

- **Physical layer**
Which pins is the controller using to communicate? The Arduino module receives data on the pin marked RX (receive), and sends data out on the pin marked TX (transmit).

- **Electrical layer**
Changes in voltage to represent bits of data. Some microcontrollers use 3.3 volts, some use 5 volts.

- **Logical layer**
A high-voltage (3.3 or 5 volt) signal represents the value 1, and a 0-volt signal represents the value 0.

- **Data layer**
Data is typically sent at 9600 bits per second. Each byte contains 8 bits, preceded by a start bit and followed by a stop bit (with which you never have to bother).

- **Application layer**
In your application, you sent one byte from the personal computer to the microcontroller, which processed it and sent back one byte to the PC.

But wait, that's not all that's involved. The electrical pulses didn't go directly to the PC. First, they went to a serial-to-USB circuit on the board that communicates using TTL serial on one side, and USB on the other. On some microcontroller boards, like the Arduino Uno, the USB-to-serial converter is a separate chip. On others, like the MKR1000 and 101, the microcontroller itself handles both USB and TTL serial communication natively.

The second protocol involved is USB, the *Universal Serial Bus* protocol. It differs from TTL serial in a few ways:

- **Physical layer**
USB sends data on two wires, Data+ and Data-. Every USB connector also has a 5-volt power supply line and a ground line.

- **Electrical layer**
The signal on Data– is always the opposite of what's on Data+, so the sum of their voltages is always zero. Because of this, a receiver can check for electrical errors by adding the two data voltages together. If the sum isn't zero, the receiver can disregard the signal at that point.

- **Logical layer**
A +5-volt signal (on Data+) or -5-volt signal (on Data–) represents the value 1, and a 0-volt signal represents the value 0.

- **Data layer**
The data layer of USB is more complex than TTL serial. Data can be sent at up to 480 megabits per second. Each byte contains 8 bits, preceded by a start bit and followed by a stop bit. Many USB devices can share the same pair of wires, sending signals at times dictated by the controlling PC. This arrangement is called a *bus* (the B in USB). As there can be many devices on the same bus, the operating system gives each one its own unique address, and sees to it that the bytes from each device on the bus go to the applications that need them.

- **Application layer**
At the application layer, the USB-to-serial converter on the microcontroller board sends a few bytes to the operating system to identify itself and its capabilities. The operating system then associates the board with a driver program that other programs can use to access data from the board.

USB: An Endless Source of Serial Ports

One of the great things about microcontrollers is that, because they're cheap, you can use many of them. For example, in a project with many sensors, you can either write a complex program on the microcontroller to read them all, or you can give each sensor its own microcontroller. If you're trying to get all the information from those sensors into a personal computer, you might think it's easier to use one microcontroller because you've got a limited number of ports. Thanks to USB, however, that's not the case. If your microcontroller speaks USB, you can just plug it in and it will show up in the operating system as another serial port. If you need more ports, you can add a USB hub.

For example, if you plug three Arduino modules into the same computer through a USB hub, you'll get three new serial ports, named something like this on macOS:

```
/dev/cu.usbmodem1441
/dev/cu.usbmodem1461
/dev/cu.usbmodem1471
```

In POSIX systems, including macOS, the ports are often listed twice, once as `/dev/tty.usbmodemXX` and once as `/dev/cu.usbmodemXX`. The difference is from dialup modem days. TTY (teletype unit) was used for incoming calls, while CU (calling unit) was used for calling out. For the USB-to-serial devices used in this book, it doesn't matter significantly which you use.

In Windows, you'd see something like `COM8`, `COM9`, `COM10`.

Most microcontroller boards have a USB-to-serial converter on board these days. For those that don't, you can buy a USB-to-serial converter for about $15 to $20. One of the more popular comes from FTDI (www.ftdichip.com). FTDI makes a USB-to-TTL-Serial chip with a breadboard connector that's handy for interfacing to TTL serial devices. You can find it in breakout boards and cables from the Maker Shed, SparkFun, Adafruit, and many other vendors. It comes in 5-volt and 3.3-volt versions. Any of these can be used for USB-to-serial connections throughout this book. The pin connections are shown in Figure 2-6 later in this chapter.

All that control is transparent to you because the computer's USB controller only passes you the bytes you need. The USB-to-serial converter on your microcontroller board presents itself to the operating system as a serial communications port, and it sends data through the USB connection at the rate you choose (9600 bits per second in the example in Chapter 1).

One more protocol: If you use a BASIC Stamp or another microcontroller with a non-USB serial interface, you might have a 9-pin serial connector connecting your microcontroller to your PC or to a USB-to-serial adapter. You can see it in Figure 2-3. This connector, called a *DB-9* or *D-sub-9*, is a standard connector for another serial protocol, *RS-232*. RS-232 was the main serial protocol for computer serial connections before USB, and it's still seen on some older computer peripheral devices:

- **Physical layer**
  A computer with an RS-232 serial port receives data on pin 2, and sends it out on pin 3. Pin 5 is the ground pin.

- **Electrical layer**
  RS-232 sends data at two levels: between 3 and 25 volts, and between −3 and −25 volts.

- **Logical layer**
  A positive voltage signal represents the value 0, and a negative voltage volt signal represents the value 1. This is *inverted logic*.

- **Data layer**
  This is the same as TTL—8 bits per byte with a start and stop bit.

So why is it possible to connect some microcontrollers, like the BASIC Stamp, directly to RS-232 serial ports? Since microcontrollers can't generate negative voltages, a separate chip was generally used to convert the TTL output of the microcontroller to RS-232 levels. RS-232 is a simpler protocol than USB but unfortunately, it's mostly obsolete. Nowadays, most microcontroller boards have an integrated USB-to-TTL-serial converter.

Most of the time, you never have to think about this kind of protocol mixing, and you can just use converters to do the job for you. It's worth knowing a little about what's happening behind the scenes, though, for troubleshooting.

X

USB-to-Serial Converters

Many of the electronics modules you'll encounter are likely to have a serial interface for communication with microcontrollers. The most common interface you'll encounter is TTL serial. For example, most Global Positioning System (GPS) modules have a TTL serial interface. You'll see one in Chapter 8. A USB-to-serial converter is an essential tool for anyone interested in making electronic devices these days.

There are several USB-to-serial converter chips on the market. One of the most popular, the FT232RL, is made by Future Technology Devices International (www.ftdichip.com). FTDI makes a cable that has this chip inside, and many other companies make adapter boards using FTDI's chips as well. It's become so common that the pin arrangement of FTDI's cable has become a de facto standard in the electronic hobbyist market. You'll see its layout on many devices. Figure 2-5 shows the cable, and Figure 2-6 shows its pin configuration.

Figure 2-5
FTDI USB-to-TTL cable.

In addition to FTDI's cable, you can buy adapters from Adafruit, SparkFun, Parallax, and others. They're also featured in many products, including Digi's XBee adapters and SparkFun's RedBoard Arduino-compatible boards.

The FT232RL chip is handy because it can handle different TTL serial voltages. You can get 5V and 3.3V versions. SparkFun sells 3.3V and 5V versions of its FTDI adapter. Adafruit's FTDI friend adapter has jumpers on the back that you solder to change the voltages of the serial transmit and receive pins. The chip also handles RS-232 voltage levels. Parallax's adapter is designed for RS-232 use, and has a DB-9 connector.

Figure 2-6
FTDI USB-to-TTL cable pin configuration. In addition to the transmit, receive, voltage, and ground connections, it has connections for hardware flow control, labeled Request to Send (RTS) and Clear to Send (CTS). Some devices use this to manage the flow of serial data.

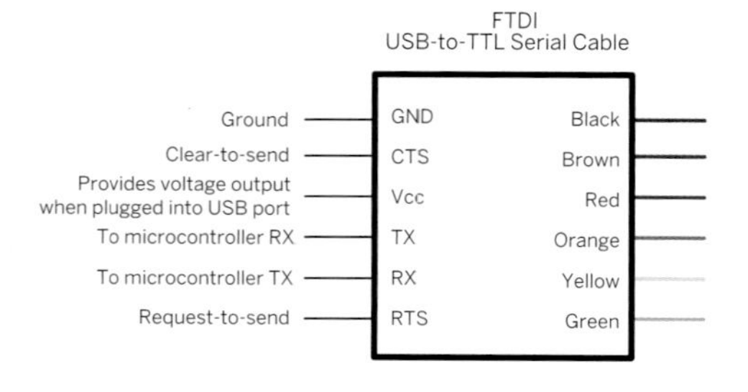

Using a USB-to-serial adapter is simple. You connect the adapter's transmit pin (TX) to your device's receive pin (RX), and vice versa. You connect their ground pins together, and if you're powering the device from the adapter, you connect the Vcc pin to the device. If you are powering your device from your adapter, make sure that the device can operate on the adapter's voltage first. Most USB-to-serial adapters supply 5 volts, which can damage a 3.3V device. In Chapter 1, you saw how to use an FTDI USB-to-serial adapter to connect to the serial terminal of a Raspberry Pi. The Pi was not powered from the adapter, though.

In order to use any USB-to-serial adapter with your personal computer, you'll need to install the drivers for the chip at the heart of the adapter. The quality of the drivers has a lot to do with the quality of a USB-to-serial adapter. Cheaper adapters usually have less compatible drivers. It's worthwhile to pay a little extra for a device with easy-to-find, well-maintained, cross-platform drivers. This is one area where FTDI excels. Their drivers are kept up to date with the current versions of macOS, Windows, and various flavors of Linux.

FTDI isn't the only company that makes USB-to-serial adapters, though. You can also find them from Prolific (PL2303), Silicon Labs (CP2102), Jiangsu Heng Qin (CP340), and others. You can find FTDI's USB-to-serial drivers at www.ftdichip.com/FTDrivers.htm. Silicon Labs' CP2102 drivers are at www.silabs.com/products/mcu/Pages/USBtoUARTBridgeVCPDrivers.aspx. Prolific's PL2303 drivers are at www.prolific.com.tw/US/ShowProduct.aspx?pcid=41. Jiangsu Heng Qin's CP340 drivers are at www.wch.cn/download/CH341SER_ZIP.html.

The Arduino Uno uses a general-purpose microcontroller, the Atmel 16U2, programmed to be a USB-to-serial converter. It doesn't need drivers for macOS or Linux, and the Windows installer installs the Windows USB driver. The source code for the USB-to-serial adapter on the Uno can be found in the Arduino github repository at github.com/arduino, under **hardware/arduino/avr/firmwares/ atmegaxxu2**. If you're an advanced user and you want more information on designing USB devices, see www.usb.org/developers/usbfaq.

If you're using a USB-to-serial adapter to program a microcontroller, you're probably going to need to use the CTS (clear-to-send) and/or RTS (request-to-send) pins as well. Many controllers that rely on a USB-to-serial adapter, like Adafruit's Huzzah! ESP8266 breakout board and SparkFun's ESP8266 Thing, have connections built in for these pins so that you don't need to think about it. See the programming guide for the microcontroller you're programming for more details.

X

Using an Arduino As a USB-to-Serial Adapter

If you don't have a USB-to-serial adapter handy, you can program an Arduino-compatible microcontroller to do the job.

The MKR1000 and the 101 and other boards based on the same processors as these are *USB-native*, so the serial port you see in the Serial Monitor is connected directly to the processor. The TX and RX pins on the board are typically addressed using `Serial1`. The following sketch will connect the transmit of the native USB connection to the RX pin and the USB native receive to the TX pin, turning any USB-native controller into a USB-to-serial converter:

```
void setup() {
  // initialize both serial connections:
  Serial.begin(9600);
  Serial1.begin(9600);
}

void loop() {
  // read from RX, send to USB:
  if (Serial1.available()) {
    char c = Serial1.read();
    Serial.write(c);
  }
  // read from USB, send to TX:
  if (Serial.available()) {
    char c = Serial.read();
    Serial1.write(c);
  }
}
```

Older boards like the Uno are not USB-native so they have a built-in USB-to-serial adapter so that the microcontroller can communicate with your personal computer serially. The USB-to-serial adapter's transmit pin (TX) is attached to the microcontroller's receive pin (RX), and vice versa. This means you can bypass the microcontroller and use the Arduino board as a USB-to-serial adapter. To do this, put a sketch on it that does nothing, like so:

```
void setup() {}
void loop() {}
```

Then connect the desired external serial device as follows (note that this is the reverse of what you'd think):

External serial device's receive pin → RX of the Arduino board

External serial device's transmit pin → TX of the Arduino board

Now your serial device will communicate directly with the USB-to-serial adapter on the Arduino. When you're ready to use the microcontroller again, just disconnect the external serial device and upload a new sketch.

❝ Saying Something: The Application Layer

Now that you've got a sense of how to make the connections between devices, it's time to build a couple of projects to understand how to organize the data you send.

 **Project 1**

Type Brighter

In this project, you'll control the output of a microcontroller with keystrokes from your computer. It's a very simple example with minimal parts, so you can focus on the communication.

MATERIALS

- **» 1 RGB LED, common cathode**
- **» 1 220-ohm resistor**
- **» 1 Arduino-compatible microcontroller (MKR100 and 101 shown)**
  - **» Features used: UART, PWM output**
- **» 1 solderless breadboard**
- **» 1 personal computer**
- **» 1 ping-pong ball**

Every application needs a communications protocol, no matter how simple it is. Even turning on a light requires that both the sender and receiver agree on how to say "turn on the light."

This project will work on any Arduino-compatible micro-controller as long as it's got three digital output pins on which you can use the `analogWrite()` command. These are called PWM pins, because you can pulsewidth modulate the pins (turn them on and off very fast) to produce something like a variable voltage.

For this project, you'll make a tiny colored lamp. You'll be able to control the brightness and color of the lamp by sending it commands from your computer. The RGB LED at the heart of the lamp is actually three LEDs in one package: one red, one green, and one blue. They share a common *cathode*, or negative terminal. Connect the cathode, which is the longest leg, to the ground of your Arduino module through a 220-ohm resistor, and connect the three other legs (called the *anodes*, one for each color) to three PWM pins of the microcontroller as shown in Figure 2-7. For the MKR1000, connect them to pins 3, 5, and 4, respectively. For the 101, connect the anodes to pins 3, 6, and 5, respectively.

When you've got the LED on the board, drill a small hole in the pong-pong ball, slightly larger than the LED. Fit the ball over the LED, as shown in Figure 2-9. It will act as a nice lampshade. If the LEDs form too harsh a spot on the ball, you can diffuse them slightly by sanding the top of the LED case.

The Protocol

Now that you've got the circuit wired up, you need to decide how you're going to communicate with the micro-controller to control the LEDs. You need a communica-tions protocol. This one will be very simple:

- To choose a color of LED to control, send the first letter of the color, in lower-case (r, g, b).
- To set the brightness for that color, send a single digit, 0 through 9.

For example, to set red at 5, green at 3, and blue at 7 (on a scale from 0 to 9), you'd send:

```
r5g3b7
```

That's about as simple a protocol as you could imagine. However, you still need to write a program for the micro-controller to read the data one byte at a time, and then decide what to do based on the value of each byte.

MKR1000 Breadboard view

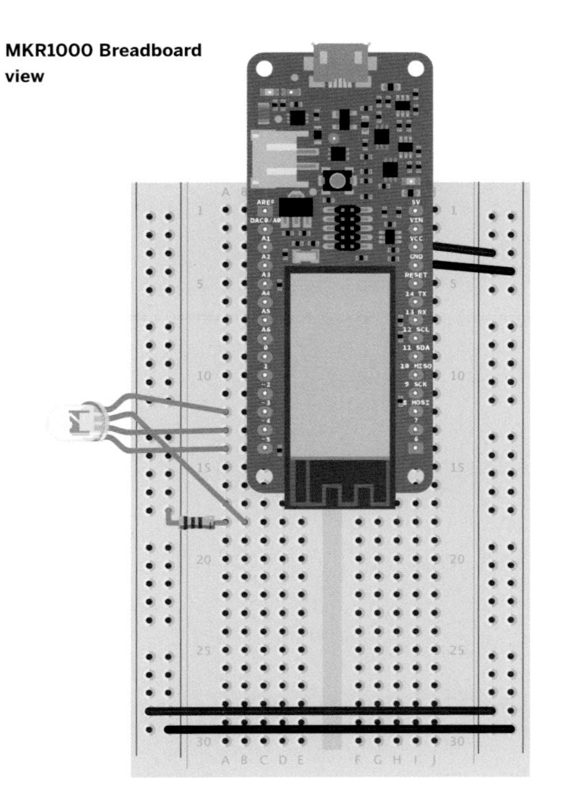

Figure 2-7
RGB LED attached to three PWM pins. On the MKR1000, use pins 3, 4, and 5 . On the 101, use pins 3, 6, and 5. The common cathode connects through a 220-ohm resistor to ground.

Schematic view
Only the pins that are used are shown.

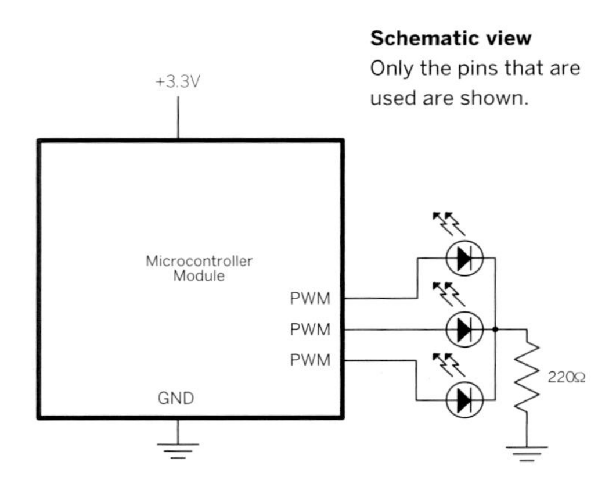

101/Uno Breadboard view

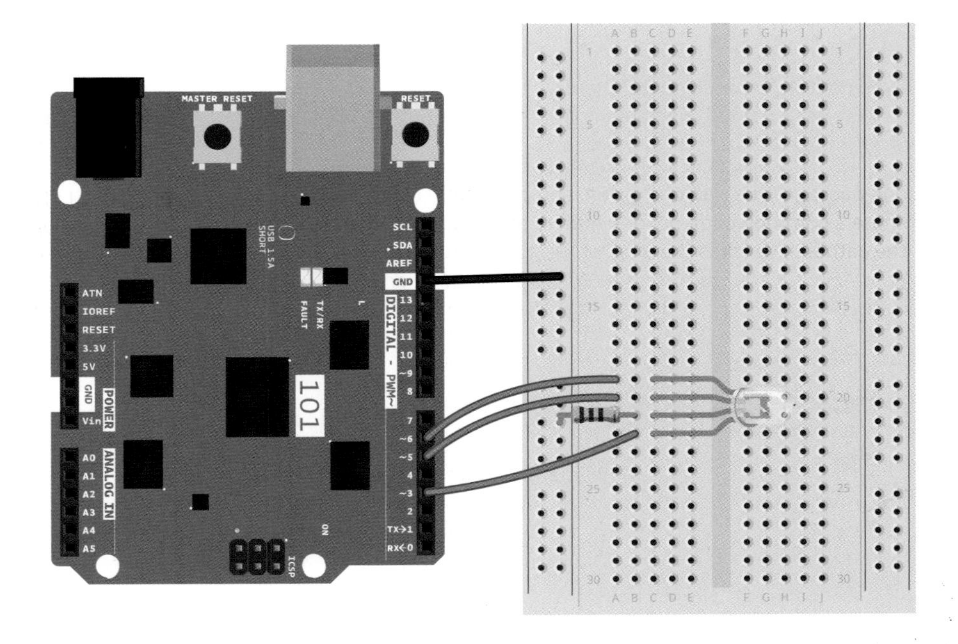

Try It To start, you need to set up constants to hold the pin numbers. Set them appropriately for your board. You'll also need a variable to hold the number of the current pin to be faded, and one for the brightness.

```
/*
  Serial RGB LED controller
  Context: Arduino

  Controls an RGB LED whose R, G and B legs are
  connected to pins 3, 5, and 4, respectively.
*/
// constants to hold the output pin numbers:
// The defaults are for the MKR1000. For the 101, use numbers in comments
const int redPin = 3;     // for 101 use 3
const int greenPin = 5;   // for 101 use 6
const int bluePin = 4;    // for 101 use 5

int currentPin = 0; // current pin to be faded
int brightness = 0; // current brightness level
```

The `setup()` method opens serial communications and initializes the LED pins as outputs.

```
void setup() {
  // initiate serial communication:
  Serial.begin(9600);

  // initialize the LED pins as outputs:
  pinMode(redPin, OUTPUT);
  pinMode(greenPin, OUTPUT);
  pinMode(bluePin, OUTPUT);
}
```

In the main loop, everything depends on whether you have incoming serial bytes to read.

If you do have incoming data, there are only a few values you care about. When you get one of the values you want, use if statements to set the pin number and brightness value.

Finally, set the current pin to the current brightness level using the `analogWrite()` command.

```
void loop() {
  // if there's any serial data in the buffer, read a byte:
  if (Serial.available() > 0) {
    int inByte = Serial.read();

    // respond to the values 'r', 'g', 'b', or '0' through '9'.
    // you don't care about any other value:
    if (inByte == 'r') {
      currentPin = redPin;
    }
    if (inByte == 'g') {
      currentPin = greenPin;
    }
    if (inByte == 'b') {
      currentPin = bluePin;
    }

    if (inByte >= '0' && inByte <= '9') {
      // map incoming byte value to the range of the analogRead() command:
      brightness = map(inByte, '0', '9', 0, 255);
      // set the current pin to the current brightness:
      analogWrite(currentPin, brightness);
    }
  }
}
```

Upload the sketch to your microcontroller, then open the Serial Monitor by clicking the icon at the top-right side of the editor, as shown in Figure 2-8.

Once it's open, type:

```
r9
```

Then click Send. You should see the ball light up red. Now try:

```
r2g7
```

The red will fade and the green will come up. Now try:

```
g0r0b8
```

The blue will come on. Voilà, you've made a tiny serially controllable lamp!

If the colors don't correspond to the color you type, you probably bought an LED of a different model than the one specified in the sketch, so the pin numbers are different. You can fix this by changing the pin number constants in your sketch.

You don't have to control the lamp from the Serial Monitor. Any program that can control the serial port can send your protocol to the Arduino to control the lamp. Once you've finished the next project, try writing your own lamp controller in Processing.
X

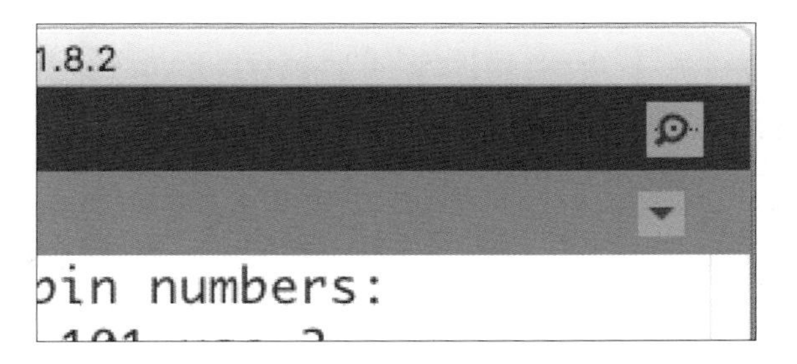

Figure 2-8
Arduino IDE toolbar, highlighting the Serial Monitor.

Figure 2-9
LED with a ping-pong ball on top to diffuse the light.

> Notice in the program how the characters you type are in single quotes? That's because you're using the ASCII values for those characters. *ASCII* is a protocol that assigns numeric values to letters and numbers. For example, the ASCII value for the letter 'r' is 114. The ASCII value for '0' is 48. By putting the characters in single quotes, you're programming the Arduino to use the ASCII value for that character, not the character itself. For example, this line:

```
brightness = map(inByte, '0', '9', 0, 255);
```

could also be written like this:

```
brightness = map(inByte, 48, 57, 0, 255);
```

because in ASCII, the character '0' is represented by the value 48, and the character '9' is represented by the value 57. Using the character value in single quotes instead of the actual values isn't essential to make your program run, but it makes it easier to read. In the first version of the previous line, you're using the ASCII characters to represent the values to map; in the second version, you're using the raw values themselves. You'll see examples in this book that use both approaches. For more on ASCII, see "What's ASCII?" on page 62.

Complex Conversations

In the previous project, you controlled the microcontroller from the computer using a very simple protocol. This time, the microcontroller will control an animation on the computer. The communications protocol is more complex as well.

Project 2

Monski Pong

In this example, you'll make a replacement for a mouse. If you think about the mouse as a data object, it looks like Figure 2-10.

MATERIALS

- » **2 flex sensor resistors**
- » **2 momentary switches**
- » **4 10-kilohm resistors**
- » **1 solderless breadboard**
- » **1 Arduino-compatible microcontroller (MKR1000 and 101 shown)**
    - » **Features used: Digital input, Analog input, UART**
- » **1 personal computer**
- » **1 small pink monkey**

What the computer does with the mouse's data depends on the application. For this application, you'll make a small pink monkey play pong by waving his arms. He'll also have the capability to reset the game by pressing a button, and to serve the ball by pressing a second button.

Connect long wires to the flex sensors so that you can sew the sensors into the arms of the monkey without having the microcontroller in his lap. Connect long wires to the buttons as well. Use flexible wire; old telephone cable works well.

Mount the buttons in a piece of scrap foam-core or cardboard until you've decided on a final housing for the electronics. Label the buttons "Reset" and "Serve." Connect the sensors to the microcontroller, as shown in Figure 2-11.

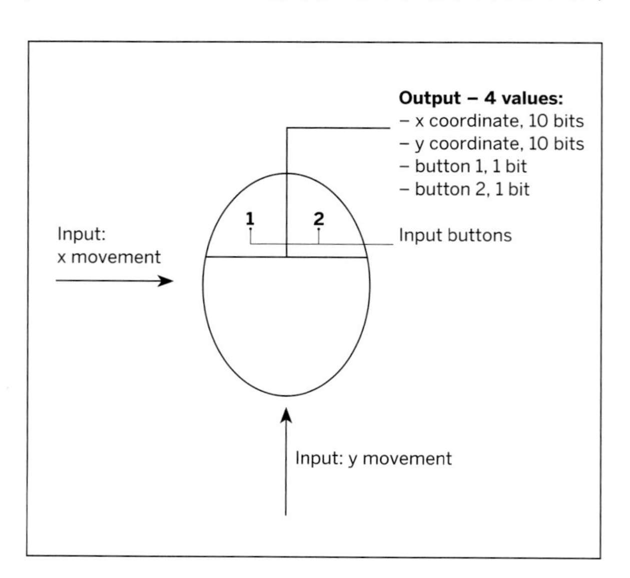

Figure 2-10
The mouse as a data object.

Figure 2-11
The Monski Pong circuit. Sensors are shown here with short wires so that the image is clear. You should attach longer wires to your sensors, though.

101/Uno Breadboard view

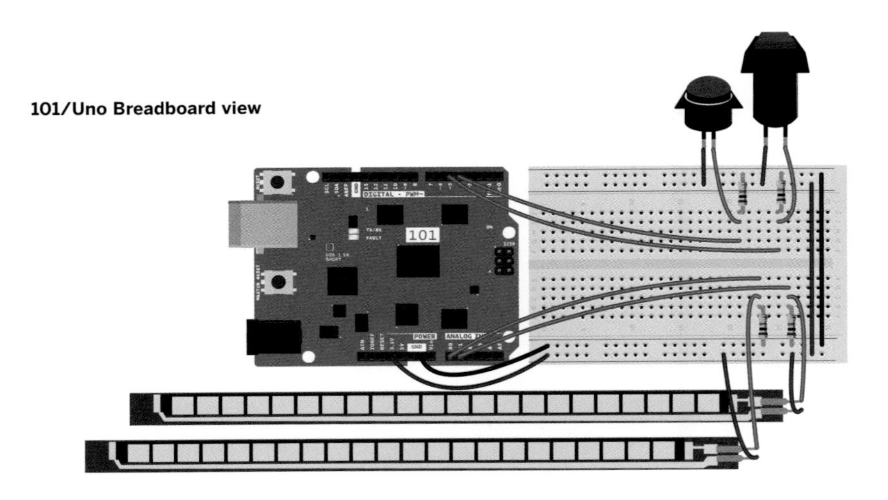

Schematic view
Only the pins that are used are shown.

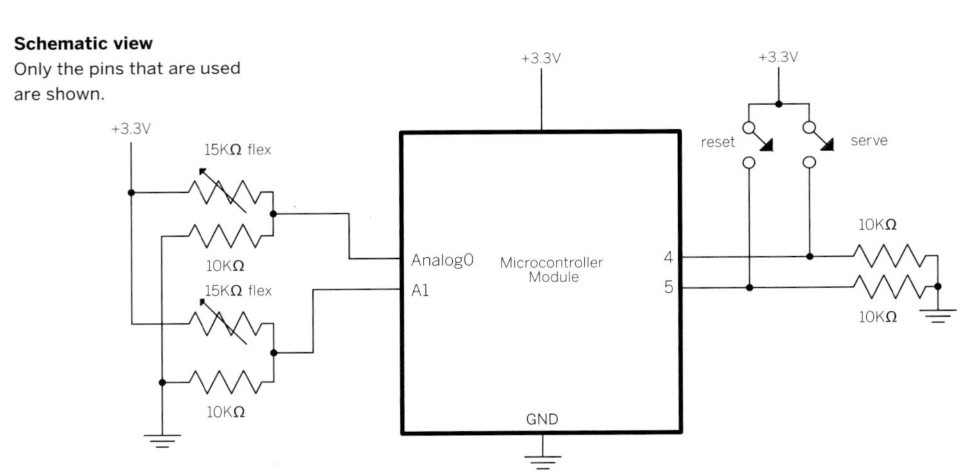

MKR1000 Breadboard view

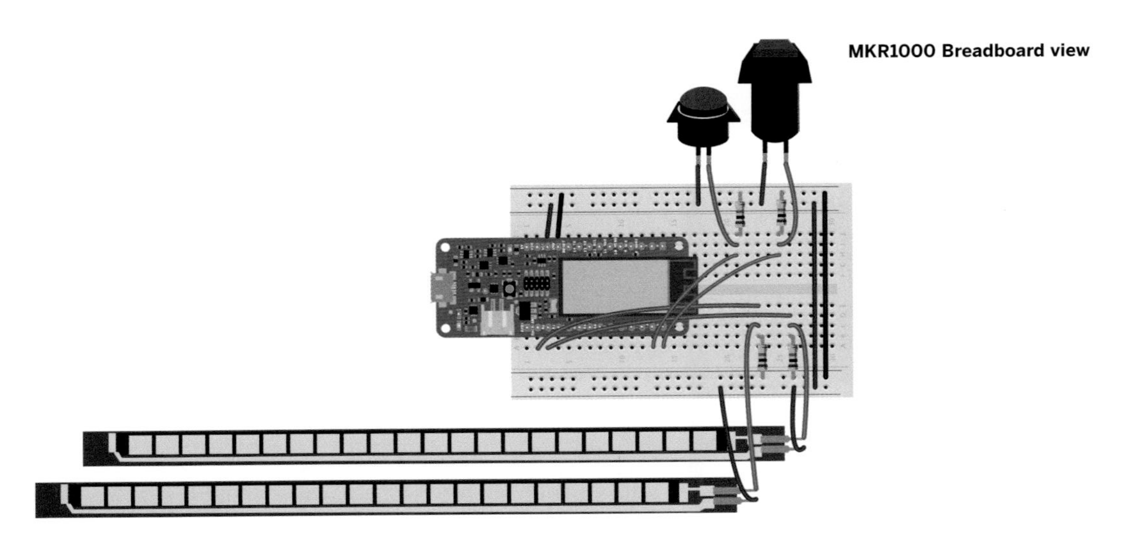

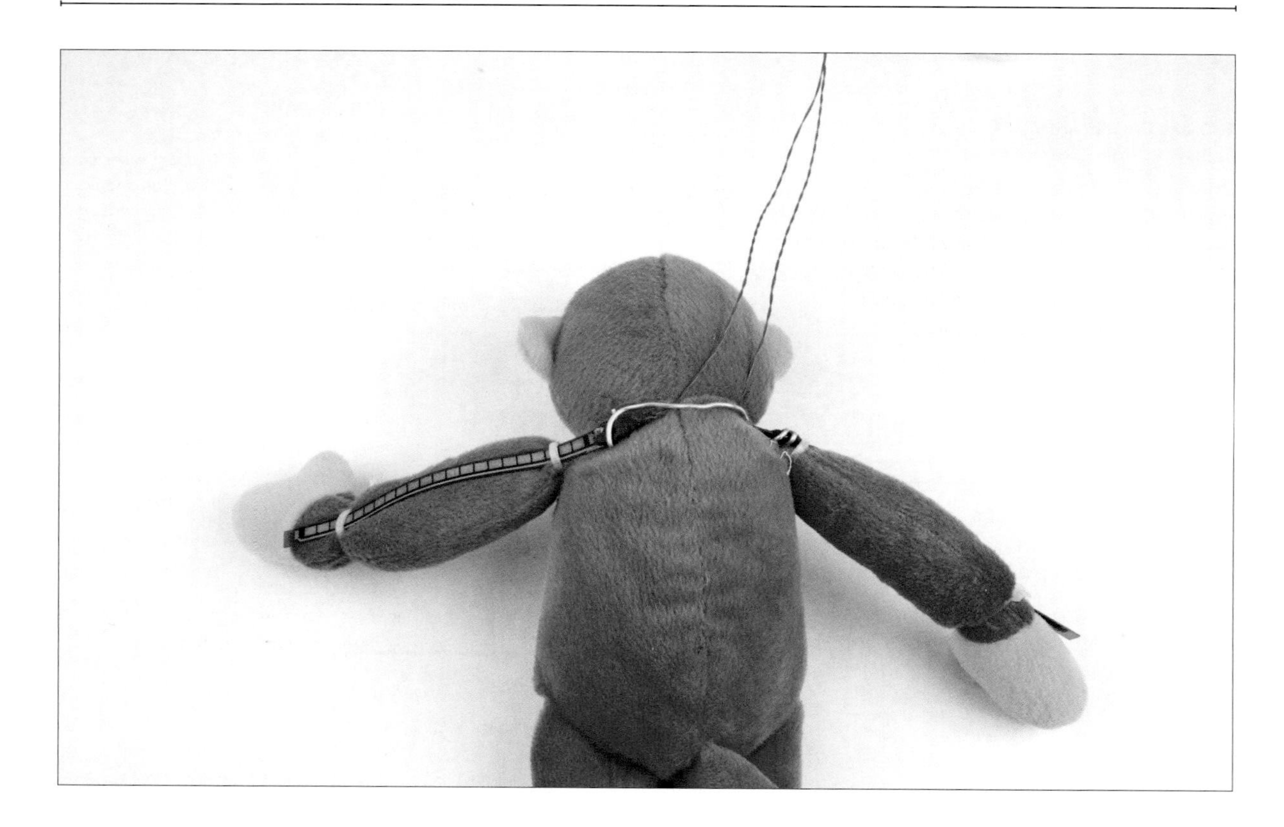

Figure 2-12
A stable support for the sensors is essential if you want good readings from them. Once you know your support works, move it inside the monkey and test it.

Cut a small slit in each of the monkey's armpits to insert the sensors. If you don't want to damage the monkey, you can use tie-wraps or tape to secure the sensors to the outsides of his arms, as shown in Figure 2-12. Position the sensors so that their movement is consistent. You should add some sort of support that keeps them in position relative to each other—a piece of flexible modeling wire will do the job nicely. Make sure that both sensors are facing the same direction, because flex sensors give different readings when flexed one direction than they do when flexed the other direction. Insulate the connections well, because the foam inside the monkey might generate
static electricity when he's moving, which would affect the sensor readings. Hot glue or electrical tape will work well as an insulator.

Make sure that the sensors and electrical connections are stable and secure before you start to work on code. Debugging is much harder if the electrical connections aren't consistent.
X

Test It Now use the following code on the Arduino module to confirm that the sensors are working.

If you open the Serial Monitor in the Arduino IDE—or your preferred serial terminal application at 9600 bits per second, as you did in Chapter 1—you'll see a stream of results like this:

```
284,284,1,1
285,283,1,1
286,284,1,1
289,283,1,1
```

Just as you programmed it, each value is separated by a comma, and each set of readings is on a line by itself.

```
/*
 Sensor Reader
 Context: Arduino

 Reads two analog inputs and two digital inputs and outputs
 their values.

 Connections:
 analog sensors on analog input pins A0 and A1
 switches on digital I/O pins 4 and 5
 */

const int leftSensor = A0;    // analog input for the left arm
const int rightSensor = A1;   // analog input for the right arm
const int resetButton = 4;    // digital input for the reset button
const int serveButton = 5;    // digital input for the serve button

int leftReading = 0;           // reading from the left arm
int rightReading = 0;          // reading from the right arm
int resetReading = 0;          // reading from the reset button
int serveReading = 0;          // reading from the serve button

void setup() {
  // configure the serial connection:
  Serial.begin(9600);
  // configure the digital inputs:
  pinMode(resetButton, INPUT);
  pinMode(serveButton, INPUT);
}

void loop() {
  // read the analog sensors:
  leftReading = analogRead(leftSensor);
  rightReading = analogRead(rightSensor);

  // read the digital sensors:
  resetReading = digitalRead(resetButton);
  serveReading = digitalRead(serveButton);

  // print the results:
  Serial.print(leftReading);
  Serial.print(",");
  Serial.print(rightReading);
  Serial.print(",");
  Serial.print(resetReading);
  Serial.print(",");
  // print the last sensor value with a println() so that
  // each set of four readings prints on a line by itself:
  Serial.println(serveReading);
}
```

Try replacing the part of your code that prints the results with the code to the right.

```
Serial.write(leftReading);
Serial.write(44);
Serial.write(rightReading);
Serial.write(44);
Serial.write(resetReading);
Serial.write(44);
// print the last sensor value with a println() so that
// each set of four readings prints on a line by itself:
Serial.write(serveReading);
Serial.write(10);
Serial.write(13);
```

When you view the results in the Serial Monitor or terminal, you'll get something that looks like this:

```
.,P,,
(,F,,
(,A,,
),I,,
```

Before you go to the next section, where you'll be writing some Processing code to interpret the output of this program, you must undo this change.

What's going on? The original example uses the `Serial.print()` command, which displays the values of the sensors as their ASCII code values: this modification sends out the raw binary values using `Serial.write()`. The Serial Monitor and serial terminal applications assume that every byte they receive is an ASCII character, so they display the ASCII characters corresponding to the raw binary values in the second example. For example, the values 13 and 10 correspond to the ASCII return and newline characters, respectively. The value 44 corresponds to the ASCII comma character. Those are the bytes you're sending in between the sensor readings in the second example. The sensor variables (`leftValue`, `rightValue`, `reset`, and `serve`) are the source of the mystery characters. In the third line of the output, when the second sensor's value is 65, you see the character "A" because the ASCII character "A" has the value 65. For a complete list of the ASCII values corresponding to each character, see www.asciitable.com.

Which way should you format your sensor values: as raw binary or as ASCII? It depends on the capabilities of the system that's receiving the data, and of those that are passing it through. When you're writing software on a personal computer, it's often easier for your software to interpret raw binary values. However, many of the network protocols you'll use in this book are ASCII-based. In addition, ASCII is readable by humans, so you may find

What's ASCII?

ASCII is the American Symbolic Code for Information Interchange. The scheme was created in 1967 by the American Standards Association (now ANSI) as a means for all computers, regardless of their operating systems, to be able to exchange text-based messages. In ASCII, each letter, numeral, or punctuation mark in the English alphabet is assigned a number. Anything an end user types is converted to a string of numbers, transmitted, and then reconverted on the other end. In addition to letters, numbers, and punctuation marks, certain page-formatting characters—like the linefeed and carriage return (ASCII 10 and 13, respectively)—have ASCII values. That way, not only the text but also the display format of the message could be transmitted along with the text. These are referred to as control characters, and they take up the first 32 values in the ASCII set (ASCII 0 31).

All the numbers, letters, punctuation, and control characters are covered by 128 possible values. However, ASCII is too limited to display non-English characters, and its few control characters don't offer enough control in the age of graphic user interfaces. Unicode—a more comprehensive superset of ASCII—has replaced ASCII as the standard for text interchange. Unicode has multiple character sets, to accommodate character sets of all languages.

it easier to send the data as ASCII. For Monski Pong, use the ASCII-formatted version (the first example); later in this chapter you'll see why it's the right choice.

If you haven't already done so, undo the changes you made on page 62 to the Sensor Reader program and make sure that it's working as it did originally. Once you've got the microcontroller sending the sensor values consistently to the terminal, it's time to send them to a program where you can use them to display a pong game. This program needs to run on a host computer that's connected to your microcontroller board. Processing will do this well.

X

Try It Open the Processing application and enter this code:

```
/*
 Serial String Reader
 Context: Processing
 */

import processing.serial.*;     // import the Processing serial library

Serial myPort;                  // The serial port
String resultString;            // string for the results

void setup() {
  size(480, 130);               // set the size of the applet window
  printArray(Serial.list());    // List all the available serial ports

  // get the name of your port from the serial list.
  // The first port in the serial list on my computer
  // is generally my microcontroller, so I open Serial.list()[0].
  // Change the 0 to the number of the serial port
  // to which your microcontroller is attached:
  String portName = Serial.list()[0];
  // open the serial port:
  myPort = new Serial(this, portName, 9600);

  // read bytes into a buffer until you get a linefeed (ASCII 10):
  myPort.bufferUntil('\n');
}

void draw() {
  // set the background and fill color for the applet window:
  background(#044f6f);
  fill(#ffffff);
  // show a string in the window:
  if (resultString != null) {
    text(resultString, 10, height/2);
  }
}

// serialEvent method is run automatically by the Processing sketch
// whenever the buffer reaches the byte value set in the bufferUntil()
// method in the setup():

void serialEvent(Serial myPort) {
  // read the serial buffer:
  String inputString = myPort.readStringUntil('\n');
```

Text or Binary?

One of the most confusing aspects of data communications is understanding the difference between a binary protocol and a text-encoded protocol. ASCII—and its more modern cousin, Unicode—allow you to convert any stream of text into binary information that a computer can read, and vice versa. Understanding the difference between a protocol that is text-based and one that's not, and why you'd make that choice, is essential to understanding how electronic things communicate.

Isn't All Data Binary?

Well, yes. After all, computers only operate on binary logic. So, even text-based protocols are, at their core, binary. However, it's often easier to write your messages in text and insist that the computers translate that text into bits themselves. For example, take the phrase "1-2-3, go." You can see it laid out in characters below, with the ASCII code for each character, and the bits that make up that code. It may seem like a lot of ones and zeroes for the computer to process, but when it's reading them at millions or billions of bits a second, it's no big deal. But what about when it's sending it to another computer? There are 80 bits there. Imagine you're sending it using TTL serial at 9600bps. If you add a stop bit and a start bit between each byte—as TTL and RS-232 serial protocols do—that's 96 bits, meaning you could send this message in 1/100th of a second. That's still pretty fast.

Character	**1**	**-**	**2**	**-**	**3**	**,**		**g**	**o**	**!**
ASCII code	49	45	50	45	51	44	32	103	111	33
Binary	00110001	00101101	00110010	00101101	00110011	00101100	01000000	01100111	01101111	01000001

What about when the messages you have to send are not text-based? Imagine you're sending a string of RGB pixel values, each ranging from 0 to 255 like this:

102,198,255,127,127,212,255,155,127,

You know that each reading is 8 bits (0 - 255 is a range of 256 values, or 2^8, or 8 bits), so in their most basic form, they take up 3 bytes per pixel. But if you're sending them in text form it's one byte per character, so the string above would be 36 bytes to send nine values. If you hadn't encoded it in text form, but just sent each value as a byte, you'd have only sent 9 bytes. When you start to send millions of pixels, that's a big difference! In cases like this, where all the data you want to send is purely numeric, sending data without encoding it as text makes sense. When you need to send text (for example, email or hypertext), encoding it as ASCII/Unicode makes sense. When the number of bytes you'd send is minimal, encoding it as ASCII/Unicode can help with debugging, because most serial or network terminal programs interpret all bytes that they receive as text, as you've seen. So:

- If there's a lot of data and it's all numeric, send as raw binary.
- If there's text, send as Unicode/ASCII.
- If it's a short data string, send as whichever makes you happy.

Interpreting a Binary Protocol

Since most of the protocols you'll deal with in this book are text-based, you won't have to interpret the value of a byte very often. Binary protocols often demand that you know which bit represents what, so it's useful to know a little about the architecture of a byte and how to manipulate it. Let's now talk for a bit about bits.

Binary protocols often show up in communications between chips in a complex device, particularly in synchronous serial protocols. Many SPI and I2C devices have small command sets. Their single-byte *operational codes*, or *opcodes*, are often combined with the parameters for the commands in the same byte. These protocols are usually written out in hexadecimal notation, binary notation, or both. You've seen hexadecimal or base-16 notation already in this book. Just as hexadecimal numbers begin with 0x, binary numbers in Arduino—and by extension in C—begin with `0b`, like so: `0b10101010`.

Which digit matters most in the number below?

$2,508

The 2, because it represents the largest amount, two thousands, or two groups of 10^3. It is the *most significant digit*. That number was in decimal, or base-10 notation. The same principle applies when you're writing in binary, or base-2. Which is the *most significant bit*:

0b10010110

The leftmost 1 is most significant because it represents 1 group of 128, or 2^7.

Usually when you see bits written out, though, you care less about their decimal numeric values than their position in the byte. For example, in the XBee API data protocol in Chapter 7, there is a 2-byte value called Channel Indicator that tells you which analog and digital inputs are active. It's a 2-byte (16-bit) value where the 9 least significant bits represent the digital inputs, the next 6 bits indicate the four analog inputs, and the most significant bit is not used. You'll see it in the pages that follow. So if you had activated all of the analog inputs and none of the digital inputs, the Channel Indicator would look like this:

0111111000000000

You can see that the lower 9 bits, representing the digital channels, are set to 0, and the next 6 bits, representing the analog channels, are set to 1, indicating that they're active. In this case, you don't care what the decimal value of this 2-byte number is, but you do care what each bit's value is, so you can tell which channels you should use.

Sometimes 2 bytes' values are combined into a single 16-bit value, or 4 bytes are combined into a 64-bit value. You can see this in the analog-to-digital I/O readings for the XBee protocol. Each ADC can read to a 10-bit resolution, so the protocol reserves 2 bytes for each reading. The maximum possible reading is 1023. If you looked at the byte values for this, the first byte would be 3, and the second would be 255. In hexadecimal notation, it would be 0x3FF. Take the first byte, multiply by 256, and add it to the second, and you get the whole value. You might think of bytes as being numbers in base-256.

Hex: What Is It Good For?

Since you can manipulate binary protocols bit by bit, you're probably wondering what hexadecimal notation is good for. Hex is useful when you're working in groups of 16, of course. For example, the Musical Instrument Digital Interface Protocol, MIDI, is grouped into banks of instruments, with 128 instruments per bank, and each instrument can play on up to 16 channels. For example. 0x9n is a Note On command, where n is the channel number, from 0 to A in hex. 0x9A means note on, channel A (or 10 in decimal). Similarly, 0x8A means note off, channel A. Once you know that MIDI is organized in groups of 16, it makes a lot of sense to read and manipulate it in hexadecimal. Similarly, web colors are represented in hexadecimal values of three bytes, red, green and blue, like so:

```
red = 0xFF0000
green = 0x00FF00
blue = 0x0000FF
```

Once you know this, it's easy to read the relative color levels in hexadecimal. For example, 0x26B0E7 has relatively little red (0x26, or 38 of a possible 255), a good amount of green (0xB0, or 176) and a lot of blue (0xE7, or 231). This is a nice shade of teal, the color of the technical terms in this book.

X

Continued from page 63.

```
    // trim the carriage return and linefeed from the input string:
    inputString = trim(inputString);
    // clear the resultString:
    resultString = "";

    // split the input string at the commas
    // and convert the sections into integers:
    int sensors[] = int(split(inputString, ','));

    // add the values to the result string:
    for (int sensorNum = 0; sensorNum < sensors.length; sensorNum++) {
      resultString += "Sensor " + sensorNum + ": ";
      resultString += sensors[sensorNum] + "\t";
    }
    // print the results to the console:
    println(resultString);
  }
```

Data Packets, Headers, Payloads, and Tails

Now that you've got data going from one device (the microcontroller attached to the monkey) to another (the computer running Processing), take a closer look at the sequence of bytes you're sending to exchange the data. Generally, it's formatted like this, with a comma between each field:

Left-arm sensor (0–1023)	Right-arm sensor (0–1023)	Reset button (0 or 1)	Serve button (0 or 1)	Return, newline characters
1–4 bytes	1–4 bytes	1 byte	1 byte	2 bytes

Each section of the sequence is separated by a single byte whose value is ASCII 44 (a comma). You've just made another data protocol. The bytes representing your sensor values and the commas that separate them are the payload, and the return and newline characters are the tail. The commas are the delimiters. This data protocol doesn't have a header, but many do.

A header is a sequence of bytes identifying what's to follow. It might also contain a description of the sequence to follow. On a network, where many possible devices could receive the same message, the header might contain the address of the sender, the receiver, or both. That way, any device can just read the header to decide whether it needs to read the rest of the message. Sometimes a header is as simple as a single byte of a constant value, identifying the beginning of the message. In this example, where there is no header, the tail performs a similar function, separating one message from the next.

On a network, many messages like this are sent out all the time. Each discrete group of bytes is called a packet and includes a header, a payload, and usually a tail. Any given network has a maximum packet length. In this example, the packet length is determined by the size of the serial buffer on the personal computer. Processing can handle a buffer of a few thousand bytes, so this 16-byte packet is easy for it to handle. If you had a much longer message, you'd have to divide the message into several packets and reassemble them once they all arrived. In that case, the header might contain the packet number so the receiver knows the order in which the packets should be reassembled.

Let's take a break from writing code and test out the sketch. Make sure you've closed the Serial Monitor or serial port in your serial terminal application so that it releases the serial port. Then run this Processing sketch. You should see a list of the sensor values in the console and in the applet window, as shown in Figure 2-13.

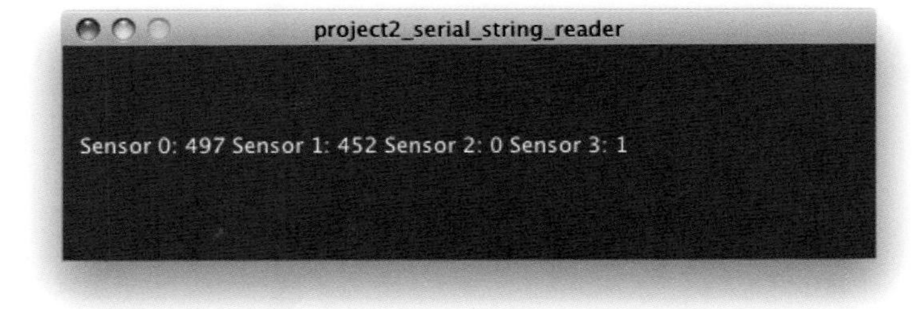

Figure 2-13
Output of the Processing sketch so far.

Next, it's time to use the data to play pong. First, add a few variables at the beginning of the Processing sketch before the `setup()` method, and change the `setup()` to set the window size and initialize some of the variables (the new lines are shown in blue).

NOTE: The variables relating to the paddle range in this example are floating-point numbers (floats), because when you divide integers, you get integer results only. For example, 480/400 gives 1, not 1.2, when both are integers. Likewise, 400/480 returns 0, not 0.8333. Using integers when you're dividing two numbers that are in the same order of magnitude produces useless results. Beware of this when using scaling functions like map().

```
float leftPaddle, rightPaddle;   // variables for the flex sensor values
int resetButton, serveButton;    // variables for the button values
int leftPaddleX, rightPaddleX;   // horizontal positions of the paddles
int paddleHeight = 50;           // vertical dimension of the paddles
int paddleWidth = 10;            // horizontal dimension of the paddles

float leftMinimum = 120;         // minimum value of the left flex sensor
float rightMinimum = 100;        // minimum value of the right flex sensor
float leftMaximum = 530;         // maximum value of the left flex sensor
float rightMaximum = 500;        // maximum value of the right flex sensor

void setup() {
 size(640, 480);             // set the size of the applet window

   String portName = Serial.list()[0];
  // open the serial port:
  myPort = new Serial(this, portName, 9600);

  // read bytes into a buffer until you get a linefeed (ASCII 10):
  myPort.bufferUntil('\n');

  // initialize the sensor values:
  leftPaddle = height/2;
  rightPaddle = height/2;
  resetButton = 0;
  serveButton = 0;

  // initialize the paddle horizontal positions:
  leftPaddleX = 50;
  rightPaddleX = width - 50;

  // set no borders on drawn shapes:
  noStroke();
}
```

Now, replace the `serialEvent()` method with this version, which puts the serial values intc the sensor variables.

```
void serialEvent(Serial myPort) {
  // read the serial buffer:
  String inputString = myPort.readStringUntil('\n');

  // trim the carriage return and linefeed from the input string:
  inputString = trim(inputString);
  // clear the resultString:
  resultString = "";

  // split the input string at the commas
  // and convert the sections into integers:
  int sensors[] = int(split(inputString, ','));
  // if you received all the sensor strings, use them:
  if (sensors.length == 4) {
    // scale the flex sensors' results to the paddles' range:
    leftPaddle = map(sensors[0], leftMinimum, leftMaximum, 0, height);
    rightPaddle = map(sensors[1], rightMinimum, rightMaximum, 0, height);

    // assign the switches' values to the button variables:
    resetButton = sensors[2];
    serveButton = sensors[3];

    // add the values to the result string:
    resultString += "left: "+ leftPaddle + "\tright: " + rightPaddle;
    resultString += "\treset: "+ resetButton + "\tserve: " + serveButton;
  }
}
```

Finally, change the `draw()` method to draw the paddles (new lines are shown in blue).

```
void draw() {
  // set the background and fill color for the applet window:
  background(#044f6f);
  fill(#ffffff);

  // draw the left paddle:
  rect(leftPaddleX, leftPaddle, paddleWidth, paddleHeight);

  // draw the right paddle:
  rect(rightPaddleX, rightPaddle, paddleWidth, paddleHeight);
}
```

> "You may not see the paddles until you flex the sensors. The range of your sensors will be different, depending on how they're physically attached to the monkey, and how far you can flex his arms. The `map()` function maps the sensors' ranges to the range of the paddle movement, but you need to determine what the sensors' range is first. For this part, it's important that you have the sensors embedded in the monkey's arms, as you'll be fine-tuning the system, and you want the sensors in the locations where they'll actually get used. Once you've set the sensors' positions in the monkey, run the Processing program again and watch the left and right sensor numbers as you flex the monkey's arms. Write down the maximum and minimum values on each arm. Then, enter them into the four variables, `leftMinimum`, `leftMaximum`, `rightMinimum`, and `leftMaximum`, in the `setup()` method. Once you've adjusted these variables, the paddles' movement should cover the screen height when you move the monkey's arms.

Finally, it's time to add the ball. The ball will move from left to right diagonally. When it hits the top or bottom of the screen, it will bounce off and change vertical direction. When it reaches the left or right, it will reset to the center. If it touches either of the paddles, it will bounce off and change horizontal direction. To make all that happen, you'll need five new variables at the top of the program, just before the `setup()` method.

```
int ballSize = 10;      // the size of the ball
int xDirection = 1;     // the ball's horizontal direction.
                        // left is -1, right is 1.
int yDirection = 1;     // the ball's vertical direction.
                        // up is -1, down is 1.
int xPos, yPos;         // the ball's horizontal and vertical positions
```

At the end of the `setup()` method, you need to give the ball an initial position in the middle of the window.

```
  // initialize the ball in the center of the screen:
  xPos = width/2;
  yPos = height/2;
```

Now, add two methods at the end of the program, one called `animateBall()` and another called `resetBall()`. You'll call these from the `draw()` method shortly.

```
void animateBall() {
  // if the ball is moving left:
  if (xDirection < 0) {
    //  if the ball is to the left of the left paddle:
    if  ((xPos <= leftPaddleX)) {
      // if the ball is in between the top and bottom
      // of the left paddle:
      if((leftPaddle - (paddleHeight/2) <= yPos) &&
        (yPos <= leftPaddle + (paddleHeight /2))) {
        // reverse the horizontal direction:
        xDirection =-xDirection;
      }
    }
  }
  // if the ball is moving right:
  else {
    //  if the ball is to the right of the right paddle:
    if  ((xPos >= ( rightPaddleX + ballSize/2))) {
      // if the ball is in between the top and bottom
```

»

Continued from previous page.

```
      // of the right paddle:
      if((rightPaddle - (paddleHeight/2) <=yPos) &&
        (yPos <= rightPaddle + (paddleHeight /2))) {

        // reverse the horizontal direction:
        xDirection =-xDirection;
      }
    }
  }

  // if the ball goes off the screen left:
  if (xPos < 0) {
    resetBall();
  }
  // if the ball goes off the screen right:
  if (xPos > width) {
    resetBall();
  }

  // stop the ball going off the top or the bottom of the screen:
  if ((yPos - ballSize/2 <= 0) || (yPos +ballSize/2 >=height)) {
    // reverse the y direction of the ball:
    yDirection = -yDirection;
  }
  // update the ball position:
  xPos = xPos + xDirection;
  yPos = yPos + yDirection;

  // Draw the ball:
  rect(xPos, yPos, ballSize, ballSize);
}

void resetBall() {
  // put the ball back in the center
  xPos = width/2;
  yPos = height/2;
}
```

You're almost ready to set the ball in motion. But first, it's time to do something with the reset and serve buttons. Add another variable at the beginning of the code (just before the `setup()` method with all the other variable declarations) to keep track of whether the ball is in motion. Add two more variables to keep score.

```
boolean ballInMotion = false;  // whether the ball should be moving
int leftScore = 0;
int rightScore = 0;
```

Now you're ready to animate the ball. It should move only if it's been served. This code goes at the end of the `draw()` method. The first `if()` statement starts the ball in motion when the serve button is pressed. The second moves it if it's in service. The third resets the ball to the center, and resets the score when the reset button is pressed.

```
// calculate the ball's position and draw it:
if (ballInMotion == true) {
  animateBall();
}

// if the serve button is pressed, start the ball moving:
if (serveButton == 1) {
  ballInMotion = true;
}

// if the reset button is pressed, reset the scores
// and start the ball moving:
if (resetButton == 1) {
  leftScore = 0;
  rightScore = 0;
  ballInMotion = true;
}
```

Modify the `animateBall()` method so that when the ball goes off the screen left or right, the appropriate score is incremented (added lines are shown in blue).

```
// if the ball goes off the screen left:
if (xPos < 0) {
  rightScore++;
  resetBall();
}
// if the ball goes off the screen right:
if (xPos > width) {
  leftScore++;
  resetBall();
}
```

To include the scoring display, add a new global variable before the `setup()` method.

```
int fontSize = 36;      // point size of the scoring font
```

Then add two lines before the end of the `setup()` method to initialize the font.

```
// create a font with the third font available to the system:
PFont myFont = createFont(PFont.list()[2], fontSize);
textFont(myFont);
```

Finally, add two lines before the end of the `draw()` method to display the scores.

```
// print the scores:
text(leftScore, fontSize, fontSize);
text(rightScore, width-fontSize, fontSize);
```

Now you can play Monski Pong! Figure 2-14 shows the game in action. For added excitement, get a second pink monkey and put one sensor in each monkey so you can play with a friend.

Figure 2-14
The completed Monski Pong Processing sketch.

Flow Control

You may notice that the paddles don't always move as smoothly onscreen as Monski's arms move. Sometimes the paddles don't seem to move for a fraction of a second,and sometimes they seem to lag behind the actions you're taking. This is becausethe communication between the two devices is asynchronous.

Although the devices agree on the rate at which data is exchanged, it doesn't mean that the receiving computer's program has to use the bytes as they're sent. Monitoring the incoming bytes is actually handled by a dedicated operating system process, and the incoming bytes are stored in a memory buffer, called the *serial buffer*, until the current program is ready to use them. Most personal computers allocate a buffer of a few thousand bytes for each serial port. The program using the bytes (Processing, in the previous example) is juggling a number of other tasks, like redrawing the screen, handling the math that goes with it, and sharing processor time with other programs through the operating system. It may get bytes from the buffer less than a hundred times a second—even though the bytes are coming in much faster.

There's another way to handle the communication between the two devices that can alleviate this problem. If Processing asks for data only when it needs it, and if the microcontroller only sends one packet of data when it gets a request for data, the two will be in tighter sync.

To make this happen, first add the following lines to the `startup()` of the **Arduino sketch** (the Sensor Reader program shown back in the beginning of the "Project 2: Monski Pong" section). This makes the Arduino send out a serial message until it gets a response from Processing.

```
while (Serial.available() <= 0) {
   Serial.println("hello");   // send a starting message
}
```

Next, wrap the whole of the `loop()` method in the Arduino sketch in an `if()` statement like this (new lines are shown in blue).

In the next step, you'll add some code to the Monski Pong Processing sketch.

```
void loop() {
   // check to see whether there is a byte available
   // to read in the serial buffer:
   if (Serial.available() > 0)   {
      // read the serial buffer;
      // you don't care about the value of
      // the incoming byte, just that one was
      // sent:
      int inByte = Serial.read();
      // the rest of the existing main loop goes here
      // ...
   }
}
```

Add the following lines at the end of `serialEvent()` in the **Processing sketch** (new lines are shown in blue).

```
void serialEvent(Serial myPort) {
   // rest of the serialEvent goes here
   myPort.write('\r');      // send a carriage return
}
```

Now, the paddles should move much more smoothly. Here's what's happening: the microcontroller is programmed to send out a "hello" string until it receives any serial data. When it does, it goes into the main loop. There, it reads the byte just to clear the serial buffer, then sends out its data once, then waits for more data to arrive. Whenever it gets no bytes, it sends no bytes.

Processing, meanwhile, starts its program by waiting for incoming data. When it gets any string ending in a newline, `serialEvent()` is called, just as before. It reads the string, and if there are commas in it, it splits the string up and extracts the sensor values, like before. If there are no commas in the string (for example, if the string is "hello"), Processing doesn't do anything with it.

The change in the Processing sketch is at the end of the `serialEvent()`. There, it sends a byte back to the microcontroller, which, seeing a new byte coming in, sends out another packet of data, and the whole cycle repeats itself. This way, the serial buffer on Processing's side never fills up, and it's always got the freshest sensor readings.

The value of the byte that the microcontroller receives is irrelevant. It's used only as a signal from the Processing sketch to let the microcontroller know when it's ready for new data. Likewise, the "hello" that the controller sends is irrelevant—it's only there to trigger Processing to send an initial byte—so Processing discards it. This method of handling data *flow control* is sometimes referred to as a *handshake* method, or *call-and-response*. Whenever you're sending packets of data, call-and-response flow control can be a useful way to ensure consistent exchange.

X

Project 3

Wireless Monski Pong

Monski Pong is fun, but it would be more fun if Monski didn't have to be tethered to the computer through a USB cable. This project breaks the wired connection between the microcontroller and the personal computer, and introduces a few new networking concepts: the *modem* and the *address*.

MATERIALS

- **1 completed Monski Pong project, including**
- **1 Arduino-compatible microcontroller (MKR1000 and 101 shown)**
  - **Features used: Digital Input, Analog input, UART**
- **1 9V battery and snap connector or appropriate battery for your board**
- **DC power plug, 2.1mm ID, 5.5mm OD**
- **1 Bluetooth Serial module**
- **1 project box**

NOTE: If your computer doesn't have built-in Bluetooth, you'll need a Bluetooth adapter. Most computer retailers carry USB-to-Bluetooth adapters.

Bluetooth: A Multilayer Network Protocol

The new piece of hardware in this project is the Bluetooth module. This module has two interfaces: two of its pins, marked RX and TX, are an asynchronous serial port that can communicate with a microcontroller. It also has a radio that communicates using the Bluetooth communications protocol. It acts as a *modem*, translating between the Bluetooth and regular asynchronous serial protocols.

The first digital modems converted data signals to audio to send them across a telephone connection. They modulated the data on the audio connection, and demodulated the audio back into data. Their descendants are everywhere, from set-top boxes that modulate and demodulate between a cable TV signal and Internet connection, to the sonar modems that convert data into ultrasonic pings used in marine research.

Bluetooth is a multilayered communications protocol, designed to replace wired connections for a number of applications. As such, it's divided into a group of possible application protocols called *profiles*. The simplest Bluetooth devices are serial devices, like the module used in this project. These implement the Bluetooth *Serial Port Profile (SPP)*. Other Bluetooth devices implement other protocols. Wireless headsets implement the audio *Headset Profile*. Wireless mice and keyboards implement the *Human Interface Device (HID) Profile*. Because there are a number of possible profiles a Bluetooth device might support, there is also a *Service Discovery Protocol*, by which radios exchange information about what they can do. Because the protocol is standardized, you get to skip over most of the details of making and maintaining the connection, letting you concentrate on exchanging data. It's a bit like how TTL serial and USB made it possible for you to ignore most of the electrical details necessary to connect your microcontroller to your personal computer, which let you focus on sending bytes in the last project.

Add the Bluetooth module to the Monski pong breadboard, as shown in Figure 2-15. The figures show an Adafruit Bluefruit EZ-Link, but the SparkFun Bluetooth Mates will also work. Connect the module's ground and Vcc to the breadboard ground and +Vin, respectively. Connect Arduino's TX to the module's RX, and vice versa, Connect the battery, and the module will start up.

Wait, Doesn't the Arduino 101 Have a Built-in Bluetooth Radio?

Yes it does. But that radio is Bluetooth 4.0, sometimes called *Bluetooth LE* or *Bluetooth Smart*. It uses the latest Bluetooth protocol, designed to save power by turning the radio off when not in use. Bluetooth 4.0 doesn't include a Serial Port protocol, however, which makes it more complicated to use. You'll see that radio in use in later chapters.

Meanwhile, Bluetooth 2.0's Serial Port Profile is very useful because there are so many TTL serial devices on the market that can connect to a Bluetooth Serial radio, just as you're about to do. So even though it may seem silly to attach a radio to a microcontroller with a built-in radio, there is value in learning about external Bluetooth serial modules like this one. Also, this project will work with any Arduino-compatible microcontroller that has a *Universal Asynchronous Receiver-Transmitter (UART)*.

Pairing Your Computer with the Bluetooth Module

To make a wireless connection from your computer to the module, you have to pair them. To do this, open your computer's Bluetooth control panel to browse for new devices.

If you're using macOS, choose the Apple menu System Preferences, then click Bluetooth. Make sure Bluetooth is turned on and Discoverable, and click "Show Bluetooth status in the menu bar." Wait a few seconds and computer will search for devices.

For Windows 10 users, click the Start menu search field and search for "Bluetooh Settings." When the control panel opens, the system will search for new devices and present you with a list.

Ubuntu Linux's Bluetooth manager for version 1.0 is a bit limited, so it's easier to install BlueMan instead. Go to the Ubuntu Software center, search for BlueMan, and install it. When it's installed, open the System control panel and you'll see Bluetooth Manager, in addition to the default Bluetooth control panel. Open Bluetooth Manager, and it will scan for available devices and show them.

At first, the devices' addresses will show up like this: `00:3A:45:6C:9A:06`. These are the *Media Access Control*, or *MAC addresses* of the radios it's finding. In a network of more than two devices, networking is essential.

Eventually the numbers will change to names. You'll see one called `Adafruit-EZ-Link-XXXX`, where `XXXX` is the last four digits of the module's address. If you have no other Bluetooth devices on, it will be the only one. Choose this device. if you're asked for a passcode, enter `1234`. Click Continue. A connection will be established, as will a serial port.

On macOS, the port will be called `Adafruit-EZ-Link-XXXX-SPP`. In Windows, it will be a variation on the usual `COMX` (mine is `COM14`).

In Ubuntu, once the module is paired, click Setup, and you'll get a dialog asking you if you want to connect to a serial port. Click Forward, and it will tell you the name of the serial port for the Bluetooth module, **/dev/rfcomm0**.

Adjusting the Microcontroller Program

You'll need to change your Arduino sketch to work with the Bluetooth module. What you change depends on the board you're using.

The processors on older Arduino and Arduino-compatible boards like the Uno and Mega, or SparkFun's RedBoards cannot communicate via USB natively, so they use a USB-to-serial converter on the board. On these boards, the processor's UART connections are connected both to the USB-to-serial converter, and to the TX and RX output pins on the board. For these boards, you don't need to make any changes to your program, but you do need to disconnect the Bluetooth module from the TX and RX pins before uploading any new code, so that the Bluetooth radio doesn't interfere with the upload.

If you're using a board that's USB-native like the 101 or the MKR1000, there is a virtual UART that communicates over the USB connection. That's what you're using when you use the Serial library. The TX and RX pins on the board are attached to a separate UART inside the microcontroller. To address them, you use `Serial1` instead of `Serial`. Any data that comes in the RX pin can be read using `Serial1.read()` and any data that you want to send out can be written using `Serial1.write()`, `Serial1.print()`, or `Serial1.println()`. If you change all occurrences of `Serial` in your current Arduino sketch to `Serial1`, the sketch will work with the Bluefruit module connected to TX and RX. You won't see any output in the Serial Monitor, though.

When your microcontroller sends data to the Bluetooth module serially, the radio's blue LED will turn on. When the computer makes a connection to the module, the red LED on the module will turn on.

Adjusting the Monski Pong Program

Once your computer has made contact with the Bluetooth module, you can connect to it like a serial port. Run the Monski Pong Processing sketch and check the list of serial ports. You should see the new port listed along with the others. Take note of which number it is, and change this line in the `setup()` method:

```
String portName = Serial.list()[0];
```

For example, if the Bluetooth port is the ninth port in your list, change the first line to open `Serial.list[8]`. Everything should work exactly as it did using the USB serial connection.

X

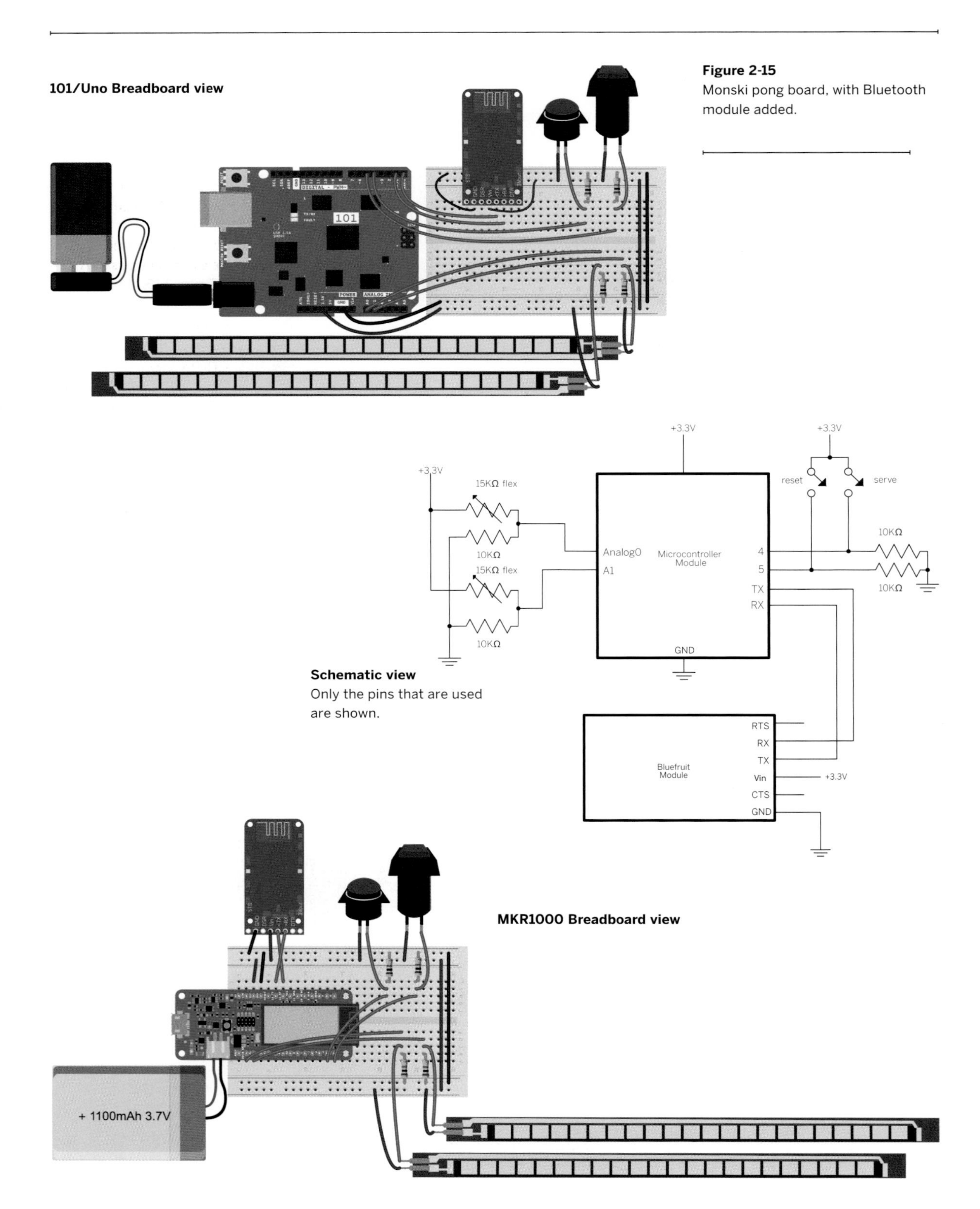

Figure 2-15
Monski pong board, with Bluetooth module added.

Finishing Touches: Tidy It Up, Box It Up

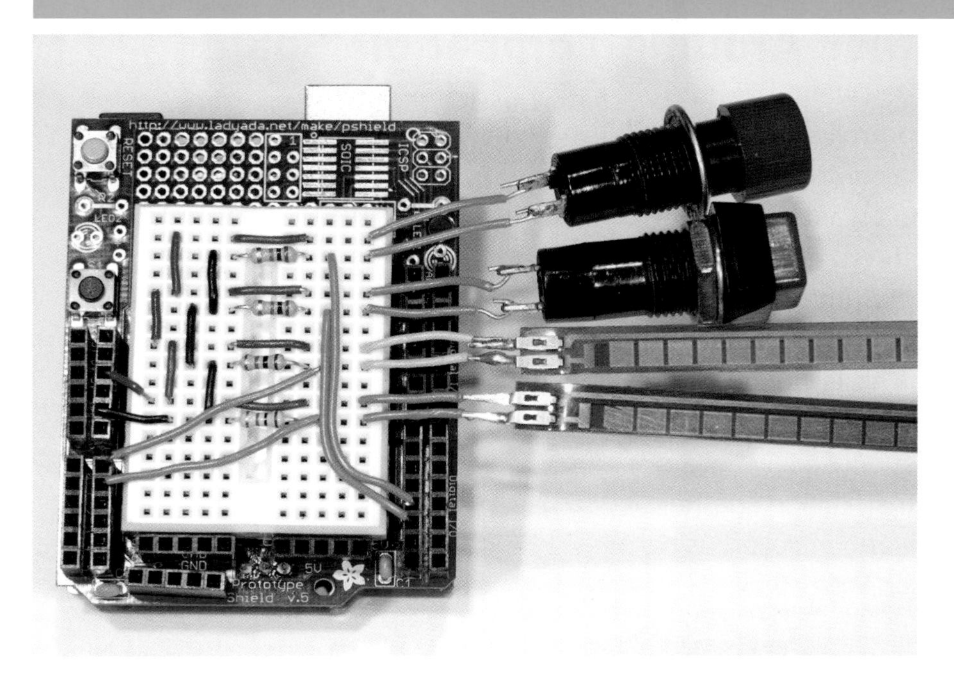

◀◀ Figure 2-16
You might want to shrink Monski Pong so it's more compact. This figure shows the Monski Pong circuit on a breadboard shield. This is the same circuit as the one in Figure 2-15.

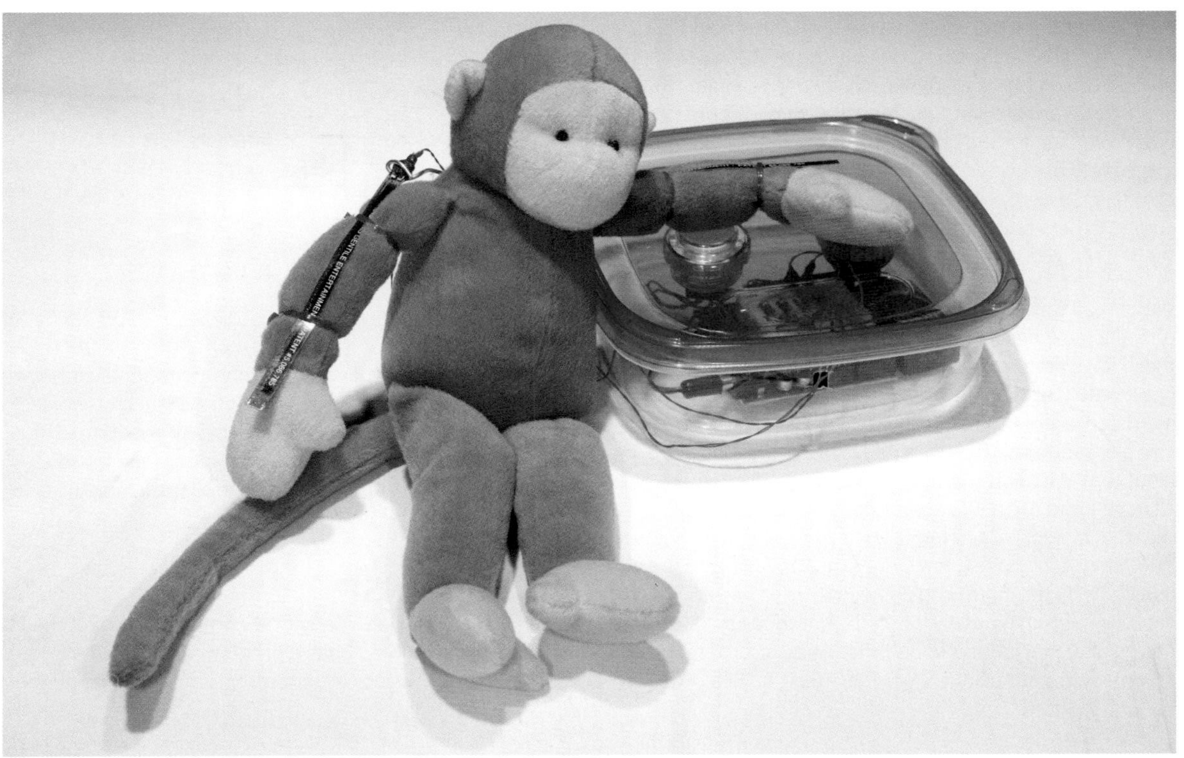

▼ Figure 2-17
Here's the Monski Pong controller with Monski attached. Kitchen storage containers make excellent project boxes. Once you've built the circuit for this, drill holes in the project box for the buttons and the wires leading to the flex sensors. Mount the breadboard and electronics in the project box.

 Project 4

Making Your Own Arduino-Compatible Board

There are times when you want to use more than one microcontroller in a project, or when you just need a controller with one or two I/O pins and don't want to spend the money on a full-featured microcontroller. There may also be times when it's simpler to have multiple processors each dedicated to one task. This is when Atmel's ATtiny microcontrollers come in handy. They're small, they cost only a few dollars, and you can program them using another Arduino-compatible microcontroller, using the SPI synchronous serial protocol.

MATERIALS

- **1 Arduino-compatible microcontroller**
  - **(Uno or MKR1000 shown)**
  - **Features used: UART, SPI**
- **1 ATtiny84 microcontroller**
- **1 solderless breadboard**
- **Jumper wires**
- **1 RGB LED, common cathode**
- **1 220-ohm resistor**

Microcontroller boards like the MKR1000, the 101, and even the Uno have a lot more parts on them besides the microcontroller itself. Most have a reset button and a *clock crystal* or *resonator* to supply the processor with a precise timing pulse. Many have a *voltage regulator* that can take a variable input voltage and supply a constant output voltage for the controller. This allows you to plug in a variety of power sources to the DC power jack, for example. Some have a USB-to-serial adapter, as explained earlier. Others have additional sensors or radios on board, like the MKR1000 and the 101.

Without all these features, the microcontroller alone can be quite inexpensive. Unfortunately, most microcontrollers these days are too small to work with on a breadboard, but it is still possible to get some of Atmel's AVR 8-bit microcontollers in a *dual inline package (DIP)* that can fit into a breadboard. The ATmega328P, which is the controller at the heart of the Uno, is one of these. Two others, the ATtiny84 and ATtiny85, are perhaps the most useful for the size and cost. Their pin diagrams are shown in Figure 2-18. They each have a few GPIO pins that can be used as analog or digital input, and a few PWM pins as well. They can't do everything the others can, but they can do the basics of digital input and output, analog input, and pseudo-analog output using PWM. They don't have a UART on board, but they can manage asynchronous serial communication using the SoftwareSerial library. When combined with a more fully featured processor, they are quite useful. This project shows you how you can reproduce the LED lamp from the first project on a controller that costs less than four dollars.

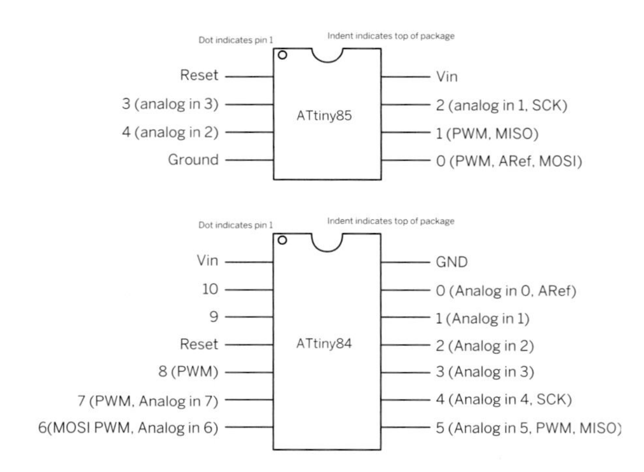

Figure 2-18
ATtiny84 and 85-pin diagrams.

Arduino as ISP

When you get an Uno or any Uno-compatible board, the microcontroller already has a program on it called a *bootloader*. The bootloader is a tiny program that stays on the

Introducing Serial Peripheral Interface (SPI)

In-crcuit serial programming uses a form of synchronous serial communication called Serial Peripheral Interface, or SPI. SPI, along with another synchronous serial protocol, Inter-Integrated Circuit or I2C (sometimes called Two-Wire Interface, or TWI), are two of the most common synchronous serial protocols you'll encounter. You'll see SPI used for other devices, like the WiFi radio on the MKR1000, or for communication with SD memory cards, as well as many sensors.

Synchronous serial protocols all feature a controlling device that generates a regular pulse, or clock signal, on one pin while exchanging data on every clock pulse. The advantage of a synchronous serial protocol is that it's a bus: you can have several devices sharing the same physical connections to one master controller.

SPI connections have three or four connections between the controlling device (or master device) and the peripheral device (or slave), as follows:

- Clock (SCK): The pin that the master pulses regularly.
- Master Out, Slave In (MOSI): The master device sends a bit of data to the slave on this line every clock pulse.
- Master In, Slave Out (MISO): The slave device sends a bit of data to the master on this line every clock pulse.
- Slave Select (SS) or Chip Select (CS): Because several slave devices can share the same bus, each has a unique connection to the master. The master sets this pin low to address this particular slave device. If the master's not talking to a given slave, it will set the slave's select pin high.

If the slave doesn't need to send any data to the master, there will be no MISO pin.

Since SPI is the standard method for programming AVR controllers, most boards using those controllers, including the Uno and earlier Arduino models, all have an ICSP header that breaks out the SPI pins. It looks like this:

White dot indicates pin 1

1: MISO | 2: +Vin
3: SCK | 4: MOSI
5: Reset | 6: Ground

If there's an ICSP header on your board, you can count on the pins having this arrangement. Different microcontrollers break the SPI functions to different pins, however. Here are the SPI pin configurations for the Uno, 101, and MKR1000:

Function	Uno	101	MKR1000
MOSI	11 or ICSP4	ICSP4	8
MISO	12 or ICSP1	ICSP1	10
Clock	13 or ICSP3	ICSP3	9
Chip Select	10	10	user choice

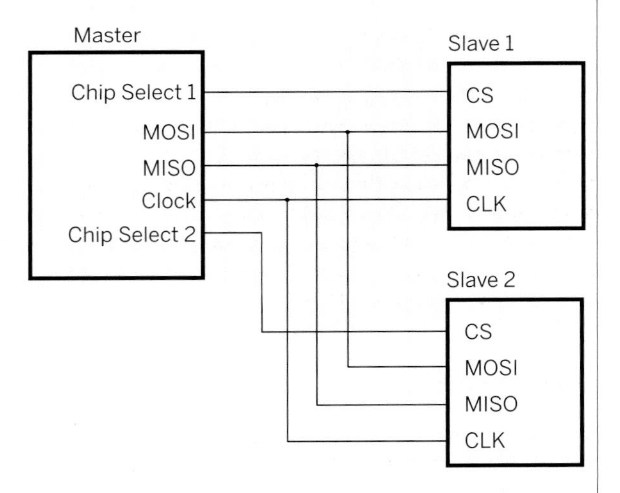

microcontroller at all times. Its only job is to read bytes in through the UART. If those bytes are formatted as a new program, the bootloader writes them to the rest of the controller's program memory. This is why you can upload a new program to your Uno via the USB serial port.

Microcontrollers don't normally come with a bootloader on board. Instead, they are programmed with a separate piece of hardware called an *in-circuit serial programmer (ICSP or ISP)*, using the SPI synchronous serial protocol. The programmer sends a clock signal to the microcontroller, and every time the clock pulses, the microcontroller reads another bit of data and writes it to program memory. When the whole program is written to memory and the controller is reset, the program runs. Programmers like Atmel's AVRISP mkII and Adafruit's USBTinyISP are popular AVR programmers. DIfferent microcontroller architectures usually require different programmers.

There's a sketch included with the Arduino IDE examples that will turn most any board into an ICSP programmer. Click on File→Examples→ArduinoISP, and upload it to the board that you want to use as a programmer. Then connect an ATtiny84 to the board as shown in Figure 2-19.

Before you can program the ATtiny84, you'll need to add its board definition to the IDE. As you did in Chapter 1 (see the "Boards Manager" sidebar), click on Preferences and look for the Additional Boards Manager URLs box. Enter the following URL in the box: http://raw.githubusercontent.com/damellis/attiny/ide-1.6.x-boards-manager/package_damellis_attiny_index.json. Then click OK and restart the IDE. When you restart, you'll find the ATtiny boards in the Tools Boards menu. For other processors in the AVR line, the process is the same. For example, the ATmega328P uses the Uno board definition. The ATmega2560 uses the Mega board definition.

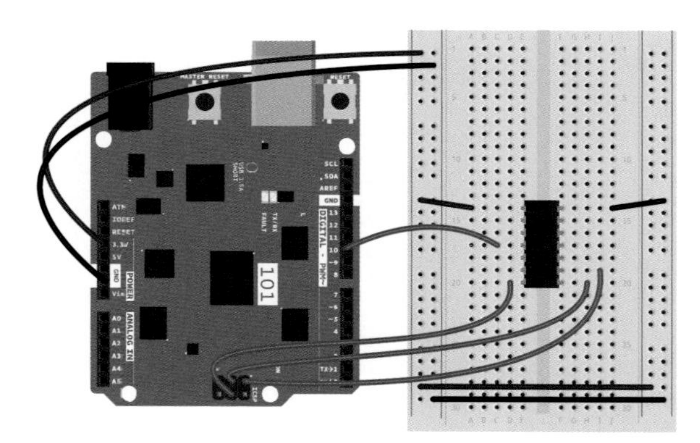

101/Uno Breadboard view

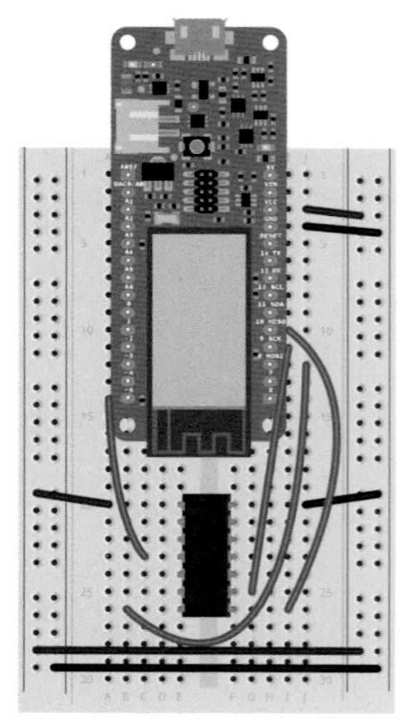

MKR1000 Breadboard view

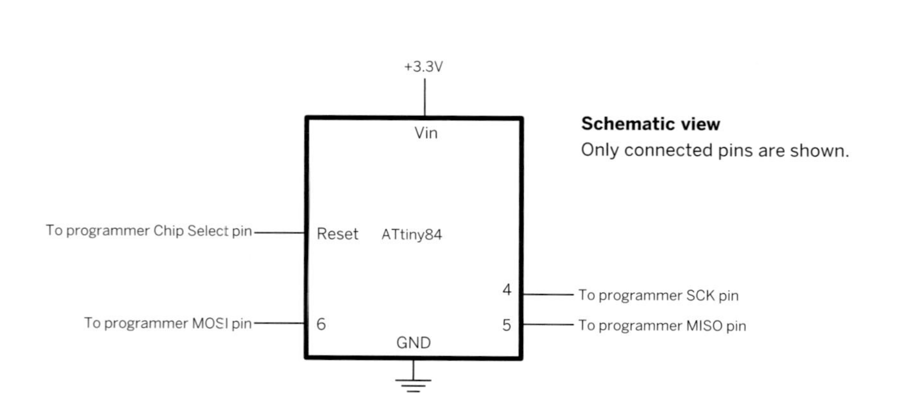

Schematic view
Only connected pins are shown.

Figure 2-19
Programming an ATtiny84 from an Arduino-compatible board. The half-circle at the top of the ATtiny indicates the top of the component, and the indent indicates physical pin number 1. This is standard on *dual-inline package (DIP)* components.

In order to program the ATtiny, you need to set the microcontroller's basic configurations. These are stored in 3 bytes of permanent memory and are called *fuses*. The fuses configure the controller's clock speed, running voltage, and so forth. In the Boards menu, set the ATtiny85 as follows:

- Board: ATtiny
- Processor: ATtiny84
- Clock: 8MHz (internal)
- Port: whatever port your programmer board is using

Click on the Tools Programmer menu and choose "Arduino as ISP." This tells the IDE to use your programmer board as the programmer for the ATtiny.

Next choose Burn Bootloader from the Boards menu. This will set the fuses. When it's done, you'll see the message "Done Burning Bootloader" in the console pane. Now you're ready to put a sketch on the ATtiny84.

You can see that the clock is set to 8MHz (internal). This means that the processor will use an internal circuit to keep its timing. With the ATtiny set like this, you need only a connection to voltage (between +3 and +5V) and ground and a pullup resistor to voltage on the reset pin to make it work. Try uploading the Blink sketch to it. First change the LED pin number from 13 to 6, then click Upload Using Programmer. The IDE will use your programmer board as the in-circuit serial programmer to program the ATtiny. Add an LED to digital I/O pin 6 and you should see the LED blinking when the ATtiny microcontroller is powered.

SoftwareSerial and ATtiny Type Brighter

The ATtiny has all the capabilities you need to run either of the projects in this chapter, so let's re-create the first, Type Brighter.

Connect the RGB LED to digital I/O pins 6 through 8 of the ATtiny84. These pins can be used for PWM, just like the PWM pins of the 101 and MKR1000 as you saw earlier. Load the Serial RGB LED controller sketch from the first project, and make the following changes:

Try It Add the following two lines to the beginning of the sketch. These lines include and make an instance of the SoftwareSerial library, which allows you to send and receive serial data on any two GPIO pins. You'll use pins 0 and 1.

Change the LED pin numbers as shown here as well.

```
#include <SoftwareSerial.h>  // include the SoftwareSerial library

SoftwareSerial swSerial(0, 1); // RX, TX

// constants to hold the output pin numbers:
const int redPin = 8;
const int greenPin = 7;
const int bluePin = 6;
```

Once you've added these lines, change every instance of `Serial` in your current Arduino sketch to `swSerial`. The SoftwareSerial library has the same `print()`, `println()`, `read()`, `write()`, and `available()` commands as the main Serial library.

Upload the changed sketch to your ATtiny84 using your programmer board just like you did with the Blink sketch. Then disconnect the ATtiny84 from the breadboard, and connect it to a USB-to-serial adapter as shown in Figure 2-20. The adapter's TX goes to pin 0 and RX goes to pin 1. In this case, you'll power the ATtiny84 from the adapter, so connect Vin and ground to the microcontroller's Vin and ground as well. Open CoolTerm or the Serial Monitor to your adapter and type commands just like you did in the first project, for example:

```
r5g3b7
```

The LED should change just like it did in the first project. It's the same project, but much simpler and less expensive!

X

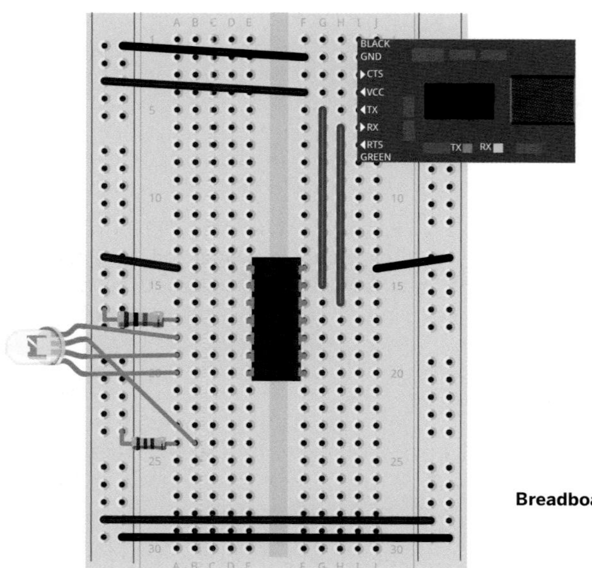

Figure 2-20
Type Brighter project on an ATtiny84. The USB-to-serial converter is supplying power for the circuit via USB from the personal computer. The ATtiny84 and the LED can operate at +3.3V or +5V, so it doesn't matter which the adapter is supplying. The 10-kilohm pullup resistor on the reset pin keeps the microcontroller from resetting spontaneously, by keeping the reset pin high. To reset the micrcontroller, connect the reset pin to ground.

Breadboard view

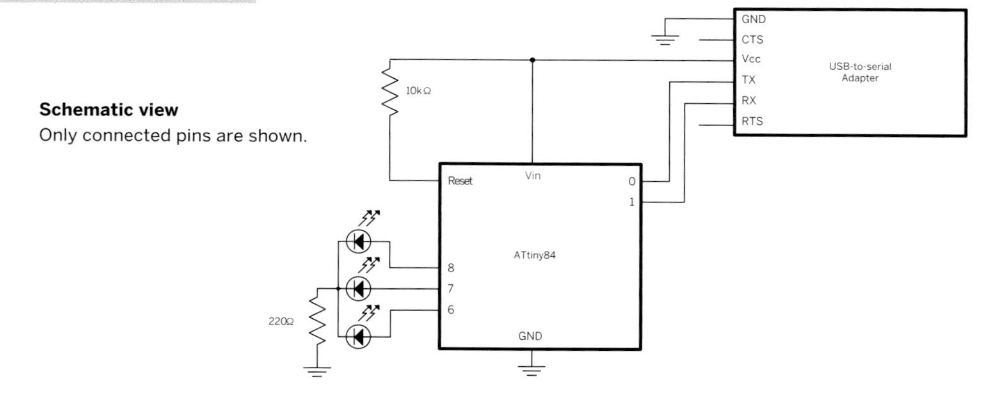

Schematic view
Only connected pins are shown.

More Serial Ports: Software Serial

A microcontroller's Universal Asynchronous Receiver-Transmitter (or UART, labeled RX and TX) allows it to send and receive serial data reliably, no matter what your code is doing, because the UART listens for serial all the time. What do you do when you need to attach more than one asynchronous serial device to your Arduino? Some microcontrollers have multiple UARTs. The Arduino Mega 2560 has four, for example. But you may not need all that a Mega has to offer just to get another serial port. SoftwareSerial allows you to use two digital pins as a "fake" UART. The library can listen for incoming serial on those pins, and transmit as well. Because it's not a dedicated hardware UART, SoftwareSerial isn't as reliable as hardware serial at very high or very low speeds, though it does well from 4800bps through 57.6kbps. When you're using limited controllers like the ATtiny processors, it's handy. It can also be useful on the Uno and other processors that have only one UART, when you want to connect to another asynchronous serial device and still want to use the main UART for communication with your personal computer.

X

Although many projects work fine with just one microcontroller, there are cases where it's helpful to have two or more with different capabilities. For example, the ESP8266 controller, which you'll see in later chapters, is great at network communications via WiFi, but it's only got one analog input, and that input has only a 1-volt range. For a project that needs multiple analog inputs, like Monski Pong, the ESP8266 wouldn't be much good on its own. But it could be connected serially to an ATtiny84 or ATtiny85 to handle the analog input.

Being able to expand the number of inputs and outputs isn't the only reason you might want two controllers. You might have a project in which you need very precise timing of an output device at the same time as you need to make a complex network exchange. By separating the tasks to two processors, you can simplify the programming by splitting it into two programs.

The Arduino IDE can program a number of controllers in the AVR microcontroller family using in-circuit serial programming, simply by selecting the right board from the Boards menu and burning the bootloader. Here's what it can do with the built-in board settings:

- ATmega328P - bootload using Uno board setting
- ATMega168 - bootload using Diecimila or Duemilanove
- ATMega8 - bootload using NG or older
- ATMega2560 - bootload using Mega2560
- ATmega32U4 - bootload using Micro

Using the board definitions from this project, it can also program these:

- ATtiny84
- ATtiny44
- ATtiny 85
- ATtiny45

Of these, the ATtinys and the ATmega328P are probably the most useful. The former are small and very inexpensive, and the latter is the same processor used in the Uno, so it's compatible with many existing examples on the web.

Figure 2-21 shows the circuit and components necessary to make your own Uno-compatble breadboard circuit. It's bootloaded just like the ATtiny controllers: connect the SPI pins to your programmer board, choose "Uno" as the board type, then burn the bootloader.

Once you've put the bootloader on the controller, you can program it over the USB-to-serial converter, but you'll need to hit the reset button right before upload each time.

It also includes a 5-volt voltage regulator and DC power jack so that you can power it from a 9–12V DC power source. The full pin diagram of the ATMega328P can be found at www.arduino.cc/en/Hacking/PinMapping168. X

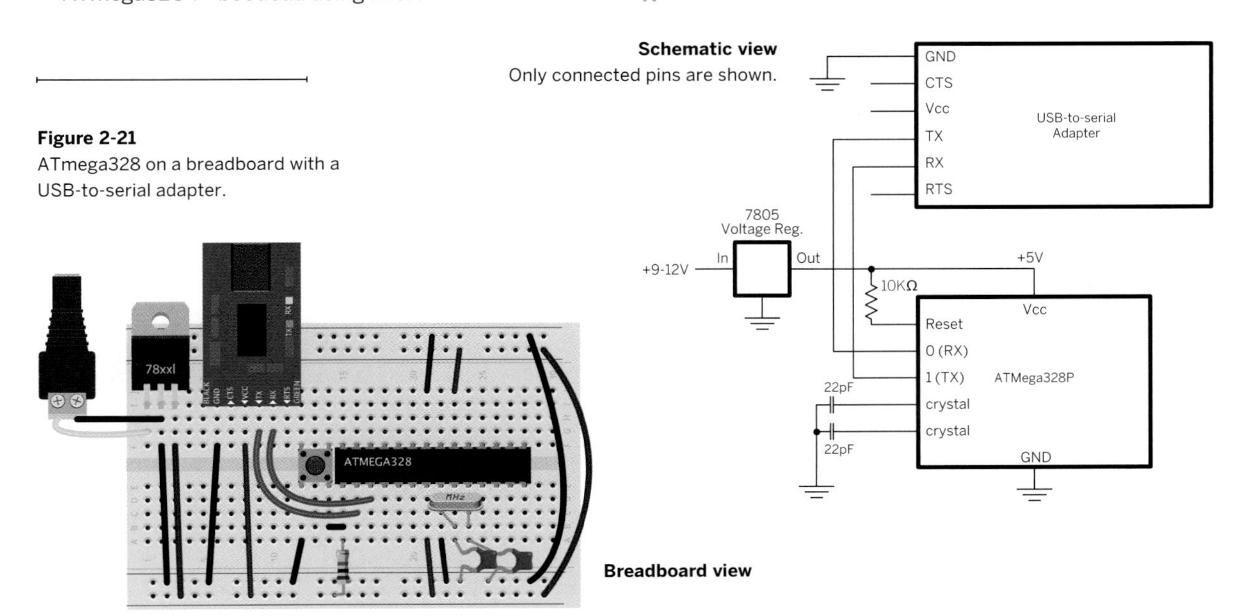

Figure 2-21
ATmega328 on a breadboard with a USB-to-serial adapter.

Conclusion

The projects in this chapter have covered a number of ideas that are central to all networked data communication. First, remember that data communication is based on a layered series of agreements, starting with the physical layer; then the electrical, the logical, the data layers; and finally, the application layer. Keep these layers in mind as you design and troubleshoot your projects, and you'll find it's easier to isolate problems.

Second, remember that serial data can be sent either as ASCII-encoded or as raw binary values, and which you choose to use depends on both the capabilities and limitations of all the connected devices. It might not be wise to send raw binary data, for example, if the modems or the software environments you program in are optimized for ASCII data transfer.

Third, when you think about your project, think about the messages that need to be exchanged, and come up with a data protocol that adequately describes all the information you need to send. This is your data packet. You might want to add header bytes, separators, or tail bytes to make reading the sequence easier.

Fourth, consider the flow of data, and look for ways to ensure a smooth flow with as little overflowing of buffers or waiting for data as possible. A simple call-and-response approach can make data flow much smoother.

Fifth, get to know the communications devices that link the objects at the end of your connection, whether they're protocol adapters like the USB-to-serial adapter or radios like the Bluefruit. Understand their addressing schemes and data protocols so that you can factor their strengths and limitations into your planning, and eliminate those parts that make your life more difficult.

Finally, think about your projects in terms of distributed computing rather than a single computer. Now that you know how to program both high-end microcontroller modules and simple microcontrollers like the ATtiny84 or ATtiny85, and how to communicate with them using asynchronous serial communication, you can assign each task in your project to a separate controller if you want.

With a little planning, you can take the computing needs of your project, and distribute them across many different computers. This is the real power of microcontrollers and serial communication.
X

▸▸ The JitterBox by Gabriel Barcia-Colombo
The JitterBox is an interactive video Jukebox created from a vintage 1940s radio restored to working condition. It features a tiny video-projected dancer who shakes and shimmies to the music. The viewer can tune the radio and the dancer will move in time with the tunes. The JitterBox uses serial communication from an embedded potentiometer tuner—which is connected to an Arduino microcontroller—in order to select from a range of vintage 1940s songs. These songs are linked to video clips and played back out of a digital projector.
The dancer trapped in the JitterBox is Ryan Myers.

54 70 90 120 150
60 80 100 130 160

3

MAKE: PROJECTS

A More Complex Network

Now that you've got the basics of device-to-device communications, it's time to tackle something more complex. The best place to start is with the most familiar data network: the internet. It's not actually a single network, but a collection of networks owned by different network service providers and linked using some common protocols. This chapter describes the structure of the internet, the devices that hold it together, and the shared protocols that make it possible. You'll get hands-on experience with what's going on behind the scenes when your web browser or email client is doing its job, and you'll use the same messages those tools use to connect your own objects to the internet.

◂◂ Networked Flowers by Doria Fan, Mauricio Melo, and Jason Kaufman
Networked Flowers is a personal communication device for sending someone digital blooms. Each bloom has a different lighting animation. The flower sculpture has a network connection. The flower is controlled from a website that sends commands to the flower when the web visitor chooses a lighting animation.

Supplies for Chapter 3

DISTRIBUTOR KEY
A Arduino Store (store.arduino.cc)
AF Adafruit (adafruit.com)
D Digi-Key (www.digikey.com)
F Farnell (www.farnell.com)
J Jameco (jameco.com)
RS RS (www.rs-online.com)
SF SparkFun (www.sparkfun.com)
SS Seeed Studio (www.seeedstudio.com)

PROJECT 5: Networked Cat

- **1 microcontroller module** This project will run on just about any Arduino-compatible board. The Arduino 101 is shown, but the Uno and MKR1000 will work as well. MKR1000: AF 3156, RS 124-0657, A ABX00004, GBX00011 (Africa and EU), D 1659-1005-ND
  Arduino 101: **D** 1660-1003-ND, **J** 2239331, **SF** DEV-13787, AF 3033, F 2520713, RS 913-9999, SS 114990575, A ABX00005, GBX00005 (Africa and EU)
  Uno: **D** 1050-1024-ND, **J** 2151486, **SF** DEV-11021, A A000099, AF 50, F 1848687, RS 715-4081, SS ARD132D2P
  1 ATtiny84 microcontroller Use the ATtiny85 if you want an even smaller board. **D** ATTINY84A-PU-ND, **SF** COM-11232, F 1455160, RS 738-0684
- **Between 2 and 4 force sensing resistors, Interlink 400 series** The model 402 is shown in this project, but any of the 400 series will work well. **D** 1027-1001-ND, **J** 2128260, **SF** SEN-09375, A 166
- **One 1-kilohm resistor** Any model will do. **D** 1.0KQBK-ND, **J** 690865, F 9339051, R 707-7666
- **1 solderless breadboard** D 438-1045-ND, J 20723 or 20601, SF PRT-12615 or PRT-12002, F 4692810, AF 64, SS 319030002 or 319030001
- **1 personal computer**
- **All necessary converters to communicate serially from microcontroller to computer** Just like the previous projects.
- **1 web camera**
- **1 cat** A dog will do if you have no cat.
- **1 cat mat**
- **2 thick pieces of wood or thick cardboard, about the size of the cat mat**
- **Wire-wrapping wire** D K386-ND, J 22577, F 09WX4670
- **Rubber bumpers** **D** 3M156065-ND, RS 120-6041, **J** 2119718, A 550, **SF** COM-10594, F 1165068
- **Wire-wrapping tool** D K445-ND or WSU-30M, J 2150361, F 441089
- **Male header pins** **D** A26509-20-ND, J 103377, SF PRT-00116, F 1593411
- **USB-to-TTL serial adapter** Use this if you're building the circuit with an ATtiny. **SF** DEV-09716 or DEV-14050, A 3309 or 284, SS 317990026

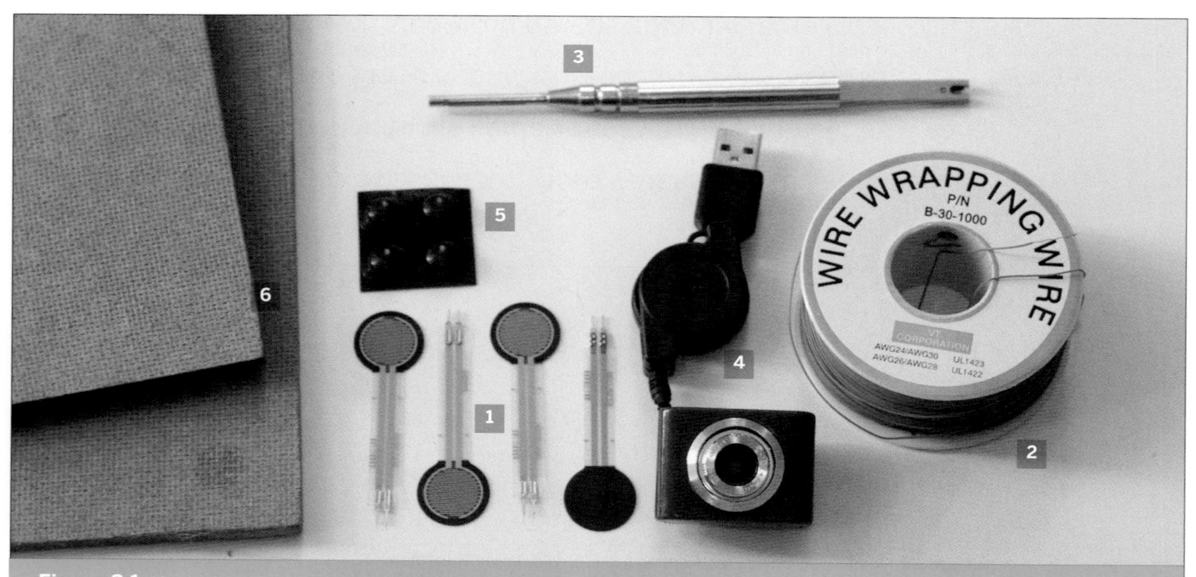

Figure 3-1. New parts for this chapter: **1.** Interlink Series 402 force-sensing resistors (FSRs) **2.** 30AWG wire-wrapping wire **3.** New tool: wire-wrapping tool. The wire stripper to the left of the tool lives inside the handle. **4.** USB camera **5.** Rubber bumpers **6.** thick wood for cat sensing base

Network Maps and Addresses

In the previous chapter, it was easy to keep track of where messages went because there were only two points in the network you built: the sender and the receiver. In any network with more than two objects—from three to three billion—you need a *map* to keep track of which objects are connected to which. You also need an *addressing scheme* to know how a message gets to its destination.

Network Maps: How Things Are Connected

The arrangement of a network's physical connections depends on how you want to route its messages. The simplest way is to make a physical connection from each object in the network to every other object. That way, messages can get sent directly from one point to another. The problem with this approach, as you can see from the directly connected network in Figure 3-2, is that the number of connections gets large very fast, and the connections get tangled. A simpler alternative to this is to put a central hub in the middle and pass all messages through it, as seen in the star network shown in Figure 3-2. This way works great as long as the hub continues to function, but the more objects you add, the faster the hub must be in order to process all the messages. A third alternative is to chain the objects together, connecting them together in a ring. This design makes for a small number of connections, and it means that any message has two possible paths, but it can take a long time for messages to get halfway around the ring to the most distant object.

In practice (such as on the internet), a multitiered model, like the one shown in Figure 3-3, works best. Each connector (symbolized by a light-colored circle) has a few objects connected to it, and each connector is linked to a more central connector. At the more central tier (the dark-colored circles in Figure 3-3), each connector may be linked to more than one other connector, so that enabling messages to pass from one endpoint to another via several different paths. This system takes advantage of the redundancy of multiple links between central connectors, but avoids the tangle caused by connecting every object to every other object.

If one of the central connectors isn't working, messages are routed around it. The connectors at the edges are the weakest points. If they aren't working, the objects that depend on them have no connection to the network. As long as the number of objects connected to each of these is small, the effect on the whole network is minimal. It may not seem minimal when your connection is through the failed connector but the rest of the network remains stable, so it's easy to reconnect when your connector is working again.

Figure 3-2
Three types of network: direct connections between all elements, a star network, and a ring network.

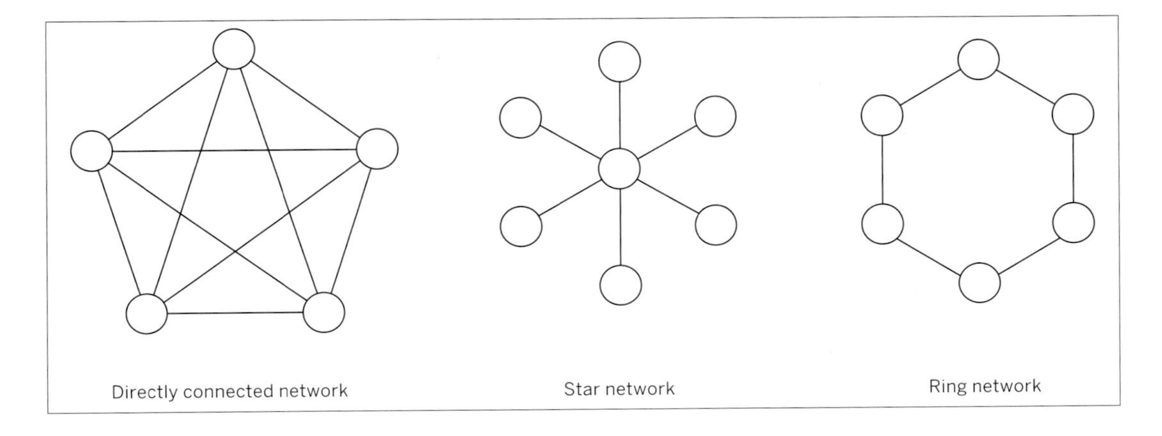

If you're using the internet as your network, you can take this model for granted. If you're building your own network, however, it's worth comparing all these models to see which is best for you. In simpler systems, one of the three networks shown in Figure 3-2 might do the job just fine, saving you some complications. As you get further into the book, you'll see some examples of these; for the rest of this chapter, you'll work with the multitiered model by relying on the internet as your infrastructure.

X

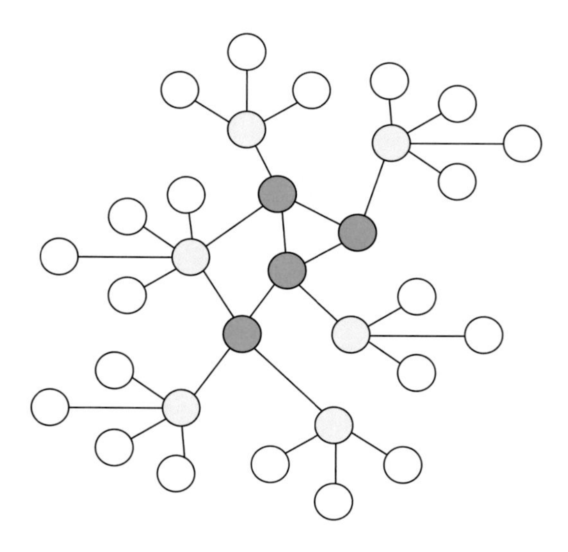

Figure 3-3
A complex, multitiered network.

Modems, Hubs, Switches, and Routers

The connectors in Figure 3-3 represent several different types of devices on the internet. The most common among these are modems, hubs, switches, and routers. Depending on how your network is set up, you may be familiar with one or more of these. There's no need to go into detail as to the differences, but some basic definitions are in order.

A modem is a device that converts one type of signal into another, and connects one object to one other object. Your home cable or DSL modem is an example. It takes the digital data from your home computer or network, converts it to a signal that can be carried across the phone line or cable line, and connects to another modem on the other end of the line. That modem is connected to your internet service provider's network. By this definition, the Bluetooth radios from Chapter 2 could be considered modems, as they convert electrical signals into radio signals and back.

A hub is a device that multiplexes data signals from several devices and passes them upstream to the rest of the network. It doesn't care about the recipients of the messages it's carrying—it just passes them through in both directions. All the devices attached to a hub receive all the messages that pass through the hub, and each one is responsible for filtering out any messages that aren't addressed to it. Hubs are cheap and handy, but they don't really manage traffic.

A switch is like a hub, but it's more sophisticated. It keeps track of the addresses of the objects attached to it, and it passes along messages addressed to those objects only. Objects attached to a switch don't get to see messages that aren't addressed to them.

Modems, hubs, and switches generally don't actually have their own addresses on the network. A router, on the other hand, is visible to other objects on the network. It has an address of its own, and it can mask the objects attached to it from the rest of the network. It can give them private addresses, meaningful only to the other objects attached to the router, and pass on their messages as if they come from the router itself. It can also assign IP addresses to objects that don't have one when they're first connected to the router. Most home connections, whether fiber, DSL, or cable modem, combine a modem and a router in one device.

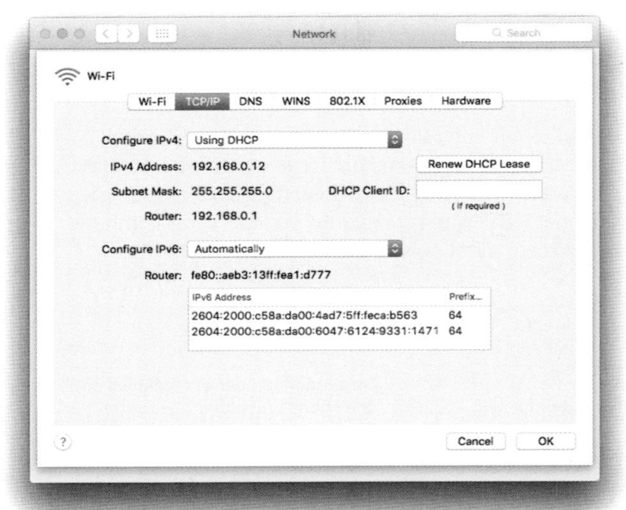

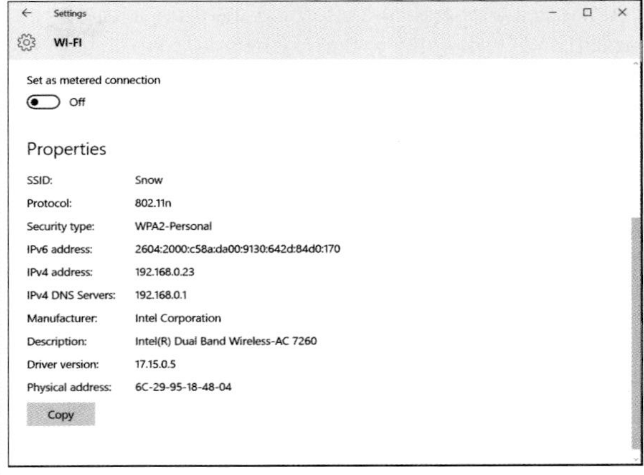

Figure 3-4
Network settings panels for macOS and Windows.

Hardware Addresses and Network Addresses

Whether you're using a simple network model where all the objects are directly connected, a multitiered model, or anything in between, you need an addressing scheme to get messages from one point to another on the network. When you're making your own network from scratch, you have to create your own addressing scheme. For the projects you're making in this book, however, you're relying on existing network technologies, so you get to use the addressing schemes that come with them. For example, when you used the Bluetooth radios in Chapter 2, you used the Bluetooth protocol addressing scheme. The devices you connect to the internet use a group of protocols from the Institute of Electrical and Electronics Engineers (IEEE), specifically the *IEEE 802.x* protocols. *Ethernet*, the wired connection your router probably uses to connect to your cable modem, is IEEE 802.3. *WiFi*, the wireless version of Ethernet is IEEE 802.11, with variations. For example current routers can handle 802.11a, b, g, or n. *Bluetooth* is based on IEEE 802.15.1. All of the 802.x protocols share a common addressing scheme. Devices are given a *hardware address*, or *Media Access Control (MAC) address*, a unique 6-byte ID number assigned by the manufacturer that differentiates that device from all the others using the protocol. Manufacturers license ranges of addresses from the IEEE. A device's MAC address typically doesn't change for the lifetime of the device.

When you connect Ethernet and WiFi devices to the internet, you use the *Internet Protocol (IP)* addressing scheme. A device's IP address can change when it's moved from one network to another, but its MAC address stays the same. When a device connects to an IP-based network, it tells the router its MAC address, and the router assigns it an IP address. The router maintains a table mapping IP addresses to the MAC addresses of currently connected devices.

You're probably already familiar with your computer's IP address and maybe even its hardware address. In macOS, click Apple Menu→System Preferences→Network Preferences to open the Network control panel. Here you'll get a list of the possible network interfaces through which your computer can connect to the internet. It's possible that you have both a built-in Ethernet interface and a WiFi interface. The built-in Ethernet and WiFi interfaces both have MAC addresses, and if you select either, you can find out that interface's address. In either interface, click on the Advanced button to get to both the Hardware tab (where you can see the MAC address), and the TCP/IP tab (where you can see the machine's IP address if you're connected to a network).

In Windows 10, click the Start Menu→Control Panel, then double-click "Network and Internet." You'll see a list of available WiFi networks, including the one to which you're connected. Scroll down to the bottom and click Advanced Options. Scroll to the bottom and you'll see the Properties section listing your Physical (hardware) address and IP address.

For Ubuntu Linux, click the System Settings menu, then Network. You'll see a list of network interfaces. Click Edit to see their details.

Figure 3-4 shows the network connection settings for macOS and Windows. No matter what platform you're on, the hardware address and the internet address will take these forms:

- The *hardware address* is made up of six numbers written in hexadecimal notation, like this: `00:11:24:9b:f3:70`
- The *IP address* is made up of four numbers written in decimal notation, like this: `192.168.1.20`

Your computer needs to know its IP address to send and receive messages, and the router to which it's connected needs the hardware address in order to give it an IP address. So whenever you begin working on a new project, note both addresses for every device you're using.

Street, City, State, Country: How IP Addresses Are Structured

Geographic addresses can be broken down into layers of detail, starting with the most specific (the street address) and moving to the most general (the country). Internet Protocol (IP) addresses are also multilayered. The most specific part is the final number, which tells you the address of the computer itself. The numbers that precede this tell you the *subnet* that the computer is on. Your router shares the same subnet as your computer, and its number is usually identical except for the last number. The numbers of an IP address are called *octets*, and each octet is like a section of a geographic address. For example, imagine a machine with this number: `217.123.152.20`. The router that this machine is attached to most likely has this address: `217.123.152.1`.

Each octet can range from 0–255, and some numbers are reserved by convention for special purposes. For example, the router is often the address `xxx.xxx.xxx.1`. The subnet can be expressed as an address range, for example, `217.123.152.xxx`. Sometimes a router manages a larger subnet or even a group of subnets, each with its own local router. The router that this router is connected to might have the address `217.123.1.1`.

Each router controls access for a defined number of machines below it. The number of machines it controls is encoded in its *subnet mask*. You may have encountered a subnet mask when configuring your personal computer. A typical subnet mask looks like this: `255.255.255.0`.

Table 3-1. The relationship between subnet mask and maximum number of machines on a network.

Subnet mask	Maximum number of machines on the subnet, including the router (accounting for reserved addresses)
255.255.255.255	1 (just the router)
255.255.255.192	62
255.255.255.0	254
255.255.252.0	1022
255.255.0.0	65,534

You can read the number of machines in the subnet by reading the value of the subnet mask. It's easiest if you think of the subnet in terms of bits. Four bytes is 32 bits. Each bit you subtract from the subnet increases the number of machines it can support. You "subtract" the value of subnet mask from its maximum value of 255.255.255.255 to get the number of machines. For example, if the subnet were 255.255.255.255, there could be only one machine in the subnet: the router itself. If the last octet is 0, as it is in Table 3-1, there can be up to 255 machines in the subnet in addition to the router. The .255 address is reserved for broadcast messages. A subnet of 255.255.255.192 would support 62 machines and the router (255 – 192 - 1 = 62), and so forth. There are a few other reserved addresses, so the real numbers are a bit lower. Table 3-1 shows a few other representative values to give you an idea.

Knowing the way IP addresses are constructed helps you to manage the flow of messages you send and receive. Normally, all of this is handled for you by the software you use: browsers, email clients, and so forth. But when you're building your own networked objects, it's necessary to know at least this much about the IP addressing scheme so you can find your router and what's beyond it.

Private and Public IP Addresses

Not every device on the internet can be addressed by every other device. Sometimes, in order to support more devices, a router hides the addresses of the devices attached to it, sending all their outgoing messages to the rest of the network as if they came from the router itself. There are special ranges of addresses set aside in the IP addressing scheme for use as private addresses. For example, all addresses in the range 192.168.xxx.xxx (as well as 10.xxx.xxx.xxx, and 172.16.xxx.xxx–172.31.xxx.xxx) are to be used for private addressing. These address ranges are used commonly in home routers. All the devices on your home network likely show up with addresses in one of these ranges. When they send messages to the outside world, those messages show up as if they came from your router's public IP address. This is called *Network Address Translation (NAT)*. **It works like this:**

My computer, with the address 192.168.1.45 on my home network, makes a request for a web page on a remote server. That request goes first to my home router. On my home network, the router's address is 192.168.1.1, but to the rest of the internet, my router presents a public address, 66.187.145.75. The router passes on my message, sending it using its public address and requesting that any replies come back to that address. When it gets a reply, it sends the reply to my computer at its local address, 192.168.1.45. Thanks to private addressing and subnet masks, multiple devices can connect to the internet through a single public IP address, which expands the total number of things that can be attached to the internet.

Address Resolution Protocol

You can use the Address Resolution Protocol to identify other devices on your network. The POSIX `arp` command allows you to read and modify the ARP table, a list of devices by hardware (MAC) address, and their associated IP address on your local network. If you're on your home network or a network to which you have administrative rights, you can also use *arp* to learn the IP addresses of other devices on your network.

If you're working on an institutional network, for example at a school, university, or corporation, the network you're on may not support the use of the ping tool. Check with your network administrator before you start using network diagnostic tools like arp and ping, as your activities might be interpreted as hostile if your network administrator doesn't know what you're doing. Make friends with your network administrator. It will make your net-connected projects simpler.

Try It Open the command-line application on your computer (Terminal on macOS and Ubuntu, the command or bash prompt on Windows). Type the command at right at the prompt (the $ is the prompt; don't type it). You'll get a response like the one shown to the right:

```
$ arp -a

? (192.168.0.1) at ac:b3:13:a1:d7:77 on en0 ifscope [ethernet]
? (192.168.0.2) at 0:17:88:a:17:45 on en0 ifscope [ethernet]
? (192.168.0.15) at 0:e0:4c:9:3b:3f on en0 ifscope [ethernet]
? (192.168.0.176) at (incomplete) on en0 ifscope [ethernet]
? (192.168.0.255) at (incomplete) on en0 ifscope [ethernet]
? (224.0.0.251) at 1:0:5e:0:0:fb on en0 ifscope permanent [ethernet]
? (255.255.255.255) at (incomplete) on en0 ifscope [ethernet]
```

Each line is a device in your computer's *address resolution protocol* (ARP) table. It's the computer's neighbors on your local network. The first address is the IP address, and the second (after the "at") is the MAC address. When you call `arp -a`, the computer looks at this table and sends a request to every address in the table. When it gets a reply, it notes the MAC address of each responding machine.

When you see (incomplete), that means the computer didn't get a response. This may mean the machine that had that address last has left the network, or it may mean that's a special address, without a device assigned to it.

Ping: Are You There?

The `arp` tool is useful for determining which other devices your computer knows the addresses of, but it doesn't tell you much information about their status. There's another command-line tool, *ping*, that can be useful in determining whether other devices are available to respond to your messages. It sends a message to another object on the network to say "Are you there?" and then waits for a reply.

Type the code to the right. On Windows, replace `-c` with `-n` (again, the $ is the prompt).

You'll get a response like on the right:

```
$ ping -c 3 127.0.0.1
PING 127.0.0.1 (127.0.0.1): 56 data bytes
64 bytes from 127.0.0.1: icmp_seq=0 ttl=64 time=0.056 ms
64 bytes from 127.0.0.1: icmp_seq=1 ttl=64 time=0.056 ms
64 bytes from 127.0.0.1: icmp_seq=2 ttl=64 time=0.072 ms
```

This sends a message to address 127.0.0.1 and waits for a reply. Every time it gets a reply, it tells you how long it took. After counting 3 packets (that's what the `-c 3` on Mac and `-n 3` on Windows means), it stops and gives you a summary, like this:

```
--- 127.0.0.1 ping statistics ---
3 packets transmitted, 3 packets received, 0.0% packet loss
round-trip min/avg/max/stddev = 0.056/0.061/0.072/0.008 ms
```

It gives you a picture of not only how many packets got through, but also how long they took. When you send multiple messages as you did here, you'll see the minimum transmission time, the maximum, and the average time of all the packets, as well as the standard deviation between these times. Ping is sometimes disabled on institutional networks, because it can also be used for malicious purposes. When used properly and carefully, though, it's a helpful tool for checking the status of remote devices on the network, and the transmission health of the network itself.

Special Addresses on a Local Area Network

127.0.0.1 is a special address called the *loopback address* or *localhost address*. Whenever you use it, the computer from which you're sending it sends the message to itself. You can also use the name `localhost` in its place. You can use localhost to test and connect applications on your computer to each other, even if it's not connected to a network. You'll see this in use later when you run one program as a server and another as a client on the same computer.

There's another special address, x.x.x.255, that you saw when you ran the `arp` command. This address is reserved as the *broadcast address*. Messages sent to that address are sent to every machine on the local network. It's useful if a device is advertising a service it offers, or if a router needs to send commands to all devices at once.

You can use the broadcast address to learn the address of all the addresses on your local area network using `ping` as follows:

Try It Type the code to the right. Again, on Windows, replace `-c` with `-n`.

You'll get a response like the one on the right:

```
$ ping -c 3 192.168.0.255

64 bytes from 192.168.0.12: icmp_seq=0 ttl=64 time=0.102 ms
64 bytes from 192.168.0.4: icmp_seq=0 ttl=64 time=1.457 ms
64 bytes from 192.168.0.1: icmp_seq=0 ttl=64 time=12.232 ms
64 bytes from 192.168.0.12: icmp_seq=1 ttl=64 time=0.061 ms
64 bytes from 192.168.0.4: icmp_seq=1 ttl=64 time=1.672 ms
64 bytes from 192.168.0.1: icmp_seq=1 ttl=64 time=10.419 ms
64 bytes from 192.168.0.12: icmp_seq=2 ttl=64 time=0.062 ms

--- 192.168.0.255 ping statistics ---
3 packets transmitted, 3 packets received, +4 duplicates, 0.0% packet loss
round-trip min/avg/max/stddev = 0.061/3.715/12.232/4.877 ms
```

> **The `ping` command may be disabled on your network if your security settings are set very strictly. If you get a message saying "Request timed out," talk to your network administrator to see if this is the case.**

Notice that none of the packets came back from 192.168.0.255, but they came from the other devices on the network that have MAC addresses. Whenever the router gets a message for 192.168.0.255, it sends that message to all the other addresses on the local network. So the replies come back from those other devices. You should see three replies from each device on the local network, including your own computer. Since every device that's connected should reply to the ping, you should end up with a complete list of the devices this way. This includes not only your personal computer, but also any tablets, phones, or home automation devices that connect to your local network via Ethernet or WiFi as well.

Sending messages to the broadcast address of a network should always be done sparingly. It can flood the network with traffic if not used properly.

If you've got computers on your local network that are asleep, they may not reply as quickly. This is why you did a ping with a count of three instead of one. If you don't see all the devices that you know are on your network, try changing the count to 5 or 7 to see what happens.

Keep in mind you can also send to a specific device using ping once you know its IP address. This is generally a better approach, since it creates less traffic on the network.

Numbers into Names

You're probably thinking this is ridiculous because you only know internet addresses by their names, like www.makezine.com or www.archive.net. You never deal with numerical addresses, nor do you want to. There's a separate protocol, the *Domain Name System (DNS)*, for assigning names to the numbers. Machines on the network called *name servers* keep track of which names are assigned to which numbers. In your computer's network configuration, you'll notice a slot where you can enter the DNS address. Most computers are configured to obtain this address from a router using the *Dynamic Host Configuration Protocol (DHCP)*, which also provides their IP address, so you don't have to worry about configuring DNS.

Packet Switching: How Messages Travel the Internet

So how does a message get from one machine to another? Imagine the process as akin to mailing a bicycle. The bike's too big to mail in one box, so first you break it into box-sized pieces. On the network, this is initially done at the Ethernet layer—also called the *datalink layer*—where each message is broken into chunks of more or less the same size, and given a header containing the packet number. Next, you'd put the address (and the return address) on the bike's boxes. This step is handled at the IP layer, where the sending and receiving addresses are attached to the message in another header. Finally, you send it. Your courier might want to break up the shipment among several trucks to make sure each truck is used to its best capacity. On the internet, this happens at the *transport layer*. This is the layer of the network responsible for making sure packets get to their destination. There are two main protocols used to handle transport of packets on the internet: *Transmission Control Protocol (TCP)*, and *User Datagram Protocol (UDP)*. You'll learn more about these later. The main difference is that TCP provides more error-checking from origin to destination, but is slower than UDP. On the other hand, UDP trades off error-checking in favor of speed.

Each router sends off the packets one at a time to the routers to which it's connected. If it's attached to more than one other router, it sends the packets to whichever router is least busy. The packets may each take a different route to the receiver, and they may take several hops across several routers to get there. Once the packets reach their destination, the receiver strips off the headers and reassembles the message. This method of sending messages in chunks across multiple paths is called *packet switching*. It ensures that every path through the network is used most efficiently.

Prior to the development of packet switching, telephone and communications networks were *circuit switched*, meaning that there had to be a dedicated circuit connecting the sender to the receiver. Banks of phone operators patched switchboards to make this happen initially, and later towers full of relays switched the circuits. The invention of routing and packet switching was a major step forward in enabling flexible communications.

X

Clients, Servers, and Message Protocols

Now you know how the internet is organized, but how do things get done on the internet? For example, how does an email message get from you to your friend? Or how does a web page get to your computer when you type a URL into your browser or click on a link? It's all handled by sending messages back and forth between devices using the addressing and transport scheme just described. Let's take a look at this in depth, using the World Wide Web as an example.

How Web Browsing Works

Figure 3-5 is a map of the route that a web page takes to reach your computer. Your browser sends out a request for a page to a web server, and the server sends the page back. Which route the request and the reply take is irrelevant, as long as there is a route.

The web server itself is just a program running on a computer somewhere else on the internet. A *server* is a program that provides a service to other programs on the internet. The computer that a server runs on, also referred to as a server, is expected to be online and available at all times so that the service is not disrupted. In the case of a web server, the server provides access to a number of HTML files, images, sound files, and other elements of a website to clients from all over the internet.

Clients are programs that take advantage of services. Your browser, a client, makes a connection to the server to request a page. The browser makes a connection to the server computer, the server program accepts the connection and delivers the files representing the page, and the exchange is made.

The server computer shares its IP address with every server program running on it by assigning each program a *port number*. For example, every connection request for port 80 is passed to the web server program. Every request for port 25 is passed to the email server program. Any program can take control of an unused port, but only one program at a time can control a given port. In this way, network ports work much like serial ports. Many of the lower port numbers are assigned to common applications, such as mail, file transfer, telnet, and web browsing. Higher port numbers are either disabled or left open for custom applications (you'll write one of those soon). A specific request goes like this:

1. Type `http://www.example.com/index.html` into your browser.

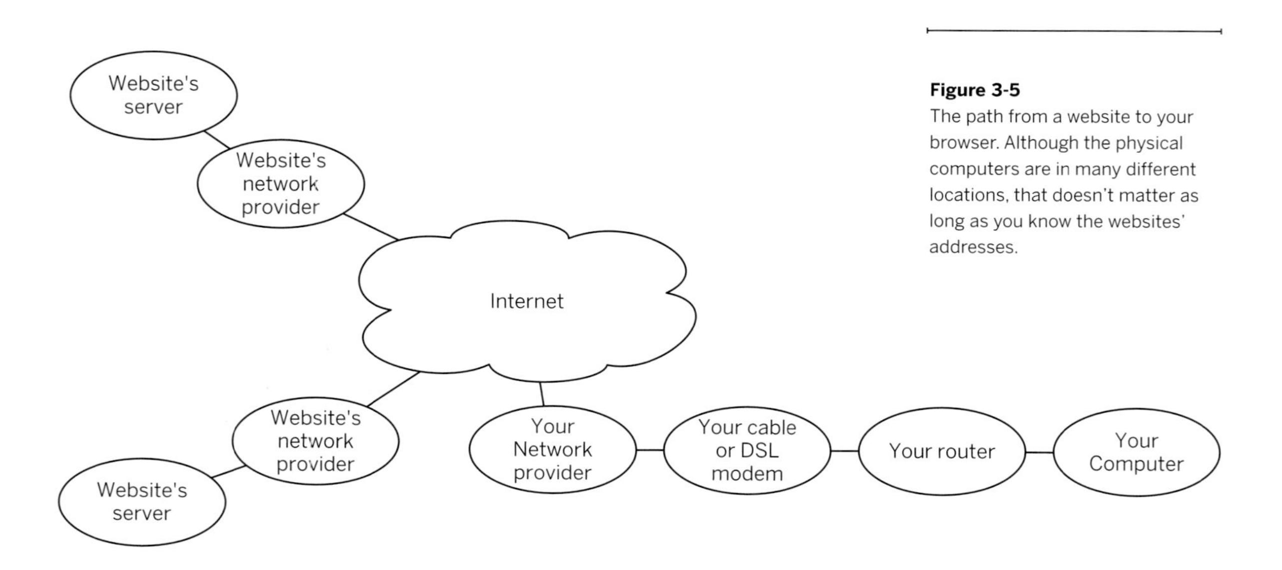

Figure 3-5
The path from a website to your browser. Although the physical computers are in many different locations, that doesn't matter as long as you know the websites' addresses.

2. The browser program contacts www.example.com on port 80.
3. The server program accepts the connection.
4. The browser program asks for a specific filename, **index.html**.
5. The server program looks up that file on its local file system, and sends the file out via the connection to the browser. Then, it closes the connection.
6. The browser reads the file, looks up any other files it needs (like images, movies, style sheets, and so forth), and repeats the connection request process, getting all the files it needs to display the page. When it has all the files, it strips out any header information and displays the page.

All the requests from browser to server, and all the responses from server to browser (except the images and movie files), are just strings of text. To see this process in action, you can duplicate the request process in the terminal window. Open up your terminal program again, just as you did for the `ping` example shown earlier (on Windows, use PuTTY).

Try It Type the code at the right.

```
$ telnet www.example.com 80
```

The server will respond as follows:

```
Trying 64.233.161.147...
Connected to www.example.com.
Escape character is '^]'.
```

The built-in Windows version of telnet is not very good. For example, you won't be able to see what you type without setting the `localecho` option, and the informative "Trying . . . Connected" prompts do not appear. Use PuTTY instead.

Now type the code to the right.

```
GET /index.html HTTP/1.1
Host: www.example.com
Connection: Close
```

Press the Enter key twice after this last line. The server will then respond with something like this:

```
HTTP/1.1 200 OK
Cache-Control: max-age=604800
Content-Type: text/html
Date: Mon, 11 Apr 2016 15:51:58 GMT
Etag: "359670651+gzip+ident"
Expires: Mon, 18 Apr 2016 15:51:58 GMT
Last-Modified: Fri, 09 Aug 2013 23:54:35 GMT
Server: ECS (lga/1384)
Vary: Accept-Encoding
X-Cache: HIT
x-ec-custom-error: 1
Content-Length: 1270
Connection: close
```

NOTE: If telnet doesn't close on its own, you may need to press Ctrl-] to get to the telnet prompt, where you can type q followed by Enter to exit.

After the header, the next thing you'll see is a lot of HTML. This is the HTML of the index page of example.com. This is how browsers and web servers talk to each other, using a text-based protocol called the *Hypertext Transport Protocol (HTTP)*. The `http://` at the beginning of every web address tells the browser to communicate using this protocol. The stuff that precedes the HTML is the HTTP header information. Browsers use it to learn the types of files that follow, how the files are encoded, and more. The end user never needs this information, but it's very useful in managing the flow of data between client and server. X

Programming a Web Server

Remember the node.js web server example from Chapter 1? Even though it doesn't do much, it is a web server and implements the Hypertext Transport Protocol (HTTP), so you can use it to learn a bit more about the exchange between a web server and a web browser.

Servers don't just return files to a client. They can do computation for the client as well, and return a result. In this section, you'll get a better sense for how a web browser responds to a client's request for data by computing a response dynamically, based on the parameters of the request.

Try It Type the following modified version of **server.js** in a text editor and save it as **dateServer.js.** Then run the script from the command line as you did in Chapter 1:

```
$ node dateServer.js
```

Now,open a second terminal window and telnet to localhost (or 127.0.0.1) on port 8080 and request the file from the command line. You should get a response with just the HTTP header and the date, within pointed brackets, like so:

```
HTTP/1.1 200 OK
X-Powered-By: Express
Content-Type: text/html
Date: Tue, 11 Apr 2017 16:43:39 GMT
Connection: keep-alive
Transfer-Encoding: chunked

< Tue Apr 11 2017 12:43:39 GMT-0400
(EDT)>
```

```
/*
    date server
    context: node.js
*/
var express = require('express');     // include the express library
var server = express();                // create a server using express

// define the callback function that's called when
// a client makes a request:
function respondToClient(request, response) {
  // write back to the client:
  response.writeHead(200, {"Content-Type": "text/html"});
  response.write("< " + new Date() + ">");
  response.end();
}

// start the server:
server.listen(8080);
 // define what to do when the client requests something:
server.get('/*', respondToClient);
```

> ⚠ **If you save it in a different folder than the server.js file from Chapter 1, make sure to install express.js using** `npm install express` **again as well.**

Even though the results of this script aren't as pretty in a browser, it's very simple to extract the date from within a client program like Processing—or even a microcontroller program. Just look for the `<` character in the text received from the server, read everything until you get to the `>` character, and you've got it.

HTTP requests aren't just for requesting files. You can add parameters to your request as well, and the server program can read those parameters and respond. To add parameters to a request, add a question mark at the end of the request and parameters after that. Here's an example:

```
http://localhost:8080/?name=tom&age=14
```

In this case, you're sending two parameters, `name` and `age`. Their values are `tom` and `14`, respectively. You can add as many parameters as you want, separating them with the ampersand (`&`). These are called *key-value pairs*.

There are data structures built into node.js's HTTP library that give you access to these parameters and other aspects of the exchange between client and server. In examples, you'll see a variable called `request`, or `req`, that contains all the aspects of the request: the address of the device that made it, the method of request, the requester's operating system and browser, parameters like the ones shown earlier, and more. You've already seen it in the previous code, along with a similar structure called `response` (sometimes `res`) that contains all the aspects of the server's response.

Following is an example that returns the parameters after the question mark in an HTTP GET request.

Try It Here's a node.js script that reads all the values sent in via a request and prints them out.

Save this script to your server as **getParameters.js.** Run the script, then view it in a browser using the URL shown earlier. You should get a page that says:

```
request: {"name":"tom","age":"14"}
```

```
/*
    parameter server
    context: node.js
*/
var express = require('express');      // include the express library
var server = express();                 // create a server using express

// define the callback function that's called when
// a client makes a request:
function respondToClient(request, response) {
  // convert the parameters to a string using JSON.stringify:
  var request = "request: " + JSON.stringify(request.query);
  // write back to the client:
  response.writeHead(200, {"Content-Type": "text/html"});
  response.write(request);
  response.end();
}

// start the server:
server.listen(8080);
 // define what to do when the client requests something:
server.get('/', respondToClient);
```

You could also request it from telnet or PuTTY like you did earlier (be sure to include the `?name=tom&age=14` at the end of the argument to GET, as in `GET /?name=tom&age=14`). You'd get something similar to the code at the right, shown here with the HTTP header:

```
HTTP/1.1 200 OK
X-Powered-By: Express
Content-Type: text/html
Date: Mon, 11 Apr 2016 17:16:48 GMT
Connection: keep-alive
Transfer-Encoding: chunked

request: {"name":"tom","age":"14"}
```

HTTP GET and POST

The ability to respond to parameters sent in an HTTP request opens all kinds of possibilities. For example, you can write a script that lets you choose the address to which you send an email message, or what message to send. You'd just add parameters to the URL after the question mark, read them in node.js, and use them to set the various mail variables.

HTTP has four main methods for making a request: GET, POST, PUT, and DELETE. Many browsers don't support PUT or DELETE, however, so stick to GET and POST for now.

In node.js, you can read the parameters from a GET request as discrete elements of the request.query. The previous example just reads the whole query as one JSON object, but the next example breaks out the individual parameters and uses them to do some computation.

Try It Of course, because you're programming this server yourself, you can do more than just print out the results. Try the script to the right. Save this version as **ageCheck.js.**

Run this script using `node ageCheck.js`, and try making a request to it with the same parameter string as the last script, `?name=tom&age=14`, and see what happens. Then try `/check/?name=tom&age=14`. Then change the age to a number greater than 21.

This version of the script has callback functions to handle two different requests. It will respond to requests for the main index of the site, `http://localhost:8080/` and it will also respond to a request for `http://localhost:8080/check`. These two different requests are called *routes* in server terminology. The route that just ends in `/` is called the *main index* or the *root* of the site.

```
/*
   Age checker
   Context: node.js

   Expects two parameters from the HTTP request:
      name (a text string)
      age (an integer)
   Prints a personalized greeting based on the name and age.
*/

var express = require('express'); // include the express library
var server = express();            // create a server using express

// define the request callback functions:
function respondToClient(request, response) {
  // convert the parameters to a string using JSON.stringify:
  var request = "request: " + JSON.stringify(request.query);
  // write back to the client:
  response.writeHead(200, {"Content-Type": "text/html"});
  response.write(request);
  response.end();
}

function checkAge(request, response) {
  var name = request.query.name;
  var age = request.query.age;
  var responseString = "";
  if (age < 21) {
    responseString = "<p>" + name
        + ", You're not old enough to drink.</p>\n";
} else {
    responseString =  "<p> Hi " +  name
        + ". You're old enough to have a drink, ";
    responseString += "but do so responsibly.</p>\n";
}
  // write back to the client:
  response.writeHead(200, {"Content-Type": "text/html"});
  response.write(responseString);
  response.end();
}

// start the server:
server.listen(8080);
// define what to do when the client requests something:
server.get('/', respondToClient);     // if the client sends /?parameters
server.get('/check', checkAge);       // if the client sends /check?parameters
```

Express.js makes it easy to program a server to respond to different routes using either GET or POST. You use the `server.get()` function to respond to GET requests, and `server.post()` to respond to POST requests. You'll learn more about routes in the REST section later on.

When you ran this script, you saw that a request to http://localhost:8080/and http://localhost:8080/check gave two different results, even with the same parameters. These are two different *routes* to which the server script can respond. Think of each different route as a function that you can ask the server to run for you.

Having to pass parameters in via the browser address line does not make for a convenient user interface. It's much easier to fill in a form, and have the browser pass the form's data in a request. POST is the HTTP method usually used to post data from a web form. Instead of adding the parameters on the end of the URL path as GET does, POST adds the parameters to the body of the HTTP request. There are a couple other parameters you need to add to a POST request too, like the content type and the content length. POST is a little more work to set up, but it's really useful for hiding the messy business of passing parameters, keeping your URLs tidy and easy to remember. Instead of the previous URL, all the user has to see is:

```
http://www.example.com/check
```

The rest can get delivered via POST. Here's how you can modify the previous example to respond to GET or POST:

To see how the script would be different for POST requests, make a couple changes to the script. Changes are shown in blue. First, add a couple of lines to the top of the script:

```
var express = require('express');// include the express library
var server = express();              // create a server using express
var bodyParser = require('body-parser');        // include body-parser
// use the parser for data that's URL-encoded:
server.use(bodyParser.urlencoded({ extended: true }));
```

Next change the checkAge function as follows. Remove the first two lines and replace with the changes shown in blue:

```
// ...no other changes before checkAge function

function checkAge(request, response) {
  var name, age;
  if (request.method === "GET") {
   name = request.query.name;
   age = request.query.age;
  } else if (request.method === "POST") {
   name = request.body.name;
   age = request.body.age;
  }
  var responseString = "";
```

Finally, add one line at the end, to define a POST method for the /check route. This method will call the same callback function as the GET method. These .get() and .post() functions are also called *listener* functions, because they direct the program to listen for incoming GET and POST requests.

```
// no other changes until the end of the script:

server.get('/check', checkAge);     // if the client sends /check?parameters
server.post('/check', checkAge);    // same for a POST request
```

Before you can run this script, you're going to have to install a new library using the node package manager (npm) like you did when you made the first server back in Chapter 1. Once you've saved this file, make sure you're in the same directory on the command line and enter the following:

```
$ npm install body-parser
```

You'll see a progress bar and some warnings as you did when you installed express, then the command prompt will return, and the body-parser library will be installed in the **node_modules** directory. Now you're ready to run this script.

Test It Telnet into to localhost port 8080 and request the file again from the command line. This time, type the POST request as shown here.

You'll get the same response as you did using GET, but now you can put all the parameters in one place, at the end.

```
POST /check HTTP/1.0
Host: example.com
Connection: Close
Content-Type: application/x-www-form-urlencoded
Content-length: 16

name=tom&age=14
```

» The content length is the length of this string, plus a linefeed. If your name is longer than three letters, or your age is greater or fewer than two digits, change the content length to match.

Serving Files

If you've done any work making web pages before, you're used to the idea that you save HTML files, images, and other content as files on a server, and the server delivers them in response to browser requests. The servers you've written here so far haven't served any files yet, but they've generated all the responses dynamically. You can serve files from node.js (called *static files* in this context, because they're not expected to change). You can also serve static files and dynamic content together from the same server.

It would be useful to have an HTML form for the script you just wrote, so that the user could enter his or her name and age. To do that, you're going to write an HTML page with a form that then calls the server's `/check` route, using a POST request. You'll also modify the `checkAge` function to respond with a link back to the HTML page.

In the previous version, you used a library called `body-parser` that parsed the parameters of your POST request. You called it using `server.use()`. That library is called *middleware*, because it sits in the middle between the client's request and the server's response. You'll use another piece of middleware in this script, `express.static`, that will set a directory from which node.js will serverstatic files. You set a directory that's in the same directory as your server script, then all static files go into that directory. You can also set a route that precedes all static files. In the following example, you'll just use `/` as the route, so that you can access any static file using the URL http://localhost:8080/filename.html.

Create a subdirectory inside the directory where your script lives, and call it **public**. Then make a new text file called **index.html** inside that directory. The text of that file is shown here.

Try It Save this file as **index.html** inside the **public** folder. You can see that the form's action is `/check`, the route that runs the `ageCheck` function, and its method is POST. So when a user clicks submit, the form will make a POST request to `/check`, and the server will respond appropriately.

```
<html>
  <body>
    <form action="/check" method="post"
      enctype="application/x-www-form-urlencoded">
      Name: <input type="text" name="name" /><br>
      Age: <input type="text" name="age" />
      <input type="submit" value="Submit" />
    </form>
  </body>
</html>
```

» Modify the **checkAge.js** script as follows (changes shown in blue), adding a line after the body-parser initialization to set up the static files directory and path:

```
// use the parser for data that's URL-encoded:
server.use(bodyParser.urlencoded({ extended: true }));
// serve static pages from public/ directory:
server.use('/',express.static('public'));
```

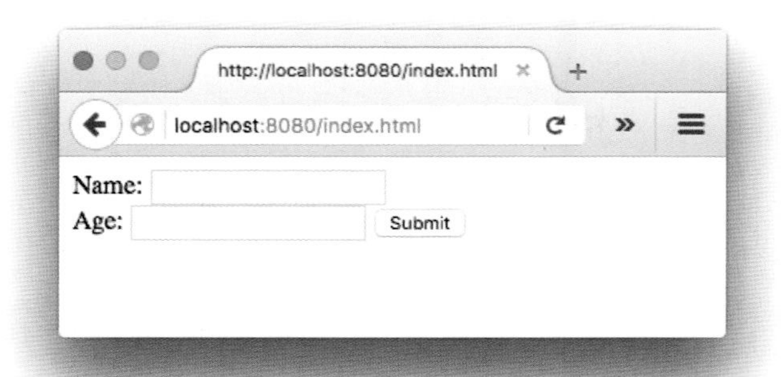

Figure 3-6
The age-checker form.

You can also remove the `respondToClient()` function from the script, and the `server.get()` call that calls it. They won't be used now. The script will run fine whether they're there or not.

Save the script, then run it again. Go to http://locaohost:8080/index.html in a browser, and you should get the page seen in Figure 3-6. The server script is responding to the client's requests, either with the static file for the GET request, or with the age check function for the POST request.

You can build a more complex web interface with media files, links, and anything else you might build in a regular site. As long as the files are in the **public** directory, they'll be served by the server. If you include links to any routes defined in the server, like `/check`, the server will run those functions, as long as the request method matches the listener function that's listening for it.

Making Requests with curl

In the previous examples, you got a firsthand look at the details of HTTP by making requests from the command line, using `telnet`. You wrote the headers of your request, and saw the headers of the server's response. It's inconvenient to type the full request every time, but the POSIX command `curl` allows you to make HTTP requests. You saw it used on the Raspberry Pi in Chapter 1. It's available on both Linux and macOS. It's not available in the Windows 10 command-line interface, but is in Cygwin if you followed the instructions in Chapter 1 and installed the Net package. Here's how you'd use `curl` to make the POST request to the server that you just wrote:

```
$ curl -v  -d "name=tom&age=14" localhost:8080/check
```

The response will be as follows:

```
*   Trying ::1...
* Connected to localhost (::1) port 8080 (#0)
> POST /check HTTP/1.1
> Host: localhost:8080
> User-Agent: curl/7.46.0
> Accept: */*
> Content-Length: 15
> Content-Type: application/x-www-form-urlencoded
>
* upload completely sent off: 15 out of 15 bytes
< HTTP/1.1 200 OK
< X-Powered-By: Express
< Content-Type: text/html
< Date: Tue, 12 Apr 2016 17:58:46 GMT
< Connection: keep-alive
< Transfer-Encoding: chunked
<
* Connection #0 to host localhost left intact
<p>tom, You're not old enough to drink.</p><a href="/index.
html">Check another person</a>
```

You can do a lot with `curl`, and you can get the full list of options from the manual page by typing `man curl` at the command prompt. Here are some of the more useful options to know. Try them out on the scripts you just wrote:

Verbose output (see all headers):

```
curl -v http://www.example.com
```

POST request with data:

```
curl -d "key=value&key2=value2" http://www.example.com
```

GET request with parameter string:

```
curl -G -d "key=value&key2=value2" http://www.example.com
```

Follow redirected links:

```
curl -L http://www.example.com
```

X

How Email Works

Transferring email also uses a client-server model. It involves four applications: your email program and your friend's (called the mail clients), and your email server (also called the mail host or mail server) and your friend's email server. Your email program adds a header to your message to say that this is a mail message, who the message is to and from, and what the subject is. Next, it contacts your mail server, which then sends the message on to your friend's mail server. When your friend checks her mail, her mail client connects to her mail server and downloads any waiting messages. The mail servers are online all the time, waiting for new messages for all users.

The transport protocol for sending mail is called the *Simple Mail Transport Protocol (SMTP)*. It's paired with two retrieval protocols: *Post Office Protocol (POP)* and *Internet Message Access Protocol (IMAP)*. Just like HTTP, these are text-based.

Try It Here's a node.js script that sends an email to you. Save this script to your server as **mailer.js**. Install the nodemailer library using `npm install nodemailer`.

> **!** You should never put a password in clear text in a program like you see here. Encrypting it is a bit more complex, and you'll see how to do it next.

```
/*
mailer script
context: node.js
*/
// include the nodemailer library:
var nodemailer = require('nodemailer');
```

Make a variable to hold the details of your mail server and account like so. Fill in the details for your server. Port 465 is the standard port number for sending mail using a *Secure Socket Layer (SSL)* connection. It ensures that your message is encrypted before it's sent, so it can't be read along the way.

```
// define your account and mail server:
var account = {
  host: 'smtp.gmail.com',
  port: 465,
  secure: true,
  auth: {
    user: 'your.mail@gmail.com',
    pass: 'your_password'
  }
};
```

▸▸ Replace with your SMTP server if you're not using Gmail.

▸▸ Fill in your email account details.

Make a variable to hold the details of the message. You can use any of the mail headers you want. The minimum you need are `to`, `from`, `subject`, and `text`. The latter is the body of your message.

```
// define the message:
var message = {
  from: account.auth.user,
  to: 'recipient@gmail.com',
  subject: 'Hello',
  text: 'Hello world',
};
```

▸▸ Fill in the recipient's email address.

When you send a message, you'll get either a response or an error message. So you need a function to handle either one.

```
// make a callback function to respond when the mail is sent:
function confirm(error, info){
  if(error){
    console.log(error);
  } else {
    console.log('Message sent: ' + info.response);
  }
}
```

Once you've defined all of these elements, make a client using the `createTransport` command, then send mail. The `sendMail` function expects a message and a callback function that it should call when it's done.

```
// make a client and use it to send the message:
var client = nodemailer.createTransport(account);
client.sendMail(message, confirm);
```

Although the examples here use the nodemailer library, there are a number of good node.js libraries for sending mail. You can use node.js as an intermediary between any local application, like the Processing sketch that will follow, or some microcontroller applications you'll see later on.

Gmail does not allow mail scripts to use plain password login by default. It uses a more complex and secure authorization system called OAuth2. To enable password login for this script, go to myaccount.google.com/lesssecureapps **and enable less secure apps. Turn it back off when you are done with this project. In fact, you may want to create a separate Gmail account just for this project. For more on writing a more secure mail client with OAuth2, see** nodemailer.com/smtp/oauth2/ **(this is for the experienced reader).**

Encrypting Your Passwords

The preceding script has your password in clear text, which is a horrible idea. Even if you carefully delete it every time you save the file, there's a good chance you'll forget sometime. At the very least, you should save the password in a separate file and scramble it using an encryption algorithm. There are standard encryption algorithms that are reasonably secure, and they are used in applications you use every day. Node.js has a library called crypto that lets you encrypt and decrypt strings of text.

To encrypt a message, you need an algorithm, a key to lock and unlock the cipher, and the message itself. The next example shows how to take a message and a key from the command line and output the message as an encrypted string in a text file. That text file can be moved anywhere, so you can read it in using a different script; for example, a modified version of the mailer script.

Try It Save this script to your server as **encrypt.js**. The libraries used are included as part of the node.js core, so there's no need to install them.

The `process.argv[]` array holds all the words you type when you call node.js from the command line (called *arguments*). You're using it here to take the message and key from the command line.

You need to set the file path and filename. In this case, you'll be saving the file in the same directory as your script.

Next, create a cipher using the key, and update it by adding the message. Then finalize it.

When you call the `writeFile()` function, you'll need a callback, so define it first.

Finally, call `writeFile()` to save your encrypted string to the file.

```
/*
encryption script
context: node.js
*/
// include the crypto and filesystem libraries:
var crypto = require('crypto');
var fs = require('fs');

// take input from the command line:
var key = process.argv[3];       // fourth element from the command line
var message = process.argv[2];   // third element from the command line

// set the location of the file that you'll output:
var fileName = 'info.txt';

// create a cipher:
var cipher = crypto.createCipher('aes-256-cbc', key);
// add the message to the cipher and finalize it:
var encryptedMsg =  cipher.update(message, 'utf8', 'hex');
encryptedMsg += cipher.final('hex')

// callback function for the writeFile function:
function success(data) {
  console.log('I wrote to the file: ' + fileName);
}

// write the encrypted message to the file:
fs.writeFile(fileName, encryptedMsg, success);
```

To run this script, type:

```
$ node encrypt.js message key
```

The two extra words, `message` and `key`, are the string you want to encrypt and the key that you want to use to recall it, respectively. If you want to use a message with more than one word, put the message in single quotes.

When you run this, you'll get a response telling you that the script wrote the file, and you'll see the file in your script's directory when you're done. It will look like a random string of numbers and letters. You'll need the decryption script that follows to read it.

Try It Save this script to your server as **decrypt.js**. It's almost identical to the previous script in structure.

```
/*
decryption script
context: node.js
*/
```

Include the same libraries as you did in the previous script.

```
var crypto = require('crypto');    // include encryption library
var fs = require('fs');            // include filesystem library
```

In this script, you're using `process.argv[]` here to take the filename and key from the command line.

```
// take input from the command line:
var key = process.argv[3];         // fourth element from the command line
var fileName = process.argv[2];    // location of the encrypted file
```

Next, create a decipherer using the key.

```
// create a decipherer:
var decipher = crypto.createDecipher('aes-256-cbc', key);
```

When you call the `readFile()` function, you'll need a callback function, so define it here first. You'll get the decrypted message from the decipher-er in this function.

```
// callback function for the readFile function:
function success(error, data) {
  if (data){
    data = data.toString();
    var decryptedMsg = decipher.update(data, 'hex', 'utf8');
    decryptedMsg += decipher.final('utf8');
    console.log('I read this from the file: ' + decryptedMsg);
  } else if (error) {
    console.log(error);
  }
}
```

Finally, call `readFile()` to read your encrypted string from the file.

```
fs.readFile(fileName, success);
```

Run this script like so:

```
$ node decrypt.js filename key
```

This time the two extra words, `filename` and `key`, are the filename you want to decrypt and the key that you want to use to recall it, respectively. Make sure that the file **info.txt** that you created with the previous script is in the same directory as this script when you run it. When you do, you should get a message like so:

```
I read this from the file: message
```

Whatever message you typed for the previous script will be the message that's output here. By combining this script with the email script, and using the **encrypt.js** script to encrypt your password, you'll have a slightly more secure way of managing your passwords for email, or for any application where you need to encrypt and decrypt messages.

Now that you've got the basics of responding to HTTP requests and mail sending, and a little encryption under your belt, it's time to put them into action in a project.

X

Networked Cat

Web browsing and email are all very simple for humans because we've developed computer interfaces that work well with our bodies. Keyboards work great with our fingers, and mice glide smoothly under our hands. It's not so easy for a cat to send email, though. This project attempts to remedy that while showing you how to build your first physical interface for the internet.

If you're a cat lover, you know how cute they can be when they curl up in their favorite spot for a nap. You might find it useful during stressful times at work to think of your cat, curled up and purring away. Wouldn't it be nice if the cat sent you an email when he lays down for a nap? It would be even better if you could then check in on the cat's website to see him at his cutest. This project makes that possible.

The system works like this: force-sensing resistors are mounted under the cat mat and attached to a microcontroller. The microcontroller and a camera are attached to a personal computer. When the cat lies down on the mat, his weight will change the sensor readings. The microcontroller then sends a signal to a Processing sketch on the personal computer, which in turn sends an HTTP GET request to the web server. The web server then sends you an email letting you know that your cat is being particularly cute. While the cat remains on the mat, the Processing sketch takes a picture with the camera every two minutes and uploads it to the web server using HTTP POST.

You'll do this project in several parts:

- Write an Arduino sketch to read sensors in the cat's mat and send the results serially to Processing.
- Write a Processing sketch that can read the serial data and make an HTTP GET request.
- Program a node.js server to serve static HTML pages, and to send mail whenever it receives an HTTP GET request.
- Make a web page for the cat cam.
- Extend the server to accept new image uploads for the web page.
- Extend the Processing sketch to take camera images and upload them to the server using an HTTP POST request.

MATERIALS

- **Between 2 and 4 force-sensing resistors, Interlink 400 series**
- **1 1-kilohm resistor**
- **1 solderless breadboard**
- **1 Arduino-compatible microcontroller module**
  - **Features used: Analog input, UART**
- **1 personal computer**
- **1 USB web camera**
- **1 cat mat**
- **1 cat**
- **2 thin pieces of wood or thick cardboard, about the size of the cat mat**
- **Wire-wrapping wire**
- **Male header pins**
- **Rubber bumpers**

Since this is a more complex project than the previous ones, with many different programs and computers, it helps to draw a system diagram showing what each program is doing and how they're communicating. Figure 3-7 is a diagram of the interaction. Figure 3-8 shows a system diagram indicating the physical computers are involved (the boxes), the programs are running on them (inside the boxes), and the communications protocols connect them (the dashed lines in between them). A diagram like this helps you plan the code and the circuit before you build anything.

Putting Sensors in the Cat Mat

First, you need a way to sense when the cat is on the mat. The simplest way to do this is to put force sensors under the mat and sense the difference when he sits on it. You can use any Arduino-compatible microcontroller to do this. All you need is an analog input and asynchronous serial communication. You can even use an ATtiny like you saw in Chapter 3, using SoftwareSerial. Figure 3-10 later in this chapter shows the circuit diagram for the Uno/101 footprint, the MKR1000 footprint, and the ATtiny84 footprint.

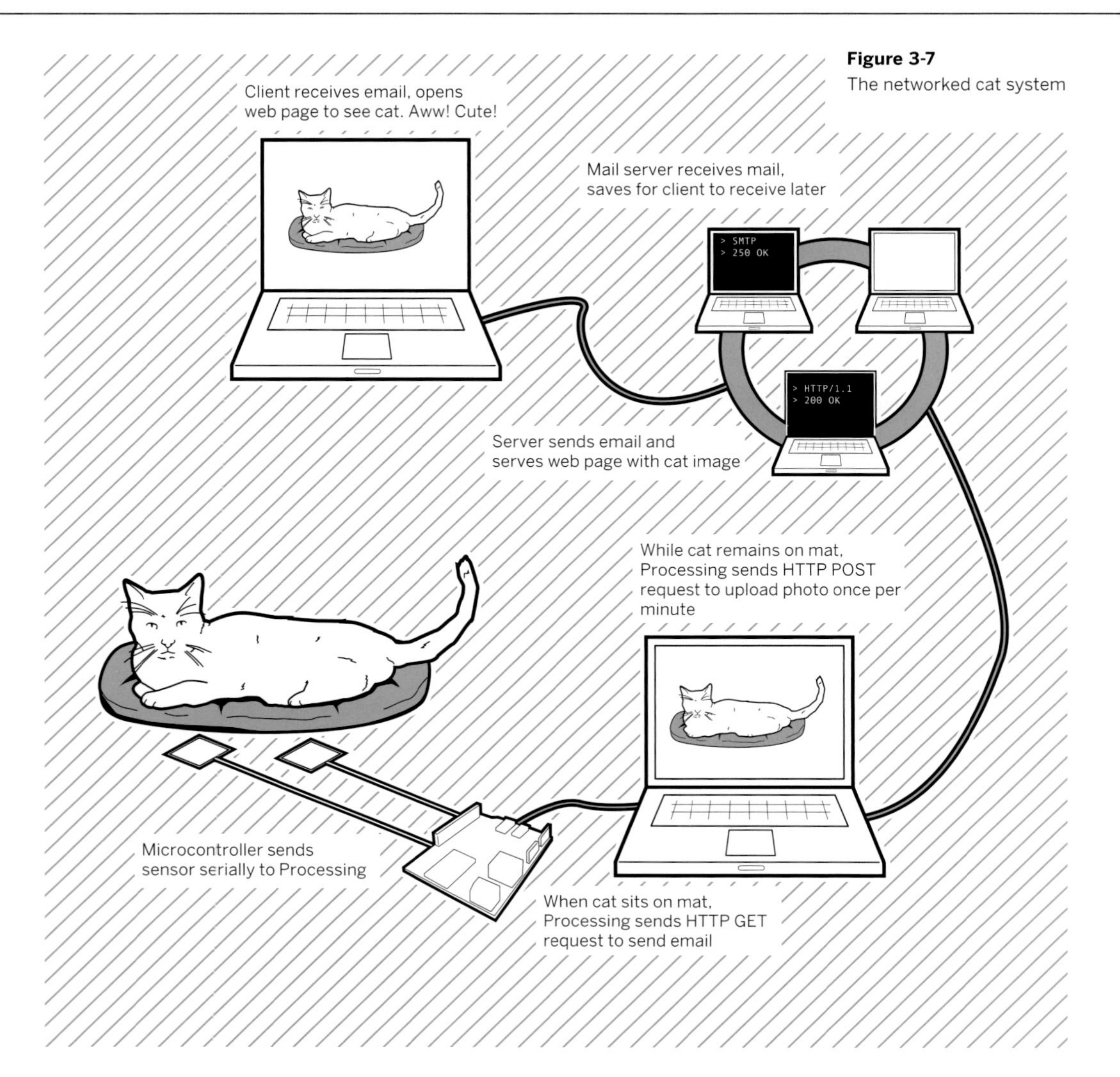
Client receives email, opens web page to see cat. Aww! Cute!
Mail server receives mail, saves for client to receive later
> SMTP
> 250 OK
> HTTP/1.1
> 200 OK
Server sends email and serves web page with cat image
While cat remains on mat, Processing sends HTTP POST request to upload photo once per minute
Microcontroller sends sensor serially to Processing
When cat sits on mat, Processing sends HTTP GET request to send email

Figure 3-7
The networked cat system

Interlink's 400 series *force-sensing resistors (FSRs)* work well for this project. They change their resistance when a force presses against them.

Many FSRs are small and round, like Interlink's 400 or 402 series, or similar models from CUI or FlexiForce. To sense an area the size of a cat, sandwich them between two firm plates, using masonite or another type of wood or firm cardboard. The plates distribute the force exerted by the cat, so you can detect when one plate is pressing on the other. Put the whole assembly underneath the cat mat, and you can sense when the cat is on the mat.

First, cut two pieces of wood or firm cardboard slightly smaller than the cat's mat. Don't use a really thick piece of wood. You just need something thick enough to provide a relatively inflexible surface for the sensors. Attach the sensors to the corners of the bottom panel. Tape the two boards together at the edges loosely, so that the weight of the cat can press down to affect the sensors. If you tape too tightly, the sensors will always be under force; too loose, and the boards will slide around too much and make the cat uncomfortable. To make sure that the weight is resting on the sensors, get some little rubber feet—available at any electronics or hardware store—and position them on the top panel so that they press down on the sensors. Position an extra rubber foot or two at the center of the panel to reduce the flex. Sandwich the sensors between the two panels. Figures 3-9, 3-10, and 3-11 show a working version of the sensor board.

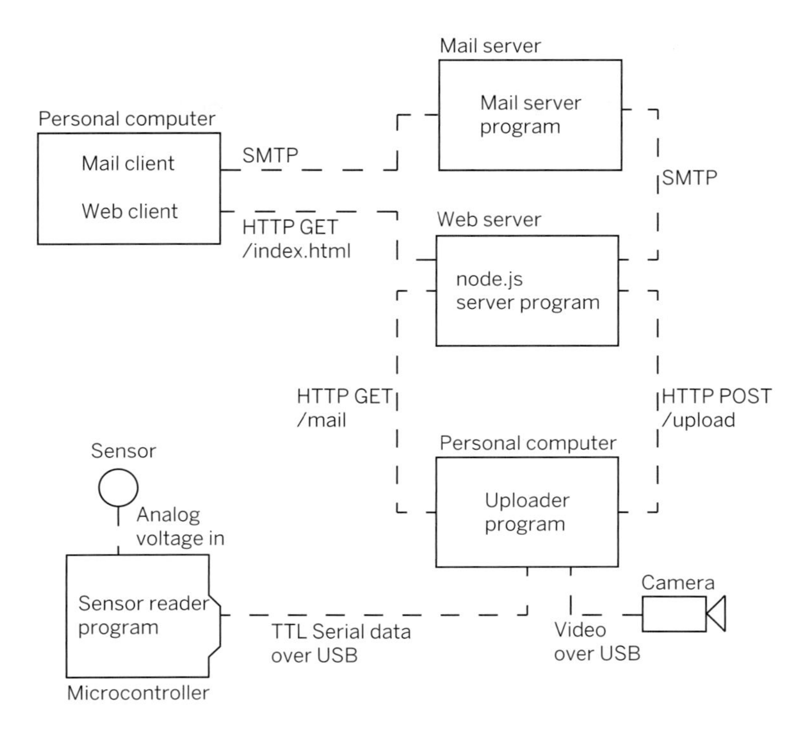

Figure 3-8
System diagram of the networked cat system.

Next, attach long wires to the force-sensing resistors to reach from the mat to the nearest possible place to put the microcontroller module.

You're using multiple sensors so you can sense a large area under the mat, but it doesn't matter which one gets triggered. Connect the sensors to an analog input of the microcontroller in parallel with each other, using the voltage divider circuit shown in Figure 3-10. This circuit combines the input from all four force-sensing resistors into one input.

You'll be connecting the microcontroller to a personal computer, serially, so if you're using an ATtiny84 or any other controller that doesn't have its own USB-to-serial adapter, you'll need to add that as well. For the Uno or 101 and the MKR1000, you can connect to the personal computer via USB, as you've done previously.
X

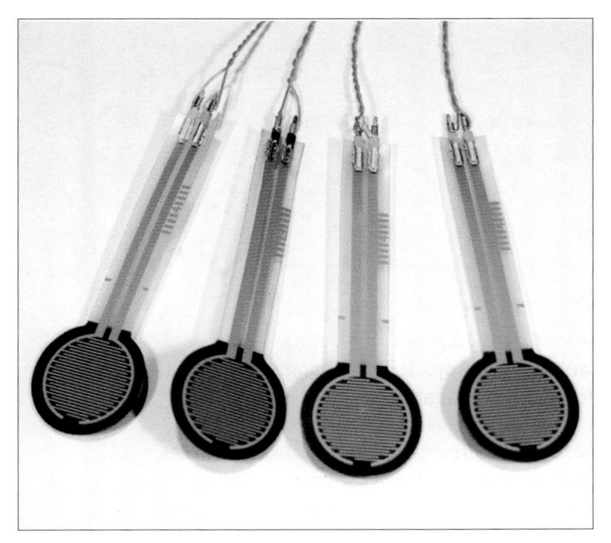

Figure 3-9
Because the force-sensing resistors melt easily, I used 30AWG wire wrap instead of solder. Wire-wrapping tools are inexpensive and easy to use, but make a secure connection. After wire wrapping, I insulated the connections with heat shrink. (Heat shrink not shown.)

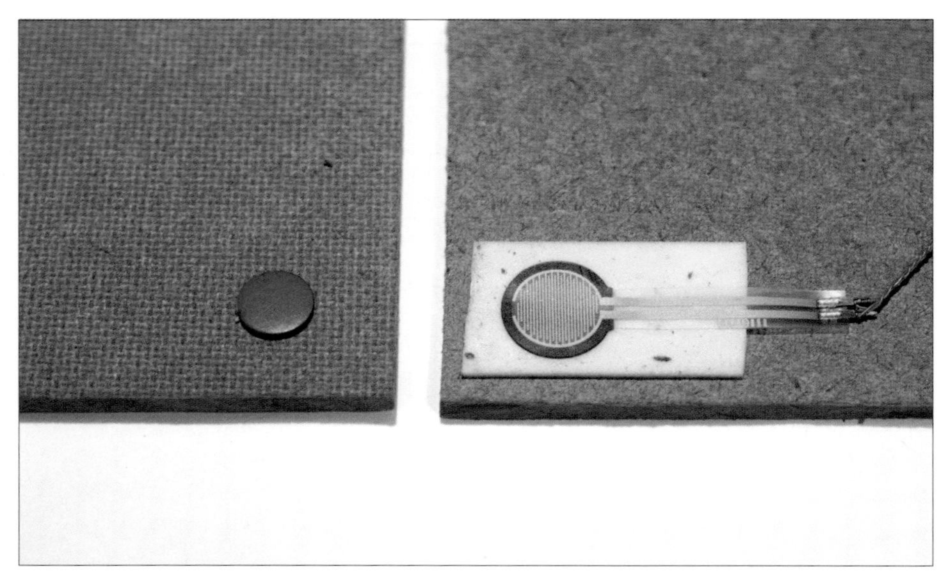

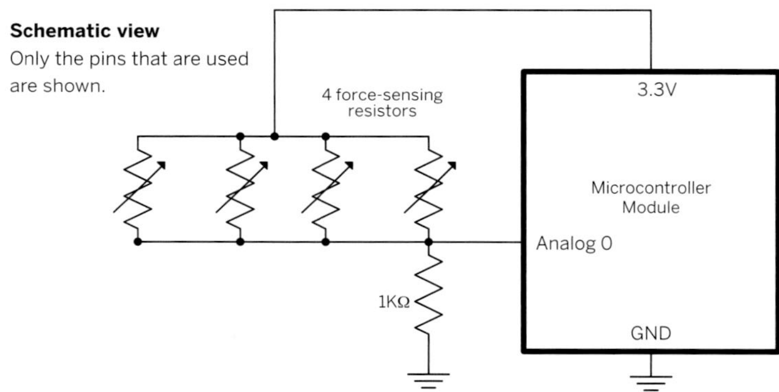

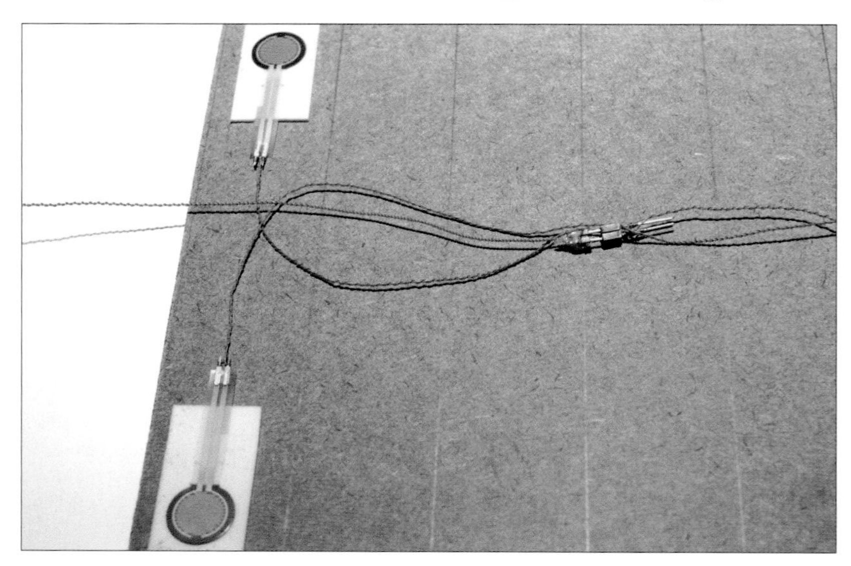

Figure 3-10
The cat-sensing panel. The four FSRs are wired in parallel. Note the rubber feet that press down more precisely on the sensors. Make sure to insulate the connections before taping the panels together. The connector is just a pair of female wire-wrap headers.

MKR1000 Breadboard view

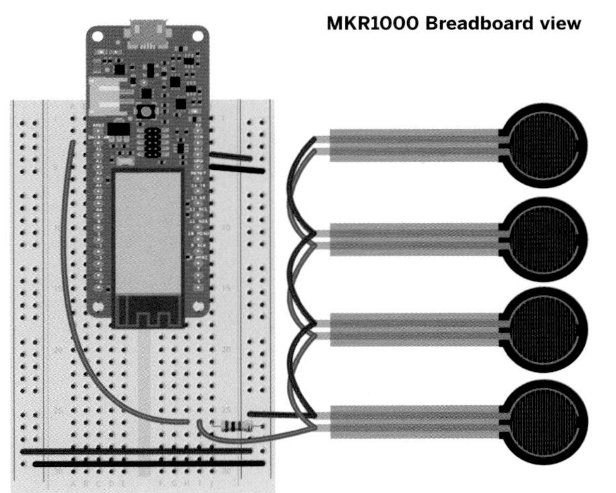

Figure 3-11

The cat-sensing circuit. Because all of the force-sensing resistors are wired in parallel, there are only two connections for all of them, voltage and analog input.

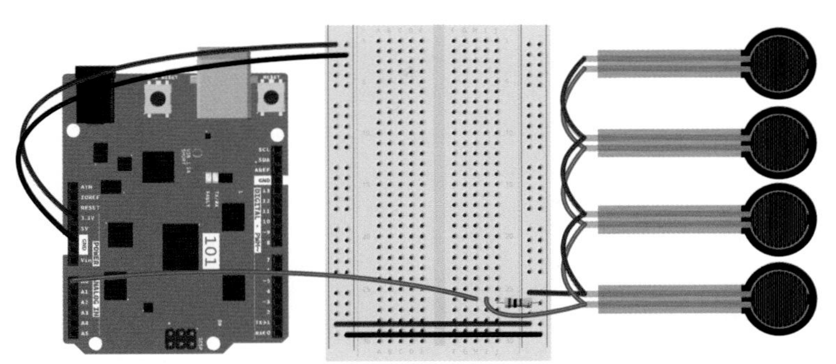

101/Uno Breadboard view

You may need a higher value base resistor depending on the resistance range of your force-sensing resistors and the weight of your cat. If a 1-kilohm resistor doesn't give you good values, try a 4.7-kilohm or a 10-kilohm resistor. The images here show a 1-kilohm resistor.

ATtiny84 Breadboard view

You'll need a 10-kilohm pullup resistor on the reset pin, and asynchronous serial connections to the computer via a USB-to-serial adapter. The adapter is also powering the microcontroller here.

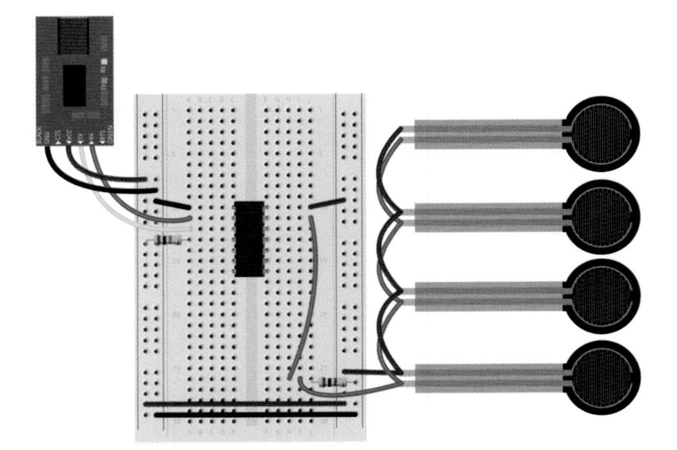

Test It Once you've got the sensor panel together and connected to the microcontroller, run this code on the Arduino board to test the sensors.

If you're using an ATtiny84 or a controller without a built-in UART, use the SoftwareSerial library as you did in Chapter 2.

To see the results, open the Serial Monitor at 9600 bits per second. Now, position the cat on the panel and note the number change. This can be tricky, as cats are difficult to command. You may want to put some cat treats or catnip on the pad to encourage the cat to stay there. When you're satisfied that the system works and that you can see a significant change in the value when the cat sits on the panel, you're ready to move on to the next step.

```
/*
   Analog sensor reader
   Context: Arduino

   Reads an analog input on Analog in 0, prints the result
   as an ASCII-formatted  decimal value.
*/
void setup() {
  // start serial port at 9600 bps:
  Serial.begin(9600);
}

void loop() {
  // read analog input:
  int sensorValue = analogRead(A0);

  // send analog value out:
  Serial.println(sensorValue);
}
```

Sensors and Events

Your project will send a mail whenever the cat steps on the mat. What does that look like in sensor terms? To find out, you need to do one of two things: get the cat to jump on and off the mat on cue (difficult to do without substantial bribery, using treats or a favorite toy), or weigh the cat and use a stand-in of the same weight. The advantage to using the cat is that you can see what happens when he's shifting his weight, preparing the bed by kneading it with his claws, and so forth. The advantage of the stand-in weight is that you don't have to herd cats to finish the project.

You can see what the output looks like graphically using the Serial Plotter in the Arduino IDE. Choose Serial Plotter from the Tools menu, and you'll see a graph of your sensor readings over time. As the cat jumps on and off the mat, you should see a change that looks like Figure 3-12. When the cat jumps on the mat, you should see a sudden increase, and when he jumps off, you'll see the graph decrease. You'll also see any small changes, which you might need to filter out. If the changes are small relative to the difference between the two states you're looking for, you can ignore them. Using the sensor values, you have enough knowledge to start defining the cat's presence on the mat as an *event*. The

Figure 3-12
Output of the sensor-graphing program.

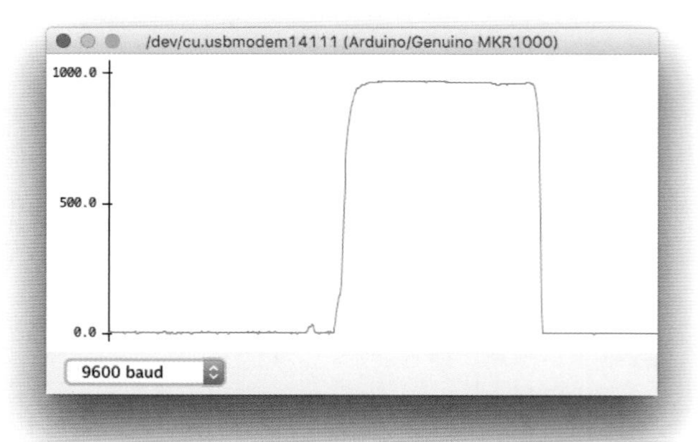

event you care about is when the sensor reading increases significantly, because that's when the cat sat on the mat. To do this, pick a threshold number in between the two states. When the sensor reading changes from being less than the threshold to greater than or equal to it, that's your first event. When the sensor value goes below the threshold, the cat has left the mat. That's your second event.

Before you write your whole program, it helps to outline in plain language what it's going to do. You know there are a few functions you'll need anyway, so you can use those functions to structure your outline. An outline like this is called *pseudocode*, and it's a good practice to use whenever you're planning a new program in any language.

Plan It Your Processing sketch is going to need to communicate using the serial library, and it's going to need to read a camera image. You'll have to import libraries for those functions.

You know it will have a `setup()` function that will initialize the program. The `draw()` function will draw the camera image to the screen.

```
/*
  CatCam uploader
 context: Processing
 */

// import libraries

void setup() {
  // initialize serial communications
  // take control of the camera
}

void draw() {
  // draw the camera image to the window
  // if there's any response from the server, print it

 //if cat remains on mat, upload picture
}
```

The sketch will listen for incoming serial data using a `serialEvent()` function. It will take action depending on the changes in the data.

```
void serialEvent(Serial myPort) {
  //read serial data
  //if cat gets on the mat:
  //  upload picture
  //  send mail
}
```

The actions that your program will take help you to define a few functions you'll need. Even if you don't know what they'll do yet, you know you need them.

Now you've got an outline to write the program. The rest is just filling in the details.

```
void sendMail() {
  // if enough time has passed since last email request:
  //   make an HTTP GET request to send an email
  //   note the time
}

void uploadPicture() {
  // if enough time has passed since last upload :
  //   take a picture
  //   make an HTTP POST request to upload the picture
  //   note the time
}
```

Connect It To start, fill in the pseudocode with real code to read the serial data. You can do this just like you read the sensor values for Monski Pong in Chapter 2.

When you run this, you'll see the sensor readings streaming down the console pane, just like you did at the beginning of the Monski Pong program. As you press on the sensor mat, you'll see the readings change. Next, you need to determine what change defines the event of the cat sitting on the mat.

```
/*
  CatCam uploader
 context: Processing
  */

// import libraries
import processing.serial.*;

Serial myPort;        // The serial port

void setup() {
  // initialize serial communications
  // change the port number to your microcontroller's serial port:
  String portName = Serial.list()[0];
  myPort = new Serial(this, portName, 9600);
  // don't generate a serialEvent() unless you get a newline character:
  myPort.bufferUntil('\n');
}

void draw() {
}

void serialEvent(Serial myPort) {
  //read serial data
  String inString = myPort.readStringUntil('\n');
  int sensorValue = 0;
  if (inString != null) {
    // trim off any whitespace:
    inString = trim(inString);
    // convert to an int:
    sensorValue = int(inString);
    println(sensorValue);
  }
}
```

Next you need to determine what defines "if the cat gets on the mat." The event you care about is when the sensor reading increases significantly, because that's when the cat sat on the mat. To do this, pick a threshold number in between the two states. When the sensor reading changes from being less than the threshold to greater than or equal to it, that's when the cat first sits on the mat. When the sensor value goes below the threshold, the cat has left the mat.

You don't just want to respond to events, though. You might have a fickle cat who jumps on and off the mat a lot, or takes his time to settle on the mat. Once you've sent a message, you don't want to send another one right away, even if the cat gets off the mat and back on. Decide on an appropriate interval, wait that long, and don't send any requests during that time. Once the interval's over, you're clear to send requests again if an appropriate event happens.

▶ To know if the sensor reading changed, you're going to need to know the previous reading and compare it to the previous one. Add three new global variables, `lastSensorReading`, `threshold`, and `catOnMat`, at the beginning of your sketch, right after you declare the serial port (changes are shown in blue):

```
Serial myPort;              // The serial port
int threshold = 400;        // above this number, the cat is on the mat.
int lastSensorValue = 0;    // the previous sensor reading
boolean catOnMat = false;   // whether the cat is on the mat
```

▶ Now add the following code at the end of the `SerialEvent` function. This will compare the current sensor reading to the last reading, and save the current reading as `lastSensorReading` for the next comparison. Comment out the `println(sensorValue)` statement, since you no longer need it.

```
void serialEvent(Serial myPort) {
  //read serial data
  String inString = myPort.readStringUntil('\n');
  int sensorValue = 0;
  if (inString != null) {
    // trim off any whitespace:
    inString = trim(inString);
    // convert to an int and map to the screen height:
    sensorValue = int(inString);
    // println(sensorValue);
  }
  // if "cat on mat" event occurs:
  if (sensorValue > threshold ) {        // if current reading > threshold
    if (lastSensorValue <= threshold) { // and last reading < threshold
      catOnMat = true;                   // then the cat just got on the mat
      println("Cat just got on mat");
      //  upload picture, then send mail:
      uploadPicture();
      sendMail();
    }
  } else {
    // if "cat off mat" occurs:
    if (lastSensorValue > threshold) {
      catOnMat = false;
      println("Cat just left mat");
    }
  }
  // save the current reading for comparison to the next reading:
  lastSensorValue = sensorValue;
}
```

▶ **NOTE: Once you've got the serial connection between the microcontroller and the computer working, you might want to add in the Bluetooth radio from the Monski Pong project in Chapter 2. It will make your life easier if your computer doesn't have to be tethered to the cat mat to finish this sketch.**

▶ Finally, add methods for sending mail and uploading pictures. For now,they'll just print messages to the console pane. After the next section, you'll write code to make it send mail for real. Add these methods to the end of your program.

```
void sendMail() {
  // this function will make an HTTP GET request to send an email
  println("This is where you'd send a mail.");
}

void uploadPicture() {
  // this function will make an HTTP POST request to upload an image:
  println("This is where you'd upload an image.");
}
```

When you run the program, you should see messages in the console area saying when the cat jumps on or off the mat, and when a mail would be sent.

The code may not work for you if the `threshold` variable is too high or too low. Determine the threshold value by watching the sensor value without the cat on the mat, and picking a number that's consistently higher. Make sure it's lower than the highest value you read when the cat is on the mat, however.

Your cat may be fickle or may take his time settling on the mat, which can result in several mail messages as he gets comfortable. To avoid this, change the sketch so that once it sends a message, it doesn't send any others for an acceptable period. You can do this by modifying the `sendMail()` and `uploadPicture()` methods to keep track of when the last message was sent. Here's how to do it.

Tame It Every time the program sends a mail message, it should take note of the time. Add a few new variables at the beginning of the program.

```
int currentTime = 0;      // the current time as a single number
int lastMailTime = 0;     // last time you sent a mail
int mailInterval = 60;    // minimum seconds between mails
int lastUploadTime = 0;   // last time you sent an upload
int uploadInterval = 120; // minimum seconds between uploads
```

Add this line to the `draw()` method to update `currentTime` continuously.

```
void draw() {
  // calculate the time as a single number:
  currentTime = hour() * 3600 + minute() * 60 + second();
}
```

Now, modify the `sendMail()` and `uploadImage()` methods as shown in the code at the right.

You can see that this function is comparing the current time to the previous time to get the difference, similar to the sensor threshold check in `serialEvent()`. The `mailInterval` and `uploadInterval` variables are set to limit the rate of emailing and uploading. This is an example of how to act on the rules of good networking: listen more than you speak. You're listening to sensor data far more frequently than you're speaking to the server.

When you run the sketch this time, trigger the sensor to cross the threshold and you'll see when the sketch will send emails, and when it will upload images.
X

```
void sendMail() {
   // calculate how long has passed since the last mail request:
  int timeDifference = currentTime - lastMailTime;
  if ( timeDifference > mailInterval) {
    println("This is where you'd send a mail.");
    // save the time you mailed for comparison next time:
    lastMailTime = currentTime;
  }
}

void uploadPicture() {
   // calculate how long has passed since the last upload request:
  int timeDifference = currentTime - lastUploadTime;
  if (timeDifference > uploadInterval) {
    println("This is where you'd upload an image.");
    // save the time you uploaded for comparison next time:
    lastUploadTime = currentTime;
  }
}
```

Sending Mail from the Cat

Now that you've got the basic structure of your Processing client in place, it's time to write a web server in node.js combining what you saw earlier in this chapter with the express.js library from Chapter 1 to send mail in response to a GET request.

Your server's going to need to do three tasks:

- Send email in response to a GET request
- Accept image uploads from Processing (you'll write that code in Processing later)
- Serve a static HTML file and the uploaded image file

The following pseudocode explains what you need node.js to do. In order to do those tasks, the script will use some libraries you've seen already, like `crypto` to encrypt and decrypt your email password; `filesystem` to read and write to and from the local filesystem; `nodemailer` to send email; and `express` to structure the web server's routes. It will introduce a new library, `multer`, that will help you manage file uploads to the server.

Set up a new directory for this node.js script called **catServer**. Then make a file in that directory called **catServer.js**.

You're going to need to give this script access to your email password, and you need to secure that password securely. You can use the encryption script you wrote earlier to generate an encrypted file. Call it **info.txt** like you did earlier, then move that file into this directory as well.

Plan It Here's the pseudocode for this server. You can see it follows the same structure as you read about in Chapter 1. You'll include libraries at the top:

```
/*
  cat server
  context: node.js
*/
// include libraries:
//     crypto, filesystem, express, multer, nodemailer
```

Next you'll set up global variables to configure decryption, the server, and the mailer:

```
// configure crypto variables:
//     decryption method
//     location of encrypted file

// configure server variables
//     where static files live
//     how to handle uploaded files, and where they go
//     what types of uploaded files you'll accept

// configure mailer variables
//     the account from which to mail
//     the message to send
```

Then you'll define callback functions to handle incoming client requests:

```
// define callbacks for server routes
//     GET route to send mail
//     POST route to upload images
```

Finally, you'll start the main code running by opening the password file, starting the server, and listening for incoming requests.

```
// open password file and decrypt
// start server
// listen for incoming requests on specific routes
```

Configure It Set up the library includes for your node.js server first. You'll install these using the command-line interface shortly.

```
// include required libraries:
var crypto = require('crypto');          // encryption/decryption tool
var fs = require('fs');                  // filesystem manager
var express = require('express');        // web server
var nodemailer = require('nodemailer');  // emailer
```

Next, set up variables to get the key for the password file from the command-line arguments, like you saw with the **encrypt.js** and **decrypt.js** scripts earlier.

```
// get the key for the password file from the command-line argument:
var key = process.argv[3];
// the file itself is in the script's directory:
var fileName = __dirname + "/info.txt";
// create a decipherer:
var decipher = crypto.createDecipher('aes-256-cbc', key);
```

Then define a server, and use `express.static` to define a subdirectory from which you'll serve static HTML pages, like you did in the **checkAge.js** script.

```
// create a server using express and set up a directory for static files:
var server = express();
server.use('/',express.static('public'));
```

Next you'll define the variables for the nodemailer library like you did before, to define the user account and the message. You'll get the username from the command line just like you did before.

▸▸ Change this to your own mail server's address.

```
// set up mail account:
var account = {
  host: 'smtp.gmail.com',    // mail server
  port: 465,                 // SSL mail port
  secure: true,              // using secure sockets for mail
  auth: {
    user: process.argv[2],  // username from the command line
    pass: ''                // password will come from decryption later
  }
};
```

▸▸ Change this email address to your own.

```
// set up the mail message content:
var message = {
  from: account.auth.user,
  to: 'cat.owner@example.com',
  subject: 'Hello from the cat',
  text: 'The cat is sitting on his mat! http://www.example.com/catcam.html'
};
```

After you've set up your configuration variables, define your callback functions. The first one will handle incoming GET requests for `/mail` by sending mail.

This function is structured like a node.js program in miniature. The real action is at the end, where you create a mail client, then send a message. There's a local callback function, `confirmMail()`, to handle the results of the call to the `sendMail()` function. In JavaScript, unlike Processing (which is Java) or Arduino (which is C), you can define functions inside other functions.

```
// callback function to handle route for mail requests:
function sendMail(request, response) {
  // callback function to confirm mail was sent and inform web client:
  function confirmMail(error, info) {
    if(error){
      console.log(error);
      response.end("Something went wrong. Check the server for details.");
    } else {
      response.send("I sent a mail to " + message.to);
    }
  }

  // create a mail client and send the mail:
  var mailClient = nodemailer.createTransport(account);
  var responseString = mailClient.sendMail(message, confirmMail);
}
```

Next, define a callback function to decrypt the password file. This is basically the same function as you saw in **decrypt.js**. When it successfully decrypts the password, it stores it in the nodemailer `account.auth.pass` variable.

```
// callback function to decipher data from file:
function decryptFile(error, data) {
  // if there's valid data from the file, decrypt it:
  if (data){
    var content = data.toString();
    var decryptedPassword = decipher.update(content, 'hex', 'utf8');
    decryptedPassword += decipher.final('utf8');
    account.auth.pass = decryptedPassword;
  // if the file produces an error, report it:
  } else if (error) {
    console.log(error);
  }
}
```

Finally, start the script actually running by opening and decrypting the password file, then starting the server.

```
// read from the password file:
fs.readFile(fileName, decryptFile);
console.log("credentials for " + account.auth.user + " obtained.");

// start the server:
server.listen(8080);
server.get('/mail', sendMail);
console.log("waiting for web clients now.");
```

Once you've saved this file, open a terminal window and change directories to the **catServer** directory. Then install the libraries using the node package manager like so:

```
$ npm install crypto express nodemailer
```

This will install all the libraries you list on the command line. You can also install them one at a time as you've done before.

After the libraries are installed, run the script:

```
$ node catServer.js your.username@yourmailhost.com key
```

Replace `your.username@yourmailhost.com` and `key` with your email address and the key you used to encrypt the **info.txt** file containing your password. The server should start up with this message:

```
credentials for your.username@yourmailhost.com obtained.
waiting for web clients now.
```

Open a browser and enter the URL http://localhost:8080/mail. You should get a response saying that the server mailed your address, and if you check your email, you'll have a message like so:

```
From: your.username@yourmailhost.com
To: your.username@yourmailhost.com
Subject: Hello from the cat
Date: Wed, 27 Apr 2016 23:30:42 +0000

The cat is sitting on his mat! http://www.example.com/
catcam.html
```

Now that you know the server's successfully sending mail, go back to the Processing sketch to modify it so that it can make this same HTTP request.

Processing doesn't have any libraries for sending and receiving email, but it does have a command to make HTTP GET requests: `loadStrings()`. The same technique can be used to call other web URLs from Processing as well. POST requests are a little more complicated; you'll see how to do that later. For now, make the following modifications to the `sendMail()` function of your Processing sketch to get it to make requests to the server.

Modify It Add a few new global variables at the top of the Processing sketch after the other global variables:

```
/*
  CatCam uploader
 context: Processing
 */
// ... no change to previous global variables...
int uploadInterval = 120; // minimum seconds between uploads
String serverAddress = "localhost";
int port = 8080;
String mailRoute = "/mail";

// ... no change to setup(), draw() or serialEvent()...
```

Replace the `println()` statement in the `sendMail()` function with the lines shown in blue at the right. These lines combine the `serverAddress`, `port`, and `mailRoute` variables to make a full URL, then use `loadStrings()` to make an HTTP GET request to that URL. The last new line will print the response. The `loadStrings()` function returns an array of strings, one for each line of the server's response.

```
void sendMail() {
  // this function will make an HTTP GET request to send an email
  // calculate how long has passed since the last mail request:
  int timeDifference = currentTime - lastMailTime;
  if ( timeDifference > mailInterval) {
    String mailUrl = "http://" + serverAddress + ":" + port + mailRoute;
    String[] response = loadStrings(mailUrl);
    println("results from server:");
    printArray(response);
    // save the time you mailed for comparison next time:
    lastMailTime = currentTime;
  }
}
```

Make sure the server script is still running on your computer, then run your Processing sketch and trigger the force sensor to send a mail like you did before. This time, Processing will contact the server and you should get a response like this in the console pane:

```
Cat just got on mat
results from server:
I sent a mail to tom.igoe@gmail.com
```

You should also get an email from the cat in your mail client. Congratulations! You've closed the loop from the physical action of the cat through the network to your email account. The first major part of the project is done. Now it's time to add a web page and embed a camera image into it.

X

Making a Web Page for the Cat Cam

You need a web page for the cat cam. Your server program is already set up to serve static files and images from a subdirectory called **public**. Any HTML files or images that are placed in this directory are accessible in your browser like so: http://localhost:8080/`filename`. Make the **public** subdirectory in the **catCam** directory. Take a picture of your cat and upload the image to this directory with the filename **catcam.jpg**. When the server's running, you can see it in the browser at http://localhost:8080/catcam.jpg. Now make the following web page to frame it.

Figure 3-13
The CatCam web page in a browser.

Save this file in the **public** directory as **index.html**.

The JavaScript in the head of the page defines a function, `update()`, that updates two page elements: an image and a text div:

Then it calls a function, `setInterval()`, that runs `update()` every minute.

The page will automatically refresh the image and the text in the user's browser every minute. Feel free to make the page as detailed as you want, but keep the JavaScript and the image and text div elements in place.

```
<!DOCTYPE html>
<html>
  <head>
    <script type="text/javascript">
    // callback function to update the image and the text on the page:
    function update() {
      // get the current date and time:
      var now = new Date();
      // get the page elements: the image and the text div:
      var timeLabel = document.getElementById('timeLabel');
      var catPic = document.getElementById('catPic');
      // create a unique URL using the date as a parameter,
      // so the page will force an update:
      catPic.src= "/catcam.jpg?" + now;
      // update the text in the text div:
      timeLabel.innerHTML = now;
    }

    // set an interval to call the update function every minute:
    setInterval(update, 60000);
    </script>
    <title>CatCam!</title>
  </head>
  <body align=center>
    <h1>CatCam!</h1>
    <img src="/catcam.jpg" id="catPic">
    <div id="timeLabel"></div>
  </body>
</html>
```

Uploading Files to a Server Using node.js

Next, you're going to expand the server script so that it accepts image uploads to the server. File uploads are generally made through HTML forms, as HTTP POST requests. When you used a library as middleware to handle form uploads earlier, the middleware expected a particular content type (`application/x-www-form-urlencoded`, in that case) and returned the form elements to you in the request, as `request.body`. Similarly, the library you'll use as middleware this time will expect a different content type, and will return the uploaded file in the request, as `request.file`. Your script will then save the file to a directory in its local filesystem. The middleware library you'll use is called `multer`.

Try It Add an extra line at the top of the **catServer.js** script, after the existing library requires (new lines are shown in blue). Make sure to install multer using npm as well, like so:

```
$ npm install multer
```

```
/*
  cat server
  context: node.js
*/
// ... previous global variables don't change ...
var nodemailer = require('nodemailer'); // emailer
var multer  = require('multer');        // middleware for file uploads
```

After you've established the express server and the static files location, add the following code to configure multer.

The `imgStore` variable establishes where uploaded files will be stored locally on the server, and the name of the callback function that will do the saving.

The `upload` variable is your instance of the middleware, just like the `server` variable is your instance of the server. Once you've declared it, you need to set the type of file you'll accept. The client can only send you files with the same type, as you'll see in Processing soon.

```
var server = express();
server.use('/',express.static('public'));

// set up options for storing uploaded files:
var imgStore = multer.diskStorage({
  destination: __dirname + '/public/', // where you'll save files
  filename: saveUpload       // function to rename and save files
});

// initialize multer bodyparser using storage options from above:
var upload = multer({storage: imgStore});
// set file type: single file, with the type "image"
// (the client must have the same file type for the uploaded file):
var type = upload.single('image');
```

Now you need to write that `saveUpload()` callback function that you declared as part of the `imgStore` variable. Put this function with the other callback functions, after the `sendMail()` function. This will save any uploaded file in the destination directory that you established as part of the `imgStore` variable, using the original name the client gave it.

```
// callback function for file upload requests:
function saveUpload(request, file, save) {
  // this calls a function in the multer library that saves the file:
  save(null, file.originalname);
}
```

You need to establish a server route through which clients can upload files. Set up a call to `server.post()` right after the call to `server.get()` in the last section of your script:

```
server.get('/mail', sendMail);              // send a mail
server.post('/upload', type, getUpload);   // upload a file
console.log("waiting for web clients now.");
```

Finally, write a function to respond to requests for that route. Place this function after the `sendMail()` function.

```
// callback function to handle route for uploads:
function getUpload(request, response) {
  // print the file info from the request:
  var fileInfo = JSON.stringify(request.file);
  console.log(fileInfo);
  response.end( fileInfo + '\n');
}
```

Now your server is ready to accept file uploads. You haven't got a client that can do that yet, however. But you can use the `curl` command that you learned about earlier in this chapter as a client. Here's how to upload a file using `curl`:

```
$ curl  -F image=@path/to/catcam.jpg 'http://localhost:8080/upload'
```

Replace `path/to/catcam.jpg` with the file path to the image you want to upload, and change the server URL as needed as well. Try uploading a new JPEG of the cat, then reload the browser window of the **index.html** page you made earlier. You should see the same page with the new image.

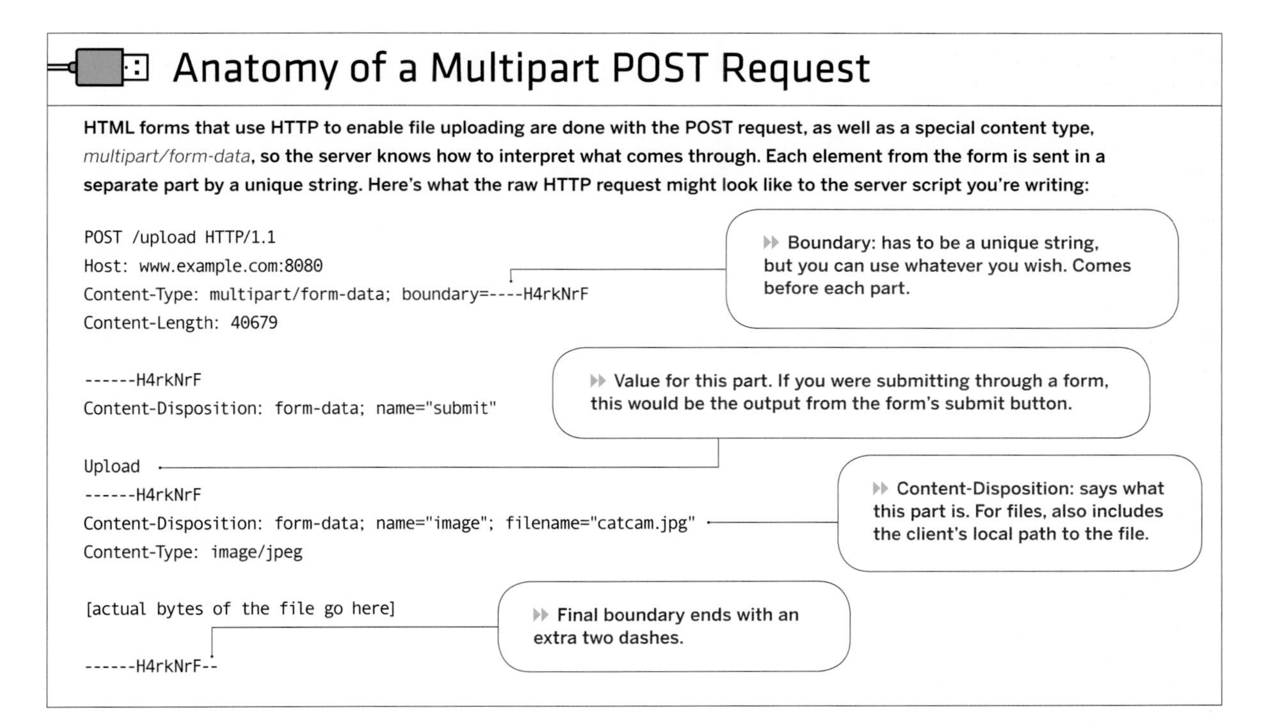

Anatomy of a Multipart POST Request

HTML forms that use HTTP to enable file uploading are done with the POST request, as well as a special content type, *multipart/form-data*, so the server knows how to interpret what comes through. Each element from the form is sent in a separate part by a unique string. Here's what the raw HTTP request might look like to the server script you're writing:

```
POST /upload HTTP/1.1
Host: www.example.com:8080
Content-Type: multipart/form-data; boundary=----H4rkNrF
Content-Length: 40679

------H4rkNrF
Content-Disposition: form-data; name="submit"

Upload
------H4rkNrF
Content-Disposition: form-data; name="image"; filename="catcam.jpg"
Content-Type: image/jpeg

[actual bytes of the file go here]

------H4rkNrF--
```

Boundary: has to be a unique string, but you can use whatever you wish. Comes before each part.
Value for this part. If you were submitting through a form, this would be the output from the form's submit button.
Content-Disposition: says what this part is. For files, also includes the client's local path to the file.
Final boundary ends with an extra two dashes.

Capturing an Image and Uploading It Using Processing

Now that you've got the server accepting uploads, you need to modify your Processing client to capture a picture of the cat and to call the script. There are several automated webcam applications on the market, but it's fun to do it on your own. This section will describe how you can do it in Processing using two external libraries: the Processing video library and the Processing net library. Both libraries work on all three major desktop operating systems now, thanks to updates in Processing 3.0.

Capturing the Image

In order to capture an image, you'll need a camera. If your computer doesn't have a built-in camera, a basic USB camera will do the job. Make sure your operating system can read the camera with the default camera application for your system. On macOS, Photo Booth will do the job. On Windows 10, try Camera. On Ubuntu and other Linux variants, Cheese works well. Once you can see yourself through the camera, you can be confident that Processing will be able to do so as well.

To install the Processing video library, open the Contribution Manager by clicking on the Sketch menu→Import Library→Add Library. Filter for "video" and install the video library by the Processing Foundation. Once it's installed, close the Contribution Manager and you're ready to use the library. It's time to fill in those sections of your sketch's pseudocode that deal with the picture from the camera.

Try It The Processing video library has some parallels in structure to the Serial library. Just like Serial, you need to import it and make an instance of it at the beginning of your sketch.

You can get a list of available video devices using `Capture.list()`. Then to initialize a particular camera, you make a new instance of Capture, just like you did with Serial.

To draw a frame of video to the window, you use the `image()` command. The first parameter, `myCam`, represents the latest frame captured. The second and third parameters are the position of the image in the window, left and top, respectively.

Whenever there's a new frame of video ready from a camera, the library will generate a `captureEvent()` for that camera. Use `myCam.read()` to read the camera image.

Save this sketch as **Camera** and run it. You should see the video image from your camera playing in the sketch window.

```
/*
 Image capture
 Context: Processing
 */
import processing.video.*;     // import the video library
Capture myCam;                  // the camera

void setup() {
  size(400, 300);  // set the size of the window
  // Get a list of cameras attached to your computer
  String[] devices = Capture.list();
  printArray(Capture.list());

  // use the first camera  in the list for capture:
  myCam = new Capture(this, width, height, devices[0]);
  myCam.start();
}

void draw() {
// draw the latest frame captured to the window:
   image(myCam, 0, 0);
}

// this event occurs whenever the camera has a new frame of video ready:
void captureEvent(Capture MyCam) {
  myCam.read();
}
```

Capture It Now that you see how the video library works, incorporate it into your **CatUploader** sketch. Start by adding the video library to the list of imports (new lines are shown in blue as usual). Then add two new global variables at the top of the sketch:

```
/*
  CatCam uploader
 context: Processing
 */
// import libraries
import processing.serial.*;
import processing.video.*;
// ... global variables from earlier sketch go here
Capture myCam;                     // the camera
String fileName = "catcam.jpg";    // local filename for the camera image
```

Add the code at the right to the beginning of the `setup()` function to take control of the camera:

```
void setup() {
   size(400, 300);  // set the size of the window
   // take control of the camera
   // list all available capture devices to the console to find your camera.
   String[] devices = Capture.list();
   printArray(devices);
   // Change devices[0] to the proper number for your camera:
   myCam = new Capture(this, width, height, devices[0]);
   myCam.start();
   // ... rest of the setup from earlier sketch goes here
}
```

Add a line to the `draw()` function to display the latest frame of video:

```
void draw() {
  // calculate the time as a single number:
  currentTime = hour() * 3600 + minute() * 60 + second();
  // draw the camera image to the window
  image(myCam, 0, 0);
}
```

Add the `captureEvent()` function to capture new frames of video:

```
// this event occurs whenever the camera has a new frame of video ready:
void captureEvent(Capture MyCam) {
  myCam.read();
}
```

In the `uploadPicture()` function, replace the line that prints "This is where you'd upload an image" with these two lines:

```
void uploadPicture() {
   // calculate how long has passed since the last upload request:
  int timeDifference = currentTime - lastUploadTime;
  if (timeDifference > uploadInterval) {
     PImage img = get();
    img.save(fileName);
    // save the time you uploaded for comparison next time:
    lastUploadTime = currentTime;
  }
}
```

Save these changes, then run your sketch again. When you trigger the sensors as if the cat is jumping on the mat, the `uploadPicture()` function will save the camera image as **catCam.jpg** in the sketch folder. Check the folder to see the image it saved. The conditional statement in `uploadPicture()` limits the sketch from taking a second picture until the `uploadInterval` has passed. Currently you've got that set to 120 seconds, or two minutes. You can shorten that by changing the value of the `uploadInterval` global variable. But don't set it too low, because you don't need a new camera image every second.

X

It is useful to have a time stamp on your image so you know what time it was taken. To make that happen, add the following lines to the end of the `draw()` function, after the `image()` call:

When you run the sketch this time, your image should have a time stamp as shown in Figure 3-14.

```
  image(myCam, 0, 0);
  // format the time as HH:DD:MM DD-MM-YYYY:
  String timeStamp = nf(hour(), 2) + ":" + nf(minute(), 2)
    + ":" + nf(second(), 2) + " " + nf(day(), 2) + "-"
    + nf(month(), 2) + "-" +  nf(year(), 4);

  // draw a dropshadow for the time text 1 pixel offset from the time:
  fill(15);
  text(timeStamp, 11, height - 19);
  // draw the main time text:
  fill(255);
  text(timeStamp, 10, height - 20);
}
```

Figure 3-14
The output of the catUploader Processing sketch.

Next, it's time to add a method that can make an HTTP POST request. For that, you'll need the Processing network library. The network library allows you to make more complex network connections than `loadStrings()` affords. The net library opens a connection to the server and lets you send through whatever you want, so it's your job to formulate the POST request that will upload the image. Your POST request will look much like what you saw in the "Anatomy of a Multipart POST Request" sidebar earlier. There will be a header with the various descriptive fields, and a body separated into parts by a boundary string. This request will have just one part, for the file, like so:

```
POST /upload HTTP/1.1
Host: localhost:8080
Content-Length: 19006
Content-Type: multipart/form-data; boundary=H4rkNrF

--H4rkNrF
Content-Disposition: form-data; name="image";
filename="catcam.jpg"
Content-Type: image/jpeg

[byte array of the file goes here]
----H4rkNrF--
```

To write this request, you need to know how many bytes are in the body of your request. The body contains both the file itself and the header and footer and boundary strings. So you need to add up all those bytes to know the content length.

Import the net library at the beginning of the sketch with the other imports, and add a few new global variables as well.

```
// import libraries
import processing.serial.*;
import processing.video.*;
import processing.net.*;

// ... global variables from earlier go here
Client thisClient;              // network client
String uploadRoute = "/upload"; // server route for uploads
String boundary = "H4rkNrF";    // boundary string for POST request
```

Now you need a function to formulate and send the actual POST request. To do this, first load the file into an array of bytes. Then open a new Client connection to the server.

```
void postFile() {
  // load the saved image into an array of bytes:
  byte[] thisFile =loadBytes(fileName);
  // open a new connection to the server:
  thisClient = new Client(this, serverAddress, port);
```

Now formulate the request. First add the header and the host address:

```
  String request = "";
  // make a HTTP POST request:
  request += "POST " + uploadRoute + " HTTP/1.1\r\n";
  request += "Host: " + serverAddress + ":" + port + "\r\n";
```

Next, form the header of the body part. This is the part that begins with the boundary, and ends with the content type of the part.

```
  // form the body part header of the request:
  String boundaryHeader = "--" + boundary + "\r\n";
  boundaryHeader +="Content-Disposition: form-data; name=\"image\"; ";
  boundaryHeader += "filename=\"" + fileName + "\"\r\n";
  boundaryHeader +="Content-Type: image/jpeg\r\n\r\n";
```

Next form the tail of the body.

```
  // form the end of the request:
  String boundaryTail ="\r\n--" + boundary + "--\r\n";
```

Now add the length of these two parts to the file length to calculate the whole content length of the request. Add that to the request, then add the content type and a blank line to end the request header.

```
  // calculate and send the length of the total request,
  // including the head of the request, the file, and the boundaryTail:
  int contentLength = boundaryHeader.length()
    + thisFile.length + boundaryTail.length();
  request += "Content-Length: " + contentLength + "\r\n";
  // tell the server you're sending the POST in multiple parts,
  // and send a unique boundary string that will delineate the parts:
  request += "Content-Type: multipart/form-data; boundary=";
  request += boundary + "\r\n\r\n";
```

Finally send the request, the boundary header, the file, and the boundary tail using `thisClient.write()`.

```
  // send the request, the boundaryHeader, the file, and the boundaryTail:
  thisClient.write(request);
  thisClient.write(boundaryHeader);
  thisClient.write(thisFile);
  thisClient.write(boundaryTail);
}
```

»

Now add a call to this new function in the `uploadPicture()` function (new lines are shown in blue).

Continued from previous page.

```
void uploadPicture() {
  // this function will make an HTTP POST request to upload a picture
  // calculate how long has passed since the last upload request:
  int timeDifference = currentTime - lastUploadTime;
  if (timeDifference > uploadInterval) {
    PImage img = get();
    img.save(fileName);
    postFile();
    // save the time you uploaded for comparison next time:
    lastUploadTime = currentTime;
  }
}
```

To see the response from the server, add the lines at the right to the `draw()` loop after the time stamp section that you just added. If `thisClient` receives any bytes from the server, this block of code will print them out as text:

Finally, add a conditional statement that will upload a picture repeatedly as long as the cat is on the mat. Even though this is being called every time through the loop, the conditional statement in `uploadPicture()` that limits the rate of uploads will ensure an upload once every two minutes as long as the cat is on the mat. Add this to the end of the `draw()` function.

```
  text(timeStamp, 10, height - 20);

  // if there is any reply from the server on the POST request,
  // print it:
  if (thisClient != null) {
    String result = "";
    while (thisClient.available() > 0) {
      result += char(thisClient.read());
    }
    if (result != "") println(result);
  }

  // if cat remains on mat, upload picture
  if (catOnMat == true) {
    // if cat remains on mat
    uploadPicture();
  }
}
```

Test the Whole System

To test this code, start the server and open the browser to the Cat Cam page, and then run the sketch. When the cat first triggers the sensor, it should take a picture and upload it, and will send you an email. Check your mail to see whether there's a new message from the cat. When the Cat Cam page refreshes itself, you'll see that image.

As long as the cat stays on the mat, the sketch will upload a new image every two minutes. Watch for a while to see the image change a few times, and then remove the cat from the mat. You should get another message in the console pane telling you the cat is off the mat. See whether the image changes once the cat is off the mat. It shouldn't; the image that remains should be the last one of the cat on the mat. The time stamp in that image will give you an idea when the cat was last on the mat. Notice that the time stamp on the image and the one in the page are not the same. The picture's stamp tells you when it was last taken, and the page's stamp tells you when it was last accessed from the server.

The final piece of the project is to move the server to a public IP address.

X

Going Public with the Server

So far, you've been developing both the Processing client and the node.js server on the same computer, which is why the address for the server is always localhost. That's not much use to you if you want to view the cat cam when you're not at home on your own network. As you saw in Figure 3-7, the server should be running on a remote web host's computer. That way, it has a public IP address and can be accessed from anywhere on the internet.

Many web hosting services feature one-click installation of popular web service frameworks, so setting up your server to run node.js should be a breeze. As mentioned in Chapter 1, your best bet is to opt for a virtual private server (VPS), where you can install anything you want. For example, on Amazon Web Services EC2 platform, there's a configuration called Elastic Beanstalk that you can use to install node.js, Python, Ruby on Rails, and many other web programming frameworks. On Digital Ocean, you create a virtual private server called a droplet, and when you do, there's a one-click install menu that allows you to automatically install a variety of frameworks. Check your web service provider's account dashboard for how to set up node.js on your particular host.

Installing Scripts on Your Server

Once node.js is installed on your server, you run scripts there just like you did on your personal computer, through the command-line interface. To run the **catServer.js** script, for example, make a new directory on your server for the script, then copy the script into that directory. Install the libraries as you saw before:

```
$ npm install crypto express nodemailer
```

The package manager will install all the required libraries. Now you can run your script just like you did on your laptop.

Connecting to a Public Server

Once your script is running on a public server, you need to know the server's address in order to access it, of course, Whether your server has a numeric address, or a named address, just change the `serverAddress` variable in your Processing sketch to the server's address and your client will be able to connect to the server. Similarly, to access the server in a web browser, enter http://server.address:8080 in the address bar (change `server.address` to your server's address).

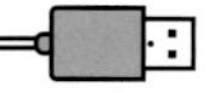

Managing Node.js Libraries using package.json

When your node.js script is dependent on many libraries, you can simplify installing them all by making a special file in your script's directory that describes the script and its dependencies called a *manifest*. The package manager npm has a command to help you generate this file. Once you've installed all your dependent libraries for the first time, type:

```
$ npm init
```

You'll be asked a series of questions, then npm will generate a file in the same directory called package.json. This is the manifest for your program. It lists the name, version, author, licenses, and any dependencies it discovers that you've installed (they live in a folder called node_modules). If you want to copy your script to another computer, you just need to copy the script and the package.json file into the same directory as each other. Then you can install all the dependencies like so:

```
$ npm install
```

The package manager will look up all the dependencies from the package.json file and attempt to install them.

Each node.js project for this book has a package.json file that you can see in the book's online code repository at www.github.com/tigoe/MakingThingsTalk2/tree/master/3rd_edition.

Running Forever

Once you've installed node.js on your web host, you might also want to install a command-line program called *forever* that can keep multiple node.js scripts running for you. You install forever using the node package manager, `npm`. To install it, log into your web host and type:

```
$ sudo npm install -g forever
```

Once npm has installed forever, then you can run your node scripts like so:

```
$ forever start catServer.js
```

You'll get a response like this:

```
warn:    --minUptime not set. Defaulting to: 1000ms
warn:    --spinSleepTime not set. Your script will exit if it does not stay up for at least 1000ms
info:    Forever processing file: catServer.js
```

Once your script is running, you can run other commands, or even log out of your server, and the script will continue to run. You can get a list of which scripts forever is running like so:

```
$forever list
info:    Forever processes running
data:        uid  command             script      forever pid   id logfile                          uptime
data:    [0] kEH1 /usr/local/bin/node catServer.js 1803    1808     /home/username/.forever/kEH1.log 0:0:2:7.956
```

Each script that's running has a process number in square brackets. To stop the **catServer.js** script without stopping others, you'd type:

```
$ forever stop 0
```

When `forever` is running a script, it will send all the output from the script to a log file. You can see all the logs it's currently logging to using:

```
$ forever logs
```

You'll get a response like this:

```
info:    Logs for running Forever processes
data:        script    logfile
data:    [0] catServer.js /home/username/.forever/9QCs.log
```

If you want to view one of the logs, you can use `less`, `tail`, `head`, and the other command-line text programs you learned in Chapter 1. For example:

```
$ tail -10 home/username/.forever/9QCs.log
```

would list the last ten lines of the log listed earlier.

As a general habit, always stop all scripts running when you log out of your server unless you intend to keep them running for a reason. It's good network hygiene not to leave things open or running when you don't have to do so.

X

Figure 3-15
The finished cat bed (at right) and a detail of the sensor pad, which sits under the cat bed itself. A bamboo jewelry box houses the electronics—and matches the furniture. The USB cable runs to the computer. Make sure to secure the wires thoroughly, or the cat may try to chew on them.

Wrap it up

When you get your server running on a web host, and your Processing client communicating with it, and you can view the changing cat cam page in a browser, take a breath and admire your work. You've got all of the pieces of the system working and it's accessible from the internet. You've built your first internet-connected device! Figure 3-15 shows the finished cat bed.

The only public-facing part of your project is the server running on your web host. This is common in internet-connected device projects. Ideally you want to keep the devices that are physically in your private space (generally clients) on a private network, and only provide one public interface through a server. In future projects, always ask yourself what parts of the project are visible on the public internet, and what parts are private. Ultimately the internet is a public space, and anything you make accessible on the internet is accessible to others, whether you realize it or not.

X

Conclusion

Now you have an understanding of the structure of the internet, and how networked applications do their business.

The internet is actually a network of networks, built up in multiple layers. Successful network transactions rely on there being at least one dependable route through the internet from client to server. Client and server applications swap strings of text messages about the files they want to exchange, transferring their files and messages over network ports. To communicate with any given server, you need to know its message protocols. When you do, it's often possible to test the exchange between client and server using a telnet session and typing in the appropriate messages. Likewise, it's possible to write programs for a personal computer or microcontroller to send those same messages, as you saw in the cat bed project. Now that you understand how simple those messages can be, you'll soon get the chance to do it without a personal computer. In the next chapter, you'll connect a microcontroller to the internet directly using an Ethernet interface for the microcontroller itself.

X

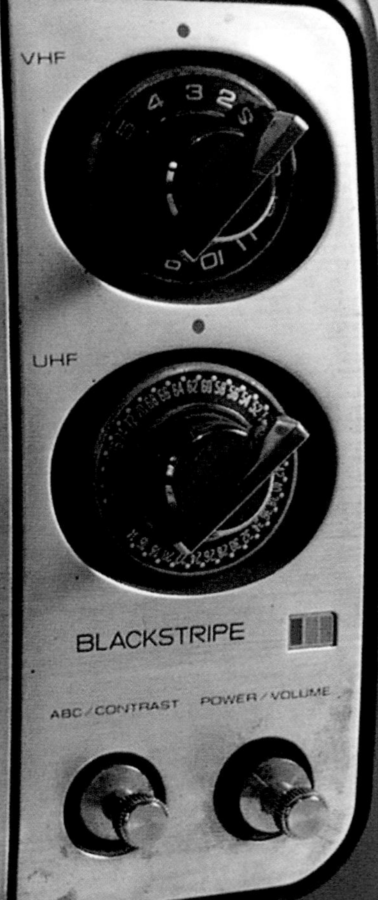
VHF
UHF
BLACKSTRIPE
ABC/CONTRAST
POWER/VOLUME

The Original Celebrated
CURIOUSLY STRONG
PEPPERMINTS
ALTOIDS
NET WT 1.76 OZ (50g)

4

MAKE: PROJECTS

Look, Ma, No Computer! Microcontrollers on the Internet

The first response that comes to many people's minds after building a project like the networked cat bed in Chapter 3 is: "Great, but how can I do this without needing to connect to my computer?" It's cumbersome to have to attach the microcontroller to a laptop or desktop computer just to enable it to connect to the internet. After all, as you saw in Chapter 3, internet message protocols are just text strings, and microcontrollers are good at sending short text strings. When combined with peripheral modules that can connect to the network, they become useful connected devices. In this chapter, you'll learn how to connect a microcontroller to the internet through a peripheral WiFi radio.

« Uncommon Projects' YBox (uncommonprojects.com/site/play/ybox-2) puts RSS feeds on your TV using an XPort serial-to-Ethernet module and a Propeller microchip. *Image courtesy of Uncommon Projects.*

Supplies for Chapter 4

DISTRIBUTOR KEY
A Arduino Store (store.arduino.cc)
AF Adafruit (adafruit.com)
D Digi-Key (www.digikey.com)
F Farnell (www.farnell.com)
J Jameco (jameco.com)
RS RS (www.rs-online.com)
SF SparkFun (www.sparkfun.com)
SS Seeed Studio (www.seeedstudio.com)

PROJECT 6: Hello Internet! Daylight Color Web Server

» **1 Arduino MKR1000** **AF** 3156, **RS** 124-0657, **A** ABX00004, GBX00011 (Africa and EU), **D** 1659-1005-ND Alternatively, an ESP8266-based board will work. **SF** WRL-13231, **AF** 2471
» **1 WiFi connection to the internet** If you access the internet at home with WiFi, you should be all set. Make sure you know the network name and password.
» **3 10-kilohm resistors** **J** 29082, **SF** COM-09939, **F** 350072, **RS** 249-9294
» **3 photocells (light-dependent resistors)** **D** PDV-P9200-ND, **J** 202403, **SF** SEN-09088, **F** 7482280
» **1 solderless breadboard** **D** 438-1045-ND, **J** 20723 or 20601, **SF** PRT-12615 or PRT-12002, **F** 4692810, **AF** 64, **SS** 319030002 or 319030001
» **3 lighting filters** One primary red, one primary green, and one primary blue. Available from your local lighting- or photo-equipment supplier.
» **1 personal computer**
» **All necessary converters to communicate serially from microcontroller to computer** Just like the previous projects.

Figure 4-1. New parts for this chapter: **1.** WiFi-enabled microcontroller. Shown here, the Arduino MKR1000 (l), SparkFun ESP8266 Thing (c), and Adafruit Feather Huzzah! ESP8266 **2.** Photocells **3.** Red, green, and blue lighting filter gels **4.** Voltmeter. You can use an off-the-shelf one, but it's better if you can find one that's antique.

PROJECT 7: Networked Air-Quality Meter

- » **1 Arduino MKR1000** **AF** 3156, **RS** 124-0657, **A** ABX00004, GBX00011 (Africa and EU), **D** 1659-1005-ND. Alternatively, an ESP8266-based board will work. **SF** WRL-13231, **AF** 2471
- » **1 WiFi connection to the internet** If you access the internet at home with WiFi, you should be all set. Make sure you know the network name and password.
- » **1 solderless breadboard** **D** 438-1045-ND, **J** 20723 or 20601, **SF** PRT-12615 or PRT-12002, **F** 4692810, **AF** 64, **SS** 319030002 or 319030001
- » **1 voltmeter** Get a nice-looking antique one if you can. Ideally, you want a meter that reads a range from 0–5V or close to that. **SF** TOL-10285, **F** 1015878, **RS** 244-890
- » **One LED** **D** 160-1144-ND or 160-1665-ND, **J** 34761 or 94511, **F** 1855510, **RS** 228-5972 or 826-830, **SF** COM-09592 or COM-09590
- » **One 220-ohm resistor** **D** 220QBK-ND, **J** 690700, **F** 9339299, **R** 707-7612
- » **1 personal computer**
- » **All necessary converters to communicate serially from microcontroller to computer** Just like the previous projects.

Introducing Network Modules

In the past decade, a wide array of commercial appliances has come on the market that can connect directly to the internet without the aid of a personal computer. All of these contain microcontrollers like the ones you've seen here, attached to a WiFi or Ethernet controller. They all engage in networked communication. Most of them handle only one transaction at a time, requesting information from a server and then waiting for a response, or sending a single message in response to some physical event. They usually act as clients to a remote server, but some can also act as servers on your local network.

Companies like D-Link, Sony, Axis, and others make security cameras with network interfaces, both Ethernet and WiFi. Nest connects your thermostat and smoke alarm to the internet. Philips makes it possible to control your lights on your home network. Belkin and others make networked power strips, garage door openers, and other connected appliances. The principles behind these products, and the controllers inside them, are the same ones you're learning in this book.

It's possible to write a program for a microcontroller that can manage all the steps of network communication, from the physical and data connections to the network address management to the negotiation of protocols like SMTP and HTTP. A code library that encompasses all the layers needed for network connections is called a *network stack*. However, it's much easier to use a network interface module to do the job. There are many such modules on the market, with varying prices and features. These are basically microcontrollers that have a network stack programmed into their firmware already, and present you with a serial interface to your own controller, either asynchronous serial or synchronous. They act basically as serial-to-WiFi or serial-to-Ethernet modems. They work much like the Bluetooth modem you used in Chapter 2, but with differing interface protocols. Earlier editions of this book used wired Ethernet modules extensively, but in recent years, wireless Ethernet (WiFi) has become so ubiquitous, and the modules have dropped so steeply in price, that WiFi is now the default.

Two WiFi Modules Compared

Many WiFi modules have libraries for Arduino-compatible microcontrollers. This and the chapters that follow feature two that have been mentioned already in Chapter 1: Atmel's WINC1500 WiFi module, and Espressif's ESP8266. The WINC1500 is the WiFi module on the MKR1000 board. This module is also featured in the Arduino WiFi101 shield, and in Adafruit's Feather M0 WiFi WINC1500 board. The ESP8266 module is at the heart of Adafruit's Huzzah! ESP8266 boards and SparkFun's Thing boards, among others. Figure 4-2 shows them all.

The WINC1500 is the more expensive of the two modules, but features built-in encryption capabilities, handy when you want to access servers using secure connections. Since the WINC1500 is running just the TCP/IP stack, it is usually interfaced to a microcontroller on the same board. The MKR1000 and the Feather M0 WINC1500 boards both have an Atmel Cortex M0 processor on board that interfaces to the WiFi radio using SPI.

The ESP8266 combines the microcontroller and radio in one. It runs its own real-time operating system (RTOS), and can be programmed using a number of different programming frameworks, including Arduino. The microcontroller doesn't expose as many GPIO pins as the MKR1000 or the Feather M0 for WINC1500, though.

The ESP8266 has no serial-to-USB interface, just a TTL serial interface. However, SparkFun's Thing Development board adds an FTDI USB-to-serial interface, and Adafruit's Feather ESP8266 Huzzah! adds a Silicon Labs CP2104 interface. You'd need the drivers for them, as mentioned in the "USB-to-Serial Converters" section in Chapter 2.

The on-board antenna for the SparkFun Thing and Thing Dev board does not appear to have as much range as the one on the Adafruit board, so it's a good idea to use an external antenna if you're using the SparkFun board. SparkFun has a good guide to the ESP8266 boards at learn.sparkfun.com/tutorials/esp8266-thing-hookup-guide/installing-the-esp8266-arduino-addon.

The default board for this chapter is the MKR1000, but you can use any of the three boards mentioned here with minimal code changes. You should consult your board's documentation to upload code from the Arduino IDE if you aren't using the MKR1000.

The WFi101 library, which can control the WINC1500 module, and the ESP8266 WiFi library for Arduino are mostly compatible with each other. This means that code you write for the MKR1000 and code you write for the ESP8266 can be almost identical. In this chapter, usually all you'll need to change are the library includes at the top of your code. Those sketches that do not work on the ESP8266, or that need modification, are noted as you go. You may have to change the inputs or outputs, however, since the ESP8266 boards do not have the same number of GPIO or analog pins as the MKR1000.

These modules do not work on 5GHz 802.11n networks and do not have software APIs for connecting to enterprise networks.

X

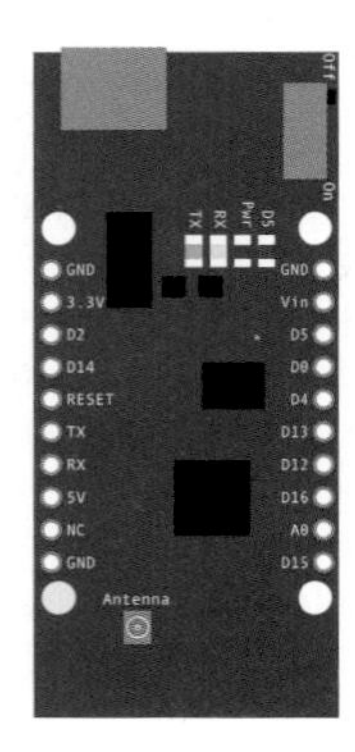

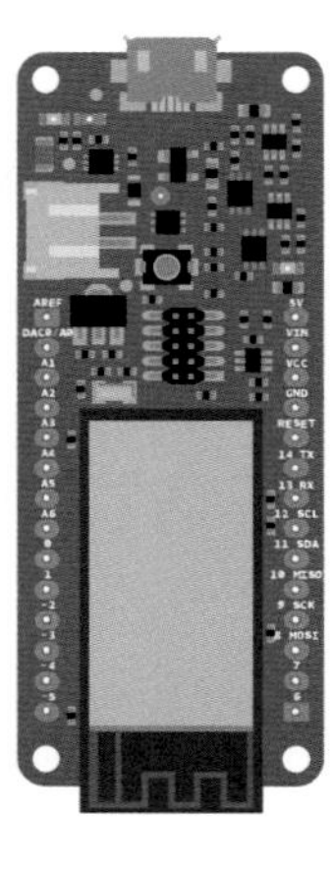

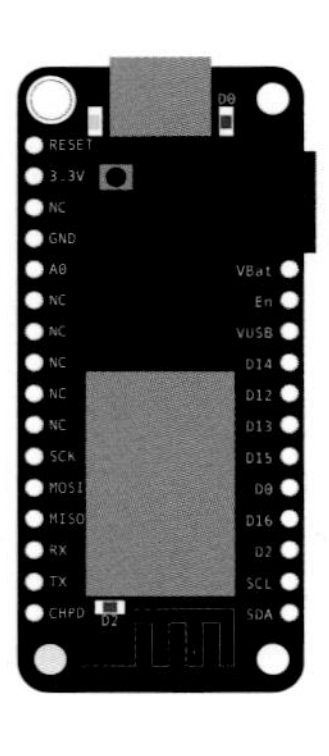

Figure 4-2
The SparkFun Thing Dev board (left), the MKR1000 (center), and the Adafruit Feather Huzzah! ESP8266 (right). You can use any of these for the WiFi projects in this chapter. The Feather Huzzah! WINC1500 should work as well. The MKR1000 has more GPIO and analog pins than the others. The MKR1000 and the Feather have on-board battery chargers.

 **Project 6**

Hello Internet!

You know from the last chapter that networked devices can act as clients or servers. In this project, you'll make a very simple web server on your microcontroller board that serves a web page whose background color changes with the color of the light where the board is located.

MATERIALS

- **1 Arduino-compatible, WiFi-enabled microcontroller:**
  - **MKR1000, or ESP8266-based board**
  - **Features used: WiFi, Analog Input, UART**
- **1 WiFi connection to the internet**
- **3 10-kilohm resistors**
- **3 photocells (light-dependent resistors)**
- **1 solderless breadboard**
- **3 lighting filters: red, green, and blue**

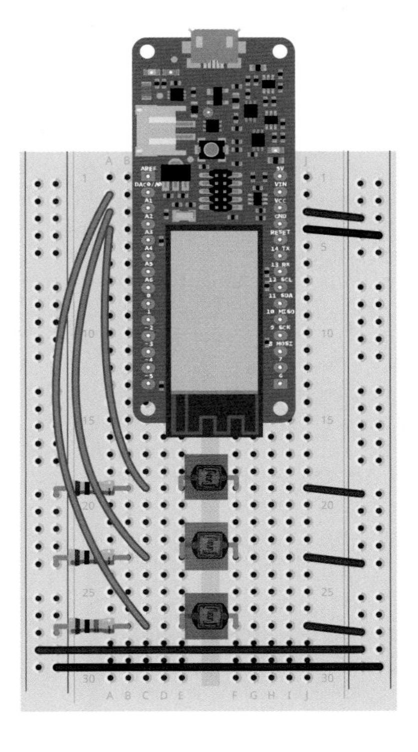

Figure 4-3
The RGB server circuit connected to an MKR1000. The schematic is shown below. Note the three color filters over the photocells.

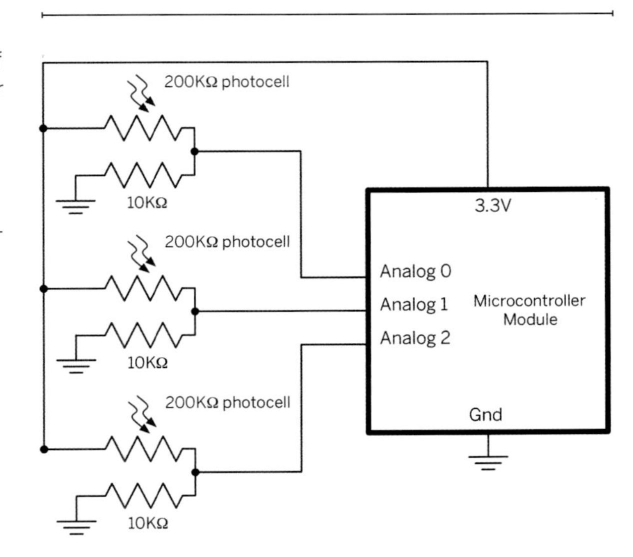

Making the Connections

The first step to this project is to make sure you have the correct board and libraries included in the Arduino IDE. Check the Tools→Boards menu to see that the MKR1000 is present, and import it through the Boards Manager if it's not. Then check the Sketch→Include Library menu to see if the WiFi101 library is included, and use the Library Manager to import it if not. If you're using an ESP8266 board, open the Arduino→Preferences menu and add the following the Additional Boards Manager URLs field: http://arduino.esp8266.com/stable/package_esp8266com_index.json. Then open the Boards Manager and look for the board definition from ESP8266 Community. This includes both the SparkFun and Adafruit boards.

To read the color of the ambient light, use three photocells and cover each one with a different color of lighting filter: red, green, and blue. You can get these at many lighting-supply stores, photo supply-stores, or art stores, or you can use any translucent colored plastic you have around the house. You don't have to be scientific here. Figure 4-3 shows the connection between the MKR1000 board and the three photocells.

How the WiFi101 Library Works

To begin, you're going to try a couple of simple programs using the WiFi101 library for Arduino (the principles described here apply to the ESP8266 WiFi library as well). This library lets you control the WINC1500 module using methods similar to those you use for printing to a serial port. There are two classes in the WiFi101 library that you'll use: `Server` and `Client`. A server waits for connections from remote devices on the internet and allows them to connect to the Ethernet module, just like regular servers. A client initiates a connection to remote devices, makes requests, and delivers the replies. Both servers and clients can be read from and written to, using the `read()`, `write()`, `print()`, and `println()` commands from the Stream library that underlies the Serial library as well as `Client` and `Server`. The `Client` class also has a `connect()` function to connect to a server, a `connected()` command that tells you whether it's connected, and a `stop()` command to disconnect. You'll see it in this project when a client connects to the server, and it will be useful later, when you're trying to connect to a remote server to get data. The ESP8266 WiFi library, which also uses Stream, includes these same classes and functions.

Before you can write your program, you need to establish some basic information about how your WiFi module will connect to the internet. You'll need to know the *Service Set Identifier (SSID)* of your WiFi access point (your router). This is sometimes called the network name, or LAN name. You also need to know its encryption type, and its password. You should also know your WiFi module's MAC address as well, since some networks require you to enter it in the router's configuration. If your router is filtering by MAC address, and you have admin access to it, look for the "MAC Address Filtering" option in the router's admin interface and add your module's MAC address to the list. You can get the address using the sketch on the following page.

If you don't know your WiFi access point's name or your device's address, you might need to scan for available networks. The WiFi101 library has a `scanNetworks()` function that lets you do that. The sketch on the following page shows you how to scan for available networks. It also prints out the MAC address of your WiFi module.

Because there's no physical connection between your WiFi module and your network, it can be tricky to diagnose when there's a problem. At the end of this chapter, you'll find some helpful troubleshooting tips and diagnostic tools. For now, if you can scan for networks and you know the network SSID and password to connect to your network, you're all set.

X

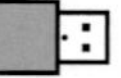

The Stream Library for Arduino

The communications protocols in the Arduino libraries, including Serial, SPI, I2C, and the WiFi `Client` and `Server` classes you're about to use are supported by a useful library that you never see directly: the *Stream* library, which lets you treat the connection between two computers as a data stream.

In Chapter 1 you read that data streams are like tubes of bytes, where the first byte in one end is the first byte out the last end. When you write a byte (for example, using `Serial.write()`), the byte is pushed into the stream. When you read a byte (for example, using `Serial.read()`) the byte is removed from the other end of the stream.

The Stream library includes the `read()`, `write()`, `available()`, `print()`, and `println()` functions, which you've seen already. They work the same in other libraries as they do in Serial.

Stream also contains functions based on Michael Margolis's TextFinder library that help you search the stream. The `readString()` function lets you read what's available in the Stream into a String. The `readBytesUntil()` function lets you read until you see a particular character, like a newline: `readBytesUntil('\n')`. The `readStringUntil()` function is similar, but returns the result in a String. Stream also includes some useful functions for parsing data. The `find()` and `findUntil()` functions let you search the stream until you find a particular substring. The `parseInt()` and `parseFloat()` functions search for the next string of numeric characters and converts them to their numeric values. These will be particularly useful for parsing the result from a network transaction like HTTP. Watch for Stream functions throughout the book. For more on Stream, see the language reference pages at www.arduino.cc/en/Reference/HomePage.

Scan It To get started, you need to include the SPI library and the WiFi101 library to control the WiFi module. Initialize serial communication so you can print out the results of the scan. If you're using an ESP8266-based board, include `<ESP8266WiFi.h>` instead of `<WiFi101.h>`.

In the `loop()` function, call two functions, one to print your WiFi module's MAC address, and one to scan for available WiFi networks. Don't scan too frequently, as that will affect the quality of the WiFi signal for others using these networks.

When you need to know the MAC address of your WiFi module, you can call the `WiFi.macAddress()` function and get the result in a 6-byte array. The `printMacAddress()` function shows how to print it out.

> **The ESP8266WiFi library has the MAC address reversed. Loop from 0 to 5 instead of 5 to 0 here.**

The `WiFi.scanNetworks()` function will return the number of networks available. You can get the network name, signal strength, and encryption type as well.

The `WiFi.encryptionType()` function returns a number corresponding to the encryption type as follows:

1. WEP
2. WPA
3. WPA2
4. None
5. Auto

The ESP8266WiFi library will only return 1 (encrypted) or 0 (no encryption).

```
/*
  network scan
  context: Arduino, with WINC1500 module
*/
#include <SPI.h>
#include <WiFi101.h>

void setup() {
  Serial.begin(9600);           //Initialize serial communications
}

void loop() {
  printMacAddress();            // Print WiFi MAC address
  listNetworks();               // scan for existing networks
  delay(10000);                 // wait 10 seconds until next scan
}

void printMacAddress() {
  Serial.print("MAC address: ");
  byte mac[6];                  // array to hold the MAC address
  WiFi.macAddress(mac);         // get MAC address
  for (int i = 5; i > 0; i--) { // loop from 5 to 1
    if (mac[i] < 0x10) {        // if the byte is less than 1 hex digit
      Serial.print("0");        // add a leading 0
    }
    Serial.print(mac[i], HEX);  // print byte of MAC address
    Serial.print(":");          // print colon
  }
  Serial.println(mac[0], HEX);  // println final byte of address
}

void listNetworks() {
  Serial.print("Scanning for available networks:");
  int numSsid = WiFi.scanNetworks();
  if (numSsid == -1)  {
    Serial.println("Couldn't get a wifi connection");
    return;     // return to the main loop
  }
  // print the number of available networks:
  Serial.println(numSsid);
  // print the network number and name for each network found:
  for (int thisNet = 0; thisNet < numSsid; thisNet++) {
    String message = WiFi.SSID(thisNet);
    message += "\tSignal Strength: ";
    message += WiFi.RSSI(thisNet);
    message += " dBm \tEncryption:";
    message += WiFi.encryptionType(thisNet);
    Serial.println(message);
  }
  Serial.println();
}
```

Plan It The most basic web server you could write needs to do the following:

- Connect to a network
- Wait for clients
- Send an HTTP response

Once your server can do that much, you can add functionality to respond to specific requests. The pseudocode for that basic server appears to the right.

```
/*
  Web Server
  Context: Arduino, with WINC1500 module
*/
// include libraries and configuration details

void setup() {
  // initialize serial communications
  // while you're not connected to a WiFi AP,
  //   try to connect
  // When you're connected, start the server
  // and print out the device's network status
}
```

As you know from the last chapter, all HTTP client requests end with a blank line. So when the client sends a blank line, your server knows it's time to respond. At the very least, your server should respond with an HTTP code followed by a blank line like so:

```
HTTP/1.1 200 OK
```

```
void loop() {
  // listen for incoming clients
  // while the client is connected,
  //   and there are incoming bytes to read,
  //   read the incoming bytes line by line
  //   print each line
  //   if the line is a blank line (\n or \r\n)
  //     send an HTTP response
  //     give the server time to get the data
  //     if the client's still connected
  //        disconnect the client
}
```

You're going to have to add your network name and password to your sketch. As you saw in the last chapter, it's never a good idea to include this information in your program. The best way to separate it in Arduino is to make a new tab and use an `#include` statement. To do this, click the new tab icon on the top-left corner of the IDE, as shown in Figure 4-4. Then give the new tab a name, such as **config.h**. This will create a new file in your sketch folder with that name. You can include it in your sketch by adding `#include "config.h"` at the top of your sketch.

Once you've made a new tab, put the following two lines in it:

```
char ssid[] = "ssid";      //  your network SSID (name)
char password[] = "s3c3r3+!"; // your network password
```

Replace `ssid` and `s3c3r3+!` with your SSID and its password. Now you can copy just your sketch and share it with others without fear of sharing your password. They will have to make their own **config.h** file, however.

Putting password information in a **config.h** file is the default in Arduino sketches for the rest of this book.

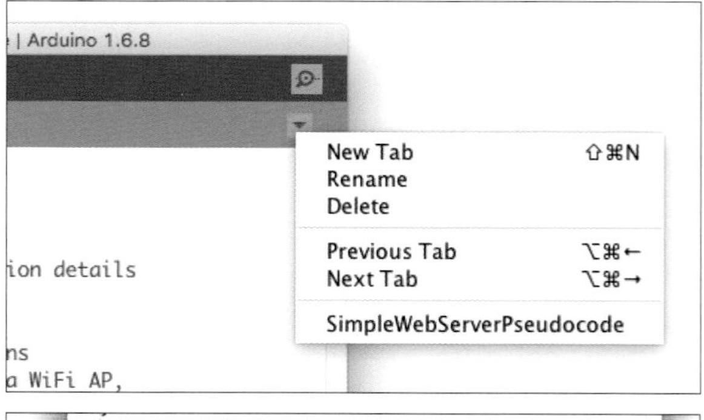

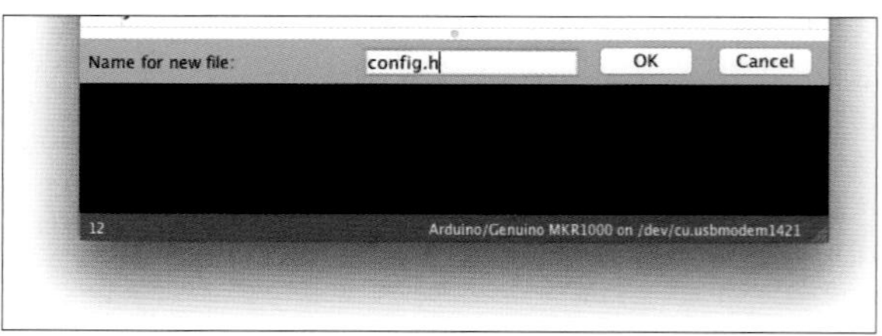

Figure 4-4
Making a new tab in the Arduino IDE.

Try It To get started, include the necessary libraries and your **config.h** file. You also need to initialize a variable to hold a server instance. The server will run on port 80, just as most web servers do.

In the `setup()`, check the WiFi module's connection status and keep trying to connect every two seconds until it's connected. The WiFi101 library can use two forms of encryption, *Wired Equivalent Privacy (WEP)*, or *WiFi Protected Access (WPA* or *WPA2)*. It can also connect to open networks. You choose the encryption scheme based on how you call the `WiFi.begin()` function. For WPA and WPA2, use `WiFi.begin(ssid, password);` as shown here. For no encryption, leave out the password. WEP is less secure and not recommended.

Once you're connected, start the server and print out the module's IP address so you know the URL to which to connect.

```
/*
  Web  Server
  Context: Arduino, with WINC1500 module
*/
#include <SPI.h>
#include <WiFi101.h>
#include "config.h"          // holds ssid[] and password[] arrays

WiFiServer server(80);        // make an instance of the server class

void setup() {
  Serial.begin(9600);// initialize serial communications
  // while you're not connected to a WiFi AP:
  while ( WiFi.status() != WL_CONNECTED) {
    Serial.print("Attempting to connect to Network named: ");
    Serial.println(ssid);
    WiFi.begin(ssid, password);  // try to connect
    delay(2000);                 // wait 2 seconds before trying again
  }

  server.begin();      // When you're connected, start the server

  // print out the device's network address
  Serial.print("To see this device's web interface, go to http://");
  IPAddress ip = WiFi.localIP();
  Serial.println(ip);
}
```

In the main loop, you'll spend all your time listening for a connection from a remote client. When you get a connection, you'll wait for the client to make an HTTP request.

When you get the blank line at the end of the request, give the client time to read the response, then disconnect the client.

```
void loop() {
  // listen for incoming clients
  WiFiClient client = server.available();
  while (client.connected()) {     // while the client is connected,
    if (client.available()) {      // and there are incoming bytes to read,
      // read the incoming data line by line:
      String request  = client.readStringUntil('\n');
      Serial.println(request);             // print the line
      // if the request is a blank line (\n or \r\n):
      if (request.length() <= 2) {
        client.println("HTTP 200 OK\n"); // send an HTTP response
        delay(10);                   // give the server time to get the data
        if (client.connected()) {  // if the client's still connected
          client.stop();           // disconnect the client
        }
      }
    }
  }
}
```

Run this sketch with the Serial Monitor open. Then open a browser window and go to the WiFi module's address as printed out in the Serial Monitor. You won't see anything in the browser, but you will see the HTTP request come through in the Serial Monitor. Now you're seeing what the server saw when you made HTTP requests in Chapter 3. A typical request will look like this (this request was made using `curl`):

```
GET / HTTP/1.1
Host: 192.168.43.184
User-Agent: curl/7.46.0
Accept: */*
```

> ⚠ **This sketch uses three analog channels, which the ESP8266 does not have.**

You can ignore most of the HTTP parameters, but two pieces are useful: the first line, where you see what the client is requesting; and the end, where it sends a blank line to finish the request. In this case, the first line shows that the client is asking for the main index page at the root of the server, signified by the / in `GET /`. In this project, you're only going to have one result to return, but in the future, you might write more complex servers that look at this part of the request and respond with different results depending on what's requested.

If you want to see something in a browser when you access this server, you're going to need to return some HTML. Here's how:

Add a function, `makeResponse()`, at the end of your sketch to make a String to send the client. The beginning of the string contains the HTTP header, and ends with a blank line (the `\n\n` represents two newline characters). Then comes the beginning of the HTML document.

The body of the HTML document prints out the values of each of the analog inputs to the MKR1000. If you're using a microcontroller with fewer analog inputs, change the value of `analogChannel` appropriately.

Finally, add the closing `<body>` and `<html>` tags and a blank line to end the HTML document, and return the String.

```
String makeResponse() {
  String result = "HTTP/1.1 200 OK\n";      // HTTP header
  result += "Content-Type: text/html\n\n"; // content type and end of header
  result += "<!doctype html>\n";           // HTML document type
  result += "<html><head><title>";         // HTML, head, title tags
  result += "Hello from Arduino</title></head>"; // end of HTML head
  result += "\n<body>\n";                   // beginning of body

  // output the value of each analog input pin
  for (int analogChannel = 0; analogChannel < 6; analogChannel++) {
    result += "analog input ";
    result += analogChannel;
    result += " is ";
    result += analogRead(analogChannel);
    result += "<br />\n";
  }
  result += "</body></html>\n\n";           // end of body and blank line
  return result;
}
```

Replace the `client.println()` statement in the `loop()` function with the following two lines, shown in blue. Then run this sketch and open its address in a browser.

```
  if (request.length() <= 2) {
    //client.println("HTTP 200 OK\n"); // send an HTTP response
    String response = makeResponse();
    client.println(response);
    delay(10);             // give the client time to get the data
```

When you enter the Arduino's address in your browser now, you'll get a web page. There's not much there, but it's legitimate enough that your browser doesn't know the difference between your Arduino and any other web server. You can add anything you want in the HTML in the `makeResponse()` method—even links to images and content on other servers (you can't easily serve images directly from the microcontroller).

Think of the Arduino as a portal to any other content you want the user to see, whether it's physical data from sensors or other web-based data. The complexity is really up to you. Once you start thinking creatively with the contents of the `makeResponse()` method, you get a wide range of ways to dynamically generate a web interface to your Arduino through the WiFi module. Here are some examples:

Reload the page, and you've got the states of the analog inputs. But how about updating them continually? Add the following line to `makeResponse()` after the line that prints the closing `</head>` tag. Then reload the page in your browser. Now the page will automatically reload itself every three seconds.

```
result += "Hello from Arduino</title></head>"; // end of HTML head
result += "<meta http-equiv=\"refresh\" content=\"3\">";
```

You can do all sorts of things in the response. Now it's time to take advantage of those three photocells you attached to three of the analog inputs. Replace the for loop in the `makeResponse()` method that prints the analog values with the code at right. This prints out a page that changes its background color with the values from the three photocells.

Now that you've got a light color server, look for colorful places to put it.

```
  result += "\n<body>\n";                  // beginning of body

  // set up the body background color tag:
  result += "<body bgcolor=#";
  // read the three analog sensors:
  int red = analogRead(A0) / 4;
  int green = analogRead(A1) / 4;
  int blue = analogRead(A2) / 4;
  // print them as one hexadecimal string:
  result += String(red, HEX);
  result += String(green, HEX);
  result += String(blue, HEX);
  // close the tag:
  result += ">";
  // now print the color in the body of the HTML page:
  result += "The color of the light on the Arduino is #";
  result += String(red, HEX);
  result += String(green, HEX);
  result += String( blue, HEX);

  result += "</body></html>\n\n";          // end of body and blank line
  return result;
}
```

Making a Private IP Device Visible to the Internet

Up until now, the internet-related projects in this book either worked only on a local subnet, or have only sent data outbound and waited for a reply. This is the first project in which you might want your device to be accessible to the internet at large. You can view it while you're on the same local network, but if it's connected to your home router and has a private IP address, it won't be visible to anyone outside your home. To get around this, you need to arrange for one of your router's ports to forward incoming messages and connection requests to your device.

To do this, open your router's administrator interface and look for controls for "port forwarding" or "port mapping." The interface will vary depending on the make and model of your router, but the settings generally go by one of these names. Port 80 on the router is used for the admin interface, so configure it so that port 8080 on your router connects to port 80 on the WiFi module (if your router allows it). Once you've done this, any incoming requests to connect to your router's public IP address on that port will be forwarded to the WiFi module's private IP address on the port that you set.

Note that the `router` reserves some ports for special purposes. For example, you won't want to port forward port 80, because the router uses it for its own interface. That's why you might use a high number, like 8080. It's no coincidence that you've been using this number for the port number in your node.js servers.

Web browsers default to making their requests on port 80, but you can make a request on any port by adding the port number at the end of the server, address like so:

```
http://www.myserver.com:8080/
```

The new public address of your WiFi module will follow this pattern, too. For example, if your home router's public address is `203.48.192.56`, you'd access your new color server at `http://203.48.192.56:8080`.

Figures 4-5 and 4-6 show the settings on an Apple AirPort Express router and a Linksys wireless router. On the Linksys router, you can find port forwarding under the Advanced tab.

> Think carefully before you enable port forwarding to any device! You're allowing anyone on the internet to access a device in your home when you do this. Depending on what that device controls (your locks, your lights, etc.), this can be dangerous.

Figure 4-5
Port mapping tab on an Apple AirPort Express router. Port mapping can be found under the Network tab.

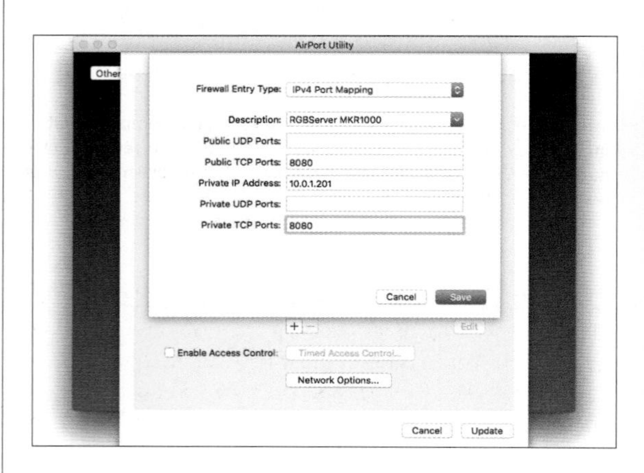

Figure 4-6
Port forwarding on a Linksys wireless router.

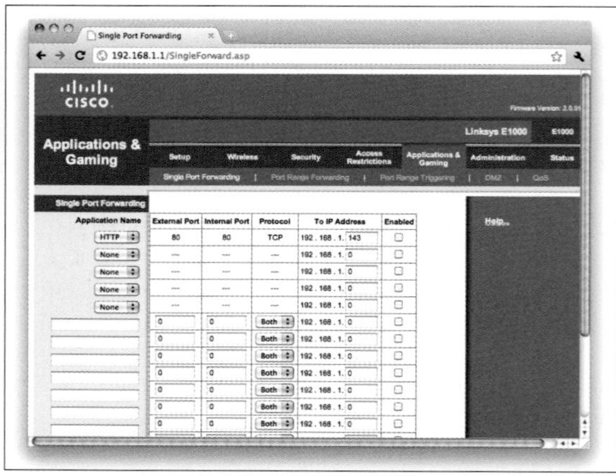

An Embedded Network Client Application

Now that you've made your first server, it's time to make a client. This project is an embedded web client. It requests data from an existing website that tracks the Air Quality Index (AQI) and uses it to affect a physical output, an analog meter.

Project 7

Networked Air-Quality Meter

Web pages aren't the only way to represent data. We have clocks, barometers, and thermometers in our houses as both decorative and information appliances. To make a physical device display data from the internet, the device must be able to connect to a server, request data, and parse the result into a number it can use to control its output devices.

MATERIALS

- **1 Arduino-compatible, WiFi-enabled microcontroller:**
  - **MKR1000, or**
  - **Uno-layout board with WiFi101 shield or ESP8266-based board**
  - **Features used: WiFi, GPIO, PWM, UART**
- **1 WiFi connection to the internet**
- **1 solderless breadboard**
- **1 voltmeter**
- **1 LED**
- **1 220-ohm resistor**

For this project, you'll need an analog voltmeter, like the kind you find in speedometers and audio VU meters. I got an antique model at a yard sale, but you can often find them in electronics surplus stores or junk shops. The model recommended in the parts list is less picturesque than mine, but it will do for a placeholder until you find one you love.

The microcontroller will make a network connection to a website through the WiFi module. Then it will read through the response for the appropriate number, and use that number to set the meter using pulsewidth modulation (PWM) via the `analogWrite()` function.

You could request the web page directly and parse through the HTML to find the number you need (this is called *web scraping*), but fortunately most data-driven websites now have an interface that lets you request the data directly, without the presentation formatting of HTML or CSS. Such an interface is called an *application programming interface (API)*.

The website you'll use is AirNow, www.airnow.gov, the U.S. Environmental Protection Agency's site for reporting air quality. It reports hourly air quality status for many U.S. cities, listed by ZIP code. You can use it to set a meter in your home or office to see the current air quality in your city. assuming you live in the United States.

Note: If you don't live in the United States, the company that built AirNow, Sonoma Technologies, is expanding the service internationally. Called AirNow-I, it's available in China, Taiwan, and parts of Mexico as of this writing. See www.sonomatech.com/project.cfm?uprojectid=1102 **or contact them for more information.**

Control the Meter Using the Microcontroller

First, you need to generate a changing voltage from the microcontroller to control the meter. Microcontrollers can't output analog voltages, but they can generate a series of very rapid on-and-off pulses that can be filtered to give an average voltage. The higher the ratio of on-time to off-time in each pulse, the higher the average voltage. This technique is called *pulse-width modulation (PWM)*. In order for a PWM signal to appear as an analog voltage, the circuit receiving the pulses has to react much more slowly than the rate of the pulses. For example, if you pulse-width modulate an LED, it will seem to be dimming because your eye can't detect the on-off transitions when they come faster than about 30 times per second. Analog voltmeters are very slow to react to changing voltages, so PWM works well as a way to control these meters. By connecting the positive terminal of the meter to an output pin of the microcontroller, and the negative pin to ground, and pulse-width modulating the output pin, you can easily control the position of the meter.

Figure 4-7 shows the whole circuit for the project. In addition to the meter, there are two LEDs in this circuit, one built into the MKR1000 module and the other on digital pin 4. You'll use these to show the status of the device.

Test It The program to the right tests whether you can control the meter.

You will need to adjust the range of `pwmValue` variable depending on your meter's sensitivity. The meters used to design this project had different ranges. The meter in the parts list responds to a 0-to-5-volt range, so the preceding program moves it through its range, since the MKR1000's maximum output voltage is 3.3V. The antique meter, on the other hand, responds to 0 to 3 volts, so it was necessary for me to limit the range of `pwmValue` to 0–230. When it was at 230, the average output voltage was about 3V and the meter reached its maximum. Note your meter's minimum and maximum values. You'll use them later to scale the air quality reading to the meter's range.

```
/*
    Voltmeter Tester
    Uses analogWrite() to control a voltmeter.
    Context: Arduino, with  WINC1500 module
*/
const int meterPin = 5;

void setup() {
  Serial.begin(9600);
}

void loop() {
  // move the meter from lowest to highest values:
  for (int pwmValue = 0; pwmValue < 255; pwmValue ++) {
    analogWrite(meterPin, pwmValue);
    Serial.println(pwmValue);
    delay(10);
  }
  delay(1000);
  // reset the meter to zero and pause:
  analogWrite(meterPin, 0);
  delay(1000);
}
```

⚠ **You will need to use a different digital out pin for the meter on the ESP8266.**

Request the Data Through an API

Figure 4-8 shows AirNow's page for New York City (www.airnow.gov/?action=airnow.local_city&zipcode=10003&submit=Go). It gives you a good idea of what data is available. You need to get the data from AirNow's site in a form the microcontroller can read, though. The microcontroller can read short strings serially, and convert those ASCII strings to a binary number. Using a microcontroller to parse through all the text of a web page is possible, but complicated. However, AirNow, like many sites, gives you a simpler way to get the data through their web-based API at docs.airnowapi.org. You'll need to create a user account, but doing so is free, and allows you up to 500 requests per hour from the API.

A typical web-based API like AirNow's is just a website that returns data in formats other than HTML. You make requests to the API using HTTP, and you state your parameters in your request, much like you saw in the age checker in Chapter 3. If the API supports different data

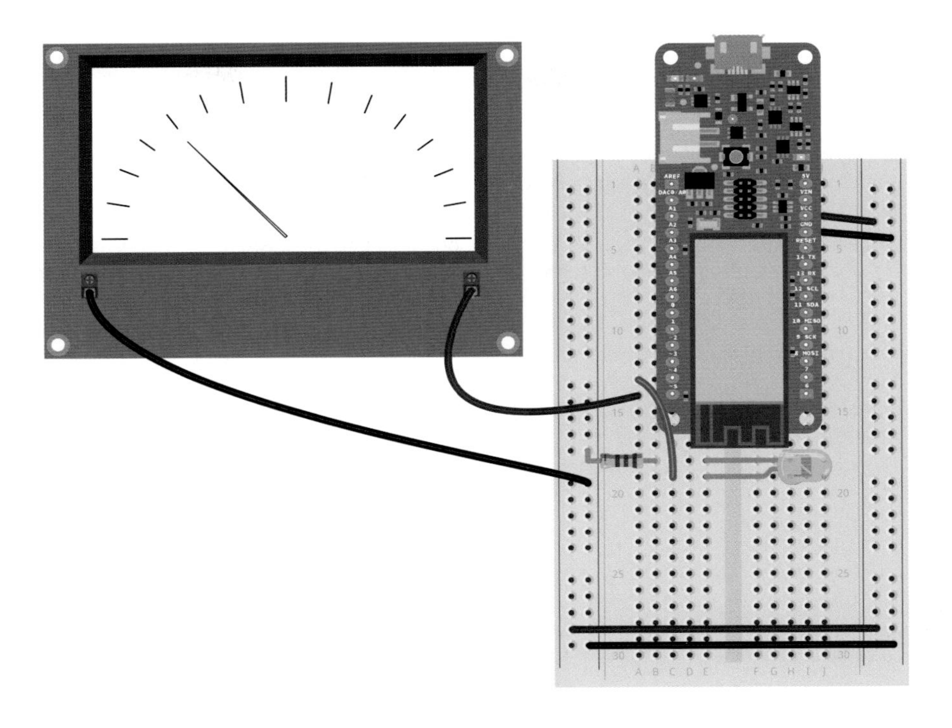

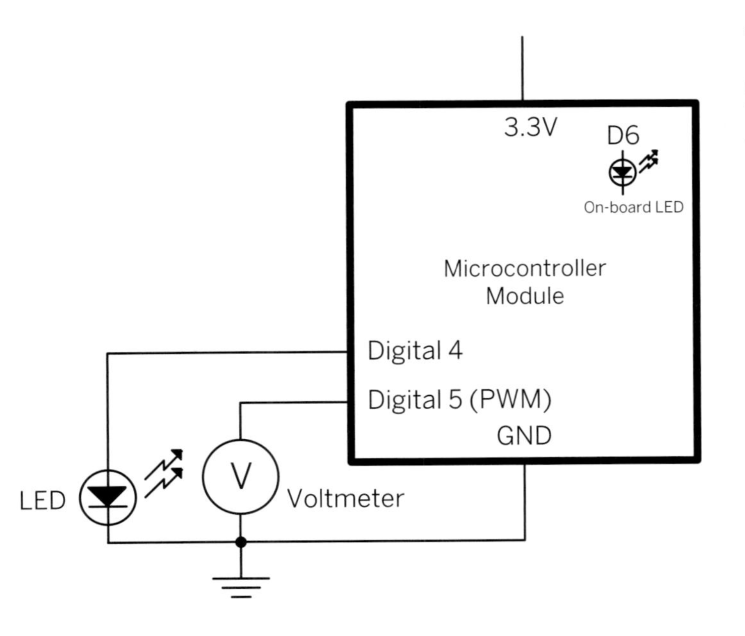

Figure 4-7
The circuit for a networked meter connected to an MKR1000. This project uses the built-in LED on pin 6 as well.

formats, your preferred format will be one of your request parameters. Many APIs also require an API key to verify your application's requests. An API key is just a long alphanumeric string that uniquely identifies your account.

Once you've set up an account for the AirNow API, click the Web Services tab, scroll down in the Current Observations by Reporting Area column to By Latitude/Longitude, and click the Query Tool link (docs.airnowapi.org/HistoricalObservationsByLatLon/query). You'll see an interactive example of how to request air quality data from the API using your latitude and longitude. There is a form that lets you build a query, adding your location, the date you want to query, the distance from the location in which you're interested, and the data format. Here's a typical query generated by that tool. Change the API key, location, and date to your own. Select "application/json" as well:

http://www.airnowapi.org/aq/observation/latLong/historical/?format=application/json&latitude=40.7496&longitude=-73.9836&date=2016-05-10T00-0000&distance=10&API_KEY=0000AAAA-0A0A-1111-A123-11223344AA55

The response to this request will look like this:


```
[{"DateObserved":"2016-05-10 ","HourObserved":0,"LocalTimeZone
":"EST","ReportingArea":"Newark","StateCode":"NJ","Latitude":40
.7496,"Longitude":-73.9836,"ParameterName":"OZONE","AQI":28,"Cat
egory":{"Number":1,"Name":"Good"}},{"DateObserved":"2016-05-10
","HourObserved":0,"LocalTimeZone":"EST","ReportingArea":
"Newark","StateCode":"NJ","Latitude":40.7496,"Longitude":-
73.9836,"ParameterName":"PM2.5","AQI":33,"Category":{"Number":1,"
Name":"Good"}}]
```


This may look complex, but it's much simpler for the microcontroller to read than the text of the HTML page would be. The parameter you'll use is the PM2.5 reading. This is a reading of the hourly average of particulate matter density. It's a measure in parts per million of particles from dust, soot, pollen, or other pollutants of 10 microns or smaller in diameter. Particles of this size can cause serious health effects when inhaled. Reading the PM2.5 value from this text is straightforward: read the text until you see "PM2.5" and then read the first numeric string after that. That's what you'll do in your next microcontroller program.

X

Figure 4-8
AirNow's page displays current air quality conditions by U.S. ZIP code. The AirNow API uses latitude and longitude instead.

Plan It The AirNow client is fairly simple, once the WiFi module is connected to the network:

1. Open a connection to the web server.
2. Send an HTTP GET request.
3. Wait for a response.
4. Process the response.
5. Set the meter with the response.
6. Wait an appropriate interval and do it all again.

Since there is no screen on your device, you need an alternate way to show whether it's connected to the network, and when it's making an HTTP request. You'll use LEDs on pin 4 and pin 6 (the built-in LED) to indicate those.

This program will use a new library, the ArduinoHttpClient library. Originally written by Adrian McEwen and expanded by Sandeep Mistry, this library can make HTTP GET, POST, PUT, and DELETE requests, and handles the details of the response for you. You can use the library to retrieve or skip the HTTP response headers, get the HTTP response codes, the content length, and other metadata about the response.

Install the HttpClient library using the Library Manager as you did other libraries in the Arduino IDE. Click the Sketch menu, choose Include Library...→Manage Libraries..., and then filter for ArduinoHttpClient and install the library.

▸▸ **There are different versions of HttpClient. You're using the one specifically called ArduinoHttpClient here.**

```
/*
  AirNow Web Client
  Context: Arduino, with WINC1500 module
*/

// include required libraries and config files
// declare I/O pin numbers and global variables

void setup() {
  // initialize serial communications
  // initialize output pins
  // while you're not connected to a WiFi AP,
  //   try to connect
  //   blink the network LED
  // When you're connected, print out the device's network status
}

void loop() {
  // make an HTTP request once every two minutes
  // set the status LEDs
}

void connectToServer() {
  // make HTTP call
  // while connected to the server,
  //   if there is a response from the server,
  //   search through for "PM2.5"
  //   read the PM2.5 value from the response
  //   throw out the rest of the response
  // If you got an AQI value,
  //  set the meter
  //  close the request
  // save the time of this HTTP call
}

void setMeter() {
  // map the PM2.5 level to a range the meter can use
  // set the meter
}

void setLeds() {
  // if the network is connected, turn on network LED
  // if the TCP socket is connected, turn on connected LED
}

void blink() {
  // Blink the LED:
}
```

Figure 4-9 is a flowchart of what happens in the microcontroller program. The major decisions (the if statements in your code) are marked by diamonds; the functions and the blocks within them are marked by rectangles. Laying out the whole program in a flowchart like this will help you keep track of what's going on at any given point. It also helps you to see what methods depend on a particular condition being true or not.

This program will check the API server every two minutes. If there's a new value for the air quality, it'll read it and set the meter. If it can't get a connection, it will try again two minutes later. Because it's a client and not a server, there's no web interface to the project, only the meter.

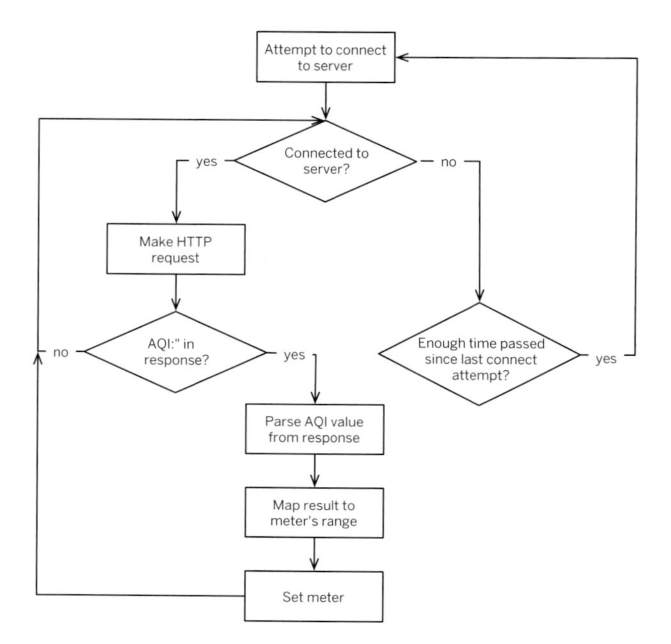

Figure 4-9
A flowchart of the Arduino program for making and processing an HTTP GET request.

Combining Different Data Types in Strings

When you're writing a web client program, you often need to combine different pieces of data into one request or response. You saw this in the earlier server program, where you appended the HTTP response elements to the response string one chunk at a time. In client programs, you might need to combine elements like the API key, the username, the latitude and longitude, various sensor readings, and other things. The Arduino String library is designed to make this easier. You can append one String to another just by adding them together. If you want to convert an `int` or a `float` data type to a `String`, you can do it by using the `String()` converter function, as you saw in the server program. You can even convert numeric data types to binary or hexadecimal notation using the BIN or HEX modifiers.

Some libraries expect arrays of characters or bytes rather than Strings. You can convert a String to a `char` array by using the String library's `toCharArray()` function. Here's how to put the contents of a String into a character array:

```
String myString = "/some/server/route";
char stringArray[myString.length() + 1];
myString.toCharArray(stringArray, myString.
length() + 1);
```

It's important to know that Strings use an extra byte of value 0 at the end, so a character array that is going to contain a String needs to be one byte longer than the String itself. You can also use the String library's `.c_str()` function to convert to the character array.

X

Try It Now that you've got the structure of the AQI server program, start with the required libraries and configuration. You'll need the SPI and WiFi101 libraries, and the ArduinoHttpClient library. Make a new file called **config.h** and store the WiFi access point SSID and password in it as you did for the server program. Add another `char` array called `APIKey` for your API key in **config.h** as well.

```
/*
  AirNow Web Client
  Context: Arduino, with WINC1500 module
*/
// include required libraries and config files
#include <SPI.h>
#include <WiFi101.h>
#include <ArduinoHttpClient.h>
#include "config.h"
```

Set up constants for the output pins (they will need to be different for the ESP8266), the meter minimum and maximum that you determined from the meter test, and the request interval.

The global variables you'll need are the WiFiClient, a character array for the server address, a String for the API route, and a long int for the last request time.

```
// declare I/O pin numbers and global variables:
const int networkLED = LED_BUILTIN; // indicates network connection
const int connectedLED = 4;         // indicates connection to server
const int meterPin = 5;             // controls voltmeter
const int meterMin = 0;             // minimum level for the meter
const int meterMax = 255;           // maximum level for the meter
const int AQIMax = 200;             // maximum level for air quality
const long requestInterval = 120000;// delay between updates to the server

WiFiClient netSocket;               // network socket to server
const char serverAddress[] = "www.airnowapi.org";  // server name
String route = "/aq/observation/latLong/current/"; // API route
long lastRequestTime = 0;           // last request time, in ms
```

In the `setup()`, initialize serial communications, configure the output pins, and connect to the WiFi network as you did in the server program. There's one change here, though: while you're waiting for the WiFi connection to happen, call a function called `blink()` to blink the network connection LED. You'll add this function at the end of the program shortly.

At the end of the `setup()`, append the rest of the parameters to the route variable so you've got the full URL that you built using the AirNow API Query Builder tool. Technically, you don't need to do this for this project. But this example shows how you might combine different variables as query parameters into one longer query string. For example, if you were getting the latitude and longitude from an external device, you could do the same with them.

```
void setup() {
  Serial.begin(9600);              // initialize serial communication
  pinMode(networkLED, OUTPUT);     // set network LED pin as output
  pinMode(connectedLED, OUTPUT);   // set connected LED pin
  pinMode(meterPin, OUTPUT);       // set voltmeter pin

  // while you're not connected to a WiFi AP,
  while ( WiFi.status() != WL_CONNECTED) {
    Serial.print("Attempting to connect to Network named: ");
    Serial.println(ssid);          // print the network name (SSID)
    WiFi.begin(ssid, password);    // try to connect
    blink(networkLED, 5);          // blink the network LED
  }

  // When you're connected, print out the device's network status:
  IPAddress ip = WiFi.localIP();
  Serial.print("IP Address: ");
  Serial.println(ip);
  //append the parameters and API key to the route:
  route += "?format=application/json&latitude=40.7296";
  route += "&longitude=-73.9936&date=2016-05-10T00-0000&distance=10";
  route += "&API_KEY=";
  route +=  APIKey;
}
```

»

The `loop()` function does only two things. First it checks the difference between `lastRequestTime` and the current time (the `millis()`) to see if the proper interval between requests has passed. If so, it calls `connectToServer()`. It also calls a function called `setLeds()` to set the status LEDs. You'll write that shortly.

Continued from previous page.

```
void loop() {
  // make an HTTP request once every two minutes:
  if (millis() - lastRequestTime > requestInterval) {
    connectToServer();
  }
  // set the status LEDs:
  setLeds();
}
```

The `connectToServer()` function is the core of the client program. First, initialize the new HttpClient. Then make the HTTP GET call, and skip the response headers.

```
void connectToServer() {
  int AQI = -1;          // AQI value

  HttpClient http(netSocket, serverAddress); // make an HTTP client
  http.get(route);                           // make a GET request
  http.skipResponseHeaders();                // ignore the HTTP headers
```

While there's a connection to the server, update the status LEDs by calling `setLeds()`, then check to see if there's a response available. If there is, use the `findUntil()` function to look for the first instance of `"PM2.5"`. Then use `parseInt()` to read the next numeric string and convert it to an `int`. If you get a valid AQI value, call the `setMeter()` function to set the meter.

Finally, close the connection to the server with the `http.stop()` function, and save the current time as the `lastRequestTime`.

```
  // while connected to the server:
  while (http.connected()) {
    setLeds();
    // if there is a response from the server:
    if (http.available()) {
      http.findUntil("PM2.5", "\n"); // parse response for "PM2.5"
      AQI = http.parseInt();         // read PM2.5 value from the response
      http.flush();                  // throw out the rest of the response
    }
  }
  if (AQI > -1) {                    // If you got an AQI value,
    Serial.print("PM2.5: ");
    Serial.println(AQI);
    setMeter(AQI);                   //  set the meter
  }
  http.stop();                       // close the request
  lastRequestTime = millis();        // save the time of this HTTP call
}
```

In the `setMeter()` function, use the `map()` function to map the AQI level to a corresponding range between the meter minimum and maximum. Then set the meter with the `analogWrite()` function.

```
void setMeter(int level) {
  // map the result to a range the meter can use:
  int meterSetting = map(level, 0, AQIMax, meterMin, meterMax);
  // set the meter:
  analogWrite(meterPin, meterSetting);
}
```

In the `setLeds()` function, turn on the network LED if the `WiFi.status()` indicates that the WiFi module is connected, and turn it off if not. Similarly, use the WiFiClient `.connected()` function to see if the module is connected to the API server, and turn on the connected LED if so.

```
void setLeds() {
  // if the network is connected, turn on network LED:
  if (WiFi.status() == WL_CONNECTED) {
    digitalWrite(networkLED, HIGH);
  } else {
    digitalWrite(networkLED, LOW);
  }
  // if the TCP socket is connected, turn on connected LED:
  int connectedToServer = netSocket.connected();
  digitalWrite(connectedLED, connectedToServer);
}
```

Finally, in the `blink()` function, blink an LED every 200 milliseconds. Pass in the LED number and number of blinks as a parameter so you can blink any LED any number of times.

```
void blink(int thisPin, int howManyTimes) {
  // Blink the LED:
  for (int blinks = 0; blinks < howManyTimes; blinks++) {
    digitalWrite(thisPin, HIGH);
    delay(200);
    digitalWrite(thisPin, LOW);
    delay(200);
  }
}
```

That's the whole client sketch. Save it as **AQIWebClient** and upload it to your MKR1000. The LEDs will indicate the network connection and the server connection, and when you get a valid AQI reading, the meter will indicate it. X

Finding a Host's IP Address

If you need to find a host's IP address instead of its name, use the `ping` command mentioned in Chapter 3. For example, if you open a command prompt and type `ping -c 1 www.makezine.com`, (use `-n` instead of `-c` on Windows), you will get the following response, listing the IP address you need:

```
PING makezine.com (104.25.44.28): 56 data bytes
64 bytes from 104.25.44.28: icmp_seq=0 ttl=52 time=38.870 ms
```

If you can't use `ping` (some service providers block it, since it can be used for nefarious purposes), you can also use `nslookup`. For example, `nslookup makezine.com` will return the following:

```
Server:   8.8.8.8
Address:  8.8.8.8#53

    Non-authoritative answer:
    Name:          makezine.com
    Address:  104.25.45.28
```

nslookup returns the Domain Name Server it used to do the lookup as well.

Any one of the addresses listed will point to its associated name, so you can use any of them.

The Finished Project

Figure 4-10
The completed networked air-quality meter.

"In building this project, you encountered a few characteristics that are typical of microcontroller-based networked appliances. It's worthwhile to note these for future designs.

First, appliances are usually not turned on all the time, so it makes sense for them to be clients rather than servers. In a system of networked devices, one of a server's most important roles is to provide continuity by being always available. Ironically, physical accessibility often works against network accessibility, because a device you can see is one you might turn off. So devices like this meter, which are designed to be physically visible, are best made as clients.

Second, whether for aesthetic reasons like this project or for cost reasons, appliances often have very simple displays, so there's not an easy way to display metadata about the device's status. LEDs like the network connection LED and server connection LED in this project are valuable as diagnostic tools. You may want to hide them in the back or inside so they're not always visible, but include them in your design so that the end user has some way of determining whether or not the device is working properly.

Third, users expect appliances to respond faster than software on a personal computer, so keep your code simple. Appliances don't usually have progress bars. If the server you're accessing does not have an API as simple as the one shown here, you might want to write a client on your own server that accesses the main server and simplifies the response for your appliance client.

The previous project, the RGB color server, acted as a server rather than a client, and you may be thinking, "That didn't look so bad, why not do that?" Running a microcontroller with no operating system as a server can be a good alternative to give an appliance a screen interface when it has no screen. Commercial devices like Philips' Hue lighting hub take this approach. However, WiFi modules like the WINC1500 or the ESP8266 can generally only accept one or two clients at a time due to the limits of their software. As long as you set users' expectations clearly, namely that this interface is intended for LAN use only, and for one or two clients at a time, it can be an effective alternative to adding a physical screen to your appliance. In commercial devices, it can reduce the cost of your device, but places the burden on the consumer to have a second device to access that server. So it's a good idea to build in a simple LED status interface as well.

X

Data Formats

You've seen a few text-based protocols so far in this book, and you've even written a few of your own. Protocols are ways of structuring data for communication. Some protocols simply relay information, and others give commands—either explicitly or implicitly—through a request. You'll see many throughout the book. It's worth summarizing the ones you've seen so far.

Simple Data Formats

In the Monski Pong game, you created comma-separated value strings. *Comma-separated values (CSV)* and *tab-separated values (TSV)* are ubiquitous, and it's common to end a CSV string with a newline, or a newline and carriage return. The protocol used by GPS receivers, called NMEA-0183, is a good example of CSV in practice. You'll see this later on in the book. You also saw a few examples of organizing data in *key-value pairs*. Any time you have a list of items, and each has a name associated with it (called a key), you've got a key-value pair. The HTTP Header properties you've seen are key-value pairs, as are the parameters of HTTP requests. They can be formatted in a variety of ways, but generally they're punctuated by one separator to separate the key from the value, and another to separate the pairs. For example, in the following data string, the colon separates key from value, and the comma separates each pair from the next:

```
"DateObserved":"2016-05-10 ","HourObserved":0,"LocalTimeZone":"
EST","ReportingArea":"Newark","StateCode":"NJ","Latitude":40.7267
,"Longitude":-74.1442
```

In the following example, the key and value are separated by an equals sign (=) and the pairs are separated by an ampersand (&):

```
?name=tom&age=14
```

Here's another you've seen, where the key is separated from the value by a colon, and a newline is used to separate the pairs:

```
Host: localhost
Content-Length: 19006
Content-Type: multipart/form-data
```

You've also used a few internet *transfer protocols*, like *HyperText Transfer Protocol (HTTP)* and *Simple Mail Transfer Protocol (SMTP)*. They share a common format as well: they open with a command like GET or POST, and then follow with properties of the transfer, in colon-separated key-value pairs. The header of the transfer is separated from the body of the message by two newlines, and the message is usually closed with two newlines as well. Any data coming from the client was sent as key-value pairs, separated by a colon for information about the transfer, or by an ampersand for data items in the content of the transfer. Take a second look at this HTTP POST request from Chapter 2 as an example:

```
POST /check HTTP/1.0
Host: example.com
Connection: Close
Content-Type: application/x-www-form-urlencoded
Content-length: 16

name=tom&age=14
```

Everything in the header is a parameter of the request: the host from which you're requesting, the connection type, the type of content to be exchanged, and the content length. Everything in the body is a parameter of the content: name and age. Having a clear separation between the header formatting and the content formatting makes it easier to separate them when writing a program, as you've already seen.

> NOTE: **Invisible characters line the newline, carriage return, or tab characters are represented in Java, JavaScript, and C using an** *escape sequence* **like so:**
> **newline: \n**
> **carriage return: \r**
> **tab: \t**
> **You've seen these already, and will see them much more. You might see other escaped characters as well.**

Structured Data Formats

When you've got data to exchange that's more complex than a list of key-value pairs, you need a structure for it. For example, imagine an array of sensors spread around your house:

Address	Location	Last read	Value
1	kitchen	12:30:00 PM	60
2	living room	05:40:00 AM	54
3	bathroom	01:15:00 AM	23
4	bedroom	09:25:00 AM	18
5	hallway	06:20:00 AM	3

You've got more than just a list of single items, and more than just a few name-value pairs. In fact, each line of the table is a list of name-value pairs (e.g. "address: 3", "value:23"). This is where a structured data format comes in handy. *JavaScript Object Notation*, or *JSON*, is a popular notation format for structured data like this.

JSON represents each cell of the table as a name-value pair. Each line is a comma-separated list of pairs, enclosed in curly braces. The whole table is a comma-separated list of lists, like so:

```
[{"Address":1,"Location":"kitchen","Last read":"12:30:00
PM","value":60},
{"Address":2,"Location":"living room","Last read":"05:40:00
AM","value":54},
{"Address":3,"Location":"bathroom","Last read":"01:15:00
AM","value":23},
{"Address":4,"Location":"bedroom","Last read":"09:25:00
AM","value":18},
{"Address":5,"Location":"hallway","Last read":"06:20:00
AM","value":3}]
```

This way of formatting data is relatively simple. The punctuation that separates each element is just a single character, so you can scan through it one character at a time and know when you're done with each element. Because it's text, it's human-readable as well as machine-readable. Spaces, newlines, and tabs aren't considered part of the structure of JSON, so you can reformat it for easier reading like so:

```
[
  {
    "Address":1,
    "Location":"kitchen",
    "Last read":"12:30:00 PM",
    "value":60
  },
  {
    "Address":2,
    "Location":"living room",
    "Last read":"05:40:00 AM",
    "value":54
  },
  {
    "Address":3,
    "Location":"bathroom",
    "Last read":"01:15:00 AM",
    "value":23
  },
  {
    "Address":4,
    "Location":"bedroom",
    "Last read":"09:25:00 AM",
    "value":18
  },
  {
    "Address":5,
    "Location":"hallway",
    "Last read":"06:20:00 AM",
    "value":3
  }
]s
```

The advantage of a data interchange format like JSON is that it's lightweight, meaning there aren't a lot of extra bytes needed to structure the information you want to send. It gives you more power than a simple list but is still efficient to send from a server to a client. You can send JSON as the body of an HTTP request, and as long as there's a program on the client that can parse it, you've got a quick way to exchange complex data.

JSON and JavaScript

As you might guess, JSON was created as a part of JavaScript, so the language makes it very easy to read, write, and parse JSON. In fact, complex JavaScript variables are written in JSON. You've used some already. Anytime you've placed a dot after a variable name, you've accessed one of the properties of a JSON object.

Since many web APIs offer a JSON interface, it's easy to parse the data from them using JavaScript. The AirNow server that you accessed in Project 7 returns a JSON response. Following is how you'd parse it in node.js.

This script prints out the response from the server after parsing it using the `JSON.parse()` function. This function takes a string that's formatted properly as JSON and converts it to a binary JSON object on which you can perform operations. There's an equivalent function, `JSON.stringify()`, that converts a binary JSON object into a string.

Try It Save this as **aqiClient.js**. The HTTP library is part of node's core, so there's no need to install any libraries. Include the http library, then set the options for the client request. Note that you're formatting the elements of `options` as a JSON object. Each element of the `options` is a key-value pair, separated by a colon: replace the API key with your own, and the location coordinates with your own.

```
/*
  AQI client
  context: node.js
*/
var http = require('http');  // include HTTP library

// set options for the request in a JSON object:
var options = {
  host: 'www.airnowapi.org',
  port: 80,
  path: '/aq/observation/latLong/current/?format=application/
json&latitude=40.7296&longitude=-73.9936&date=2016-05-10T00-
0000&distance=10&API_KEY=0000AAAA-0A0A-1111-A123-11223344AA55'
};
```

The callback function for the request, called `handleResponse()` here, has two local functions inside it because the response might come back in several chunks. The variable `result` collects the chunks when they come in. The `response.on('end')` function converts the result to a JSON object and prints it out, formatted nicely.

Finally, you call the request using `http.request()`, then end it with `request.end()`. Run this on the command line like so:

```
$ node aqiClient.js
```

```
function handleResponse(response) {
  var result = '';                          // string to hold the response

  response.on('data', function (data) {  // when you get a response,
    result += data;                         // add it to the overall result
  });

  response.on('end', function () {        // when the response finally ends,
    var response = JSON.parse(result);   // parse the whole result
    console.log(response);                // and print it
  });
}

var request = http.request(options, handleResponse);  // start the request
request.end();                                        // end it
```

When you run this script, try filtering the response by treating it as a JSON object and printing only its properties. It will appear as an array, so try printing `response[0]` first. Then try `response[0].AQI` or `response[0].Longitude`. You can use any of the property names to get just that property. This script, used with any host and path, can be used as a valuable test client to learn the elements of a given web API's data structure, when it gives you a JSON response. You can also use it to make secure HTTPS requests, by changing the first line to `require('https');` and then using port 443.
X

Representational State Transfer and Web APIs

Underlying the HyperText Transfer Protocol (HTTP) is a principle known as *Representational State Transfer*, or *REST*. It's not a protocol—it's more of an architectural style for information exchange. It's used commonly in application programming interfaces (APIs) on the web. Though it started with the web, it has applications in other areas. Understanding a little about REST will not only help you understand other systems, it will design your own communications protocols as well. Once you know about it, you can never get enough REST in your life.

The basic idea of REST is this: there is a thing somewhere on a network (called a *resource*). Maybe it's a database, or maybe it's a microcontroller controlling your household appliances. You either want to know what state it's in, or you want to change its state. REST gives you a way to describe, or represent, the state of the resource on the internet, and to transfer that representation to a remote user.

You've been using REST already—it's what HTTP is all about. When you want to know about a remote device or system, you make a request to GET that information or to POST changes to it. The remote thing responds to your request, either with a representation of the state of affairs (for example, an HTML page like the AirNow.gov page in Figure 4-8, which includes representations of your local weather), or by changing the state of affairs (for example, by updating the light settings in a networked lamp).

In RESTful thinking, URLs are nouns that describe things, and requests are the verbs that act on nouns. The properties of things are described in the URL, separated by slashes. For example, imagine a RESTful description of the RGB lamp that you built in Chapter 2. If you want to know the state of the red channel, you might say to the Arduino:

```
GET /color/r/
```

The lamp would then reply with the red level, a number between 0 and 255. Or, perhaps you want to change the state of the green channel. You might do this:

```
POST /color/g/255
```

The lamp would know that it should set the green channel to 255. In this scenario, the lamp is the server—serving up representations of its state—and you're the client.

There are three important elements to this exchange:

1. The representation and its properties are separated by slashes.
2. The verb is the request: whether you want to get some information or change the state of the thing, you start with the verb. The verbs are HTTP's verbs: GET, PUT, POST, and DELETE. GET retrieves the state of a property; PUT adds a new property; POST updates an existing property; and DELETE removes an existing property. GET and POST are the most commonly used.
3. The description is technology-independent. It doesn't matter whether the resource is delivered to you as an HTML page or XML from a web server, or whether it's printed by a node.js, PHP or Ruby script on a server, a C/C++ program running on an Arduino, or sung by a choir of pixies. It doesn't matter how the result is generated; you just want to know what the state of things is, and it's the server's job to do that.

You can use REST for just about any control interface. For example, Open Sound Control, or OSC, a protocol for musical controllers that's designed to supersede MIDI, is RESTful. It's also independent of the transport mechanism, and it can be sent over Ethernet, serial, or any other physical data channel. It's simple enough that you can read it with a limited processor, and general enough to describe most anything you want information about or want to control.

The fact that REST doesn't describe the technology means that if you change the programming tools you use to run your site, you don't have to change the URLs. The client never sees file extensions like .html, so it doesn't know or care what programming tools or file types you're using. If you want to change programming langauges or formats, go ahead! The site structure can stay the same.

To design RESTfully for the web, describe your resource (e.g., your web service or connected device) in terms of its properties and come up with routes to view and change those properties appropriately. The basic RESTful style is `/object/attribute` to get the value of a parameter, and `/object/attribute/newValue` to set it. Properties with simple values, like a number or string, are passed at the end of the route using PUT or POST. More complex values are usually passed in the body of the HTTP request.

Following are two examples: one is a traditional website and the other is a physical device on a network.

A Traditional Web Service

Perhaps you're making a social media site for runners to compare their daily runs. As part of it, each runner has a profile with various attributes for each day's run: the date (day, month, and year), the distance of the run, and the time of the run. For a run on January 31 , 2012, the URLs to address those attributes might look like this.

To create a new run for 31 January 2012:

```
PUT /myrun.example.com/runnerName/31/1/2012
```

To get the distance of a run:

```
GET /myrun.example.com/runnerName/31/1/2012/distance/
```

To set the distance to 12.56km:

```
POST /myrun.example.com/runnerName/31/1/2012/distance/12.56
```

You can tell a request that gets the value of a parameter from one that sets the value because the GET requests tend to end with the name of the parameter, while the PUT and POST requests tend to end with the new value of the parameter.

A Web-Based Device

Imagine you're building a device that controls the window blinds in your office. There are 12 windows, each with its own blind. The blinds can be positioned variably, in a range from 1 to 10. The client can get the state of any blind, or you can set it. The URLs to see the state of the windows might look like this.

To see them all at once:

```
GET /mywindows.example.com/windows
```

To see an individual window (window number 2):

```
GET /mywindows.example.com/window/2/
```

To set a window's blind halfway down:

```
POST /mywindows.example.com/window/2/height/5
```

To close them all:

```
POST /mywindows.example.com/windows/height/0
```

What's actually returned to the client's request is up to you. You might choose to make a graphic that shows images of the windows with blinds, or you might show the result in text only. RESTful architecture just determines where to find something, not what you'll find there.

RESTful addresses are designed to be easy for a computer to parse. All you have to do is to separate the incoming request into substrings on the slashes, look at each substring to see what it is, and act appropriately on the substrings that follow it. They're also easy for people to read. A URL like the shown earlier is more comprehensible than this one:

http://myrun.example.com/?runnerName=George&day=31&month=1&year=2012&distance=12.56

REST is a style, not a formal protocol, so many systems use REST where it's useful and diverge from it when it's not. Not all web APIs are fully RESTful. The AirNow.gov site that you used for the air quality meter project wasn't fully RESTful, though it uses some RESTful techniques You'll also see REST used in many of the projects that follow in this book.

In the next section you'll learn a few methods to read RESTful HTTP requests in node.js and Arduino.

X

REST in node.js

Node.js and express.js were designed with REST in mind, so reading incoming requests when they're RESTfully formatted is simple. Using express.js, you can format your server routes to include parameters that you want to read, like so:

```
server.post('/check/age/:age', checkAge);
```

Using this route, your script will expect a POST request like this:

http://www.example.com/check/age/21

When you read the incoming request variable, you can check `request.params.age`, and it will return the value 21. Whatever comes in the request string after `/age/` will be read as such. You can even combine multiple parameters in a single line like so:

```
server.post('/check/name/:name/age/:age', checkAge);
```

Some would argue that combining multiple parameters on one line like this is not strictly RESTful, but that's for you as the programmer to decide. Express.js will parse your request correctly if you format it this way.

Try It Here's a version of the **ageCheck.js** script from Chapter 3, using RESTful methods. Changed lines are in blue. Save this as a new project in its own directory as **restAgeCheck.js** and then install express.js using npm as usual, like so:

```
$ npm install express
```

In this version, you're reading `request.params.name` and `request.params.age` to get the name and age. You should also check to see if age is undefined, in case the client doesn't send the age.

```
/*
RESTful age checker
Context: node.js
*/

var express = require('express');// include the express library
var server = express();              // create a server using express

function checkAge(request, response) {
  var name = request.params.name;
  var age = request.params.age;
  var responseString = "";
  if (typeof age === 'undefined') {
    responseString = "<p>Please tell me your age.</p>\n"
  } else {
    if (age < 21) {
      responseString = "<p>" + name
        + ", You're not old enough to drink.</p>\n";
    } else {
      responseString =  "<p> Hi " +  name
        + ". You're old enough to have a drink, ";
      responseString += "but do so responsibly.</p>\n";
    }
  }
  // write back to the client:
  response.writeHead(200, {"Content-Type": "text/html"});
  response.write(responseString);
  response.end();
}
```

You'll set up three routes here, so that the client can send name, age, or both at the same time.

```
// start the server:
server.listen(8080);
// define what to do when the client requests something:
server.get('/check/name/:name', checkAge);
server.get('/check/age/:age', checkAge);
server.get('/check/name/:name/age/:age', checkAge);
```

You're probably wondering why you used a GET request when you're sending in parameter values. Doesn't RESTful style say that you should be sending POST or PUT requests? Technically that's true, but sometimes GET requests are used so that you can enter the whole request from the address bar of a browser. Once you know it works, you can easily change those last lines from `server.get()` to `server.post()`.

Run this script on the command line like so:

```
$ node RestAgeCheck.js
```

Then you can test it by making requests from the address bar of your browser like so:

```
http://www.myserver.com:8080/check/name/tom/age/42
```

Or by using curl from the command line like so:

```
$ curl http://www.myserver.com:8080/check/name/tom/age/42
```

In either case, replace `www.myserver.com` with your own server's address, or `localhost` if you're running it locally.

REST in Arduino

Reading REST requests in Arduino is a bit trickier than in node.js. To do it, you need write a server like you did in Project 6, Hello Internet!, earlier in this chapter. Then you'll read the incoming text stream from the client and parse it. You can do this using commands from the Stream library mentioned earlier in this chapter (remember, the WiFi101 `Client` and `Server` classes are instances of the Stream library, just like Serial).

The `setup()` for a RESTful server would look just like it did for the Hello Internet! project. The `loop()` is shown here.

Try It All you care about is the first line of the HTTP request, which might look like this:

```
GET /check/age/21 HTTP/1.1
```

Read until the first space to see if the request type is GET or POST.

Next, set up a while loop to read each token of the route, stopping at each slash (/). If the previous token that you read was the name of the parameter you want, then the current token is the value.

At the end of the while loop, save the current token as the last token for checking next time through.

Once you've got what you want, send the client a response, and close the connection.

You can test this just as you did the node.js script. When you do, you should see the age value print out in the Serial Monitor.

```
void loop() {
  // listen for incoming clients
  WiFiClient client = server.available();
  while (client.connected()) {    // while the client is connected,
    if (client.available()) {     // and there are incoming bytes to read
      // read the first line up to the first space to get the request type:
      String request  = client.readStringUntil(' ');
      // check if the request is GET or POST:
      if (request == "GET" || request == "POST") {
        // continue reading the stream, looking for  / characters:
        String lastToken = "";            // the last token you read
        while (!lastToken.endsWith("HTTP")) {
          String currentToken = client.readStringUntil('/');
          if (lastToken == "age") {
            int age = currentToken.toInt();  // the parameter you want
            Serial.print("age: ");
            Serial.println(age);
          }
          // check other tokens the same way
          lastToken = currentToken;
        }
        client.println("HTTP 200 OK\n\n"); // send an HTTP response
        if (client.connected()) {        // if the client's still connected
          client.stop();                 // disconnect the client
        }
      }
    }
  }
}
```

Programming and Troubleshooting Tools for Embedded Modules

You may have hit a number of problems when making the connections in the projects in this chapter. Probably the most challenging thing about troubleshooting them was that the WiFi module gave little or no indication that a problem occurred. This is the norm when you're working with embedded modules that you build yourself. This section covers a few things you should always check, and a few tools that will help you solve problems. These principles apply whether you're using the WiFi modules or some other network or communications module. You'll use these methods over and over again in the rest of the book and beyond.

The Three Most Common Mistakes

Check Power and Ground

Always check whether you have made the power and ground connections correctly. This is less of an issue when you've got an all-in-one module like the MKR1000 or the ESP8266 devices. But if you're using the WiFi101 shield or another plug-in module, you can get a pin misaligned and not notice. If you're lucky, the module you're using will have indicator LEDs that light up when it's working properly. Whether it does or not, check the voltage between power and ground with a meter to make sure you've got it powered correctly.

Check the Connections

When you're wiring a module to a microcontroller by hand, it's fairly common to get the wires wrong the first time. Make sure you know what each pin does, and double-check that the pin that one transmits on connects to the pin that the other receives on, and vice versa. If it's a synchronous serial connection, make sure the clock is connected.

Check the Configuration

If you're certain about the hardware connections, check the device's configuration to make sure it's all correct. Did you get the IP address correct? Is the SSID correct? The password? The server address? Do you have WiFi signal strength from the WiFi access point?

Diagnostic Tools and Methods

Once you know the device is working, you have to program the sequence of messages that constitutes your application. Depending on the application's needs, this sequence can get complex, so it's useful to have a few simple programs around to make sure things work as desired.

It's a good idea to test your module first using sample code—if it's provided. Every good communications module or library maker gives you a starting example, like the Scan Networks example that starts this chapter. Use it, and keep it handy. When things go wrong, return to it to make sure that basic communications still work.

Physical Debugging Methods

Writing code makes things happen physically. It's easy to forget that when you're working on an exchange of information like the examples shown here. So, it's helpful to put in physical actions that you can trigger; this way, you can see that your code is working. In the last project, you saw LEDs turn on when the client connected or disconnected, when it made a successful request, and when it reset. Triggering an LED is the most basic and reliable thing you can do from a microcontroller—even more basic than sending a serial debugging message—so use LEDs liberally to help make sure each part of your code is working. You can always remove them later.

Serial Debugging Methods

Besides their physical form factor, the WiFi modules you've seen here make a number of things easy for you. For example, the fact that they use synchronous serial communication to talk to the microcontroller means you can still use the asynchronous serial port to get messages about what's going on—as you saw in the previous example. Following are a few tips for effective serial debugging of networking code using serial messages.

It's important to keep in mind that serial communication takes time. When you have serial statements in your code, they slow it down. The microcontroller has to save each bit of information received, which takes time away from your program. Even though the time is miniscule for each bit, it can add up. After a few hundred bytes, you might start to notice it. So, make sure to take out your serial transmission statements when you're done with them.

Debug It One way to manage your debugging is to have a variable that changes the behavior of your program, like this. When every debugging statement is preceded by the `if (DEBUG)` conditional, you can easily turn them all off by setting `DEBUG` to false.

```
const boolean DEBUG = true;

void setup() {
  Serial.begin(9600);
}

void loop() {
  if (DEBUG) Serial.println("this is a debugging statement");
}
```

Announce It In the Air Quality Index client project, you had a lot of different functions in your code. It's not easy to tell whether every function is being called, so during troubleshooting, make it a habit to "announce" serially when every function is being called. To do so, put a `Serial.print()` statement at the beginning of the method, like the code shown on the right.

```
void connectToServer() {
  if (DEBUG) Serial.print("running connectToServer()...");
  // rest of the method goes here
}

void setMeter() {
  if (DEBUG) Serial.print("running setMeter()...");
  // rest of the method goes here
}

void setLeds(int thisLevel) {
  if (DEBUG) Serial.print("running setLeds()...");
  // rest of the method goes here
 }
```

Check Conditionals

It's common in an application like these to have multiple nested conditions that affect a particular result. As you're working, you can easily make mistakes that prevent the right conditions from happening. As you work, check that they're still coming true when you think they are by announcing it, as shown here.

```
while (http.connected()) {
 if (DEBUG) Serial.print("http.connected...");
  setLeds();
  // if there is a response from the server:
  if (http.available()) {
   if (DEBUG) Serial.print("http.available...");
    http.findUntil("PM2.5", "\n"); // parse response for "PM2.5"
    AQI = http.parseInt();         // read PM2.5 value from the response
    http.flush();                  // throw out the rest of the response
  }
}
```

Check Nesting Make sure that you don't make one action dependent on another unless it's necessary. When you get conditional within conditional within conditional, your code gets complicated. Decouple your dependencies when possible. For example, setting the meter depends on whether you have a good AQI value, not on whether you are connected. So separate the two as shown on the right.

```
int AQI = -1;                             // set an invalid AQI value
HttpClient http(netSocket, serverAddress); // make an HTTP client
http.get(route);                          // make a GET request
while (http.connected()) {                // while connected to the server
  if (http.available()) {                 // see if there is a response
    // ...read AQI value from the response
    // ...but don't set the meter here....
  }
}
if (AQI > -1) {                          //  if you have a valid AQI value,
  setMeter(AQI);                         //  set the meter here instead
}
```

Separate It Programmers like efficiency, so they will often combine several commands into one line of code. Sometimes the problem is in the combination. For example, this line attempts to connect using `client.connect()`, and then it checks the result in a conditional statement. It gave me all kinds of trouble until I separated the connection attempt from the check of the connection.

```
  if (client.connect()) {
    // if you get a connection, report back via serial:
    Serial.println("connected");
  }
  else {
    // if you didn't get a connection to the server:
    Serial.println("connection failed");
  }
```

» Here's the less efficient code that solved the problem. Checking the connection immediately after connecting was apparently too soon. The delay stabilized the whole program.

```
  client.connect();    // connect
  delay(1);            // wait a millisecond
  if (client.connected()) {
    // if you get a connection, report back via serial:
    Serial.println("connected");
  }
  else {
    // if you didn't get a connection to the server:
  Serial.println("connection failed");
  }
```

Just Watch It Sometimes it helps to step back from your complicated program and just watch what you're receiving from the other end. One technique I return to repeatedly while developing progams like these is to simply print out all of what the client module is receiving. It usually reveals the problem. Familiarity with your code can blind you to what you need to see. So, remind yourself of what you are actually receiving, as opposed to what you think you are receiving.

```
// if you're connected, save any incoming bytes
// to the input string:
if (client.connected()) {
  if (client.available()) {
    char inChar = client.read();
    Serial.write(inChar);
  }
}
```

Write a Test WiFi Status Program

Before you can connect to any server, you need to connect to your WiFi access point. You might need to know the MAC address of your WiFi module,and it helps to know the signal strength of your connection, to see if you've got a solid connection between the two. The following sketch prints out the stats of your WiFi module, attempts to connect to the WiFi access point, and when it connects, it prints the IP address and MAC address of both the WiFi module and the access point. Using this, you should be able to get all the credentials you need to get your device on the network.

Test It You'll need to make a **config.h** file for this sketch with your WiFi SSID and password just like you did earlier.

```
/*
  WiFi Status check
  Context: Arduino, with WINC1500 module
*/

#include <SPI.h>
#include <WiFi101.h>
#include "config.h"

void setup() {
  Serial.begin(9600);
  Serial.println("Starting");
  // while you're not connected to a WiFi AP,
  while ( WiFi.status() != WL_CONNECTED) {
    Serial.print("Attempting to connect to Network named: ");
    Serial.println(ssid);           // print the network name (SSID)
    WiFi.begin(ssid, password);     // try to connect
    delay(2000);                     // wait 2 seconds before next attempt
  }
}

void loop() {
  printWiFiStatus();
  delay(10000);
}
```

The `printWiFiStatus()` function prints out the properties of your WiFi access point first:

```
void printWiFiStatus() {
  // print the SSID of the WiFi AP to which you're attached:
  Serial.print("SSID: ");
  Serial.println(WiFi.SSID());

  // print the gateway address of the WiFi AP to which you're attached:
  IPAddress ip = WiFi.gatewayIP();
  Serial.print("Gateway IP Address: ");
  Serial.println(ip);

  // print the subnet mask of the WiFi AP to which you're attached:
  IPAddress subnet = WiFi.subnetMask();
  Serial.print("Netmask: ");
  Serial.println(subnet);
```

»

Continued from previous page.

> ⚠ This sketch is not identical to the ESP8266WiFi library version. Check that library's online examples for changes to the `macAddress()` and `BSSID()` functions.

```
// print the MAC address of the WiFi AP to which you're attached:
byte apMac[6];
WiFi.BSSID(apMac);
Serial.print("BSSID (Base station's MAC address): ");
for (int i = 0; i < 5; i++) { // loop from 0 to 4
  if (apMac[i] < 0x10) {      // if the byte is less than 16 (0x0A hex)
    Serial.print("0");        // print a 0 to the string
  }
  Serial.print(apMac[i], HEX);// print byte of MAC address
  Serial.print(":");          // add a colon
}
Serial.println(apMac[5], HEX);// println final byte of address
```

After the access point properties are printed, it prints out the properties of your WiFi module:

> ⚠ The ESP8266WiFi library has the MAC address reversed. Loop from 0 to 5 instead of 5 to 0 here.

```
// print your MAC address:
byte mac[6];
WiFi.macAddress(mac);
Serial.print("Device MAC address: ");
for (int i = 5; i > 0; i--) { // loop from 5 to 1
  if (mac[i] < 0x10) {        // if the byte is less than 16 (0x0A hex)
    Serial.print("0");        // print a 0 to the string
  }
  Serial.print(mac[i], HEX);  // print byte of MAC address
  Serial.print(":");          // add a colon
}
Serial.println(mac[0], HEX);  // println final byte of address

// print your  IP address:
IPAddress gateway = WiFi.localIP();
Serial.print("IP Address: ");
Serial.println(gateway);

// print the received signal strength from the WiFi AP:
long rssi = WiFi.RSSI();
Serial.print("signal strength (RSSI):");
Serial.print(rssi);
Serial.println(" dBm");
Serial.println();
}
```

Write a Test Client Program

It's easiest to work through the steps of the program if you can step through the sequence of events. More complex development environments allow you to step through a program one line at a time. Arduino doesn't give you that ability, but you can use test programs. The following program simply makes an HTTP request and prints everything the server sends in response, headers and all. It's a good way to check that what you're looking for is coming back from the server.

Test It The handy thing about this program is that you can test the exchange of messages manually before you try to parse the results. If you get no response, or a bad response, you can then start to look for what you're doing wrong in your client.

You'll need to make a **config.h** file for this sketch with your WiFi SSID and password just like you did earlier.

```
/*
  Test HTTP Client
  Context: Arduino, with WINC1500 module
*/
// include required libraries and config files
#include <SPI.h>
#include <WiFi101.h>
#include <ArduinoHttpClient.h>
#include "config.h"

WiFiClient netSocket;                  // network socket to server
const char server[] = "myserver.com";  // server name
String route = "/foo";                 // API route

void setup() {
  Serial.begin(9600);                 // initialize serial communication
  while ( WiFi.status() != WL_CONNECTED) { // while you're not connected,
    Serial.print("Attempting to connect to Network named: ");
    Serial.println(ssid);             // print the network name (SSID)
    WiFi.begin(ssid, password);       // try to connect
    delay(2000);
  }

  // When you're connected, print out the device's network status:
  IPAddress ip = WiFi.localIP();
  Serial.print("IP Address: ");
  Serial.println(ip);
}

void loop() {
  HttpClient http(netSocket, server, 8080);     // make an HTTP client
  http.get(route);                              // make a GET request

  while (http.connected()) {         // while connected to the server,
    if (http.available()) {          // if there's a response from the server,
      String result = http.readString();  // read it
      Serial.print(result);               // and print it
    }
  }
  // when there's nothing left to the response,
  http.stop();                       // close the request
  delay(10000);                      // wait 10 seconds
}
```

Write a Test Server Program

The previous program allowed you to connect to a remote server and test the exchange of messages. The remote server was beyond your control, however, so you can't say for sure that the server ever received your messages. If you never made a connection, you have no way of knowing whether the module can connect to any server. To test this, write your own server program to which it can connect.

Here is a short node.js script that you can run on your PC. It listens for incoming connections, and prints out any messages sent over those connections. It sends a basic HTTP response. This is similar to the servers you've written in node.js already, and will print out a few details of the client's request. You can change the details as needed. When your client can successfully connect to this server, try changing the client's host address to connect to other servers.

X

```
/*
  test web server
  context: node.js
*/
// include libraries and declare global variables:
var express = require('express');      // include the express library
var server = express();                // create a server using express

// define the callback function that's called when
// a client makes a request:
function respondToClient(request, response) {
  console.log(request.connection.remoteAddress);
  console.log(request.headers);
  console.log(request.query);
  // write back to the client:
  response.write("Hello, client!\n");
  response.end();
}

// start the server:
server.listen(8080);
// define what to do when the client requests something:
server.get('/', respondToClient);
```

Conclusion

The activities in this chapter show a model for networked objects that's very flexible and useful. The object is basically a browser or a server, requesting information from the web and extracting the information it needs, or delivering information to a client. You can use these models in many different projects.

The advantage of these models is that they don't require a lot of work to repurpose existing web applications. At most, you need to write a node.js script to summarize the relevant information from an existing website. This flexibility makes it easier for microcontroller enthusiasts who aren't experienced in web development to collaborate with web programmers, and vice versa. It also makes it easy to reuse others' work if you can't find a willing collaborator.

The model has its limits, though, and in the next chapter, you'll see some ways to get around those limits with a different model. Even if you're not using this model, don't forget the troubleshooting tools mentioned here. Making simple mock-ups of the programs on either end of a transaction can make your life much easier. This is because they let you see what should happen, and then modify what actually is happening to match that.

X

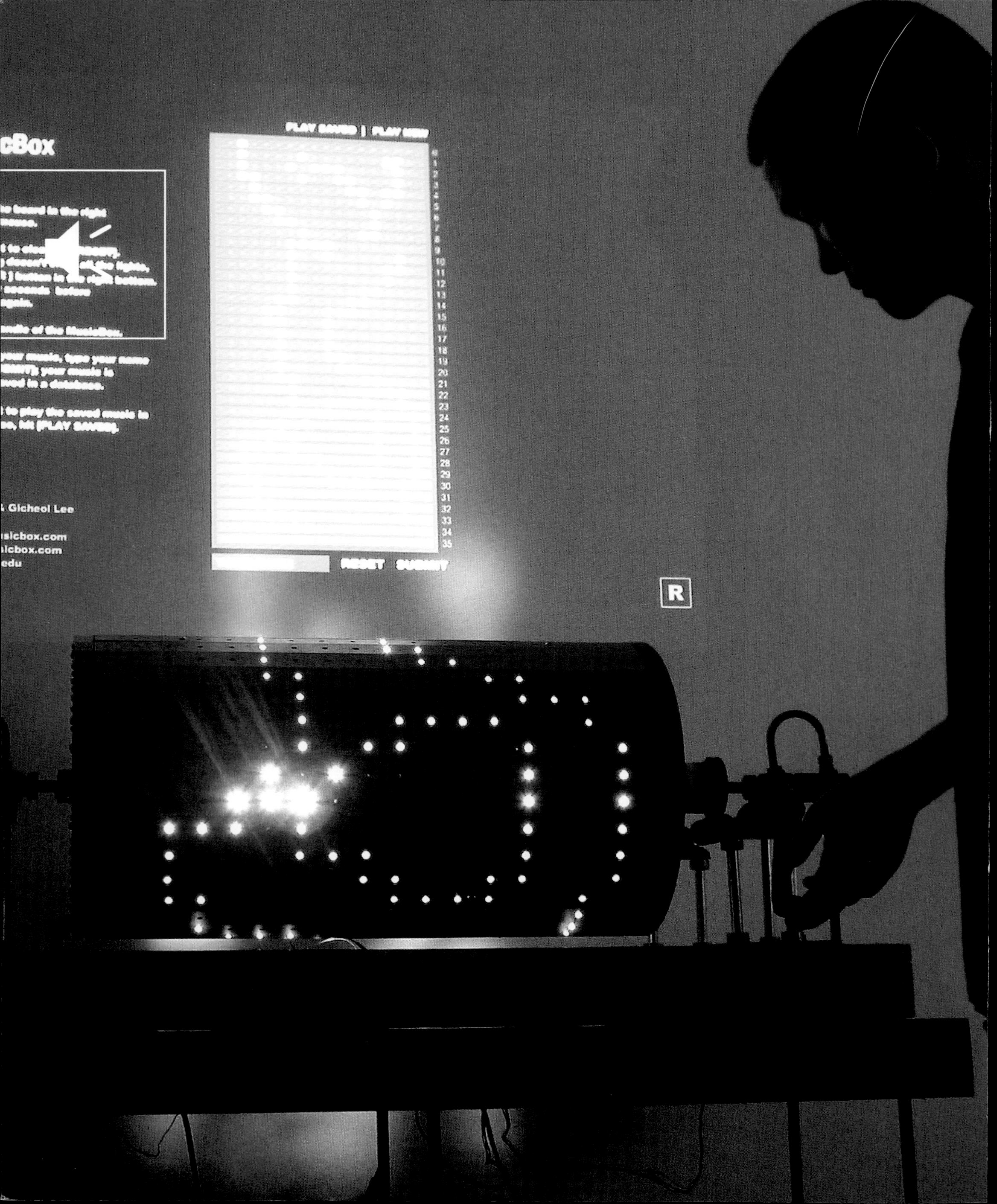

PLAY SAVED | PLAY NEW
RESET SUBMIT
Gicheol Lee
R

5

MAKE: PROJECTS 

Communicating in (Near) Real Time

So far, most of the networked communications you've seen worked through a web browser. Your device made a request to a remote server, the server ran a program, and then it sent a response. The client made a connection to the web server, exchanging some information, and then the connection was broken. In this chapter, you'll learn how to keep that connection open. You'll write two different server programs that allow you to maintain the connection in order to facilitate a faster and more consistent exchange between the server and client.

◂◂ Musicbox by Jin-Yo Mok (2004)
The music box is connected to a composition program over the internet using a serial-to-Ethernet module. The composition program changes the lights on the music box and the sounds it will play. Real-time communication between the two gives the player feedback. *Photo courtesy of Jin-Yo Mok.*

Supplies for Chapter 5

DISTRIBUTOR KEY

A Arduino Store (store.arduino.cc)
AF Adafruit (adafruit.com)
D Digi-Key (www.digikey.com)
F Farnell (www.farnell.com)
J Jameco (jameco.com)
RS RS (www.rs-online.com)
SF SparkFun (www.sparkfun.com)
SS Seeed Studio (www.seeedstudio.com)

PROJECT 8 & 9: Video Controller

» **1 Arduino MKR1000** **AF** 3156, RS 124-0657, **A** ABX00004, GBX00011 (Africa and EU), **D** 1659-1005-ND
Alternatively, an Uno-layout board with the WiFi101 shield or WINC1500-based board. **AF** 3033 with 2891
» **1 WiFi connection to the internet** If you access the internet at home with WiFi, you should be all set. Make sure you know the network name and password.
» **1 solderless breadboard** **D** 438-1045-ND, **J** 20723 or 20601, **SF** PRT-12615 or PRT-12002, **F** 4692810, **AF** 64, **SS** 319030002 or 319030001
» **5 220-ohm resistors** **D** 220QBK-ND, **J** 690700, **F** 9339299, **R** 707-7612
» **1 perforated printed circuit board** **AF** 1609, **D** V2018-ND, **J** 616673, **F** 4903213, **RS** 159-5420
» **4 3/4" hex standoffs, 4-40 threads female-female** You may need different part numbers to match your enclosure's particular height, but these will get you started on a search. **D** 36-2204-ND, **RS** 123-6835, **F** 2301244
» **8 1/4" 4-40 thread pan head machine screws** **D** 36-9300-ND, **RS** 274-5086, **F** 2500400
» **3-5 LEDs** **D** 160-1144-ND or 160-1665-ND, **J** 34761 or 94511, **F** 1855510, **RS** 228-5972 or 826-830, **SF** COM-09592 or COM-09590
» **1 rotary encoder with pushbutton** **AF** 377, **SF** COM-10982 with BOB-11722, **SS** 311130001
» **1 pushbutton** **D** GH1344-ND or SW400-ND, **J** 2231822 or 119011, **SF** COM-09337, **F** 1634684, **RS** 718-2213
» **Male headers** **D** A26509-20-ND, **J** 103377, **SF** PRT-00116, **F** 1593411
» **Female headers** **D** ED7102-ND, **F** 1122344, **SF** PRT-00115
» **1 project enclosure** See text.

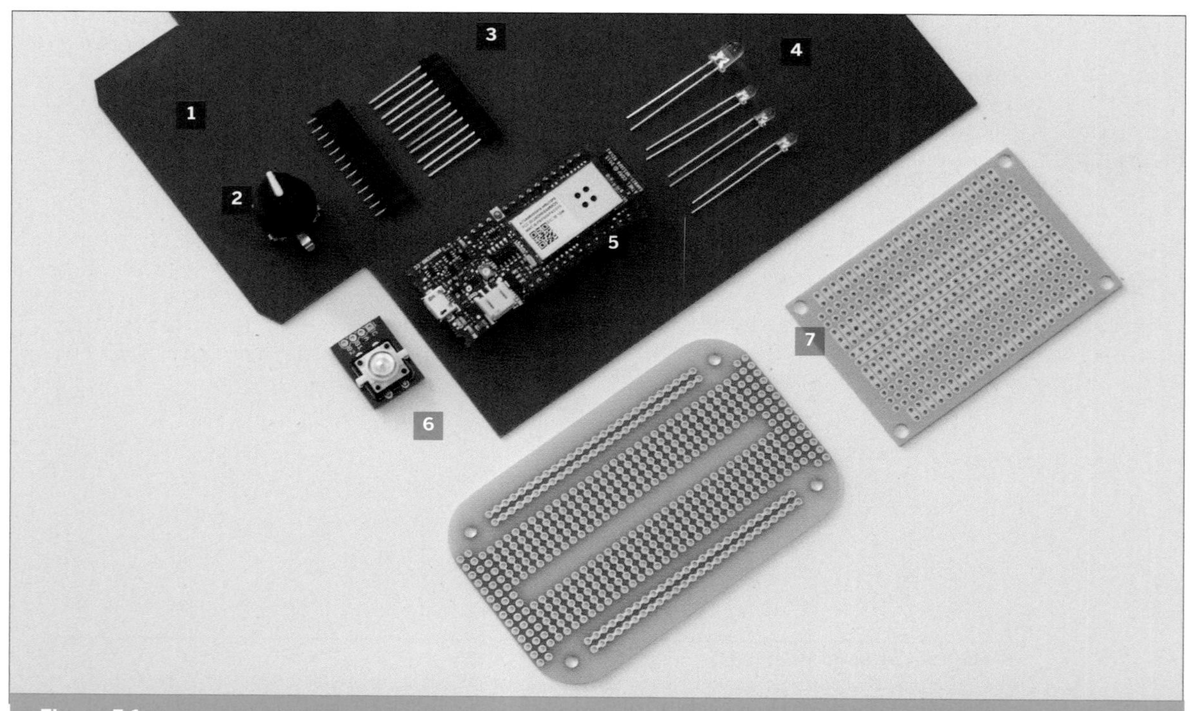

Figure 5-1. New parts for this chapter: **1.** Mat board for enclosure **2.** Rotary encoder **3.** Female headers **4.** LEDs **5.** MKR1000 **6.** Pushbutton with built-in LED **7.** Perforated circuit board

Interactive Systems and Feedback Loops

In every interactive system, there's a feedback loop: you take action, the system responds, you see the response—or a notification of it—and you take another action. In some systems, the timing of that loop can be very loose, with substantial delays between action and response. In other applications, the timing must be tight.

For example, in the cat bed application in Chapter 3, there's no need for the system to respond in less than a few seconds, because your reaction is not very time-sensitive. As long as you get to see the cat while he's on the bed (which may be true for several minutes or hours), you're happy. Monski Pong in Chapter 2 relies on a reasonably tight feedback loop in order to be fun. If it took a half-second or longer for the paddles to move when you move Monski's arms, it would be no fun. The timing of the feedback loop depends on the shortest time that matters to the participant.

Any system that requires coordination between action and reaction needs a tight feedback loop. Consider remote control systems, for example, such as a robot operated over a network. In that case, you'd need not only a fast network for the control system, but also a fast response from the camera or sensors on the robot (or in its environment) that are giving you information about what's happening. You need to be able to both control it quickly and see the results quickly. Networked action games also need a fast network. It's no fun if your game console reacts slowly, allowing other players with a faster network connection to get the jump on you. For applications like this, an exchange protocol that's constantly opening and closing connections (like HTTP does) wouldn't be very effective.

When there's a one-to-one connection between two networked objects, it's easy to establish a tight feedback loop. When multiple objects are involved, though, it gets harder. To begin with, you have to consider how the network of connections between all the objects will be configured. Will it be a star network, with all the participants connected through a central server? Will it be a ring network? Will it be a many-to-many network, where every object has a direct connection to every other object? Each of these configurations has different effects on the feedback loop timing. In a star network, the objects on the edge of the network aren't very busy, but the central one is. In a ring network, every object shares the load more or less equally, but it can take a long time for a message to reach objects on opposite sides of the ring. In a direct many-to-many network, the load is distributed equally, but each object needs to maintain a lot of connections.

In most cases where you have a limited number of objects in conversation, it's easiest to manage the exchange using a central server. The most common program example of this is a text-based chat server like the text-based chat in videoconferencing applications like Skype or Google Hangout. Server programs that accept incoming clients and manage text messages between them in real time are often referred to as *chat servers*. The server programs you'll write in this chapter are variations on the chat server model. These servers will listen for new connections and exchange messages with all the clients that connect. Because there's no guarantee how long messages take to pass through the internet, the exchange of messages can't be instantaneous. As long as you've got a fast network connection for both clients and server, though, the feedback loop will be faster than human reaction time. X

Transmission Control Protocol: Sockets & Sessions

Each time a client connects to a web server, the connection that's opened uses a protocol called *Transmission Control Protocol*, or *TCP*. TCP specifies how objects on the internet open, maintain, and close a connection that will involve multiple exchanges of messages. The connection made between any two objects using TCP is called a *socket*. A socket is like a pipe joining the two objects. It allows data to flow back and forth between them as long as the connection is maintained. Both sides need to keep the connection open in order for it to work.

For example, think about the exchanges between a web client and server that you saw in the previous two chapters. The pipe is opened when the server acknowledges the client's contact, and it remains open until the server has finished sending the file. If there are multiple files needed for a web page, such as images and style sheets, then multiple socket connections are opened and closed.

There's a lot going on behind the scenes of a socket connection. The exchange of data over a TCP connection can range in size anywhere from a few bytes to a few terabytes or more. All that data is sent in discrete packets, and the packets are sent by the best route from one end to the other.

NOTE: "Best" is a deliberately vague term: network hardware calculates the optimal route is differently, which involves a variety of metrics (such as the number of hops between two points, as well as the available bandwidth and reliability of a given path).

The period between the opening of a socket and the successful close of the socket is called a *session*. During the session, the program that maintains the socket tracks the status of the connection (open or closed) and the port number; counts the number of packets sent and received; notes the order of the packets and sees to it that packets are presented in the right order, even if the later packets arrive first; and accounts for any missing packets by requesting that they be resent. All of that is taken care of for you when you use a TCP/IP stack like the Net library in Processing or the firmware on the WiFi modules that you first saw in Chapter 4.

The complexity of TCP is worthwhile when you're exchanging critical data. For example, in an email, every byte is a character in the message. If you drop a couple of bytes, you could lose crucial information. The error-checking of TCP does slow things down a little, though, and if you want to send messages to multiple receivers, you have to open a separate socket connection to each one.

There's a simpler type of transmission protocol that's also common on the internet: *User Datagram Protocol (UDP)*. Whereas TCP communication is based on sockets and sessions, UDP is based only on the exchange of packets. You'll learn more about it in Chapter 7.

TCP Sockets vs. WebSockets

When you used telnet to connect to a web server in Chapter 3, you were opening a TCP socket connection to the server. HTTP works over TCP sockets. If you're familiar with web development at all, you may have heard of a new protocol introduced with HTML5 called *webSockets*. The webSocket protocol allows HTTP clients and servers to establish longer sessions for sustained exchanges, such as you might see in a game or other application that requires a tight interactive loop.

The second server program you'll write in this chapter is a webSocket server. The first is a TCP socket server. Understanding the first will make the second easier to comprehend.

X

A Video Control Application

Media control is a great way to learn about real-time connections. This project is a networked video application that you can control from a physical controller. The server will be a Processing program, and the client will be a physical device that is using a WiFi-connected microcontroller. The clients and the server's screen have to be physically close so that the operator can see the screen. In this case, you're using a network for its flexibility in handling multiple connections, not for its ability to connect remote places.

There are a few well-known controls on any video controller: a play/pause button, fast-forward and rewind buttons, and perhaps a knob to allow you to shuttle frame-by-frame through the video. The controller you make for this video server will be even simpler. It will have a play/pause button, a shuttle knob, and a button to connect or disconnect from the server.

Here's how the system will work:

- The video server shows a video.
- The controller clients connect to the server through a TCP connection.
- When a player connects, the server replies with a greeting string.
- The client can send the following commands:
  - start or stop
  - scroll forward or back `n` frames
  - disconnect
- The server doesn't send any messages to the client other than the hello message.

The communications protocol doesn't define anything about the physical form of the client object. As long as the client can make a TCP connection to the server and can send and receive the appropriate ASCII messages, it can work with the server. You can attach any type of physical inputs to the client, or you can write a client that sends all these messages automatically, with no physical input from the world at all (though that would be boring). Your client doesn't have to use pushbuttons and a scroll knob if you prefer to use other sensors, as long as you send the messages.

You'll need to define the specific bytes you want to use for these messages, however, and the format in which you want to send them. Since you've learned about key-value pairs, let's use them, with colons to separate the key and the value. By doing so, it will be easy to use standard programming tools that can parse that format. The messages will look like this:

- Server greeting: `connect: n` where n is client's number on the server
- Client video play: `playing: n` where n is 1 or 0 (for true or false)
- Client scroll: `position: n` where n is a positive or negative frame count. Positive numbers scroll forward, negative numbers scroll back.
- Client disconnect: `exit:n` if n=1, this 1 tells the server to disconnect the client
- All messages will be terminated with a newline character (`\n`).

A Test Chat Server

You need a server to get started. You don't need code to control the video right now; you just want to confirm that the clients can connect and send messages. Following is a server written in Processing with all the elements to handle network communications. It will let you listen for new clients, and then send them messages by typing in the applet window that appears when you run the program.

X

Try It Start with the includes and variable declarations and definitions. Include the net library, which gives you the ability to make socket connections, The main variables are an instance of the Server class, a port number on which to serve, and an ArrayList to keep track of the clients. You can make the new Server and ArrayList instances here when you initialize the variables, or you can do it in the `setup()` function.

```
/*
  Test Server Program
 Context: Processing

 Creates a server that listens for clients and prints
 what they say.  It also sends the last client anything that's
 typed on the keyboard.
 */

// include the net library:
import processing.net.*;

int port = 8080;                          // the port the server listens on
Server myServer = new Server(this, port); // the server object
ArrayList clients = new ArrayList();       // list of clients
```

»This program uses a Processing data type you may not have seen before: *ArrayList*. Think of it as a super-duper array. ArrayLists don't have a fixed number of elements to begin with, so you can add new elements as the program continues. It's useful when you don't know how many elements you'll have. In this case, you don't know how many clients you'll have, so you'll store them in an ArrayList, and add each new client to the list as it connects. Processing inherits ArrayLists from Java. There is an introduction to ArrayLists on the Processing website at www.processing.org. JavaScript arrays are expandable like ArrayLists as well. Unfortunately, Arduino and C do not have expandable arrays.

» The `setup()` method sets the window size.

```
void setup() {
  size(640, 360);                    // set the window size
}
```

» The `draw()` method listens for new messages from clients and calls a function called `readMessage()` to deal with them.

```
void draw() {
  // listen for clients:
  Client currentClient = myServer.available();

  // if a client sends a message, read it:
  if (currentClient != null ) {
    readMessage(currentClient);
  }
}
```

» The `readMessage()` function handles incoming messages. It skips empty messages (strings with just a newline, for example) and prints the message and the sender's IP address. If a client sends any message containing the substring "exit," the server disconnects it and removes it from the list of clients.

```
void readMessage(Client thisClient) {
  // read available text from client as a String and print it:
  String message = thisClient.readStringUntil('\n');
  // if there's no message, skip the rest of this function:
  if (message == null) return;
  // print the message and who it's from
  println(thisClient.ip() + ": " + message);

  if (message.contains("exit")) {    // if it's a disconnect message,
    myServer.disconnect(thisClient); // disconnect client
    clients.remove(thisClient);      // delete client from the clientList
  }
}
```

The `serverEvent()` message is generated by the server when a new client connects to it. `serverEvent()` announces new clients and adds them to the client list. It sends the new client the greeting message telling it its place in the `clients` ArrayList.

```
// ServerEvent occurs when a new client connects to the server:
void serverEvent(Server myServer, Client thisClient) {
  println("New client: " + thisClient.ip()); // print client's IP
  clients.add(thisClient);                  // add it to the clientList
  thisClient.write("client:" + clients.size() + "\n");     // say hello
}
```

Finally, the `keyReleased()` method sends any keystrokes typed on the server to all connected clients. This is just for testing.

```
void keyReleased() {
    myServer.write(key);
}
```

Run the server and open a telnet connection to it. Remember, it's listening on port 8080, so if your computer's IP address is, say, 192.168.1.45, you'd connect like so: `telnet 192.168.1.45 8080`. If you're telnetting in from the same machine, you can use: `telnet localhost 8080` or `telnet 127.0.0.1 8080`. (Windows 10 users: telnet may ask for password permission to connect). Whatever you type in the telnet window will show up in Processing's console pane when you press Enter, and whatever you type in the server's applet window will show up at the client's command line. Try sending the messages as described previously: playing, position, and exit. Sending `exit:1` will cause the server to close the connection to the client.

Once you understand the server, it's time to make a microcontroller-based client to connect to it.

X

The Controller Client

The video controller client listens to local input and remote input. The local input is from you, the user. The remote input is from the server. The client is constantly listening to you, but it only listens to the server when it's connected.

To listen to you, the client needs:

- An input for instructing it to connect. The same input can be used to send a disconnect message.
- An input for sending a scroll message.
- An input for sending a play/pause message.

To let the user know what the client device is doing, add:

- An output to indicate whether the client is connected to the server.
- An output to indicate whether the video should be playing or not
- Outputs to indicate when the buttons or scroll wheel are active

It's always a good idea to put outputs on the client to give local feedback when it receives input from the user. This is what the last outputs are for. Even if there is no connection to the server, local feedback lets the user know that the client device is responding to her actions. For example, pressing the connect button doesn't guarantee a connection, so it's important to separate the output that acknowledges a button push from the one that indicates successful connection. If there's a problem, this helps the user determine whether the problem is with the connection to the server, or with the client device itself. The circuit for this project includes LEDs triggered directly by the sensors, so no code is needed to activate them.

In the code for this client, you'll control two other LEDs, which are used to indicate whether the client is connected or disconnected, and whether the remote video should be playing or stopped. If this client had a more complex interaction, you'd need more status indicators.

X

The Circuit

The circuit for the client uses an MKR1000 as its microcontroller. It uses several of the GPIO pins for the pushbuttons, the scroll wheel, and the indicator LEDs. The ESP8266 boards don't have enough GPIO pins to build this project, so if you want to use one of them, you'll need to use a second microcontroller connected serially to it in order to control the physical inputs and outputs. You'd also need to break the code into the network code and the physical I/O code as well.

This client uses a *rotary encoder* as the scroll wheel. A rotary encoder, also called a *quadrature encoder*, is a sensor that has two output pins and a knob that can turn a full 360°. When the knob turns, the output pins pulse. Their pulsing is out of phase with each other so that by watching the sequence of pulses, you can tell which direction the encoder is turning. The sequence of pulses produced is called *gray code*. A program to read an encoder must read and interpret gray code. Different encoders have different resolutions. The ones recommended here both produce 24 pulses per revolution, meaning that you'd see 24 changes of the I/O pins every time you turn the encoder one full revolution.

Rotary encoders often incorporate a pushbutton in the shaft, as the ones recommended here do. The pushbutton is handy, as it allows you to combine the play/pause button and the scroll wheel into one physical control. If you're using the SparkFun encoder, it also has an RGB LED in it, so you could use the red, green, and blue channels of the LED as three of your indicator LEDs if you wish. I used the Adafruit encoder in the illustrations shown here.

This project uses an illuminated pushbutton as well, for the connect button. It's a convenient way to show the connection state to the server right on the control that affects it. If you're not using an illuminated pushbutton, you'll need a separate LED to replace the one inside it.

The pushbuttons are wired differently from a normal switch. First, they're wired to ground instead of to voltage. When you push one of the buttons, the input pin to which it's connected will read as LOW. This is called *inverted logic*, and allows you to read the buttons without a separate pulldown resistor. In the program, you'll initialize them using `pinMode(pinNumber, INPUT_PULLUP)`. Most microcontrollers have internal pullup resistors on their GPIO pins so that you can wire switches this way.

MATERIALS

- **1 Arduino-compatible, WiFi-enabled microcontroller:**
  - **MKR1000, or Uno-layout board with WiFi101 shield or WINC1500-based board**
  - **Features used: WiFi, Digital I/O, UART**
- **1 WiFi connection to the internet**
- **1 solderless breadboard**
- **5 220-ohm resistors**
- **1 perforated printed circuit board**
- **3–5 LEDs (see text)**
- **1 pushbutton**
- **1 rotary encoder with pushbutton**
- **1 project enclosure (see text)**

The pushbuttons also have LEDs connected to them from voltage. This way, when you push the button, the LED lights up, because its cathode connects to ground through the button. There's no need to add any code to make this happen—the pushbutton will ground the LED circuit automatically, and you'll get local feedback on when the button is being pressed. The encoder pins also have LEDs wired similarly.

Figures 5-2 and 5-3 show the circuit on a solderless breadboard as usual. Figures 5-4 and 5–5 show the device laid out on a printed circuit prototyping board (known as a *perfboard*). The perfboard makes it possible for the circuit to be a little smaller and more physically robust. It's harder to assemble than a solderless breadboard circuit, though, so you should build the circuit on a breadboard first. Both layouts are shown so you can choose the method you like.

You'll need an enclosure for the circuit. You can buy a project case from your favorite electronics retailer, but you can also get creative. For example, a pencil box from your friendly neighborhood stationery store will work well. Drill holes in the lid for the controls and the LEDs, add a battery inside, and you're all set. If you're skilled with a mat knife, you can also make your own box from cardboard or mat board. Figures 5-6 and 5–7 show the box I made, and Figure 5-8 shows the template from which I made it. If you're lucky enough to have access to a laser cutter, use it; if not, a mat knife and a steady hand can do the job.

X

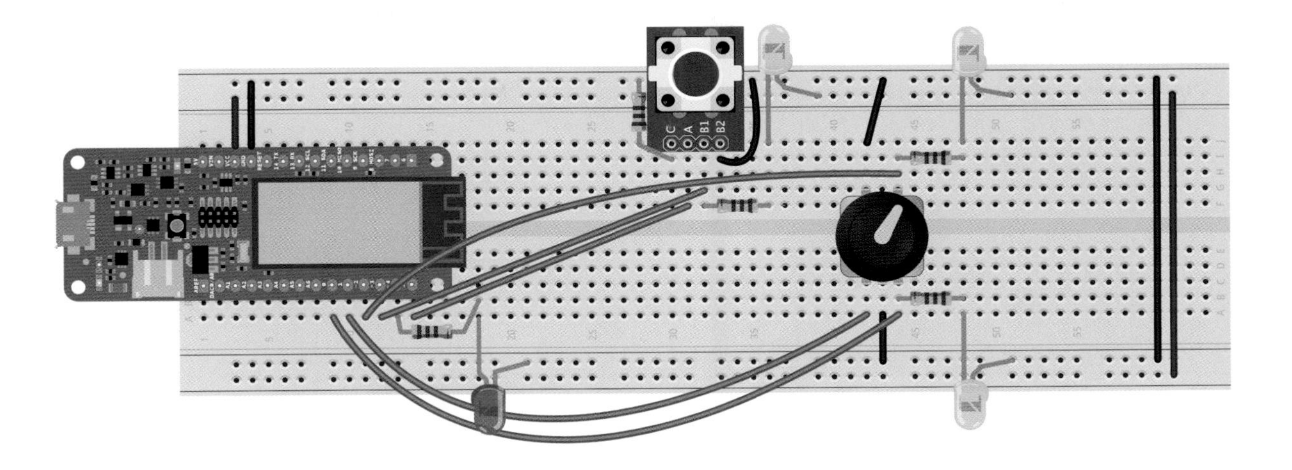

Figure 5-2
The video controller client, breadboard layout. Although it's shown on a full-size breadboard for clarity, it will fit on a half-size breadboard.

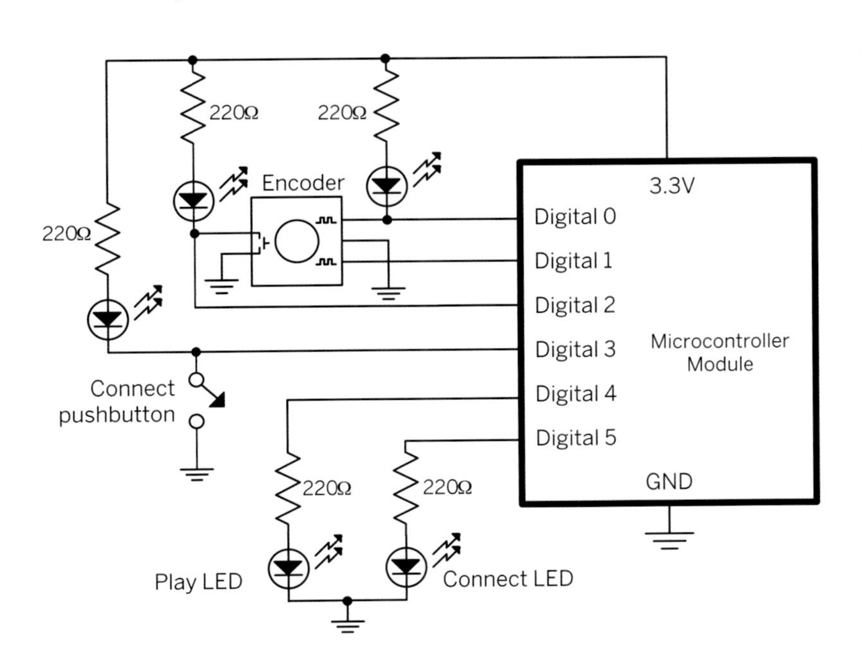

Figure 5-3
The video controller client schematic. The following page shows the layout for the circuit on a printed circuit prototyping board (a perfboard).

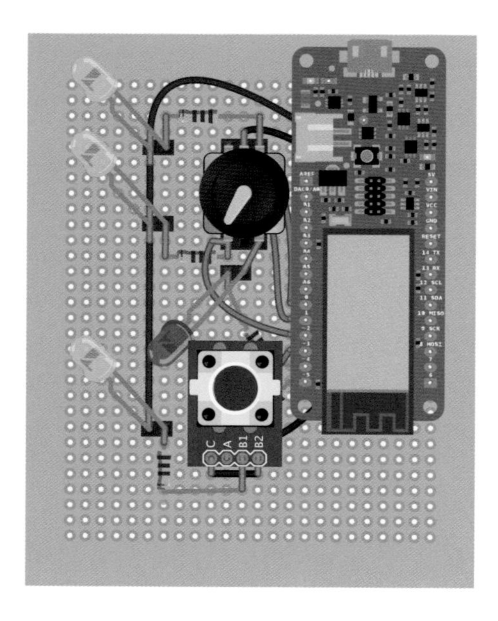

Figure 5-4
The circuit as laid out on a printed circuit prototyping board (perfboard), with all components on board. All the components except the resistors are all attached to the board via socket headers so they can be removed easily.

Figure 5-5
The circuit board layout showing where the solder joints (in gray) connect wires, resistors, and sockets underneath the board.

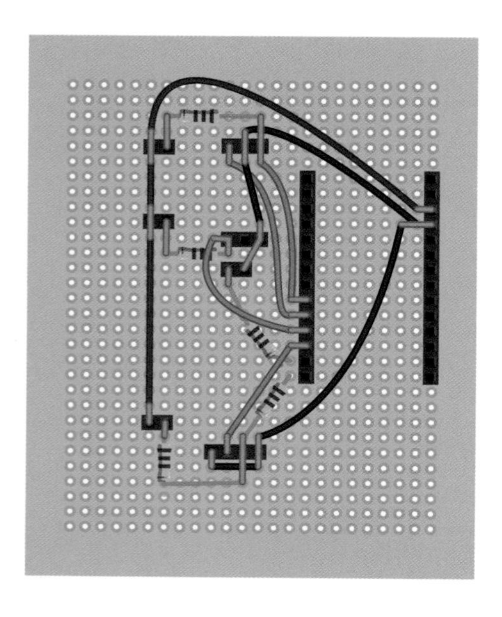

Figure 5-6
The assembled video controller. The circuit inside the housing was designed to be just tall enough so that the LEDs and encoder would stick through the top. Assemble the circuit first so you know how much space you need for the housing.

Figure 5-7
Detail of the assembled circuit board. Use female headers to mount the LEDs, the encoder, and the pushbutton. Additional extension headers make the job of getting the components to the right height much easier. You can trim the LED legs bit by bit until they are the right height, then stick them in the headers. The hex spacers lift the circuit board from the box, leaving room for a Lithium-poly battery underneath, which plugs conveniently into the MKR1000. Make sure to use at least a 1000mA battery.

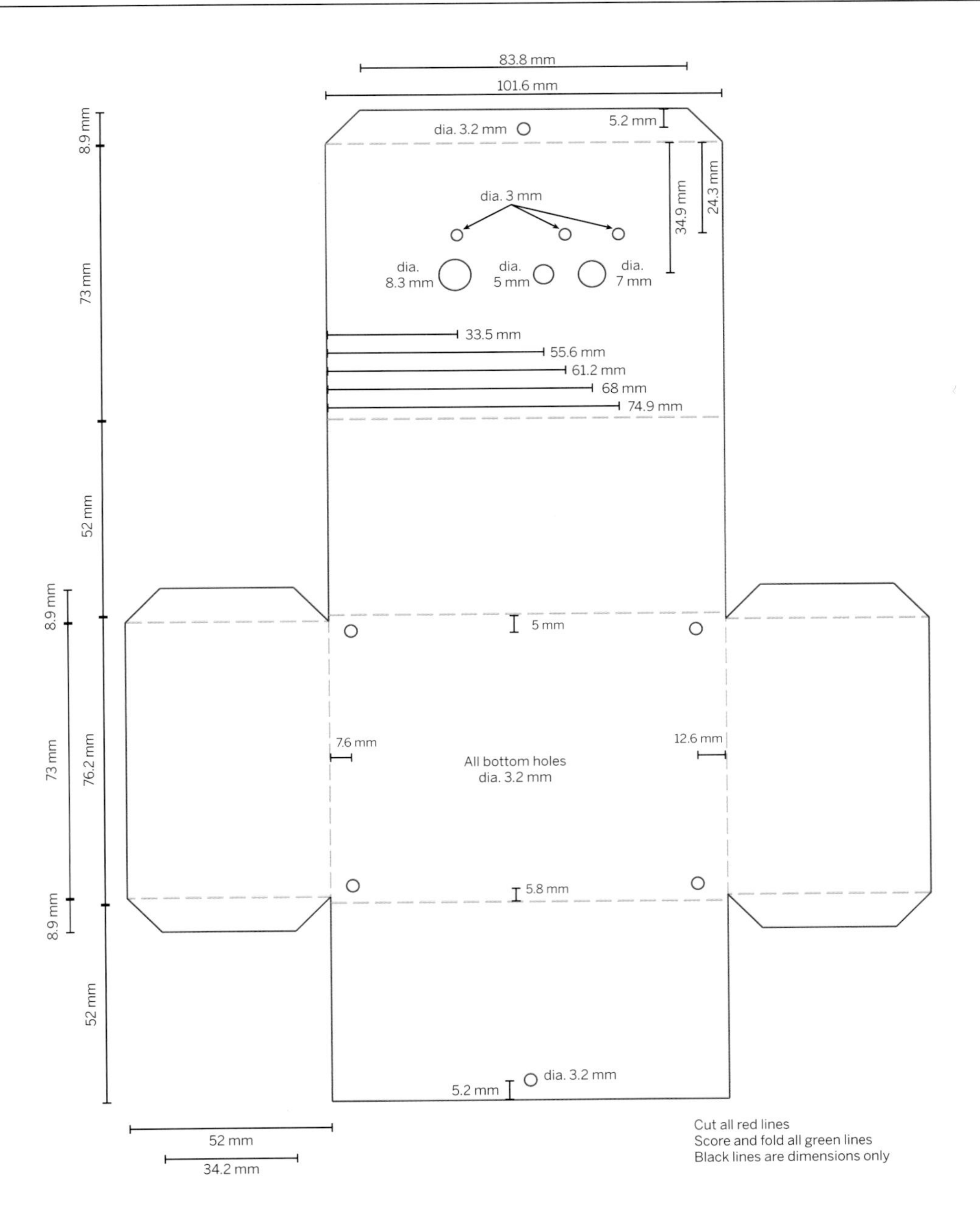

Figure 5-8
Template for the housing. This template can be cut out of poster board or mat board and then folded to make a housing for the controller. The dimensions will depend on how you assemble your circuit, so modify them as needed. Although they look precise here, they were measured with a caliper after the fact.

The screw holes in the bottom and the front are drilled for 4-40 size machine bolts, to hold the box on its standoffs.

The first two prototypes were drawn by hand and cut out of paper, then scrap cardboard. I didn't move to a final version cut out of mat board until I had tested the fit of the sides with a paper version. Score the folds on the outside side of the board so they fold better.

The Code

This sketch uses two new libraries, Encoder, by Paul Stoffregen and Button, by Michael Adams. You can find and install both using the Library Manager as you did for previous libraries. The Encoder library simplifies the reading of a rotary encoder. It reads the changes in the I/O pins of the encoder and returns the number of steps and direction that the encoder has moved. Similarly, the Button library simplifies the reading of a pushbutton attached to an I/O pin. It reads the state of its I/O pin and keeps track of whether it's toggled, pressed, or released.

Plan It The video controller client code reads the sensors, and listens for data from the server. If it's connected to the server, it sends changes from the sensors to the server.

In the code that follows, the sensor reading and sending is broken out into separate functions.

```
/*
Video client
context: Arduino
*/

// include necessary libraries
// initialize global variables

void setup() {
  // connect to WiFi
  // initialize inputs and outputs
  // initialize serial communications
}

void draw() {
  // read scroll wheel
  // read buttons

  // if connected to server,
  //    listen for incoming data from server
  //    if scroll wheel has changed, send change to server
  //    if connect button changed, toggle connect state
  //    if play button changed, toggle playing state

  // update status output LEDs
}
```

Try It First, make a new file called **config.h** and store the WiFi access point SSID and password in it as you did for the Arduino programs in Chapter 4.

```
// config.h
char ssid[] = "ssid";      //  your network SSID (name)
char password[] = "s3c3r3+!"; // your network password
```

At the beginning of your code, import the libraries you need.

```
// include required libraries and config files
#include <SPI.h>
#include <WiFi101.h>
#include <Encoder.h>
#include <Button.h>
#include "config.h"
```

Set up global variables to hold the server address (your personal computer's address, where the server runs) and make an instance of the WiFiClient library. This is how you'll create a TCP socket connection to the server, and read from and write to the server.

```
const char serverAddress[] = "192.168.0.12"; // server address
int port = 8080;                              // port number
WiFiClient tcpSocket;                         // server socket
```

Change this to match your server"s IP address

Then make instances of the Encoder and Button libraries to read the encoder and buttons. Set up constants for the LED pin numbers to make them easier to remember as well. Finally, you need a Boolean variable to track the playing state, and a long integer to hold the last read encoder position.

```
Encoder myEncoder(0, 1);         // instance of the encoder library
Button playButton(2);            // instances of the button library
Button connectButton(3);
const int playLED = 4;           // pin numbers for the LEDs
const int connectLED = 5;

boolean playing = false;         // what the video state should be
long lastPosition  = 0;          // last read position of the encoder
```

The `setup()` function sets the states of the I/O pins and starts the button instances. Notice that the pinMode for the encoder pins is `INPUT_PULLUP`. This tells the microcontroller that you want to use the internal pullup resistors on these pins.

Once the pins are initialized, you'll connect to your WiFi network, just like you did in the sketches in Chapter 4.

```
void setup() {
  Serial.begin(9600);              // initialize serial communication
  pinMode(0, INPUT_PULLUP);        // initialize encoder pins
  pinMode(1, INPUT_PULLUP);
  pinMode(connectLED, OUTPUT);     // initialize LED pins
  pinMode(playLED, OUTPUT);
  connectButton.begin();           // initialize buttons
  playButton.begin();

  // while you're not connected to a WiFi AP,
  while ( WiFi.status() != WL_CONNECTED) {
    Serial.print("Attempting to connect to Network named: ");
    Serial.println(ssid);          // print the network name (SSID)
    WiFi.begin(ssid, password);    // try to connect
    delay(2000);
  }

  // When you're connected, print out the device's network status:
  IPAddress ip = WiFi.localIP();
  Serial.print("IP Address: ");
  Serial.println(ip);
}
```

The Encoder library relies on the interrupt function of the microcontroller to read the encoder pins. ***Hardware interrupts*** **allow the controller to interrupt whatever is going on in the code when a given input pin changes. If interrupts interfere with the communication between the main processor on the MKR1000 and the WiFi radio, you can turn off the use of interrupts by using the `#define ENCODER_DO_NOT_USE_INTERRUPTS` directive before you include the library.**

Begin the `loop()` function by calling two new functions, `readEncoder()` and `readButtons()`, which you'll write shortly. These will do what their names imply. Then check to see if there's a socket connection to the server, and if so, check if there's any data available to be read from the server. If there is, read it and print it out to the Serial Monitor.

Finally, update the status LEDs. The Connect LED depends on whether there's a socket connection, so you can use the value returned by the `tcpSocket.connected()` function, which returns 1 or 0 for true or false. Likewise, you can turn on or off the Playing LED using the `playing` variable, which you'll set to true or false as well.

```
void loop() {
  // read the sensors:
  readEncoder();
  readButtons();

  // check for incoming data from the server:
  if (tcpSocket.connected()) {     // if connected to the server,
    if (tcpSocket.available()) {  // if there is a response from the server,
      String result = tcpSocket.readString();  // read it
      Serial.print(result);       // and print it (for diagnostics only)
    }
  }
  // update the status LEDs:
  digitalWrite(connectLED, tcpSocket.connected());
  digitalWrite(playLED, playing);
}
```

Start the `readEncoder()` function by calling `myEncoder.read()`. It will return the number of steps the encoder's taken. Positive values indicate clockwise movement and negative values indicate counterclockwise movement. Compare the position to the last position read, and if there's a socket connection to the server open, send the difference to the server. Then save the current position for comparison next time you read.

```
void readEncoder() {
  long position = myEncoder.read();        // read the encoder
  long difference = position - lastPosition;  // compare to last position
  if (difference != 0) {                   // if it's changed,
    if (tcpSocket.connected()) {           // if the socket's connected,
      tcpSocket.print("position:");        // send the key
      tcpSocket.println(difference);       // send the value
    }
    lastPosition = position;               // update lastPosition
  }
}
```

Start the `readButtons()` function by checking if the Connect button is toggled (changed), and if so, read it to see if it's pressed. If so, and there's no socket connection to the server, attempt to connect. If you're connected, send `exit:1` and the server will disconnect.

By determining if `.toggled()` is true and the button is low, you ensure that you're only taking action when the button changes state. If you only checked if it was pressed, you'd take action repeatedly as long as it's held down.

```
void readButtons() {
  if (connectButton.toggled()) {           // if connect button has changed
    if (connectButton.read() == LOW) {     // and it's pressed
      if (!tcpSocket.connected()) {        // if you're not connected,
        connectToServer();                 // connect to server
      } else {                             // if you're already connected,
        tcpSocket.println("exit:1");       // disconnect
      }
    }
  }                                        // end of connectButton.toggled
```

Finish the `readButtons()` function by checking the Play button. Check both toggled and pressed conditions just like you did with the Connect button. If its state has changed, change the variable playing to its opposite, and if there's a socket connection to the server, send a message with the current state of this variable.

```
  if (playButton.toggled()) {              // if the play button has changed
    if (playButton.read() == LOW) {        // and it's LOW
      playing = !playing;                  // toggle playing state
      if (tcpSocket.connected()) {         // if you're connected,
        tcpSocket.print("playing:");       // send the key
        tcpSocket.println(playing);        // send the value
      }
    }
  }                                        // end of playButton.toggled
}                                          // end of readButtons()
```

Finally, you need the `connectToServer()` function that's called in the first half of `readButtons()`. This calls the `.connect()` function from the WiFiClient library to attempt a connection to the server. This is a *blocking function*, meaning that it stops your whole program until a connection is made. You'll probably notice a delay between when you press the button and when the Serial Monitor reports the result.

```
void connectToServer() {
  Serial.println("attempting to connect");
  // attempt to connect to the server on the given port:
  if (tcpSocket.connect(serverAddress, port)) {
    Serial.println("connected");
  } else {
    Serial.println("failed to connect");
  }
}
```

Upload the client code to the microcontroller, then run the Processing chat server that you wrote previously. Open the Serial Monitor so you can see the status messages. Once your controller's connected to WiFi, press the Connect button, and you should see it connect to the server. Click the Play button or turn the encoder, and you should see the messages come through in Processing's console pane. If you type in the Processing sketch window and hit the Enter key, the message should show up in the Arduino Serial Monitor as well. Congratulations, you've got a working client!

Send a few messages to the server by triggering the encoder and the buttons to get a sense how both the server and the controller client device work. Once you're comfortable, it's time to add video to the server application.
X

Hardware Interrupts

Hardware interrupts **allow a microcontroller to interrupt whatever is going on in the code when a given input pin changes. The interrupt function in this library's examples (called an *interrupt service routine* or *ISR*) is internal to the Encoder library. You can read more about them on the Arduino reference site at www.arduino.cc/en/Reference/AttachInterrupt. The hardware interrupt pin numbers vary from processor to processor. The ATMega328 processor on the Uno has two interrupts, numbered 0 and 1, attached to digital pins 2 and 3, respectively. On the 101 all the digital pins can be used as interrupts. On the MKR1000, pins 0, 1, 4, 5, 6, 7, 8, 9, A1, and A2 are interrupts. The `digitalPinToInterrupt(pin)` function translates the digital pin number to the interrupt number. For example, on the Uno, `digitalPinToInterrupt(3)` returns 1. This command doesn't work on the 101, though, since it's not needed on that controller.**
X

Adding Video to the Server

The main difference between the video server application and the chat server you've already written is that the video server parses the messages from the client and uses them to control the video. The network connection management is the same as the chat server, so you can start with the previous Processing code and add to it to make the server. The changes will be as follows:

- **Global variables:** add video library and variables to manage the video
- **Setup function:** add code to set the window size and initialize the video
- **Draw function:** add code to display a frame of video and text overlay
- **serverEvent function:** no changes needed
- **readMessage function:** add code to parse messages into key-value pairs and take appropriate actions

New functions:

- **movieEvent function:** called by sketch every time a new video frame is ready to read. Use it to read the video.
- **scrub function:** jumps to new position in the movie

You'll also need a video for your sketch to run. The video library for playing movies is the same as the one you used to capture video in Chapter 3. It can run a wide variety of video formats, but the Processing foundation recommends encoding your video using the H.264 format. Keep the size small, 640x480 or thereabouts. If you don't have a video handy, the video library examples provide a few you can use. Save your video in a subdirectory of your sketch's directory, called **data**.

X

Code It At the beginning of your sketch, import the video library and add a few new global variables for managing the video and client messages. New lines are shown in blue.

```
/*
  Video server
 context: Processing
 */

// include necessary libraries
import processing.net.*;
import processing.video.*;

// set global variables:
int port = 8080;                        // the port the server listens on
Server myServer;                        // the server object
ArrayList clients = new ArrayList(); // list of clients
Movie myVideo;                          //  movie player
boolean playing = false;               // state if the video: playing/paused
String lastMessage = "";               // last message received from client
```

In the `setup()` function, add a call to the `size()` function to set the window size. The size parameters should be the width and height of your video, in pixels, and the new movie instantiation should use the name of your movie. You'll add the `scrub()` function shortly; it will jump to a given frame of the movie.

```
void setup() {
  size(640, 360);
  myVideo = new Movie(this, "movie.mov"); // initialize the video
  myServer = new Server(this, port);      // start the server
  scrub(0.0);                             // set the video to the start
}
```

In the `draw()` function you'll add a line to display the latest frame of the video in the window. After that, you'll add text overlay as you did in the cat server in Chapter 3, to display the last client message received, and who it's from.

```
void draw() {
  // listen for clients:
  Client currentClient = myServer.available();

  // if a client sends a message, read it:
  if (currentClient != null ) {
    readMessage(currentClient);
  }
  // display the latest video frame:
  image(myVideo, 0, 0, width, height);

  // display the client that sent the last message:
  fill(15);                              // dark gray fill
  text(lastMessage, 11, height-19);     // drop shadow
  fill(255);                             // white fill
  text(lastMessage, 10, height-20);     // main text
}
```

The two new events you'll add are `movieEvent()` and `scrub()`. The former is called automatically when there's a new frame of video ready. It just calls the video library's `.read()` command to get the new frame.

The `scrub()` function jumps to a given position in the movie. You have to play or loop first, then use the `.jump()` command to move the position. If you're playing state is paused, then use the `pause()` command to do so.

```
void movieEvent(Movie myVideo) {
  myVideo.read();
}

void scrub(float newPosition) {
  myVideo.loop();                   // play the video
  myVideo.jump(newPosition);         // jump to the calculated position
  if (!playing) myVideo.pause(); // if the movie should be paused, pause it
}
```

Raspberry Pi as a Networked Display

You probably don't want to dedicate your personal computer to a display for this project. This is where a Raspberry Pi or other embedded computer can come in handy. You can attach a Pi to any HDMI-capable monitor or TV and then run Processing on it to act as the display for this sketch. A Pi 2 or 3 can run this sketch reasonably well. Connect your Pi to the monitor or TV with an HDMI cable, and add a keyboard and mouse via USB. Then plug in the Pi and it'll boot into its graphic user interface.

To install Processing on the Pi, connect to it through the command-line interface, and make sure it's networked and that you've installed Processing as you saw in Chapter 1. Processing will now be available through both the graphic user interface and the command-line interface. You'll find it in the Programming submenu of the Raspberry menu in the GUI, or you can launch it by typing `processing` **on the command line.**

Processing on the Pi has a series of quirks and exceptions you'll need to deal with, including using a Pi-specific variation on the video library. There's a helpful guide to running Processing on the Pi at github.com/processing/processing/wiki/Raspberry-Pi.

Finally, make the following changes to the `readMessage()` command, which reads incoming client messages. The new code replaces the conditional statement from the chat server sketch that begins with `if (message.contains("exit"))`.

First, use the `trim()` command to remove any extraneous tabs, newlines, or carriage return bytes from the end of the string. These characters are called *whitespace*. They're added by some systems (for example, the Arduino `println()` statement adds `\r\n`), and they can interfere with converting a substring to a different data format.

Next use the `split()` command to split the string at the colon into two substrings. The first substring is the message's property name, and the second is the property value.

Once you have the property name and value, use three conditional statements (`if` statements) to take the appropriate action. The position message converts the `position` value and moves the video. One video frame is approximately 0.033 second. The `playing` property starts or pauses video. The `exit` property causes the server to disconnect the client.

At the end of the function, save the client message to a global variable so that the `draw()` function can use it to display it onscreen.

```
void readMessage(Client thisClient) {
  // read available text from client as a String and print it:
  String message = thisClient.readStringUntil('\n');
  // if there's no message, skip the rest of this function:
  if (message == null) return;
  // print the message and who it's from
  println(thisClient.ip() + "\t" + message);

  message = message.trim();                      // trim whitespace
  String[] decodedMsg = split(message, ":");     // split message at the colon
  String property = decodedMsg[0];               // first part is the key
  int value = int(decodedMsg[1]);                 // second part is the value

  //    if it's a play/pause message, change the video state:
  if (property.equals("playing")) {
    playing = boolean(value);        // convert value to a boolean
    if (playing) {                   // if it's true,
      myVideo.loop();                // play the video
    } else {                         // if not,
      myVideo.pause();               // pause it
    }
  }

  // if it's a position message, change the video time:
  if (property.equals("position")) {
    float frames = value * 0.033;
    float videoTime = myVideo.time() + frames;
    scrub(videoTime);       // set the video to the appropriate position
  }

  // if it's a disconnect message, disconnect the client:
  if (property.equals("exit") && value == 1) {
    myServer.disconnect(thisClient); // disconnect client
    clients.remove(thisClient);      // delete client from the clientList
  }
  // save the last client message received for printing on the screen:
  lastMessage = thisClient.ip() + ": " + message;
}
```

From the client's point of view, this server will respond exactly the same as the chat server. Run the sketch and try connecting to it using your controller client, or from a telnet connection and you'll see, However, now it will display and control video in reaction to the client's messages. Figure 5-9 shows what the server displays, using video shot from a moving train.

You can log into the server with as many clients as you wish. The server will respond to the last client's message. One side effect of this is that the server's playing state can get out of sync with the clients' states. To correct this, you could add code to broadcast the server's playing state whenever it changes. You'd also need to add client code to receive and act on these messages. See if you can work out how to do this.

X

Figure 5-9
Screenshot from the Processing video server. You can see the client IP address and last message in the lower-left corner.

 **Project 9**

A WebSocket Video Controller

In the server sketch, you probably noticed some lag between when you triggered your sensors and when the video moved in Processing. That's because video playback is computationally intensive. It would be better if you could separate the server, which manages the network transactions, from the display, which controls the video, so that the server doesn't have to do the most computation. In this project, you'll do that by writing a web server in node.js to replace Processing as the server, and a web page to replace it as the display client. Your controller client device will control video on the display client, with the server managing the exchange. You'll maintain a persistent connection between clients and server using webSockets.

You might think, "Why not make the server an HTTP server like we did in Chapter 4? You can put video in a web page, right?" You can, but controlling the video in the client remotely would be tricky using normal HTTP requests While HTTP is convenient because of its ubiquity, it's not so effective for real-time or near-real-time applications like this video server, because you're opening and closing a socket for each exchange between the server and client.

The *webSocket protocol* addresses the need for realtime client-server interaction in HTTP. It's an extension to HTTP that allows clients to upgrade their connection so that the socket remains open and the server and client can exchange messages continually. Instead of separate TCP sessions for every transaction, the client can now establish one session with the server. This protocol was designed to support applications requiring a tighter interactive loop between clients and server.

Figure 5-10 shows the system diagram for the new version of this project. The server can support multiple clients, both controller and display clients. With a more complex protocol, each controller client could even target a specific display client.

WebSockets are more complex than normal TCP sockets, however. First a client has to make an HTTP request, and in the header of the request, it must include the following headers:

```
Upgrade: websocket
Connection: Upgrade\r\n
Sec-WebSocket-Key: dGhlIHNhbXBsZSBub25jZQ==
Sec-WebSocket-Version: 13
```

The `Upgrade` and `Connection` headers tell the server that the client wants to establish a websocket, the `Sec-WebSocket-Key` is a unique string to handshake with the server, and the `Sec-WebSocket-version` says what version of the webSocket protocol the client is using. The server will respond like so:

```
HTTP/1.1 101 Switching Protocols
Upgrade: websocket
Connection: Upgrade
Sec-WebSocket-Accept: s3pPLMBiTxaQ9kYGzzhZRbK+xOo=
```

The `Sec-WebSocket-Accept` is a version of the client's key encrypted with a string that's standard to the webSocket protocol. This allows the client to verify the server.

Once they're connected, the client and server can exchange bytes continually. However, the server expects the client to encode its messages using a formula specific to the webSocket protocol. There are webSocket clients for most platforms that will take care of the protocol management for you. WebSockets are built into HTML5 and client-side JavaScript now, so it's easy to use them in an HTML page. There are a few webSocket client and server libraries for node.js as well. In this project, you'll use one called `ws`. The Arduino-based library you'll use is the `webSocket` class from the ArduinoHttpClient library you used in the previous chapter.

X

Many webSocket libraries implement the protocol incompletely, or add proprietary features that make it difficult to interoperate with other libraries written in other programming languages. The ws library for node.js was the most interoperable with Arduino and HTML5 at the time of this writing. If you're interested in learning more, explore other libraries, but be prepared for some incompatibilities.

Figure 5-10
System diagram for a websocket video server. Server supplies video HTML page and video assets to browsers via HTTP, then connects both controller and display clients' webSockets via HTTP Upgrade. This server can support multiple clients of both types.

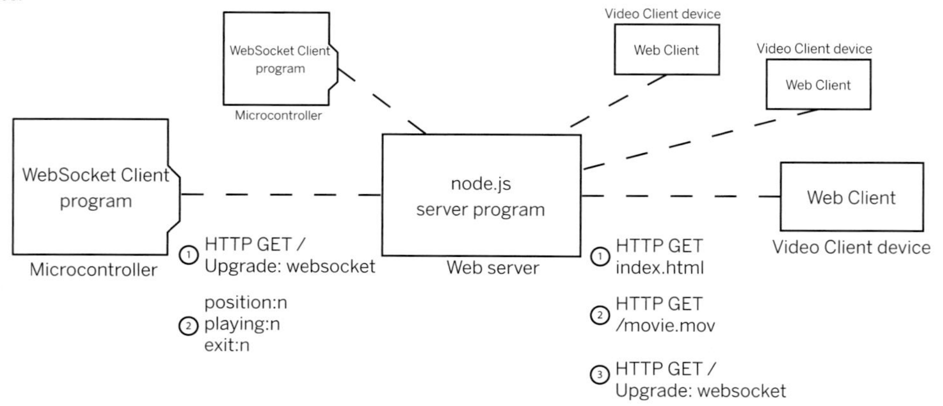

The Server and Browser Client

WebSocket servers are built on top of HTTP servers, so your server for this project will start with a node.js express-based HTTP server like the one you saw in Chapter 4. The browser client will be an HTML page that displays and controls the video using the JavaScript library called p5.js.

Try It Start with the basics for a server using the express.js library. Save this file in its own project directory as **wsExpressServer.js**, then add the express.js library using npm as you did before like so:

```
$ npm install express
```

You can see this server serves static files from a subdirectory called `public` like the one in Chapter 4 did. Shortly, you'll make a new subdirectory in the project directory called **public**.

The webSocket library will need access to an instance of node.js's http library, so this program makes that instance from the Express server, then uses it to listen for clients. This server will run now, but it won't do much so far. You'll need a web page to serve. To make it, you're going to use p5.js.

```
/*
Express-based webSocket Server
context: node.js
*/

// Include libraries:
var express = require('express');           // the express library
var http = require("http");                 // the http library

var server = express();                     // the express server
var httpServer = http.createServer(server); // the http server

// serve static files from /public:
server.use('/',express.static('public'));

// start the server:
httpServer.listen(8080);                    // listen for http connections
```

`express.static()` is called middleware. It's a function that intercepts all HTTP requests before passing them to listener functions like `http.get()` and `http.post()`. You'll see other middleware in later chapters. The `.use()` function routes all requests that match the path in the first parameter through the middleware that you specify in the second.

Adding Interactive Elements with p5.js

In Chapter 1 you got a brief introduction to p5.js, a JavaScript framework for making interactive web pages in a Processing-like style. This will be your first project with it. Make a new subdirectory in your project directory called **public**, and copy the files from the p5.js **empty-example** project into it. The files in it should be as follows:

```
public/
   index.html
   sketch.js
   libraries/
      p5.js
      p5.dom.js
      p5.sound.js
```

If you installed the command-line tool p5-manager, you can install the whole p5.js project on the command line like so:

```
$ p5 g -b public
```

Then change directories into the **public** directory and get the latest versions of the p5.js libraries like so:

```
$ p5 update
```

You'll see this pattern frequently in the chapters that follow: a node.js project that's a server, with a public subdirectory that's a p5.js project. Together, node.js and p5.js are a handy combination.

Try It When you open the **index.html** file from the p5.js empty project template, it will look similar to this:

```
<!DOCTYPE html>
<html>
  <head>
    <meta charset="utf-8">
    <script type="text/javascript" src="p5.js"></script>
    <script type="text/javascript" src="libraries/p5.dom.js">
    <script type="text/javascript" src="sketch.js">
    <title></title>
  </head>
```

The JavaScript code you'll write will all be in the **sketch.js** file. Open it and replace the default code with the code to the right.

Copy your video file into the **public** directory as well, so the server can serve it.

Now you can run your server program and see some results. On the command line, run the server like so:

```
$ node wsExpressServer.js
```

Then open a browser and go to http://localhost:8080. You should see your video playing. Change the video name and video size to match the width and height of your particular movie.

```
/*
  Video client
  context: p5.js
*/
var myVideo;                               // movie player

function setup() {
  myVideo = createVideo("movie.mov"); // initialize the video
  myVideo.size(640, 360);              // set its size
  myVideo.position(10, 30);            // position it
  myVideo.loop();                      // loop the video
}
```

Adding WebSocket Functionality

Now that you've got the HTTP server and the HTML client started, you need to give both the capability to make webSocket connections. So far, you've worked on the center and the right side of the diagram in Figure 5-10. You've got steps 1 and 2 of that side working, in that you can serve the **index.html** page and the movie. Now you'll add the third step on the right side. It's best to get the webSocket functionality working in software alone before you add the microcontroller to the system.

The communications protocol for this server is the same as the one in the previous project. The messages will look like this:

- Server greeting: `connect: n` where n is client's number on the server
- Client video play: `playing: n` where n is 1 or 0 (for true or false)
- Client scroll: `position: n` where n is a positive or negative frame count. Positive numbers scroll forward, negative numbers scroll back.
- Client disconnect: `exit:n` if n=1, this 1 tells the server to disconnect the client
- All messages will be terminated with a newline character (`\n`).

This time, since the server and the browser client are written in JavaScript, you'll send the messages in JSON, because it's easy for JavaScript to parse it that way. That means that you need to surround your messages with curly braces, make sure the key names are in quotation marks, and separate the key-value pairs with commas as needed. Then you can use `JSON.parse()` and `JSON.stringify()` to convert them from strings to data objects and back again.

Add It At the beginning of your server script, you need to include the ws library and make an instance of it. New lines are shown in blue as usual. You need to install the ws library using npm as well, like so:

```
$ npm install ws
```

It may seem strange to have three server objects in your code. The first variable, `server`, is the instance of express.js that handles the server routing and static files. The second, `httpServer`, is the part of `server` that handles the HTTP protocol. The third, `wss`, is the part that handles the webSocket protocol.

```
/*
Express-based webSocket Server
context: node.js
*/
var express = require('express');          // the express library
var http = require("http");                 // the http library
var WebSocketServer = require('ws').Server; // ws library's Server class

var server = express();                                 // express server
var httpServer = http.createServer(server);             // http server
var wss = new WebSocketServer({ server: httpServer }); // websocket server
```

At the end of your code, you already start the HTTP server. Add an event listener to the webSocket server to listen for incoming webSocket connections.

```
// start the servers:
httpServer.listen(8080);                // listen for http connections
wss.on('connection', connectClient);    // listen for webSocket messages
```

In the body of your script, after you configure the Express server to serve static files from the **public** directory, you'll add a new function to handle incoming webSocket connections. `connectClient()` is the callback function for the event listener you just added on the previous page. When a new connection is made, the event listener will pass the new client to this function.

A webSocket client can generate a few possible events: send a message, close the socket, or trigger an error. You'll set up three event listeners for those events inside the `connectClient()` function, because they can only occur if a client connects. You'll put their callback functions, `readMessage()`, `readError()`, and `disconnect()`, here as well.

At the end of `connectClient()`, you'll define what happens when the client first connects: define a greeting message and send it. You're using `.size`, not `.length`, because wss.clients is a data type called a *Set*. Sets in JavaScript similar to arrays, but each element has to be unique. The ws library stores clients in a Set called `wss.clients`.

```
// serve static files from /public:
server.use('/',express.static('public'));

function connectClient(newClient) {
  // when a webSocket message comes in from this client:
  function readMessage(data) {
    // you'll see more on this function shortly
  }

  // if there's a webSocket error:
  function readError(error){
    console.log(error);
  }

  // when a client disconnects from a webSocket:
  function disconnect() {
    console.log('Client ' + newClient.clientName + ' disconnected');
  }

  // set up event listeners:
  newClient.on('message', readMessage);
  newClient.on('error', readError);
  newClient.on('close', disconnect);

  // when a new client connects, send the greeting message:
  var greeting = {"client": wss.clients.size};
  newClient.send(JSON.stringify(greeting));
}
```

The `readMessage()` function is the most complex of the three callback functions in `connectClient()`, because it has to read incoming messages and decide what to do with them. Replace the earlier stub of code with this function.

When the incoming message has a property called `playing` or `position`, you need to broadcast it to the other clients so they can adjust the video. You'll write the `broadcast()` function shortly. If the incoming message has an `exit` property and the value is 1, then close the webSocket to this client.

```
  // when a webSocket message comes in from this client:
  function readMessage(data) {
    var result = JSON.parse(data);                  // parse the JSON
    if (result.hasOwnProperty('clientName')) {    // if there's a clientName,
      newClient.clientName = result.clientName;  // name the client
    }
    if (result.hasOwnProperty('playing') ||        // if there's a playing prop
       result.hasOwnProperty('position')) {       // or a position property
      broadcast(newClient, result);                // broadcast it
    }
    if (result.exit === 1) {                        // if it's an exit message,
      console.log("client " + result.clientName + " logging out");
      newClient.close();                            // close the webSocket
    }
    console.log(result);                            // print the message itself
  }
```

The `broadcast()` function goes after `connectClient()`. It runs through the list of connected webSocket clients and checks to see if they are the client that sent the message. If not, it sends them the message. You're using the `.forEach()` function to call the same function, `sendToAll()`, for each element of the wss.clients Set.

```
// broadcast data to connected webSocket clients:
function broadcast(thisClient, data) {
  function sendToAll(client) {
    if (client !== thisClient) {
      console.log('broadcasting from:' + client.clientName);
      client.send(JSON.stringify(data));
    }
  }
 // run the sendToAll function on each element of wss.clients:
  wss.clients.forEach(sendToAll);
}
```

The webSocket server is now different than your Processing server in two ways. First, the webSocket server keeps its own list in a property called `clients`, unlike Processing, where you had to maintain your own ArrayList of clients. Second, you're not really doing anything with the incoming messages; you're just checking that they have useful information and sending them out to the other clients. The web client code for handling incoming messages will look more like the Processing sketch's `readMessage()` function, because they both manage the video. In this case, each client controls its own copy of the video. By contrast, the Processing server controlled the video, not the clients.

You've added all the code you need for the server to handle webSockets. Make sure you installed the ws library and run this script from the command line. You won't see any difference when you reload the web page yet, though, since the script in that page doesn't yet have code to make a websocket connection. Next you'll add some code to the **sketch.js** script to make that happen.

Add It At the beginning of your client-side script, you'll need a few new global variables to keep track of the socket connection, the last received message, the host's address, and the outgoing message. In the `message` variable, set a client property so the server knows what type of client this is. Although the server's not checking this, it can be useful for diagnostics.

```
/*
  Video client
  context: p5.js
*/


var myVideo;                               // movie player
var socket;                                // the websocket
var lastMessage;                           // last server message received
var host = document.location.host;        // address of the server
var message = {"client": "browser"};     // what you'll send to the server
```

In the `setup()` function, add some code to create a new text division in the HTML using the `createDiv()` function. You'll use this to show messages from the server. Then create a new webSocket connection back to the server. Set up two listener functions, one to listen for a successful opening of the socket, and the other to read messages.

```
function setup() {
  myVideo = createVideo("movie.mov"); // initialize the video
  myVideo.size(640, 360);             // set its size
  myVideo.position(10, 30);           // position it
  myVideo.loop();                      // loop the video
  lastMessage = createDiv('');        // create a div for text messages
  lastMessage.position(10, 10);       // position the div
  socket = new WebSocket('ws://' + host); // connect to server
  socket.onopen = sendIntro;           // socket connection listener
  socket.onmessage = readMessage;      // socket message listener
}
```

As mentioned above, you won't have a `draw()` function, because everything in this script is driven by events generated by the webSocket. The `sendIntro()` function just sends a greeting to the server when the socket's open.

The `readMessage()` function is the heart of the script. It reads incoming messages and converts them to JSON objects. Then it looks for the appropriate properties, `playing` or `position`. If those properties are present, the client uses them to control the video, just like the `readMessage()` function in the Processing server did. At the end of the `readMessage()` function, it converts the latest message into a string and places it in the text division.

Save these changes to **sketch.js**, make sure the server is still running, and reload the page in the browser. You should see the video load, and get a message from the server sending the greeting as you see in Figure 5-11. In the server console, you should see the client login and send the message: `{ client: 'browser' }`. Now it's time to write the controller client.
X

```
function sendIntro() {
  // convert the message object to a string and send it:
  socket.send(JSON.stringify(message));
}

function readMessage(event) {
   // read  text from server:
  var msg = event.data;            // read data from the onmessage event
  var videoTime = myVideo.time();  // get the current video time
  var message = JSON.parse(msg);   // convert incoming message to JSON

  // if it's a play/pause message, change the video state:
  if (message.playing) {
    myVideo.loop();
  } else {
    myVideo.pause();
  }

  // if it's a position message, change the video time:
  var value = parseFloat(message.position);
  if (!isNaN(value)) {           // if it's a numeric value,
    var frames = value * 0.033;  // 1 frame in seconds at ~30fps
    videoTime += frames;         // add the change to the current time
    myVideo.time(videoTime);     // set the video position
  }

  // save the last client message received for printing on the screen:
  lastMessage.html(JSON.stringify(message));
}
```

Figure 5-11
The video browser client

The WebSocket Controller Client

The video controller client for this project is physically identical to the controller client for the last project. You can use the same circuit; only the firmware on the microcontroller needs to change. The only difference between this microcontroller client and the last one is that this one needs to make webSocket connections instead of TCP socket connections. To do that, you'll use the ArduinoHttpLibrary that you've already been using. That library features a `webSocket` class that makes it easy to make the connection.

Try It Start by making a copy of the finished Arduino program from the previous project. You'll need to add an extra line at the top of the program to include the ArduinoHttpLibrary. You'll also need to add a line at the top to make a webSocket instance. New lines are shown in blue.

There are no changes needed to the rest of the global variable section or `setup()` function, so you can leave those as is.

```
/*
  Video Controller WebSocket Client
  Context: Arduino, with WINC1500 module
*/
#include <SPI.h>
#include <WiFi101.h>
#include <ArduinoHttpClient.h>
#include <Encoder.h>
#include <Button.h>
#include "config.h"

const char serverAddress[] = "192.168.0.12";  // server address
int port = 8080;                              // port number
WiFiClient tcpSocket;                         // server socket

// make a websocket instance
WebSocketClient webSocket = WebSocketClient(tcpSocket, serverAddress, port);
```

▸▸ Change this to match your server's address.

▸▸ The `loop()` function is very similar. You're just swapping out the webSocket for the TCP socket; the rest is the same. The `webSocket` class has a function called `parseMessage()` that parses the message and returns how many bytes are in the incoming message. Assuming there are bytes available, you can use the usual Stream methods like `read()` or `readString()` to get the message.

You'll also need to change the `connectLED` update, as it relies on whether the webSocket's connected, not the TCP socket.

```
void loop() {
  readEncoder();
  readButtons();

  // while websocket is connected, listen for incoming messages:
  if (webSocket.connected()) {
    int msgLength = webSocket.parseMessage();  // get message length
    if (msgLength > 0) {                       // if it's > 0,
      String message = webSocket.readString(); // read it
      Serial.println(message);                 // print it
    }
  }
  // update the status LEDs:
  digitalWrite(connectLED, webSocket.connected());
  digitalWrite(playLED, playing);
}
```

The readEncoder() function has only one change: instead of checking that the TCP socket's connected, you're now checking that the webSocket's connected. If it is, you'll call a new function, sendJsonMessage(), to format and send the message out the webSocket, which you'll see shortly.

```
void readEncoder() {
  long position = myEncoder.read();             // read the encoder
  long difference = position - lastPosition;    // compare to last position
  if (difference != 0) {                        // if it's changed,
    if (webSocket.connected()) {                // and webSocket's connected,
      sendJsonMessage("position", difference);  // send a JSON message
    }
    lastPosition = position;                    // update lastPosition
  }
}
```

The readButtons() function's changes are similar to those for the readEncoder() function. When the connectButton changes, you'll replace the TCP socket check with the webSocket check, and call connectToServer() if you're not connected. If you are connected, you'll send the exit message so the server can disconnect you.

When the play button changes, check the webSocket connection and send the playing state if you have a connection.

```
void readButtons() {
  if (connectButton.toggled()) {          // if the connect button has
changed
    if (connectButton.read() == LOW) {    // and it's LOW
      if (!webSocket.connected()) {       // if you're not connected,
        connectToServer();                // connect to server
      } else {                            // if you're already connected,
        sendJsonMessage("exit", 1);       // disconnect
      }
    }
  }                                       // end of connectButton.toggled

  if (playButton.toggled()) {              // if the play button has changed
    if (playButton.read() == LOW) {        // and it's LOW
      playing = !playing;                  // toggle playing state
      if (webSocket.connected()) {         // if you're connected,
        sendJsonMessage("playing", playing); // send a message
      }
    }
  }                                       // end of playButton.toggled
}                                         // end of readButtons()
```

The connectToServer() function attempts to connect just like the previous one. However, webSocket.begin() returns an error 1 if it fails to connect, so the logic is reversed from the previous code. This version also sends the client name by way of greeting to the server, using the sendJsonMessage() function that you'll see next.

```
void connectToServer() {
  Serial.println("attempting to connect");
  boolean error = webSocket.begin();    // attempt to connect
  if (error) {
    Serial.println("failed to connect");
  } else {
    Serial.println("connected");
    sendJsonMessage("", 0);  // send the client name and nothing else
  }
}
```

The only new function, called `sendJsonMessage()`, forms a JSON-encoded message to send to the server. The `webSocket` class needs to form a message packet before sending, using `webSocket.beginMessage()`. This function lets you specify the type of message, using the webSocket protocol types (these include text, binary, ping, and others). Then you can `print()` or `println()` to the packet. Finally, the `webSocket.endMessage()` function sends the whole message that you've assembled.

```
// This function forms a JSON message to send,
// from a key-value pair:
void sendJsonMessage(String key, int val) {
  webSocket.beginMessage(TYPE_TEXT);    // message type: text
  webSocket.print("{\"clientName\":\"MKR1000\"");
  if (key != "") {           // if there is no key, just send name
    webSocket.print(",\""); // comma, opening quotation mark
    webSocket.print(key);   // key
    webSocket.print("\":"); // closing quotation mark, colon
    webSocket.print(val);   // value
  }
  webSocket.print("}");
  webSocket.endMessage();
}
```

Upload this sketch to your board and open the Serial Monitor. Make sure your server's still running, then press the connect button. On the server side, you should see a message like this: `{ clientName: 'MKR1000' }`. The connect LED should light and in the Serial Monitor, you should see a greeting message like this: `{"client":1}`. If you've got a browser client logged in, turn the encoder or press the play button and you should be able to control the video now.

With this version of the video server, you can have both multiple controller clients and multiple display clients, and the server won't see a perceptible change in performance.

This version doesn't allow you to target specific controller clients to specific display clients, but now that you have the idea, see if you can modify the system to make that happen.

X

Figure 5-12
One controller can control many browser clients simultaneously.

Conclusion

The basic structure of the clients and servers in this chapter can be used any time you want to make a system that manages synchronous connections between several devices on the network. The servers' main jobs are to listen for new clients, keep track of the existing clients, and make sure that the right messages reach the right clients. They must place a priority on listening at all times.

The clients should also place a priority on listening, but they have to balance listening to the server with listening to the physical inputs. They should always give clear and immediate responses to local input, and should indicate the state of the network connection at all times.

When you're planning the communications between client and server, keep it simple, and leave the system open for more commands, because you never know when you might decide to add a feature. Make sure to build in responses where appropriate, like the "client" and "exit" messages you saw here. Keep the messages unambiguous and, if possible, keep them short as well. Consider using formats that are native to your programming tools, like JSON was used here.

You've seen two different approaches to client-server systems with tight interactive loops here. The first used plain TCP sockets connection between the server and its clients, and the second used webSockets. The plain TCP socket is a simpler approach if both your client and your server can make plain TCP socket connections. The webSocket approach, on the other hand, works well with existing HTTP applications and lets you take advantage of the web browser as client.

Now you've seen examples of client-server exchanges with loose interactive loops (the HTTP system in Chapter 4) and tight interactive loops, as you saw in this chapter. With those two tools, you can build almost any application in which there's a central server and a number of clients. For the next chapter, you'll step away from the internet and take a look at various forms of wireless communication.

X

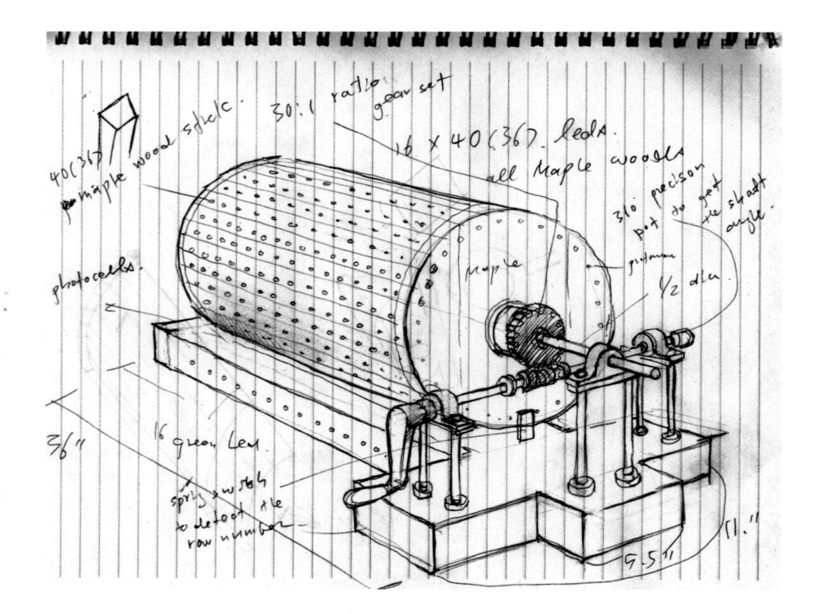

At left

◀◀ Jin-Yo Mok's original sketches of the music box.

At right

▶▶ The music box composition interface.

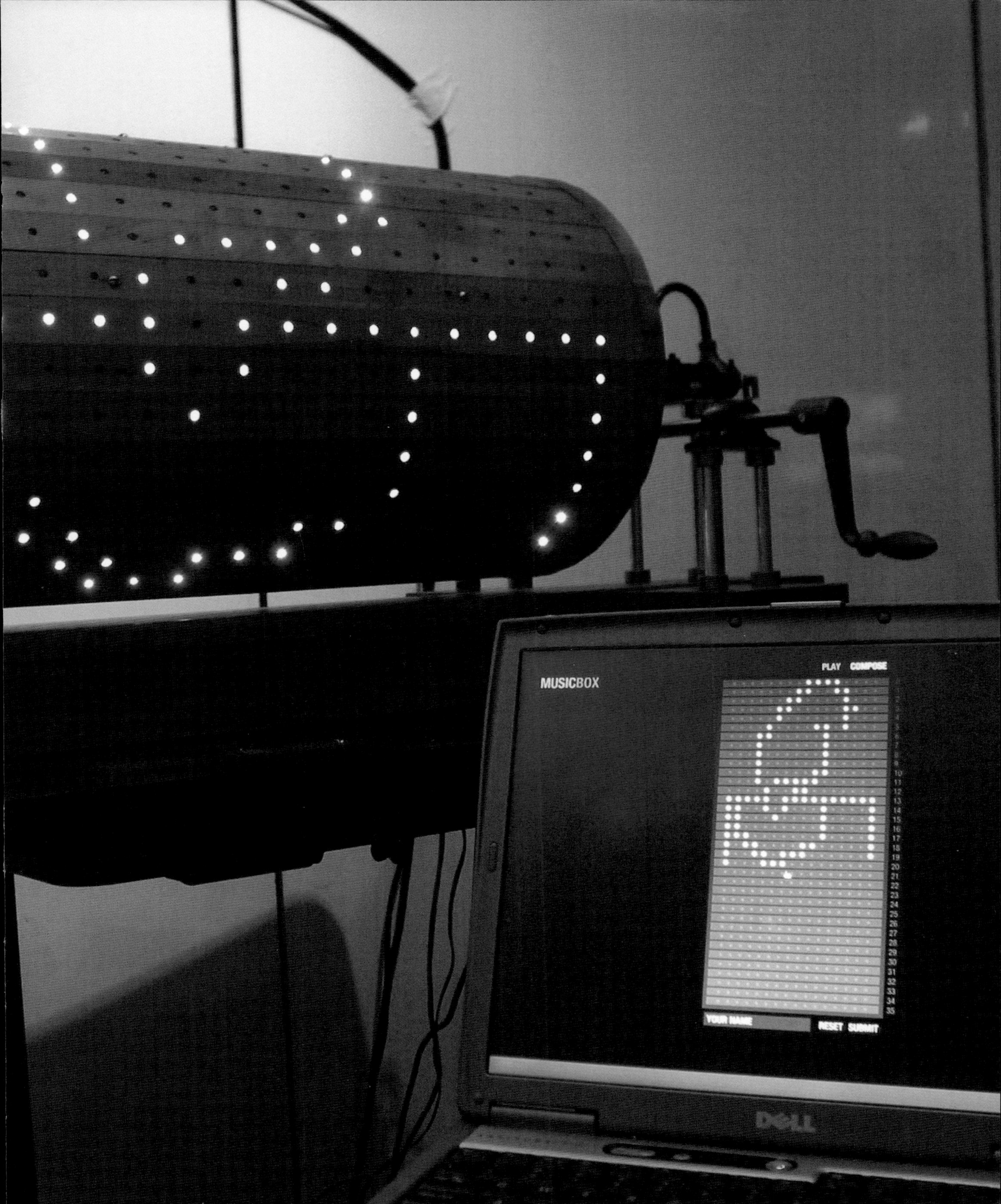
MUSICBOX
PLAY
COMPOSE
YOUR NAME
RESET
SUBMIT
DELL

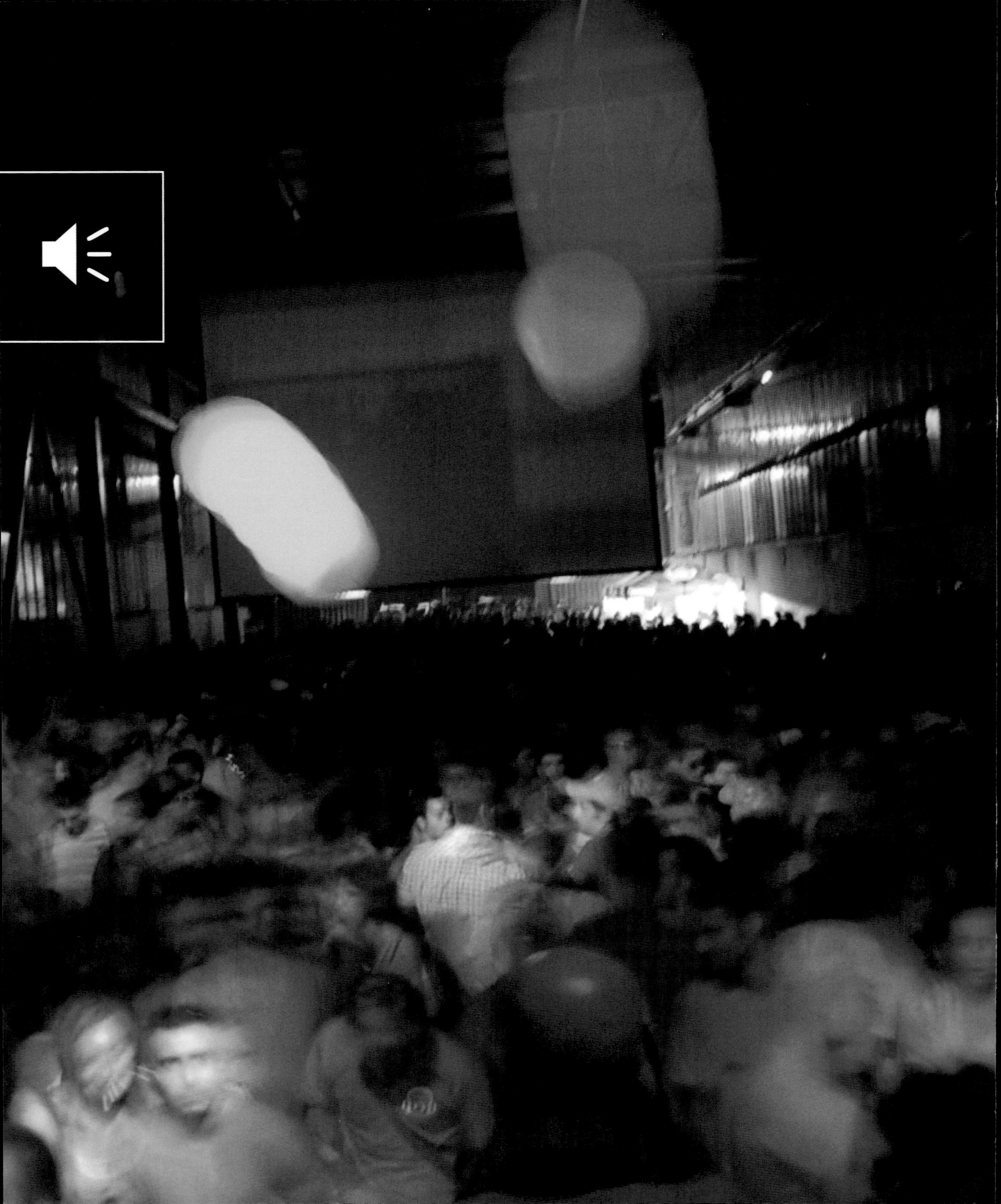

Wireless Communication

If you're like most people interested in this area, you picked up this book because you had ideas for building devices that talk to each other wirelessly. Perhaps you're so eager that you just skipped straight to this chapter. If you did, go back and read the rest of the book! The earlier chapters introduce common forms of wireless, WiFi and Bluetooth, but more importantly they cover some of the fundamentals of digital communication on which wireless methods depend. In particular, if you're not familiar with serial communication between computers and microcontrollers, you'll want to read Chapter 2 before reading this chapter. This chapter explains wireless communication in more detail. In it, you'll learn about two types of wireless communication, and build some working examples.

◂◂ **Alex Beim's Zygotes** (www.tangibleinteraction.com) are lightweight, inflatable rubber balls lit from within by LED lights. The balls change color in reaction to pressure on their surface, and they use ZigBee radios to communicate with a central computer. A network of zygotes at a concert allows the audience to have a direct effect not only on the balls themselves, but also on the music and video projections to which they are networked. *Photo courtesy of Alex Beim.*

Supplies for Chapter 6

DISTRIBUTOR KEY
A Arduino Store (store.arduino.cc)
AF Adafruit (adafruit.com)
D Digi-Key (www.digikey.com)
F Farnell (www.farnell.com)
J Jameco (jameco.com)
MS Maker SHED (www.makershed.com)
RS RS (www.rs-online.com)
SF SparkFun (www.sparkfun.com)
SS Seeed Studio (www.seeedstudio.com)

PROJECT 10: Infrared Control of a Digital Camera

» **1 microcontroller module** This project will run on just about any Arduino-compatible board. The Arduino 101 is shown, but the MKR1000 or Uno will work as well. **MKR1000**: **AF** 3156, **RS** 124-0657, **A** ABX00004, GBX00011 (Africa and EU), **D** 1659-1005-ND **Arduino 101**: **D** 1660-1003-ND, **J** 2239331, **SF** DEV-13787, **AF** 3033, **F** 2520713, **RS** 913-9999, **SS** 114990575, **A** ABX00005, GBX00005 (Africa and EU) **Uno**: **D** 1050-1024-ND, **J** 2151486, **SF** DEV-11021, **A** A000099, **AF** 50, **F** 1848687, **RS** 715-4081, **SS** ARD132D2P
» **1 infrared LED** **J** 106526, **A** 387, **SF** COM-09469, **F** 1716710, **RS** 577-538, **SS** MTR102A2B
» **1 pushbutton** **D** GH1344-ND or SW400-ND, **J** 2231822 or 119011, **SF** COM-09337, **F** 1634684, **RS** 718-2213
» **1 220-ohm resistor** **D** 220QBK-ND, **J** 690700, **F** 9339299, **R** 707-7612
» **1 10-kilohm resistor** **D** 10KQBK-ND, **J** 691104, **F** 9339060, **R** 707-7745
» **Oscilloscope** **SS** 109990013, **SF** TOL-11702, **AF** 468
» **1 IR camera remote** See text
» **1 IR phototransistor** **D** 365-1068-ND, **RS** 654-8542
» **1 solderless breadboard** **D** 438-1045-ND, **J** 20723 or 20601, **SF** PRT-12615 or PRT-12002, **F** 4692810, **AF** 64, **SS** 319030002 or 319030001
» **1 personal computer**

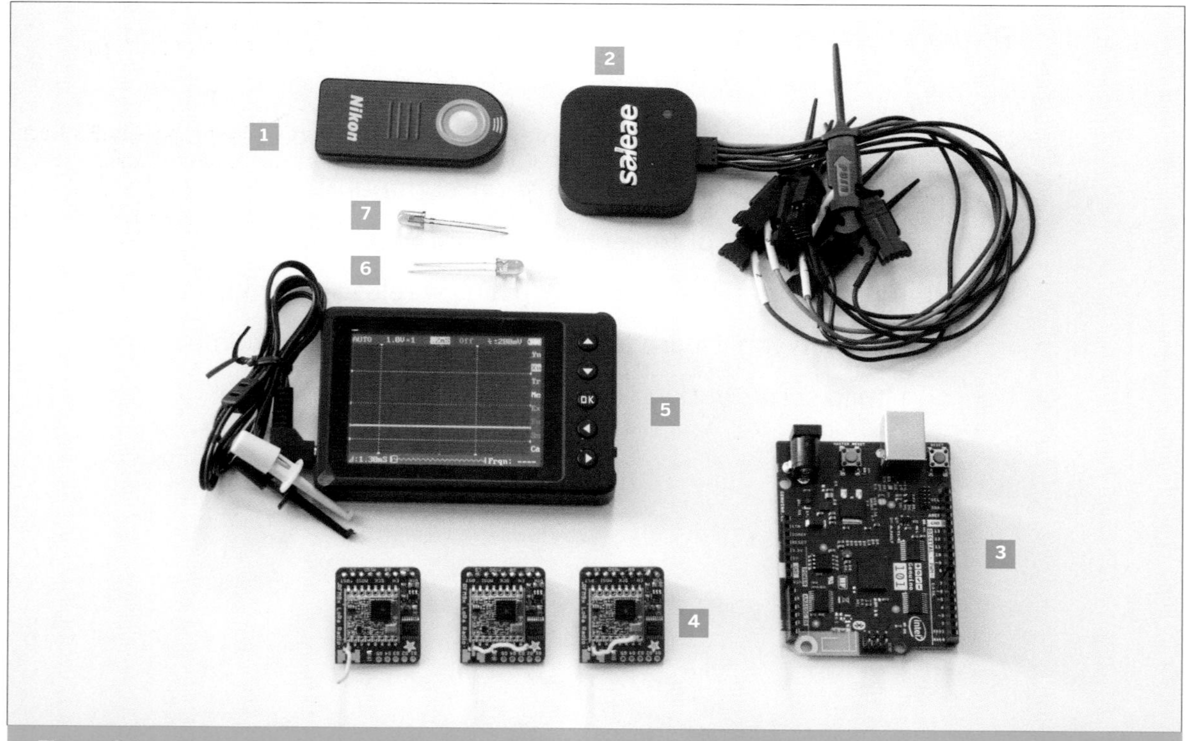

Figure 6-1. New parts for this chapter: **1.** DSLR infrared remote **2.** Saleae Logic Probe (optional) **3.** Arduino 101 **4.** RFM95W LoRa radios **5.** DSO Nano Oscilloscope **6.** Infrared LED **7.** Infrared phototransistor

- **All necessary converters to communicate serially from microcontroller to computer** Just like the previous projects.

PROJECT 11: Duplex Radio Transmission

- **2–3 solderless breadboards** **D** 438-1045-ND, **J** 20723 or 20601, **SF** PRT-12615 or PRT-12002, **F** 4692810, **AF** 64, **SS** 319030002 or 319030001
- **1 microcontroller module** This project will run on just about any Arduino-compatible board.
  MKR1000: **AF** 3156, **RS** 124-0657, **A** ABX00004, GBX00011 (Africa and EU), **D** 1659-1005-ND
  Arduino 101: **D** 1660-1003-ND, **J** 2239331, **SF** DEV-13787, **AF** 3033, **F** 2520713, **RS** 913-9999, **SS** 114990575 , **A** ABX00005, GBX00005 (Africa and EU)
  Uno: **D** 1050-1024-ND, **J** 2151486, **SF** DEV-11021, **A** A000099, **AF** 50, **F** 1848687, **RS** 715-4081, **SS** ARD132D2P
- **2–3 HopeRF RFM95W or Semtech SX1276 radio modules** **AAF** 3072, **SS** 113060006
- **2–3 pushbuttons** **D** GH1344-ND or SW400-ND, **J** 2231822 or 119011, **SF** COM-09337, **F** 1634684, **RS** 718-2213
- **2–3 220-ohm resistors** **D** 220QBK-ND, **J** 690700, **F** 9337792, **RS** 707-7612
- **2–3 LEDs** **D** 160-1144-ND or 160-1665-ND, **J** 34761 or 94511, **F** 1855510, **RS** 228-5972 or 826-830, **SF** COM-09592 or COM-09590
- **1 personal computer**
- **All necessary converters to communicate serially from microcontroller to computer** Just like the previous projects.

PROJECT 12: Bluetooth LE Camera Control

- **One Arduino 101** or other Arduino-compatible board paired with a Nordic nRF8001 or nRF51822 Bluetooth LE radio
  MKR1000: **AF** 3156, **RS** 124-0657, **A** ABX00004, GBX00011 (Africa and EU), **D** 1659-1005-ND
  Arduino 101: **D** 1660-1003-ND, **J** 2239331, **SF** DEV-13787, **AF** 3033, **F** 2520713, **RS** 913-9999, **SS** 114990575 , **A** ABX00005, GBX00005 (Africa and EU)
  Uno: **D** 1050-1024-ND, **J** 2151486, **SF** DEV-11021, **A** A000099, **AF** 50, **F** 1848687, **RS** 715-4081, **SS** ARD132D2P
  Bluetooth LE radio: AF 1697
  Alternatively, **RedBear BLE Nano and MK20 USB board** Sold together **MS** MKRBL5, **SF** WRL-14071
- **1 solderless breadboard** **D** 438-1045-ND, **J** 20723 or 20601, **SF** PRT-12615 or PRT-12002, **F** 4692810, **AF** 64, **SS** 319030002 or 319030001
- **1 infrared LED** **J** 106526, **A** 387, **SF** COM-09469, **F** 1716710, **RS** 577-538, **SS** MTR102A2B
- **1 220-ohm resistor** **D** 220QBK-ND, **J** 690700, **F** 9339299, **R** 707-7612
- **1 Bluetooth LE-enabled tablet, phone, or computer** Many devices made in the last 5 years have BLE. Check the product documentation if you're uncertain.
- **All necessary converters to communicate serially from microcontroller to computer** Just like the previous projects.

Why Isn't Everything Wireless?

The advantage of wireless communication seems obvious: no wires! This makes physical design much simpler for any project where the devices have to move and talk to each other. Wearable sensor systems, digital musical instruments, and remote control vehicles are all simplified physically by wireless communication. However, there are some limits to this communication that you should consider before going wireless.

• Wireless communication is not as reliable as wired communication.
You have less control over the sources of interference with wireless. You can insulate and shield a wire carrying data communications, but you can never totally isolate a radio or infrared wireless link. There will always be some form of interference, so you must make sure that all the devices in your system know what to do if they get a garbled message, or no message at all, from their counterparts.

• Wireless communication is never just one-to-one communication.
The radio and infrared devices mentioned here broadcast their signals for all to hear. Sometimes that means they interfere with the communication between other devices. For example, Bluetooth, most WiFi radios (802.11b, g, and n), and ZigBee (802.15.4) radios all work in the same frequency range: 2.4 gigahertz (802.11n will also work at 5GHz). They're designed to not cause each other undue interference, but if you have a large number of ZigBee radios working in the same space as a busy WiFi network, for example, you'll get some interference.

• Wireless communication is less private by default.
It's easier to eavesdrop on a wireless conversation than a wired one. The broadcast nature of wireless means that any compatible receiver in range of your transmitter can pick up the energy you're sending, whether you want it to or not. Nowadays most wireless communications systems mitigate this through the use of addressing and encryption, but it's important to remember that wireless communication is by nature easier to listen in on than wired.

• Wireless communication does not mean wireless power.
You still have to provide power to your devices, and if they're moving, this means using battery power. Batteries add weight, and they don't last forever. The failure of a battery when you're testing a project can cause all kinds of errors that you might attribute to other causes. A classic example of this is the "mystery radio error." Most radios consume more power when they transmit than when they're communicating with their host controller over wires. This causes a slight dip in the voltage of the power source. If the radio's power supply circuit isn't built to account for this, the voltage can dip low enough to make the radio turn off or reset itself. The radio may appear to function normally when you're sending it serial messages from your controller, but it may not have enough power to transmit, and you won't know why. When you start to develop wireless projects, it's good practice to first make sure that you have the communication working using a regulated, plugged-in power supply, and then create a stable battery supply once you've got the communications working.

• Wireless communication generates electromagnetic radiation.
This is easy to forget about, but every radio you use emits electromagnetic energy. Radio and infrared light are forms of radiation on the electromagnetic spectrum. The same energy that cooks your food in a microwave sends your mp3 files across the internet. And while there are many studies indicating that it's safe at the low operating levels of the radios used here, why add to the general level of radio energy if you don't have to?

• Make the wired version first.
The radio and IR transceivers discussed here can replace the communications wires used in previous chapters, but they will require code and circuitry changes. Before you decide to add wireless to any application, make sure you've got the basic exchange of messages between devices working over wires first.

X

Flavors of Wireless: Infrared and Radio

There are two common types of wireless communication in most people's lives: infrared light communication and radio communication. The main differences between them, from a user's or developer's position, is their *directionality* and their *range*.

Television remote controls typically use infrared (IR) communication. Unlike radio, IR is dependent on the orientation between transmitter and receiver. There must be a clear line of sight between the two. Sometimes IR can work by bouncing the beam off a reflective surface, but this technique is not always reliable. Ultimately, the receiver is an optical device, so it has to "see" the signal. Car door openers, mobile phones, garage door remote controls, and many other devices use radio. These work regardless of whether the transmitter and receiver are facing each other. They can even operate through walls, in some cases. In other words, their transmission is *omnidirectional*. Radio devices can usually transmit over a greater distance as well, depending on the transmitter power and the receiver sensitivity. Generally, IR is used for short-range line-of sight applications, and radio is used for everything else; Figure 6-2 illustrates this difference.

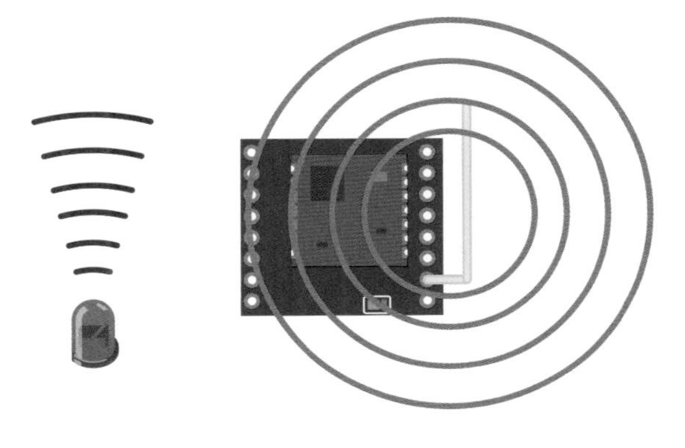

Figure 6-2
The signal from the LED at left radiates out in a beam from the LED, while the signal from a radio antenna like on the radio at right radiates omnidirectionally.

Transmitters, Receivers, and Transceivers

There are three types of devices common to both IR and RF systems: *transmitters*, which send a signal but can't receive one; *receivers*, which receive a signal but can't send one; and *transceivers*, which can do both. You may wonder why everything isn't a transceiver, as it's the most flexible device. It's more complex to make a transceiver than it is to make the other two. You have to make sure the transceiver is not receiving its own transmission, or it will cause feedback. For some applications, it's cheaper to use a transmitter-receiver pair and handle any errors by re-transmitting the message until the receiver gets it. That's how TV remote controls work, for example. It makes the components much cheaper.

It's increasingly common in radio applications to just make every device a transceiver, and incorporate a microcontroller to manage the transmitter-receiver filtering. All Bluetooth and WiFi radios work this way. However, it's still possible to get transmitter-receiver pair radios, and they are still cheaper than their transceiver counterparts.

Keep in mind the distinction between transmitter-receiver pairs and transceivers when you plan your projects, and when you shop. Consider whether the communication in your project must be two-way, or whether it can be one-way only. If it's one-way, ask yourself what happens if the communication fails. Can the receiver operate without asking for clarification? Can the problem be solved by transmitting repeatedly until the message is received? If the answer is yes, you might be able to use a transmitter-receiver pair and save some money.

How Infrared Works

IR communication works by pulsing an IR LED at a set data rate, and receiving the pulses using an IR photodiode. It's serial communication transmitted using infrared light. Since there are many everyday sources of IR light (the sun, incandescent light bulbs, any heat source), it's necessary to differentiate the IR data signal from other IR energy. To do this, the serial output is sent to an oscillator before it's sent to the output LED. The wave created by the oscillator, called a *carrier wave*, is a regular pulse that's modulated by the pulses of the data signal. The receiver picks up all IR light but filters out anything that's not vibrating at the carrier frequency. Then it filters out the carrier frequency, so all that's left is the data signal. This method allows you to transmit data using infrared light with less interference from other IR light sources.

The directional nature of infrared makes it more limited, but it's cheaper than radio, and requires less power. As radios have gotten cheaper, more power-efficient, and more robust, though, it's less common to see an IR port on a computer. However, it's still both cost-effective and power-efficient for line-of-sight remote control applications.

Data protocols for the IR remote controls of home electronics vary from manufacturer to manufacturer. To decode them, you need to know both the carrier frequency and the message structure. Most commercial IR remote control devices operate using a carrier wave usually between 38 and 40kHz. The carrier wave's frequency limits the rate at which you can send data on that wave, so IR transmission is usually done at a low data rate, typically between 500 and 2,000 bits per second. It's not great for high-bandwidth data transmission, but if you're only sending the values of a few pushbuttons on a remote, it's acceptable. Unlike the serial protocols you've seen so far in this book, IR protocols do not all use an 8-bit data format. For example, Sony's Control-S protocol has three formats: 12 bit, 15 bit, and 20 bit. Philips' RC5 format, common to many remotes, uses a 14-bit format.

If you have to send or receive remote control signals, there are many good sites on the web that explain the various protocols. SparkFun, Adafruit, Seeed Studio, and other hobbyist electronics retailers sell IR demodulators. SparkFun and Adafruit have some useful tutorials as well. EPanorama has a number of useful links describing many of the more common IR protocols at www.epanorama.net/links/irremote.html.

If you're building both the transmitter and receiver, your job is fairly straightforward. You just need an oscillator through which you can pass your serial data to an infrared LED, and a receiver that listens for the carrier wave and demodulates the data signal. If you're building a receiver to match an existing transmitter, your job is a bit more complex.

X

Making Infrared Visible

Even though you can't see infrared light, cameras can. If you're not sure whether your IR LED is working, one quick way to check is to point the LED at a digital camera and look at the resulting image. If it's working, you'll see the LED light up. Figure 6-3 shows the IR LED in a home remote control, viewed through a personal computer's camera. You can even see this in the LCD viewfinder of a digital camera. If you try this with your IR LED, you may need to turn the lights down or close the curtains to see this effect. Some cameras have a built-in IR filter, however, so it's good to first check your camera with a working IR remote that you know is working before using the camera to test your custom IR project.

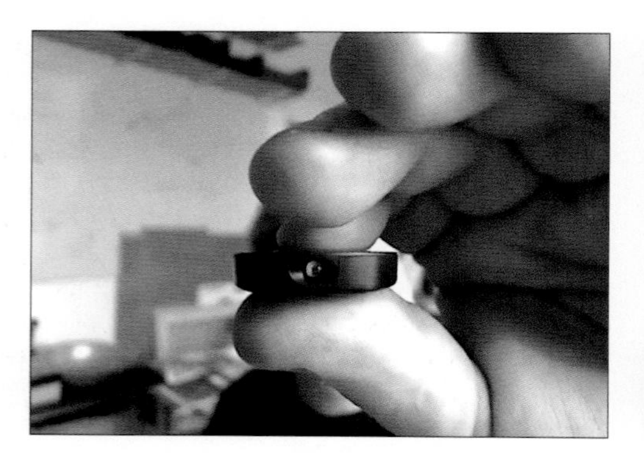

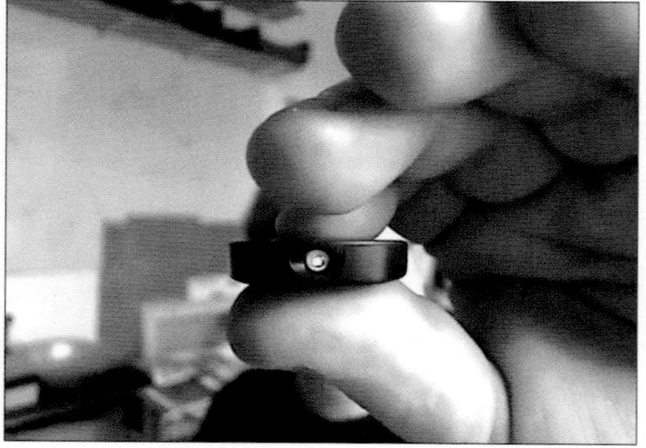

Figure 6-3. Having a camera at hand is useful when troubleshooting IR projects.

 Project 10

Infrared Control of a Digital Camera

This example uses an infrared LED and a microcontroller to control a digital camera. It gives you an idea how IR remote controls work.

MATERIALS

- **» 1 Arduino-compatible microcontroller**
  - **» Features used: Digital I/O**
- **» 1 infrared phototransistor**
- **» 1 IR camera remote**
- **» 1 oscilloscope**
- **» 1 infrared LED**
- **» 1 pushbutton**
- **» 1 220-ohm resistor**
- **» 1 10-kilohm resistor**
- **» 1 solderless breadboard or prototyping shield**

Most digital SLR cameras on the market today can be controlled remotely via infrared. Each brand uses a slightly different protocol, but they all tend to have the same basic command: trigger the shutter. There are a number of good tutorials on how to sense the signal from a DSLR remote online, and this project borrows from two in particular. For further reference, see Adafruit's tutorial at learn.adafruit.com/ir-sensor, or look for Sebastian Setz's now-obsolete MultiCameraIR library that, though it no longer works with the Arduino IDE, contains the timing codes for a number of popular cameras.

This project is based on the signals from the Canon RC-3 and the Nikon ML-L3 IR remotes for DSLR cameras. If you've got a Nikon or a Canon DSLR, you won't need to reverse-engineer the signal, because the following code will do the job for you. If you're using a different camera, the next section will explain a little about how to capture and decode the IR signal using a logic probe or scope.

Detecting the Signal

Reading the IR signal from a remote requires an IR phototransistor and an oscilloscope or logic probe. The DSO Nano is an inexpensive scope that works well for this job. For about the same price, Saleae's Logic 4 logic probe (www.saleae.com) will also work. The DSO Nano and other standalone scopes allow you to watch the voltage change in a circuit in real time. Logic probes like the Logic 4 connect to your computer and capture the signal for a few seconds, then display it on your screen for you.

To see the remote's signal on an oscilloscope, connect the scope probes to ground and to the phototransistor's emitter, and fire the remote at the phototransistor. Once you see activity, adjust the voltage and time divisions on the scope until you see readable activity (refer to your scope's instructions on how to do this). Most scopes will tell you the frequency of the signal automatically. Putting your scope in single-shot trigger mode will help you capture the actual signal. Once you can see the timing of each signal's pulse, you can work out how to duplicate it by generating your own pulses on an IR LED.

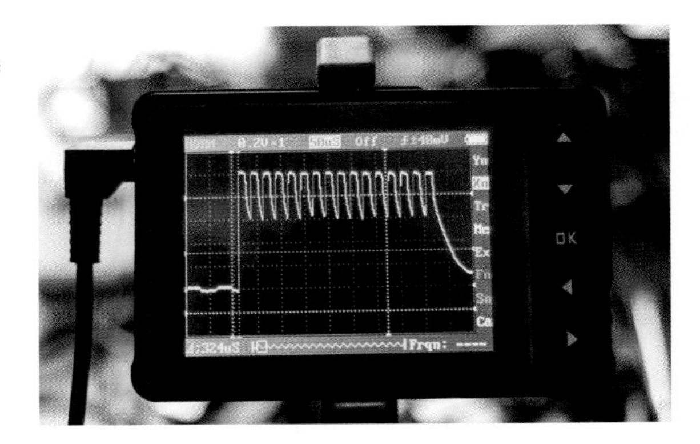

Figure 6-4. The beginning of the Nikon IR remote signal shown on a DSO Nano oscilloscope.

What you'll typically see in the output of the remote is a series of timed pulses separated into groups. The timing of the pulses will be at the remote's carrier frequency. The Canon remote operates at 32.7kHz, and the Nikon one operates at 38kHz. The timing between the pulses determines what the camera should do. Table 6-1 shows the timing patterns for the Canon and Nikon remotes, discovered through experiment and comparison to others' findings. You may have to adjust for your own remote. Armed with this information, you're ready to build the circuit and write some code.

The Circuit

All you need for this project is an IR LED and a pushbutton. It's one of the simplest circuits you'll build for this book.

Figure 6-5 shows the circuit. It's shown on the Arduino 101, but it's been tested successfully on the M0, AVR, and Curie processors. Once you've built it and written the code, you might want to check that it's working by replacing the IR LED with a visible LED, or looking at it with a camera before you test with the DSLR.

Once you've programmed the microcontroller, you're ready to click some pictures. The code follows.
X

Table 6-1
IR signal to trip the shutter, from two different DSLR remotes.

Nikon (38kHz)	Canon (32.7kHz)
2ms on	16 pulses on
27ms off	7.33ms off
0.4ms on	16 pulses on
1.5ms off	100ms before repeat
0.5ms on	
3.44ms off	
0.5ms on	
65ms off before repeat	

Figure 6-5
The microcontroller with an IR LED and pushbutton attached.

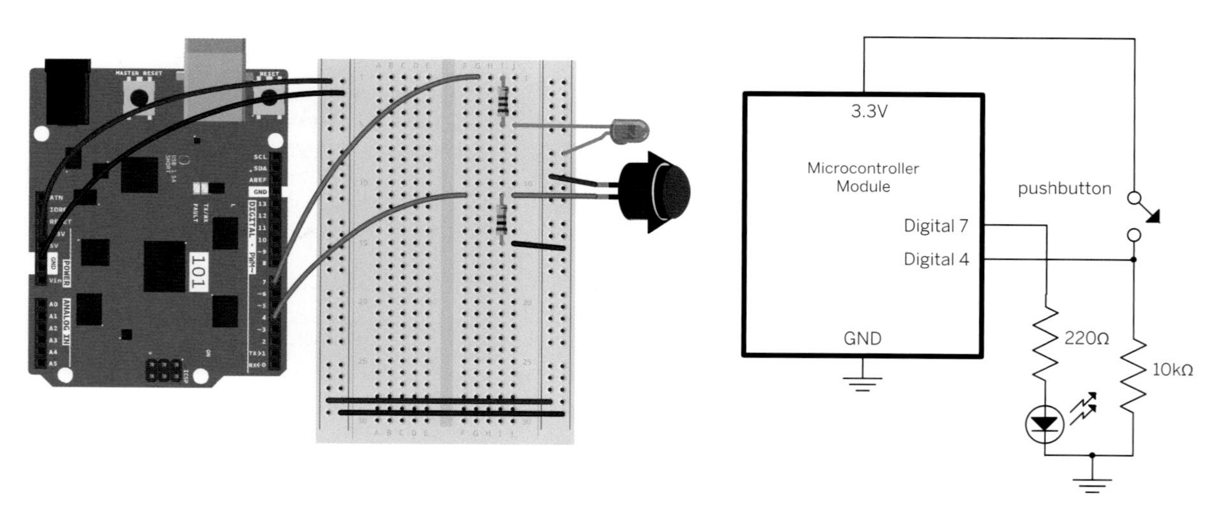

Try It To start your sketch, set up global constants for the pin numbers and the camera types, and variables for the current and previous button states.

```
/*
  IR Camera control
  Context: Arduino

  This sketch controls a digital camera via an infrared LED.
*/

const int pushButtonPin = 4;      // pushbutton I/O pin
const int IRPin = 7;              // IR LED I/O pin
const int CAM_NIKON = 1;          // DSLR camera types
const int CAM_CANON = 2;
int buttonState = 0;              // current button state
int lastButtonState = 0;          // previous button state
```

In the `setup()`, set the pinMode for the input and the output.

```
void setup() {
  pinMode(pushButtonPin, INPUT);        // initialize pushbutton
  pinMode(IRPin, OUTPUT);               // initialize IR pin
}
```

In the `loop()`, read the pushbutton and see if it's changed since the previous read. If so, then send the shutter control signal using a function you'll call `shutterClick()`. Change the parameter of this to `CAM_CANON` if you're using a Canon instead of a Nikon. Then save the current button state for comparison next time through.

```
void loop() {
  // read the pushButtonPin input pin:
  buttonState = digitalRead(pushButtonPin);

  if (buttonState != lastButtonState) { // if button state changes,
    if (buttonState == HIGH) {          // and button is pressed
      shutterClick(CAM_NIKON);          // send IR shutter signal
    }
  }
  // save the current state for next comparison:
  lastButtonState = buttonState;
}
```

The `shutterClick()` function sends the appropriate shutter signal for the camera you're using. IR remotes generally send the signal twice, as this code does.

In order to pulse the IR LED at the right frequency, `shutterClick()` calls a function called `IRPulse()`. This function uses Arduino's built-in `tone()` function to generate the carrier frequency on the IR LED pin, then pauses for the appropriate amount of microseconds, then turns the tone off. The `tone()` command can affect the timing on the `delayMicroseconds()` command, which is why you're not using the latter in this function.

```
void shutterClick(int cameraType) {
  if (cameraType == CAM_NIKON) {        // if it's a Nikon
    unsigned long irFrequency = 38000;  // IR freq. = 38K
    for (int i = 0; i < 2; i++) {       // send message twice
      IRPulse(2000, irFrequency);       // 2ms on
      delay(27);                        // 27ms off
      IRPulse(400, irFrequency);        // 0.4ms on
      delayMicroseconds(1500);          // 1.5ms off
      IRPulse(500, irFrequency);        // 0.5ms on
      delayMicroseconds(3500);          // 3.44ms off
      IRPulse(500, irFrequency);        // 0.5ms on
      delay(65);                        // 65ms delay before repeat
    }
  }
  if (cameraType == CAM_CANON) {        // If it's a Canon
    unsigned long irFrequency = 32700;  // IR freq. = 32.7K
    for (int i = 0; i < 2; i++) {       // send message twice
      IRPulse(489, irFrequency);        // 16 pulses @ IR freq.
      delayMicroseconds(7330);          // delay 7.33 ms
      IRPulse(489, irFrequency);        // 16 pulses @ IR freq.
      delay(100);                       // 100ms delay before repeat
    }
  }
}

// send an IR pulse at a frequency for a time:
void IRPulse(unsigned long interval, unsigned long frequency) {
  unsigned long now = micros();        // get current time in micros
  tone(IRPin, frequency);              // pulse IRPin at frequency
  while (micros() - now < interval);// do nothing until interval passes
  noTone(IRPin);                       // turn off pulse
}
```

Figure 6-6
This intervalometer was built using the method shown on the previous page. An Arduino in the box senses a change from the PIR sensor and sends an IR signal to the camera to take a picture.

How Radio Works

Radio relies on the electrical property called *induction*. Any time you vary the electrical current in a wire, you generate a corresponding magnetic field that emanates from the wire. This changing magnetic field induces an electrical current in any other wires in the field. The frequency of the magnetic field is the same as the frequency of the current in the original wire. This means that if you want to send a signal without a wire, you can generate a changing current in one wire at a given frequency, and attach a circuit to the second wire to detect current changes at that frequency. That's how radio works.

The distance that you can transmit a radio signal depends on the signal strength, the sensitivity of the receiver, the nature of the antennas, and any obstacles that block the signal. The stronger the original current and the more sensitive the receiver, the farther apart the sender and receiver can be. The two wires act as antennas. Any conductor can be an antenna, but some work better than others. The length and shape of the antenna and the frequency of the signal all affect transmission. Antenna design is a whole field of study on its own, so I can't do it justice here, but a rough rule of thumb for a straight wire antenna is as follows:

- **Antenna length =** 5,616 in. / frequency in MHz = 14,266.06 cm. / frequency in MHz

This formula, from the **American Radio Relay League Handbook**, 1929 edition, is the formula for a *half-wave antenna*, meaning that the antenna length is roughly a half-wavelength long. It derives from the fact that radio signals, which travel at the speed of light in free space, travel about 5% lower in a wire. Wavelength is the inverse of frequency, and the speed of light is 3.0 * 10^8m/sec. So, for example, at 915MHz, a half-wave antenna's length would be:

$$1/915*10^6 * 3.0*10^8 * 0.95 /2 = 0.1557382m$$

Or about 15.6 cm. It's also common to see full-wave and quarter-wave antennas. For more information, consult the technical specifications for the specific radios you're using. Instructions on making a good antenna are common in a radio's documentation.

Radio Transmission: Digital and Analog

As with everything else in the microcontroller world, it's important to distinguish between digital and analog radio transmission. Analog radios simply take an analog electrical signal, such as an audio signal, and superimpose it on the radio frequency in order to transmit it. The radio frequency acts as a carrier wave, carrying the audio signal. Digital radios superimpose digital signals on the carrier wave, so there must be a digital device on either end to encode or decode those signals. In other words, digital radios are basically modems, converting digital data to radio signals, and radio signals back into digital data.

Radio Interference

Though the antennas you'll use in this chapter are omnidirectional, radio can be blocked by obstacles, particularly metal ones. A large metal sheet, for example, will reflect a radio signal rather than allowing it to pass through. This principle is used not only in designing antennas, but also in designing *radio frequency (RF) shields*. If you've ever cut open a computer cable and encountered a thin piece of foil wrapped around the inside wires, you've encountered an RF shield. Shields are used to prevent random radio signals from interfering with the data being transmitted down a wire. A shield doesn't have to be a solid sheet of metal, though. A mesh of conductive metal will block a radio signal as well—if the grid of the mesh is small enough. The effectiveness of a given mesh depends on the frequency it's designed to block. It's possible to block radio signals from a whole space by surrounding the space with an appropriate shield and grounding the shield. You'll hear this referred to as making a *Faraday cage*. The effect is named after the physicist Michael Faraday, who first demonstrated and documented it.

Sometimes radio transmission is blocked by unintentional shields. If you're having trouble getting radio signals through, look for metal that might be shielding the signal. Transmitting from inside a car can sometimes be tricky because the car body acts as a Faraday cage. Putting the antenna on the outside of the car improves reception. This is true for just about every radio housing. Bodies of water block RF effectively as well.

All kinds of electrical devices emit radio waves as side effects of their operation. Any alternating current can generate a radio signal, even the AC that powers your home or office. This is why you hear a hum when you lay speaker wires in parallel with a power cord. The AC signal is inducing a current in the speaker wires, and the speakers are reproducing the changes in current as sound.

WiFi operates at frequencies in the gigahertz range, commonly called the *microwave range*, because the wavelength of those signals is very short compared to lower frequency signals. To cook food, microwave ovens generate energy in this range to excite (heat up) the water molecules in food. Some of that energy leaks from the oven at low power, which is why you get all kinds of radio noise in the gigahertz range around a microwave.

Motors and generators are especially insidious sources of radio noise. A motor also operates by induction; specifically, by spinning a pair of magnets around a shaft in the center of a coil of wire. By putting a current in the wire, you generate a magnetic field, which attracts or repulses the magnets, causing them to spin. Likewise, by using mechanical force to spin the magnets, you generate a current in the wire. So, a motor or a generator is essentially a little radio, generating noise at whatever frequency it's rotating.

Because there are so many sources of radio noise, there are many ways to interfere with a radio signal. It's important to keep these possible sources of noise in mind when you begin to work with radio devices. Knowledge of common interference sources, and knowing how to shield against them, is a valuable tool in radio troubleshooting.

Multiplexing and Protocols

When you're transmitting via radio, anyone with a compatible receiver can receive your signal. There's no wire to contain the signal, so if two transmitters are sending at the same time, they will interfere with each other. This is the biggest weakness of radio: a given receiver has no way to know who sent the signal it's receiving. In contrast, consider a wired serial connection: you can be reasonably sure when you receive an electrical pulse on a serial cable that it came from the device on the other end of the wire. You have no such guarantee with radio. It's as if you were blindfolded at a cocktail party and everyone else there had the same voice. The only way you'd know who was talking to you was if each person clearly identified himself at the beginning and end of his conversation, and no one interrupted him during this time. In other words, it's all about protocols.

The first thing everyone at that cocktail party would have to do is agree on who speaks when. That way they could each have your attention for awhile. Sharing in radio communication is called *multiplexing*, and this form of sharing is called *time-division multiplexing*. Each transmitter gets a given time slot in which to transmit.

Of course, it depends on all the transmitters being in sync. When they're not, time-division multiplexing can still work reasonably well if all the transmitters speak much less than they listen (remember the first rule of love and networking from Chapter 1: listen more than you speak). If a given transmitter is sending for only a few milliseconds in each second, and if there's a limited number of transmitters, the chance that any two messages will overlap, or *collide*, is relatively low. This guideline, combined with a request for clarification from the receiver (rule number four), can ensure reasonably good RF communication.

Back to the cocktail party. If every person spoke in a different pitch, you could distinguish each individual by her pitch. In radio terms, this is called *frequency-division multiplexing*. It means that the receiver has to be able to receive on several frequencies simultaneously. But if there's a coordinator handing out frequencies to each pair of transmitters and receivers, it's reasonably effective.

Various combinations of time- and frequency-division multiplexing are used in every digital radio transmission system. The good news is that most of the time you never have to think about it because the radios handle it for you.

Multiplexing helps transmission by arranging for transmitters to take turns and to distinguish themselves based on frequency, but it doesn't concern itself with the content of what's being said. This is where data protocols come in. Just as you saw how data protocols made wired networking possible, you'll see them come into play here as well. To make sure the message is clear, it's common to use a data protocol on top of using multiplexing. For example, Bluetooth, ZigBee, and WiFi are nothing more than data networking protocols layered on top of a radio signal. All three of them could just as easily be implemented on a wired network (and, in a sense, WiFi is: WiFi is wireless Ethernet). The principles of these protocols are no different than those of wired networks, which makes it possible to understand wireless data transmission even if you're not a radio engineer. Remember the principles and troubleshooting methods you used when dealing with wired networks, because you'll use them again in wireless projects. The methods mentioned here are just new tools in your troubleshooting toolkit. You'll need them in the projects that follow.

Radio Transmitters, Receivers, and Transceivers

How do you know whether to choose a radio transmitter-receiver pair, or a pair of transceivers? The simplest answer is that if you need feedback from the device to which you're transmitting, then you need a transceiver. Most of the time, it's simplest to use transceivers. In fact, as transceivers have become cheaper to make (and therefore sell), transmit-receive pairs are getting harder to find.

There are many different kinds of data transceivers available. The simplest digital radio transceivers on the market connect directly to the serial transmit and receive pins of your microcontroller. Any serial data you send out the transmit line goes directly out as a radio signal. Any pulses received by the transceiver are sent into your microcontroller's receive line. They're simple to connect, but you have to manage the whole conversation yourself. If the receiving transceiver misses a bit of data, you'll get a garbled message. Any nearby radio device in the same frequency range can affect the quality of reception. As long as you're working with just two radios and no interference, transceivers like this do a decent job. However, this is seldom the case.

Nowadays, most transceivers on the market implement networking protocols, handling the conversation management for you. The Bluetooth modem in Chapter 2 ignored signals from other radios with which it wasn't associated, and handled error-checking for you. The other radios you'll see in following projects will do the same and much more. They require you to learn a bit more in terms of networking protocols, but the benefits you gain make them well worth that minor cost.

The biggest difference between networked radios and simple transceivers is that every device on a network has an address. That means you have to decide which other device you're speaking to, or whether you're speaking to all the other devices on the network.

X

Radio Networks

The distance that a message can be carried using a radio network depends primarily on the power of the transmitter and the sensitivity of the receiver. Transmission power costs energy. A message can be carried further using a network of radios, however. Messages can be passed along from one radio to another, or they can be stored until the receiver is within transmission range. The characteristics of a radio network are determined by how it manages the trade-off of energy, range, and message delivery.

Radio Network Topologies

The network topologies you saw in Chapter 3 apply to radio networks as well. The most common patterns for radio network topologies are *point-to-point* networks, in which nodes in the network exchange messages directly, and *star networks*, in which a central node routes messages to and from the other nodes. Sometimes you'll see *mesh networks*, in which all devices in the network can route messages to and from each other.

The simplest radio networks are broadcast networks. Radios in this kind of network do not have addresses. All traffic is one-way. Transmitters and receivers share the same frequency, and all receivers listening on the same frequency that are in the range of the transmitter can pick up the signal. These are only useful if you don't want to send any private messages. Before you add a communications protocol, all radio networks are broadcast networks.

Point-to-point networks feature radios that share a common addressing scheme. Radios in this type of network typically broadcast a request to see what other radios are in range, and when they receive a response, they attempt to connect directly with one other radio in range. Bluetooth radios use a point-to-point topology.

Star networks have a central device that assigns addresses to the other devices, and routes messages from one to another, and sometimes to other networks. WiFi works in this way. The router in a WiFi network acts as the central node.

In mesh networks, some or all nodes can route messages for the other nodes. If a node receives a message that isn't addressed to it, it passes the message on. This passing can get complicated when several nodes are within range of each other, so mesh network systems usually include router nodes and a central coordinator node that manages the routers. When confgured this way, the mesh is a hybrid between a mesh and a star network. The router nodes in

Figure 6-7
Types or radio network: point-to-point, star, or mesh. In the first, all nodes are connected directly. In the second, it's a maximum of one hop from sender to receiver. In the third, it can be multiple hops.

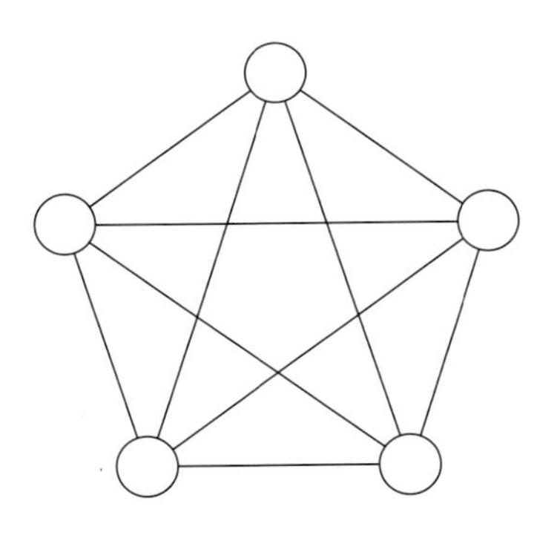

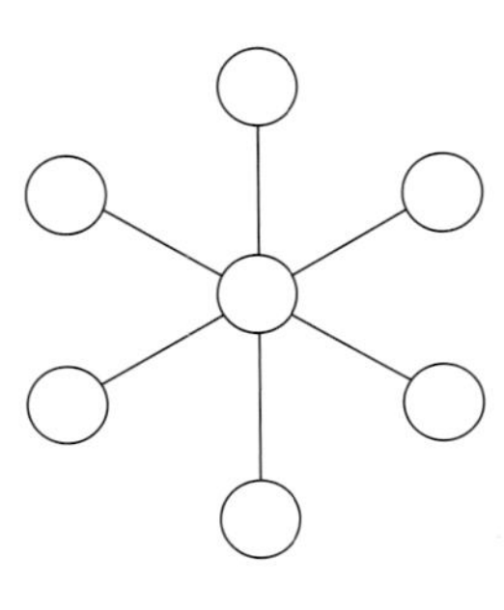

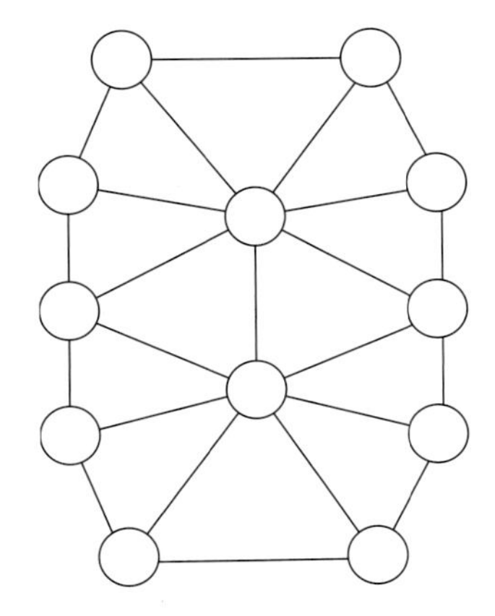

a mesh network can store messages from other nodes in case the receiver node is not currently active on the network, or they can forward them to other routers or a coordinator. This saves power for the transmitter and receiver nodes, as they don't have to be active at the same time. However, it complicates message management.

Mesh networking often seems attractive because of its flexibility, but it is the most complex of the three schemes to use. Unless you know your system will involve many radios and requires the flexibility of a mesh, you're usually better off with one of the simpler topologies.

Messages and Sessions

The connection between two radio devices may be *session-based*, like those you saw in the previous chapter. In such a system, a given device can communicate with only one partner at a time. The connection between two devices has to be broken before they can communicate with other devices. Alternately, the connection may be *message-based*, or *packet-based,* in which each message has a destination address. In this kind of system, a given device may send several messages to different receivers without having to establish a dedicated connection to any of them.

Asynchronous serial port connections that you've used so far are like session-based systems in that only one application on a computer at a time can use a given serial port. The Bluetooth serial radios that you saw in Chapter 2 are session-based in that they have to be paired with each other before communication can happen. The WiFi radios you've used can support both message-based communication using UDP, although so far you've only used them for session-based communication using TCP.

Session-based radio networks afford more reliability, like TCP does relative to UDP, but they generally use more energy in managing the connection. Message-based networks are usually more energy efficient, but less reliable. Some message-based networks support acknowledgment from the receiver as a way of increasing reliability. When a system affords acknowledgments and re-transmission of messages, the sender has to keep track of who it's sent messages to until each receiver acknowledges receipt of its message.

Wireless Protocol Stacks

The details of digital radio communication are managed by firmware on each radio. This firmware is sometimes called a *protocol stack*, like the network stacks described in Chapter 4, because it manages the layers of communication, from the physical control of the radio frequencies to the arrangement of messages into packets to the management of transmission and acknowledgment of packets to the management of point-to-point sessions. The work of managing communication can take a few milliseconds, so you can't count on the fact that a radio transmission is delivered to your microcontroller the instant that it's received by the radio. This will be significant when you learn about using radio signals to estimate distance and location in Chapter 8.

Protocol stacks also help you to manage radio interference. Remember, a radio transmitter is omnidirectional, so radios operating on the same frequency can receive transmissions from each other even when they don't use the same protocol. Protocols that are compatible with one another tend to cause less interference with each other. For example, WiFi, Bluetooth, and ZigBee are all wireless protocols which organize data and transmission similarly at lower levels. They're all variations of the IEEE 802 specification from which Ethernet derives. Ethernet is 802.3, WiFi is 802.11, Bluetooth is 802.15.1, and ZigBee is 802.15.4. They operate on some of the same frequencies, and their physical addressing and packet frames are similar. So even though you can't read WiFi messages with a Bluetooth radio, they are pretty good at ignoring each other's messages effectively.

Radio Physical Interfaces

Digital radios interface with computers or microcontrollers through wired connections using the same serial communications protocols you've already seen: asynchronous serial UARTs, SPI, and I2C for radios designed to work with microcontrollers, and USB for radios designed to work with personal computers or embedded processors. When you're investigating a new radio for use, consider the computer to which will be physically connected, and make sure they both communicate using the same wired protocol.

X

Buying Radios

There are many different digital radio communications protocols in use now, for a variety of industries and applications, and the number of protocols seems to grow very year. They all serve slightly different needs, and have different limitations. Before you start on a radio-based project, you should weigh your options carefully. Here are a few things to consider when choosing your radios.

Wireless protocols are developed to serve specific needs. Sometimes those needs overlap, and the protocols for them can seem very similar as a result. WiFi and Bluetooth are the best-known radio protocols in current digital systems because they're popular for communication between consumer computing devices (laptops, tablets and phones), appliances (printers, mice, keyboards, speakers, etc.), and the internet. ZigBee is popular in appliance-to-appliance applications like home lighting and automation. ANT is a low-power protocol popular in short-range communications, for example, between fitness devices and mobile phones. LoRa is a low-power, long-range protocol being developed for device-to-device communication over longer distances.

Though it's tempting to want to try out the latest and greatest new protocol, you should ask yourself whether the devices you plan to use in your project support this protocol. If so, then you'll simplify your work by using it. If your computer already contains one of the radios you need, you may only need to buy one more.

Physical interface affects how your radio will connect to your computer or processor. When looking for radios, look for something that works with the communications output of your microcontroller, whether asynchronous serial, SPI, or I2C. Make sure the radio operates at the same voltage levels as well.

Interoperability Do the radios from one company work with radios from another company? Quite often, the answer is no. For example, Digi's XBee series of ZigBee and 802.15.4 modules are difficult to use with radios from other companies that implement ZigBee because Digi layers an additional proprietary interface on top of the protocol stack. They're very useful devices, but easiest used in a network of all Digi radios. Interoperability is the main reason WiFi and Bluetooth have seen wide adoption. Manufacturers using these protocols have paid the greatest attention to interoperability, so a Nordic or a Texas Instruments Bluetooth module can easily communicate with the Bluetooth radio in your laptop or tablet. Radio protocols that have been adopted in personal computing devices tend to interoperaperate best.

Table 6-2
A few popular radios used with microcontroller projects. You've seen the WiFi ones in use already.

Radio	Protocol	Frequency	Topology	Wired Interface	Message or Session-based?
ESP8266	WiFi	2.4GHz	Star	Programmable	Either (TCP or UDP)
WINC1500	WiFi	2.4GHz	Star	SPI	Either (TCP or UDP)
Digi Xbee 802.15.4	802.15.4, Proprietary	2.4GHz	Point-to-point, Mesh	UART	Message-based (can also be configured for UART connection)
HopeRF RFM69	Proprietary	433, 868, 915MHz	Point-to-point	SPI	Message-based
HopeRF RFM95W	LoRa	433, 868, 915MHz	Point-to-point	SPI	Message-based
Nordic 51822	Bluetooth LE	2.4GHz	Point-to-point	SPI	Message-based
Cambridge Silicon CSR8510	Bluetooth LE	2.4GHz	Point-to-point	USB	Message-based
RF Link	TTL Serial	434MHz	Broadcast	UART	Session-based (UART connection)

The wisest thing you can do when buying your radios is to buy them as a set. Matching a transmitter from one company to a receiver from another is asking for headaches. They may say that they operate in the same frequency range, but there's no guarantee. Likewise, trying to hack an analog radio—such as that from a baby monitor or a walkie-talkie—may seem like a cheap and easy solution, but in the end, it'll cost you time and energy. If you're building all of the devices yourself, make sure the radios match. If you're using prebuilt devices, make sure your custom radios can talk to them before you buy.

Distance and energy also affect your decision. How far do you need to transmit? How much energy does the radio you're considering consume? Can your power supply give you the energy you need for the time you need? In a network where messages can make multiple hops, like a star network or a mesh network, you can potentially transmit further than in a point-to-point network.

You should also consider the **number of devices** that your project needs, as this will affect the **topology of the network** you choose. For example, if you've only got two devices that need to talk to each other in your project, then a star network like WiFi may be overkill. A point-to-point system may make more sense, or maybe even a simple broadcast transmitter-receiver pair. In a project that includes multiple radios and multiple two-way conversations, a point-to-point network or a star network may make the most sense.

Think about the **communication pattern** as well. In a project where messages are short and infrequent, a message-based protocol is great, but in one where you need a continuous connection, then a session-based protocol makes sense.

Consider the **data rate** you need for your application—more specifically, for the wireless part of it. You may not need high-speed wireless communication. For example, one common use for wireless communication in the performance world is to get data off the bodies of performers, in order to control media devices, musical instruments, and lighting dimmers. You might think that you need your radios to work at high data rates to do this, but you don't. You can send the sensor data from the performers wirelessly at a low data rate to a stationary computer, then connect to a high-bandwith network for transmitting the media at a higher data rate.

Price is of course a factor, but don't let it dicate everything you do. Sometimes, in picking a less expensive radio, you encounter so many complications that you offset the expense with workarounds to compensate for the radio's shortcomings. Count the number of devices you plan to use when you calculate your costs.

Table 6-2 covers a few popular radios used in microcontroller projects. The market for these radios has changed significantly in recent years, as the range of protocols has increased and the cost of radios has dropped. The Digi XBee radios were very popular when the first edition of this book came out, because they are reliable, flexible, and at the time, one of the most affordable choices. In recent years, they've been eclipsed by inexpensive WiFi and Bluetooth for projects that connect to a personal computer or the internet. Bluetooth is still more expensive than WiFi, in part because the Bluetooth SIG charges manufacturers more to license the Bluetooth brand than the WiFi Alliance does for WiFi.

HopeRF makes a series of radios that have gained popularity because they are low-power, long-range, and inexpensive. Their LoRa-based radios, compatible with the recently developed LoRa Alliance standard, may become very popular if LoRa takes off in consumer products, and if their radios can communicate with those products.

Transmitter-receiver pairs like the RF-link radios available from SparkFun or Seeed Studio are very inexpensive, and have a simple UART interface, but they are very prone to interference from other radio sources, and they are not addressible, so communication is less flexible than the other radios mentioned. They are mostly obsolete.

Processor manufacturers are increasingly embedding radios in the processor chip itself, as is the case with the Nordic Bluetooth radio in the Intel Curie chip at the heart of the 101, or the WiFi radio in the ESP8266. Given that, it's useful to know about many protocols, so you can use whatever radio is attached to your processor.

In the two projects that follow, you'll see microcontroller-to-microcontroller duplex radio communication using LoRa radios, and microcontroller-to-personal computer and mobile device communication using Bluetooth LE.

X

Duplex Radio Transmission

In this example, you'll connect an RF transceiver and a pushbutton to each of two or three microcontrollers. Each microcontroller will send a signal to the other when its button is pressed. When each one receives a message, it will light up an LED to indicate that it got a message. It will output the message serially as well.

MATERIALS

- **2–3 Arduino-compatible microcontrollers**
  - **Features used: Digital I/O, SPI**
- **2–3 solderless breadboards**
- **2–3 HopeRF RFM95W or Semtech SX1276 radio modules**
- **2–3 pushbuttons**
- **2–3 LEDs**
- **2–3 220-ohm resistors**

Bluetooth and WiFi are useful radio protocols when you want to talk to existing networks or consumer devices, but there are projects where you need to connect a few microcontrollers directly, without a central router or computer. This is where point-to-point radios come in handy. HopeRF and Semtech make a number of low-cost, low-power radios that serve as good examples of this kind of radio. In this project you'll make a small point-to-point network.

The HopeRF RFM9x and Semtech SX127x series radios use the LoRa radio protocol. LoRa is a relatively new protocol, designed for low-power devices and long-range communication. It works by using *spread spectrum* communication, in which data bits are sent on multiple adjacent frequencies in order to maximize the use of available radio bandwidth. The protocol is message-based, so you won't make a dedicated connection like you've done with sockets and serial ports.

LoRa doesn't mandate an addressing scheme, so you can make up your own. It's designed to have addressing schemes layered on top of it. The LoRaWAN protocol implements an addressing protocol on LoRa. For this project, though, you'll make your own simple addressing scheme by assigning each radio a one-byte address, and setting a common *broadcast address*. Your code will filter out messages whose destination doesn't match its address or the broadcast address. This scheme is typical of point-to-point networks.

When using these radios, you set a network ID, called a *sync word* in LoRa parlance, and the radio only listens for messages starting with that sync word. You can also affect which radios receive which messages by changing the radio's *spreading factor*. The spreading factor affects how the bits of a message are distributed across the different frequencies in the band that your radio is using. For example, the default bandwidth you'll use is 125kHz. The base frequency of the radio is 915MHz. The spreading factor determines how the bits of your message will be spread across a 125kHz range of frequencies centered at 915MHz. The spreading factor for these radios ranges from 6 to 12, with a default of 7. A higher spreading factor can compromise the reliability of transmission when you're transmitting continuously, so it's better to use a lower factor when possible.

The Circuit

You'll need at least two radios and two microcontrollers. You might want to get the parts to make a third device, to experiment with the networking dynamics once you have communication working between two radios. Adafruit's RFM95W LoRa Radio Transceiver Breakout was used for this example. The code should also work with the Dragino LoRa shield, the Modtronix Wireless SX1276 module, and Seeed Studio's modules. These radios come in two frequency ranges, 868–915MHz and 433MHz. The code to follow is adaptable for either frequency range, but defaults to 915MHz. These radios communicate with a microcontroller via SPI, so you'll use the SPI pins of your controller. In addition to the MOSI, MISO, clock (SCK), and chip select (CS) pins, you'll also connect to the radio's reset pin (RST) and its interrupt request pin (IRQ or G0) that the radio will use to let the microcontroller know when it has received new data. Note that the IRQ pin is connected differently on the MKR1000 than on the Uno and 101 boards, because of where the available hardware interrupts are on each processor.

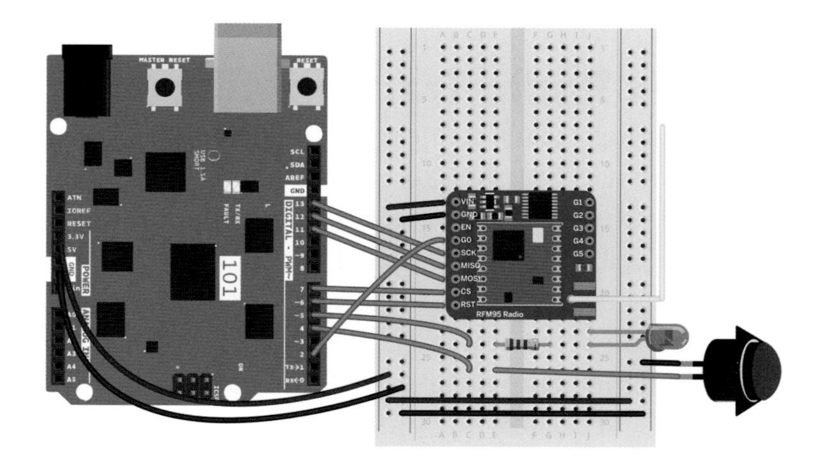

Figure 6-8
The circuit for connecting the RFM95W breakout board to a 101. You can use the same pin connections for an Uno. Note the antenna (yellow wire), which should be approximately 7.8cm for 915MHz or 16.5cm for 433MHz. These are quarter-wave antennas.

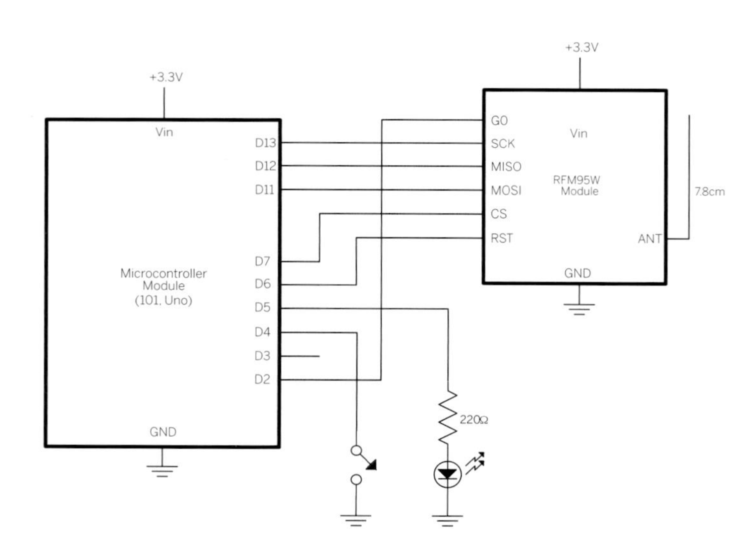

Figure 6-9
The schematic view of the circuit for connecting the RFM95W breakout board to a 101.

You'll need to attach an antenna to your radios to get a decent signal. The simplest way is to cut a wire of the appropriate length and attach it to the antenna connection. See the antenna formula in the "How Radio Works" section earlier. For 915MHz, a 7.8cm antenna will do the job. This length is one quarter of the wavelength, so this is called a *quarter-wave monopole antenna*, meaning it's only connected to one pole (the antenna pin). You can also buy prebuilt antennas for the correct wavelength. With a good antenna, you can supposedly transmit 2km or more with these radios, depending on obstructions. In my own tests, I've reached 30–50m indoors consistently with a simple wire antenna, but haven't tested longer distances.

The Code

There are a couple of different Arduino libraries that are compatible with these radios. The only one based on the Stream library like other communications libraries that you've seen is the Arduino LoRa library. Using the Arduino IDE's Library Manager, search for LoRa by Sandeep Mistry and install it. You might also find it handy to keep the library's documentation page open in a browser for reference: github.com/sandeepmistry/arduino-LoRa

X

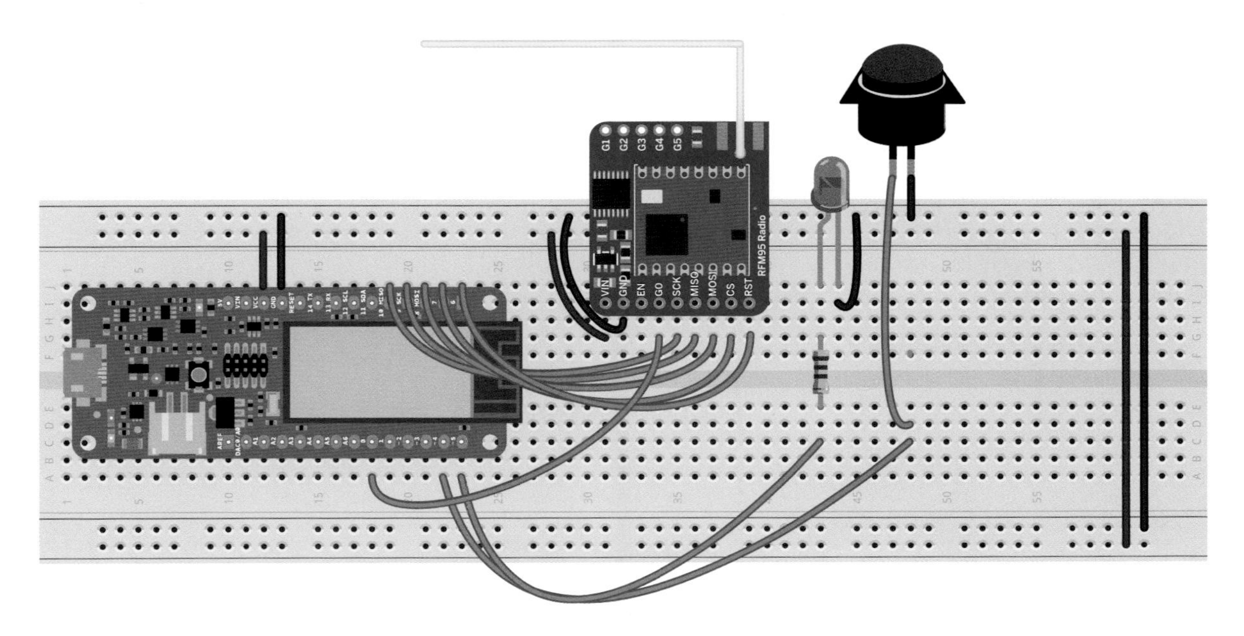

Figure 6-10
The circuit for connecting the RFM95W breakout board to an MKR1000. Note the antenna, which should be approximately 7.8cm for 915MHz or 16.5cm for 433MHz. These are quarter-wave antennas.

Figure 6-11
The schematic view of the circuit for connecting the RFM95W breakout board to an MKR1000.

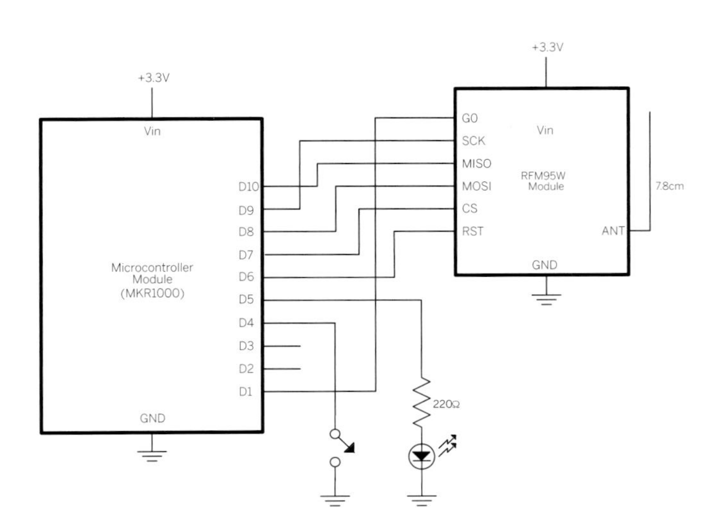

Try It Your program starts as usual by including the libraries and defining the I/O pin numbers. Change the interrupt request pin (irqPin) to a number appropriate to your processor. Chapter 5's sidebar on interrupts detailed the interrupt pin numbers. For the Uno and 101, use pin 2. For ARM M0-based boards like the MKR1000, use pin 1.

Next, set the properties of the push-button state and of the radio. Change the local address for each radio.

```
/*
  LoRa Duplex communication
  context: Arduino
*/
#include <SPI.h>              // include libraries
#include <LoRa.h>

const int buttonPin = 4;
const int receiveLED = 5;
const int csPin = 7;          // LoRa radio chip select
const int resetPin = 6;       // LoRa radio reset
const int irqPin = 1;         // interrupt request pin

int lastButtonState = HIGH;   // initial pushbutton state
byte msgCount = 0;            // count of outgoing messages
byte localAddress = 0xBB;     // address of this device
byte destination = 0xFF;      // destination to send to
byte syncWord = 0xB4;         // Sync word (network ID)
byte spreadingFactor = 8;     // spreading factor (6-12);
```

In the `setup()` function, initialize serial communication, set the pins for the radio, and initialize the radio at 915MHz. The notation `915E3` is a standard way of writing exponents in the C programming language. It means 915 * 10^3. Then set the sync word and spreading factor for the radio. Finally, set the I/O pins for the pushbutton and LED.

```
void setup() {
  Serial.begin(9600);                     // initialize serial
  LoRa.setPins(csPin, resetPin, irqPin);// set CS, reset, IRQ pin
  if (!LoRa.begin(915E6)) {              // initialize ratio at 915MHz
    Serial.println("LoRa init failed. Check your connections.");
    while (true);                         // if failed, do nothing
  }
  LoRa.setSyncWord(syncWord);
  LoRa.setSpreadingFactor(spreadingFactor);
  LoRa.setTimeout(10);                    // set Stream timeout of 10ms
  Serial.println("LoRa init succeeded.");
  // set the I/O pin modes:
  pinMode(buttonPin, INPUT_PULLUP);
  pinMode(receiveLED, OUTPUT);
}
```

ISM Radio Bands

You may be wondering why these radios come in multiple frequency ranges. These ranges are part of the *Industrial, Scientific, and Medical (ISM)* bands for the regions of the world. ISM bands are frequency ranges that do not require a license in which to operate, as long as you're transmitting below a given frequency. The ISM bands are agreed upon by the *International International Telecommuncations Union (ITU)*, and vary from country to country. For LoRa applications, the bands are:

Europe: 863–870MHz and 433MHz bands
US: 902–928MHz
Australia: 915–928MHz
China: 779–787MHz and 470–510MHz

▶ Begin the `loop()` function by reading the pushbutton. If it's changed since last read, and it's low, then you know the user just pressed the button. Note that you're using the internal pullup resistor on this pin. The delay between checking if it's changed and checking if it's low is a *debounce delay*. It ensures that the change is stable. If the user just pressed the button, send a message using a function called `sendMessage()`, which you'll write next. Then save the button state in `lastButtonState` for comparison next time through the loop.

At the end of the loop, call a new function called `onReceive()`. You'll write it shortly. This function will read and print incoming messages. Use `LoRa.parsePacket()` to determine if there's a valid packet to read. `.parsePacket()` is part of the `Stream` class, which you learned about in Chapter 4. It's used on message-based communications. You'll see it again in Chapter 7 for use with UDP messages.

```
void loop() {
  int buttonState = digitalRead(buttonPin);// read the pushbutton
  if (buttonState  != lastButtonState) {    // if it's changed
    delay(3);                               // debounce delay
    if (buttonState == LOW) {               // if it's low
      String message = "HeLoRa!";           // send a message
      sendMessage(message);
      Serial.println(message);             // print the message sent
    }
    lastButtonState = buttonState;          // save button state
  }

  // parse for a packet, and call onReceive with the result:
  onReceive(LoRa.parsePacket());
}
```

▶ In the `sendMessage()` function you'll begin a new packet to send, then you'll make up a message format. Write a header composed of the destination (1 byte), the sender (1 byte), the message count (1 byte), the message length (1 byte), and the message itself. Then increment the message count. Although this is a made-up format, it includes elements standard to many communications protocols. In the `onReceive()` function that you'll write next, you'll see how you can use it to filter messages.

```
void sendMessage(String outgoing) {
  LoRa.beginPacket();                       // start packet
  LoRa.write(destination);                 // add destination address
  LoRa.write(localAddress);                 // add sender address
  LoRa.write(msgCount);                     // add message ID
  LoRa.write(outgoing.length());            // add payload length
  LoRa.print(outgoing);                     // add payload
  LoRa.endPacket();                         // finish packet and send it
  msgCount++;                               // increment message count
}
```

The last function you need to write is `onReceive()`. First, check to see that there is an incoming message at all, by checking its packet size. If it's 0, quit the function and return to the loop. If there is a message, turn on the LED to indicate that. Then read the header bytes: recipient, sender, incoming message id (message count), message length, and finally the incoming message.

Next, check the message length against the reported length. If it's wrong, quit the function and report an error. Similarly, check to see that the recipient address is either the address of this radio, or the broadcast address. If not, quit the function and return.

Next, print out all the elements of the message. In addition to the items you put in your own header, the LoRa library gives you two other useful items: received signal strength indicator, or RSSI, and signal-to-noise ratio, or SNR.

Finally, turn off the LED to indicate the end of reception. That's the end of the sketch.

```
void onReceive(int packetSize) {
  if (packetSize == 0) return;        // if there's no packet, return

  digitalWrite(receiveLED, HIGH);     // turn on  the receive LED
  // read packet header bytes:
  int recipient = LoRa.read();        // recipient address
  byte sender = LoRa.read();          // sender address
  byte incomingMsgId = LoRa.read();   // incoming msg ID
  byte msgLength = LoRa.read();       // incoming msg length
  String incoming = LoRa.readString();// payload of packet

  if (msgLength != incoming.length()) { // check length for error
    Serial.println("error: message length is wrong.");
    return;           // skip rest of loop
  }
  // if the recipient isn't this device or broadcast,
  if (recipient != localAddress && recipient != 0xFF) {
    Serial.println("This message is not for me.");
    return;           // skip rest of loop
  }
  // if message is for this device, or broadcast, print details:
  Serial.print("Received from: ");
  Serial.println(sender, HEX);
  Serial.print("Sent to: ");
  Serial.println(recipient, HEX);
  Serial.print("Message ID: ");
  Serial.println(incomingMsgId);
  Serial.print("Message length: ");
  Serial.println(msgLength);
  Serial.print("Message: ");
  Serial.println(incoming);
  Serial.print("RSSI: ");
  Serial.println(LoRa.packetRssi());
  Serial.print("Snr: ");
  Serial.println(LoRa.packetSnr());
  Serial.println();
  digitalWrite(receiveLED, LOW);      // turn off receive LED
}
```

Once you've uploaded this code to two boards (with appropriate address changes), press the button on one and you should see the LED light up briefly on the other. Then press the second, and you should see the LED light on the first. If you open a serial terminal connection to both boards at once (you can open two windows to two different serial ports in CoolTerm), you should see messages transmitted from one to the other as well.

These radios offer features that are common in networks, both wired and wireless, so it's useful to experiment with them. Once you've seen reliable transmission from one to the other and back, there are a few extra things you can do to learn more about these radios and the network they form.

Broadcast Messages

If you have the components to build three radio boards, do so and set the third board's `localAddress` to something unique and its `destination` to 0xFF. When the destination is set to 0xFF, the `onReceive()` function will report the message even if it's not addressed to this radio.

With three, try setting two radios to send just to each other's addresses instead of to the broadcast address and see what happens. Both of them should still get messages from the third radio, which is still sending to the broadcast address, but the third radio won't report their messages. It's still receiving them, just filtering them out by address. You can change this by removing the address check in `onReceive()` and simply reporting all messages. In networks with addresses, a device listening to messages that are not addressed to it like this is said to be in *promiscuous mode*. It's a useful diagnostic tool, and a privacy vulnerability.

Sync Words and Spreading Factor

Every radio sharing the same sync word receives every message. If a third radio has a different sync word, however, it will filter out the messages from the radios that don't share its sync word. This can be useful when you need to set up multiple LoRa networks in the same space. The sync word functions as a form of network ID. Try changing the sync word on one of your radios to see the effect.

You can also use the spreading factor as a form of filtering. Two radios using different spreading factors will not receieve each others' messages. Try changing the spreading factor on one of your radios to see the effect. Even one point makes a big difference.

Spreading factor isn't intended to be used as a method for message filtering, though it does the job well. Normally it's used in conjunction with two other factors, bandwidth and coding rate, to optimize the transmission and reception of signals. The data sheets for the HopeRF and Semtech radios discuss this in further depth. For reference, find them at www.hoperf.com/upload/rf/RFM95_96_97_98W.pdf (HopeRF RFM95W) and www.semtech.com/images/datasheet/sx1276.pdf (Semtech SX1276). The two radios are functionally identical.

Received Signal Strength

One of the characteristics that shows up in the serial terminal window is the *received signal strength indicator, or RSSI*. RSSI is a measure of the strength of the last signal received. You'll learn more about it in Chapter 8. Radio transmission, like light and other forms of electromagnetic energy, decreases with the distance squared. Try moving your radios apart and pressing the button from different distances, to see the difference.

Signal-to-Noise Ratio

Another characteristic reported in your sketch is the signal-to-noise ratio, or SNR. The signal-to-noise ratio indicates how much of what the radio received was a clear data signal, and how much was unintelligble, and therefore noise. A higher SNR indicates a clearer transmision. Generally, you get a higher SNR at lower data rates, because the receiver has more time to read the incoming message. If the SNR gets low, you can try adjusting the spreading factor, bandwidth, and coding rate to improve the signal.

Cyclic Redundancy Check

Many communications protocols feature a cyclic redundancy check (CRC), and this one is no different. In a CRC, a calculation is done on the payload and the result is placed in one or two bytes at the end of the packet. The receiver then does the same calculation on the payload, and if the result is the same, the payload is valid. If not, the packet is rejected. The payload length check you're doing is a very crude CRC. You can enable the radio's internal CRC by using the command `LoRa.crc()`. This will improve reliability and filter out invalid packets.

LoRa radios offer a flexible and reliable way to add point-to-point radio to a project. With a proper antenna, you can get a long range out of them too, making them useful for outdoor applications.

X

Project 12

Bluetooth LE Camera Control

In Chapter 2, you learned how to connect a microcontroller to your personal computer using a Bluetooth radio. That project used the Bluetooth Serial Port Profile from the Bluetooth 2.0 spefication. In this project you'll use Bluetooth 4.0, also known as Bluetooth LE, to control the IR camera controller from the infrared project.

MATERIALS

- **1 Arduino 101 or compatible microcontroller with Nordic nRF8001 or nRF51822 Bluetooth LE radio**
  - **Features used: Digital I/O, Bluetooth LE**
- **1 infrared LED**
- **1 220-ohm resistor**
- **1 solderless breadboard or prototyping shield**
- **1 Bluetooth LE-enabled tablet, phone, or computer**

The IR camera controller you built in the previous project is a good example of a project that could benefit from a radio interface as well. Imagine if, instead of pushing the pushbutton, you could control the IR remote from a smartphone or a tablet. A Bluetooth connection could make that possible. Bluetooth 4.0, also known as Bluetooth LE, was designed with applications like this in mind.

The Arduino 101 has a built-in Bluetooth LE radio, so it's perfect for this project. If you want to do it with another board, however, you can. The Bluetooth library used here will work with any Bluetooth radio based on Nordic Semiconductor's nRF8001 or nRF58122 radios. You can get these as add-on modules to existing microcontroller boards, or you can find them built into boards like RedBear Labs' BLE Nano. There are other Bluetooth LE radios on the market that use their own libraries, if you want to investigate other options.

Bluetooth version 4.0 was a big change to the Bluetooth specification. It aimed to make the protocol more power-efficient, among other things, and introduced a whole new model for interacting with Bluetooth devices. Instead of connecting to a Bluetooth device using the Serial Port Profile (SPP), you now use the General Access Profile (GAP), and then connect to services it offers using the Generic Attribute Profile (GATT). For an excellent in-depth introduction to Bluetooth LE, see **Make: Bluetooth** by Alasdair Allan, Don Coleman, and Sandeep Mistry (Maker Media).

Bluetooth LE: Central and Peripheral Devices

Bluetooth LE devices are divided into two rough groups, *central* devices and *peripheral* devices. A peripheral device advertises its services using the Generic Access Profile (GAP). Central devices can listen for advertised peripherals and then browse the services they offer. Every *service* has one or more *characteristics*, and every characteristic has a *value*. The protocol for describing services and characteristics is called the Generic Attribute profile (GATT). For example, a Bluetooth LE-enabled light might offer a single service, perhaps called Light, with two characteristics: Brightness and Hue. Separating the two characteristics would allow you to change the overall brightness without changing the hue. If you wanted to enable a preset brightness, your light might have a third characteristic, On. When On is set to true, the light would be turned on to the level of the Brightness characteristic, using the Hue characteristic to set the color.

So how does the General Attribute profile (GATT) differ from Bluetooth Serial Port Profile (SPP) in practice? When you're using the SPP, you start by pairing with a Bluetooth device and exchanging a passcode in order to pair. Then the device presents you with a data stream that looks like a serial port to your computer. Beyond that, the SPP offers no structure. Any bytes you push down the stream from either end come out the other end, first in, first out, and you have to define the structure of the data you send through that stream. By contrast, the GATT is a bit more structured. You pair with the device (no passcode needed, usually), and then you can see a list of services. Each of these has a sub-list of characteristics. You can either read from these, write to them, or subscribe to them. If you subscribe to a characteristic. the peripheral device will automatically update you when it chages. Rather than having to read all the data in sequence from a stream, you pick the particular characteristics that you care about, and read to or write from them.

Typically Bluetooth 2.0 devices offer only one service. The BlueFruit radios in Chapter 2 offer SPP, for example; Bluetooth headsets offer the Audio Profile. A Bluetooth keyboard offers the Human Interface Device (HID) profile. In contrast, Bluetooth 4.0 devices can offer multiple services. Most commercial Bluetooth LE devices will offer at least a battery level service and one other, so you can check the battery level in addition to using the device.

A peripheral's characteristics can have a few properties. They can be readable or writable. They can also offer notifications to update central devices that subscribe to that property when the property changes. You'll see how a characteristic's properties are set in the code that follows.

The Circuit

If you're using the 101, you can use the same circuit diagram as Figure 6-5, shown earlier in this chapter. The Bluetooth radio is built into the board. Figures 6-14 and 6-15 show how to connect a Nordic NRF8001 radio to an Uno. The Nordic radio's interface with the processor via SPI, with one extra pin. If you're using another board based on the Uno's Atmega328 processor or the MKR1000's Cortex M0, you should be able to use the same code, as long as you change the SPI pins to match your board.

Any Arduino-compatible board that uses the Nordic nRF8001 or nRF51822 should work with the library used with this project, though only the 101, the Uno, and the MKR1000 were tested with this code.

For this project, your IR remote control for the camera will be a peripheral. It will offer one service, namely the camera control for the camera. You'll call it `cameraService`. The service will have one characteristic, the shutter control, called `shutter`,which tells whether it's currently sending a signal to the camera or not. When the characteristic's value is 1, the controller will send the command to the camera

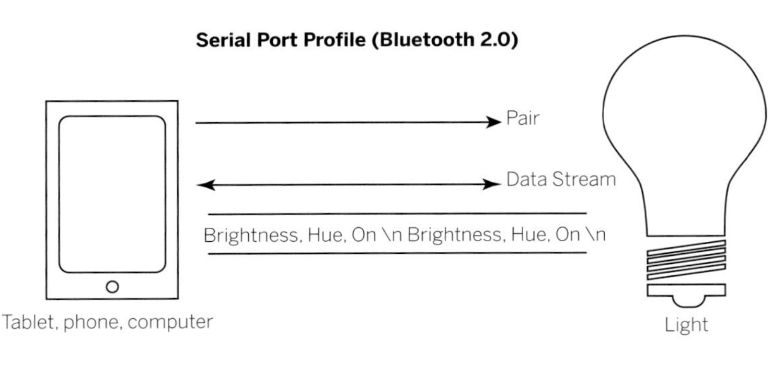

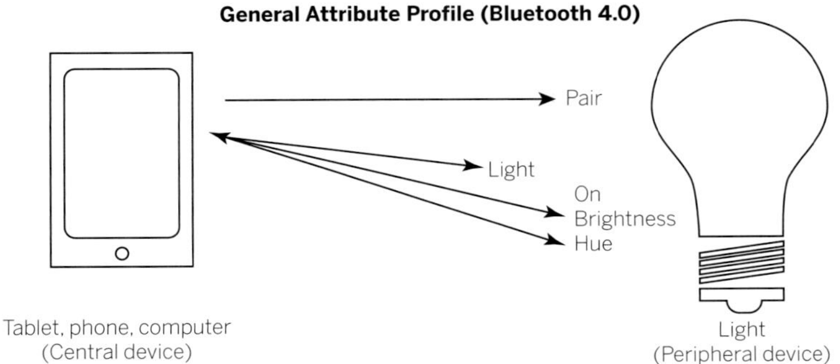

Figure 6-12
A Bluetooth 2.0 light would offer the Serial Port Profile. When you connect to it, all data is sent through the serial port stream. Your computer (tablet, phone, laptop) would read all three data points to get what you want.

In Bluetooth 4.0 (aka Bluetooth LE), the light would offer a service called Light, with three characteristics. Your computer, called a central in Bluetooth LE terms, could read, write, or subscribe to any or all of them, depending on their properties.

via the IR LED. When it's finished sending, the controller will reset the shutter characteristic's value to 0.

Bluetooth LE devices, characteristics, and services are identified by 128-bit numbers called *Universally Unique Identifiers*, or *UUIDs*, for example `2f45c1ee-5048-11e6-beb8-9e71128cae77` (UUIDs are written in hexadecimal notation). Some well-known services, like the battery service, have shorter 16-bit UUIDs that are pre-assigned by the Bluetooth SIG. The battery service is `0x180F`, and any peripheral that offers a service with that UUID is expected to report the battery level using it. In any commercial device you should use 128-bit UUIDs in your code so as not to cause conflict with existing services. I'll use 16-bit UUIDs in this project for brevity. For a full list of the well-known service UUIDs, go to developer.bluetooth.org/gatt/services.

The Code

The software library for this project is called BLEPeripheral, and it's by Sandeep Mistry. If you're using a board other than the 101, install it using the Library Manager. The version for the 101 is called CurieBLE, and it installs automatically when you installed the 101 board via the Boards Manager. The documentation at www.arduino.cc/en/Reference/CurieBLE for CurieBLE and github.com/sandeepmistry/arduino-BLEPeripheral for the BLEPeripheral version. Both libraries have similar commands, so the code differences are minimal, as shown in the following pages. Differences are shown for the non-101 boards in blue.

X

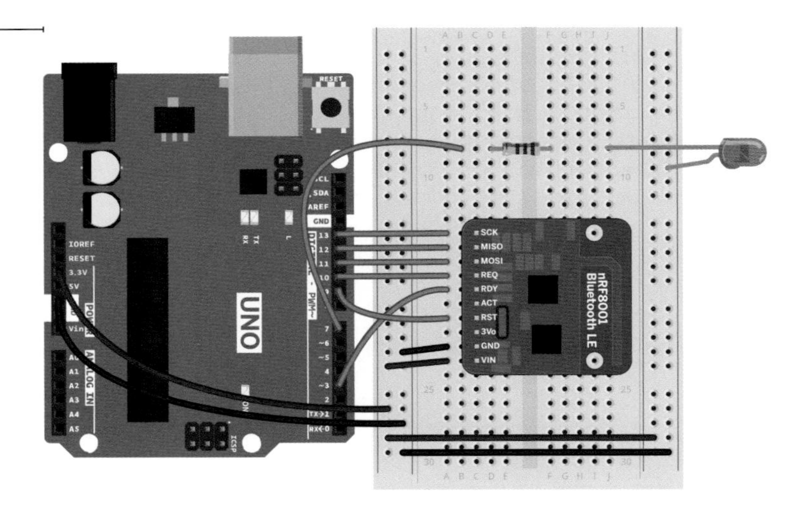

Figure 6-13
Breadboard view of the circuit for the Bluetooth LE Camera control project, using an Arduino Uno and a Nordic nFR 8001 radio. If you're using an Arduino 101, you can use the same circuit as the IR project earlier in the chapter instead, as the radio is built into the 101.

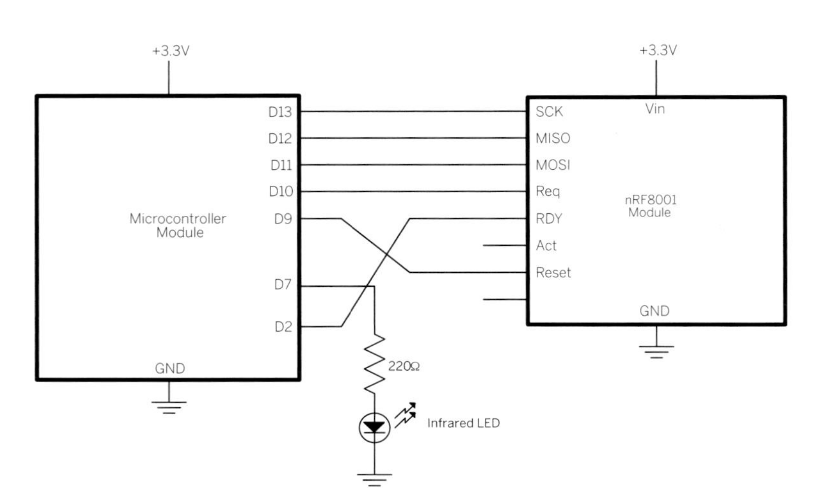

Figure 6-14
Schematic view of the circuit for the Bluetooth LE Camera control project, using an Arduino Uno and a Nordic nFR 8001 radio.

Try It Your sketch starts with the usual library imports. **If you're using the 101, you'll include the CurieBLE library** and instantiate a peripheral like so:

```
/*
Bluetooth LE IR Camera Control
context: Arduino 101
/*
#include <CurieBLE.h>
```

Open the Boards Manager and check that your Arduino 101 Board definitions are version 2.0.1 or later for this code to work on the Arduino 101. If you're using an earlier version of the board definition, update it to 2.0.1 or later.

For other boards, you'll include the SPI library and the BLEPeripheral library instead. Then you'll need to set the pin numbers for the Request, Ready, and Reset pins. Then you'll instantiate the peripheral using those pin numbers like so:

```
/*
Bluetooth LE IR Camera Control
context: Arduino Other
/*
#include <SPI.h>        // use SPI and BLEPeripheral for non-101 boards
#include <BLEPeripheral.h>
// define pins for BLE Request, Ready, and Reset pins
// (needed only for non-101 boards)
const int BLE_REQ = 10;
const int BLE_RDY = 2;
const int BLE_RST = 9;

// create peripheral instance:
BLEPeripheral blePeripheral = BLEPeripheral(BLE_REQ, BLE_RDY, BLE_RST);
```

Next, give the camera service and the shutter characteristic each a UUID. You initialize different characteristics by data type: `BLEIntCharacteristic`, as shown here, if you plan to read it as integers; `BLEFLoatCharacteristic` if you plan for floating point numbers; `BLECharCharacteristic`, if you need ASCII characters, and so forth. See the library documentation for details.

Following that, add the global variables from the IR camera control sketch in Project 10.

```
// Give the camera service a UUID:
BLEService cameraService("F01A");
// Give the shutter characteristic a UUID
BLEIntCharacteristic shutter("F01B", BLERead | BLEWrite);

const int IRPin = 7;             // IR LED I/O pin
const int CAM_NIKON = 1;         // DSLR camera types
const int CAM_CANON = 2;
int buttonState = 0;             // current button state
int lastButtonState = 0;         // previous button state
```

If you're using the 101: In the `setup()` function, you'll set a local name for the peripheral, then you'll set which service it advertises to central devices. Since it has only one service, that is what you'll advertise. Then set an initial value for the shutter characteristic using the `.setValue()` function, and finally, begin advertising with the `.begin()` function.

```
void setup() {
  Serial.begin(9600);
  BLE.begin();
  // set  local name for the peripheral and advertise the camera service:
  BLE.setLocalName("irRemote");
  BLE.setAdvertisedServiceUuid(cameraService.uuid());

  // add the service and characteristic as attributes of the peripheral:
  BLE.addService(cameraService);
  cameraService.addCharacteristic(shutter);

  BLE.advertise();              // begin advertising camera service
  shutter.setValue(0);          // set initial value for shutter
  Serial.println("Starting");
}
```

For other boards with the BLEPeripheral library: The functions are called from the `blePeripheral` instance, and the function for adding a service or a peripheral is called `.addAttribute()`. In this library, `blePeripheral.begin()` also starts advertising the service.

```
void setup() {
  Serial.begin(9600);
  // set  local name for the peripheral and advertise the camera service:
  blePeripheral.setLocalName("irRemote");
  blePeripheral.setAdvertisedServiceUuid(cameraService.uuid());

  // add the service and characteristic as attributes of the peripheral:
  blePeripheral.addAttribute(cameraService);
  blePeripheral.addAttribute(shutter);

  blePeripheral.begin();        // begin advertising camera service
  shutter.setValue(0);          // set initial value for shutter
  Serial.println("Starting");
}
```

The `loop()` function for this sketch listens for any activity from central devices using the library's `.poll()` function (again, note the difference between libraries). This enables you to then use the `.written()` function to see when a central device connects and changes the value of a writable characteristic like the shutter characteristic. If that happens, call the IR camera control `shutterClick()` function, then reset the characteristic value for next time.

Add the `shutterClick()` and `IRPulse()` functions from the IR camera control project to finish the sketch.

X

```
void loop() {
  // poll the peripheral for activity:
  BLE.poll();                // use this for the 101
  // blePeripheral.poll();   // use this instead for BLEPeripheral library

  // if a central wrote to the shutter characteristic:
  if (shutter.written()) {
    // and the characteristic's value is 1:
    if (shutter.value() == 1) {
      shutterClick(CAM_NIKON);     // send IR shutter signal
      Serial.println("click");
      shutter.setValue(0);         // reset the value to 0
    }
  }
}
```

Bluetooth LE Central Diagnostic Applications

Now comes the fun part: connecting to the Bluetooth LE peripheral you just made from a central device. Before you start writing your own, it's useful to have a working application that you can use for diagnostic purposes. There are a number of applications for Bluetooth LE scanning on mobile device platforms, but fewer options on the desktop operating systems. Three useful applications for this are available for free: LightBlue for macOS and LightBlue Explorer for iOS, both by Punch Through Designs, and nRF Connect, by Nordic, for Android. If you search the app store for macOS, iOS, or Android respectively, you'll find them. Install the one for your platform and you're ready to continue. This section will show you how to use them to scan for your IR remote control peripheral and control it.

LightBlue for macOS

LightBlue for macOS is a simple Bluetooth LE scanner. It presents you with four columns and begins scanning automatically as shown in Figure 6-15. The first column shows the devices it's found. Click on any device to see its services in the second column. Click a service to see its characteristics, and click a characteristic to see its value and whether it's readable, writable, or set to notify you when you subscribe.

The peripheral you just built will show up with the name `irRemote`, which you set as the local name. Alternately, it might show up with the name `ARDUINO 101-XXXX` where XXXX is the last four digits of the Bluetooth radio's MAC address. That's the default name the library uses if you don't give your peripheral a name. When you click on its service, notice that it's the UUID that you gave the `cameraService`. Click on the characteristic (with the UUID you gave to `shutter`). Since it's a readable and writable characteristic, you can either click the Read button to see the value, or enter a value in the text field to write a value to it. LightBlue expects either hex values or ASCII strings. Write the hex value 0x01 and then hit Enter, LightBlue will set the value of the characteristic on your peripheral device, and the device will send the IR signal to open the camera's shutter. The sketch running on your peripheral will then reset the value to 0, so you'll always get that value when you read it.

LightBlue Explorer for iOS

This app is similar to the desktop app in its functionality, though different in its appearance. You'll get a list of devices first, and clicking on your device will reveal its services. Clicking on the service brings up the characteristic screen seen in Figure 6-16. From there, you can read or write values to your characteristic just as you can with the desktop app. However, with the iOS version, you don't need to add the preceding "0x" to your hex values. If

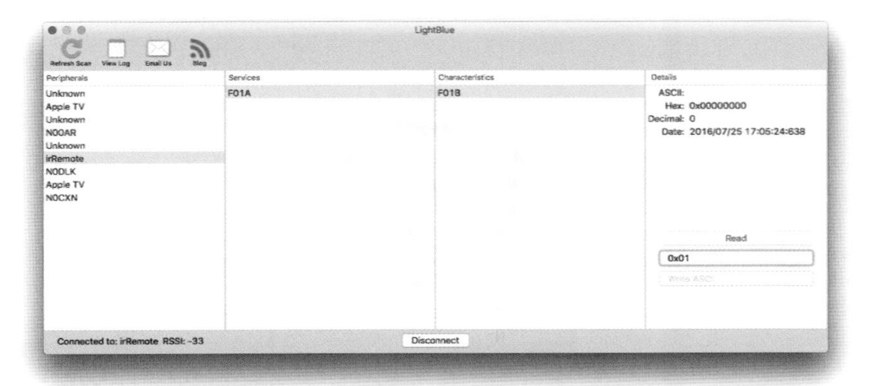

Figure 6-15
LightBlue for macOS, a Bluetooth LE scanner. This application scans for Bluetooth LE devices and lets you browse their services and characteristics.

you want to change the data type, tap the word `Hex` in the upper-right corner of the characteristic viewer screen to get to a menu where you can set the data type.

nRF Connect for Android

Nordic Semiconductor's nRF Connect app is one of several Bluetooth LE scanners available for Android devices. It works much like LightBlue does on macOS and iOS. The first screen presents a device list. The second presents that device's service list. The third presents the details of a service, and lets you write to the writable ones. You can write any data type, but you have to pick your data type from an option menu. For the peripheral you just built, pick the UINT 8 data type. Then enter 1 as shown in Figure 6-17. Tap send, and the peripheral should send the IR signal to the camera.

Whichever Bluetooth LE central app you use, keep it around for diagnostic purposes. As you work more with Bluetooth LE, these apps will help you to test connectivity while you're developing your own apps.

X

Figure 6-16
LightBlue for iOS. This view shows the Characteristic browser screen.

Figure 6-17
nRF Connect for Android.

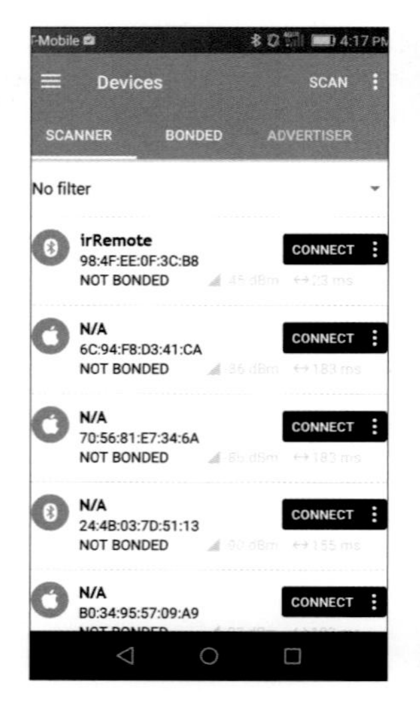

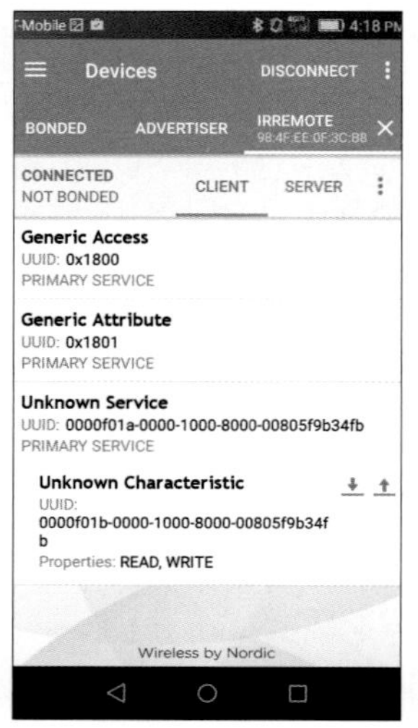

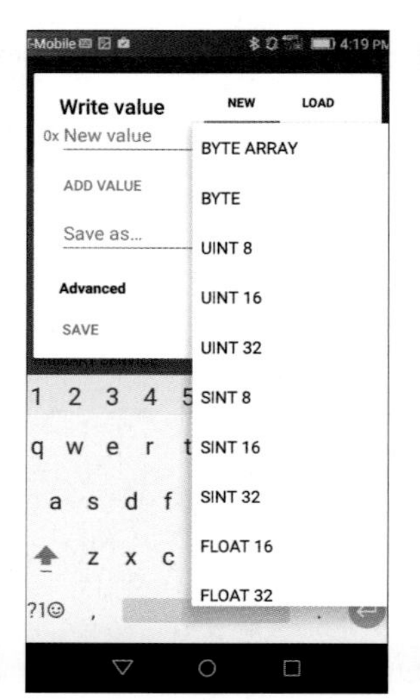

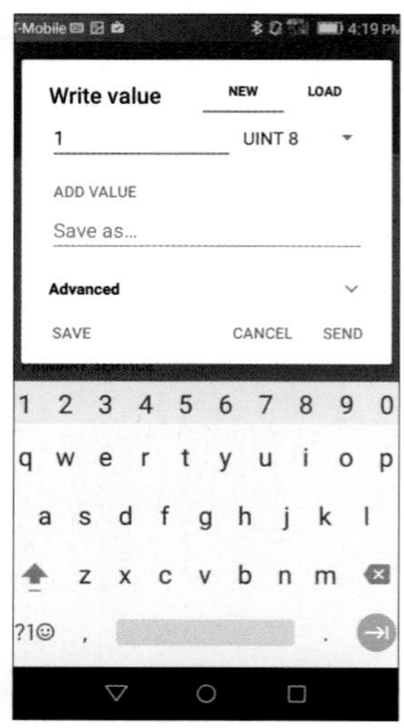

A node.js Bluetooth LE Central Application

Much of the development effort for Bluetooth LE so far has been focused on mobile operating systems rather than desktop ones. Although there are development tools for it in Windows, for example, it's not fully supported there as of this writing. There are libraries for macOS, but not many higher level ones. However, there is a library for node.js that works on macOS, Linux, and Windows. It's called noble, and it's also by Sandeep Mistry. In this final part of the project, you'll build a server in node.js that connects to your peripheral device and serves a web page with a button you can click to fire the shutter. You'll write the client side of this app in p5.js, which you saw in the previous chapter.

Installing noble and express

Installing the noble library itself is just like installing any other node.js library. You'll need the express library as well, for the server part of the application. Make a directory for your project in your documents directory (I called mine **NobleCameraControl**), then use npm to install them both like so:

```
$ mkdir NobleCameraControl
$ cd NobleCameraControl
$ npm install noble express
```

Depending on your operating system, you'll have to do a few extra steps. For macOS, you've got nothing extra to do, though if your machine was made prior to 2012, its Bluetooth radio is not Bluetooth LE-compatible. For Windows and Linux, if your computer doesn't already have a Bluetooth LE-compatible radio, you can use the Cambridge Silicon CSR8510 radio mentioned at the beginning of this chapter. The drivers for it are built into recent versions of Windows. For Linux, including the Raspberry Pi, you can install it and its dependencies like so:

```
$ sudo apt-get install bluetooth bluez libbluetooth-dev
libudev-dev
```

Windows users may also need to change the Bluetooth USB drivers in order for noble to recognize your radio, even if you're using the built-in radio. There's a software tool called Zadig that will do this for you. You can find it at zadig.akeo.ie, and you can find installation instructions on the noble library repository page, github.com/sandeepmistry/noble. You can also find more detailed Linux installation instructions there. Once you've set up a project directory and got noble installed, you're ready to write some code.

The Server Code

This program is an HTTP server that runs on your personal computer to connect a web page to Bluetooth LE devices through your computer.

Plan It You saw the process for connecting to a peripheral in the previous section: scan for all peripherals, find yours, scan for services, scan for characteristics, then read, write, or subscribe to them. This program takes advantage of the asynchronous nature of node.js, and it can get a bit confusing, so it helps to see the pseudocode version first.

In the code to follow, you'll see the usual node.js pattern, where functions are defined at the beginning of the program and called at the end of it. Similarly, functions that are local to another function are defined at the beginning of the parent function and then called at the end of it.

```
// include libraries:
// set up global variables

// if the radio's on, scan for peripherals
// if you find a peripheral,
//    stop scanning for other peripherals
//    connect to it
//       read its services and characteristics

// if you have a peripheral's services and characteristics:
//    list them
//    scan through the characteristics for the shutter characteristic
//       save it in a global variable

// if an HTTP client makes a GET request for /click:
//    if you have a valid shutter characteristic
//       change the shutter characteristic's value
//       respond to the HTTP client with the result
```

Try It Save the following file as **server.js**. in the **NobleCameraControl** directory. Start by requiring the libraries you'll need, noble and express. Then make a server variable from express as you've done before.

```
/*
  Noble Bluetooth LE server
  context: node.js
*/
var noble = require('noble');
var express = require('express');
var server = express();              // create a server using express
```

» You'll need global variables to hold the UUIDs of the service and characteristic that you'll be looking for on your peripheral. You'll also need one for the characteristic itself. Make a final one to keep track of how many times the shutter's been clicked too.

Establish the **public** directory as the one from which you'll serve those static files using `express.use()`. You'll make this directory later.

```
var cameraUuid = 'f01a';          // uuid for the camera service
var shutterUuid = 'f01b';         // uuid for the shutter characteristic
var shutter = null;               // the shutter characteristic
var clickCount = 0;               // count of camera clicks from the client
var device;                       // the peripheral

// serve static files from /public:
server.use('/',express.static('public'));
```

» At the end of your program, you'll need two event listeners from the noble library. The first tells noble what to do when the radio turns on (`.on('stateChange')`), and the other tells it what to do when it discovers a peripheral (`.on('discover')`). Each has a callback function, and you'll write those next.

Start the express server listening and set up a GET request listener for the `/click` request at the end of the program as well.

```
// Scan for peripherals with the camera service UUID:
noble.on('stateChange', scanForPeripherals);
noble.on('discover', readPeripheral);

server.listen(8080);        // start the express server
server.get('/click', click); // GET request listener
```

» Here's the callback handler for the `stateChange` event. It will get called when the `stateChange` event happens. Put this and the rest of the functions that follow before the listeners you just wrote.

Rather than scan for every known peripheral, you can scan for a peripheral that's advertising a particular service. That's what the `.startScanning()` function is doing when you pass it the camera service's UUID.

```
//  callback function for noble stateChange event:
function scanForPeripherals(state){
  if (state === 'poweredOn') {              // if the Bluetooth radio's on,
    noble.startScanning([cameraUuid], false); // scan for the camera service
    console.log("Started scanning");
  } else {                                   // if the radio's off,
    noble.stopScanning();                   // stop scanning
    console.log("Bluetooth radio not responding. Ending program.");
    process.exit(0);                        // end the program
  }
}
```

When a peripheral is discovered that's got a service matching the one you scanned for, the `readPeripheral()` callback function gets called. In this function, you'll stop scanning, attempt to connect, and when you succeed, you'll read services using the `readServices()` callback function. You're saving the peripheral in the global variable called `device` so that you can check its state later on.

```
// callback function for noble discover event:
function readPeripheral (peripheral) {
  console.log('discovered ' + peripheral.advertisement.localName);
  console.log('signal strength: ' + peripheral.rssi);

function readServices() {
    device = peripheral;    // save the peripheral to a global variable
    console.log('Checking services: ' + peripheral.advertisement.localName);
    // Look for services and characteristics.
    // Call the explore function when you find them:
    peripheral.discoverAllServicesAndCharacteristics(explore);
  }

  noble.stopScanning();                 // stop scanning
  peripheral.connect();                 // attempt to connect to peripheral
  peripheral.on('connect', readServices);  // read services when you connect
}
```

Once you've got a list of all the services and characteristics, you can explore them. The `explore()` function is the callback for the `.discoverAllServicesAndCharacteristics()` function that you just wrote. In this function, you'll go through all the discovered characteristics to find the one that matches the shutter characteristic's UUID. You don't know when you'll need to write to the characteristic, because that's determined by the user through the web browser. So save the characteristic in a global variable, so you can access it at any time.

```
// the service/characteristic explore function.
// depends on the services & characteristics, not the peripheral,
// so it doesn't have to be local to readPeripheral():

function explore(error, services, characteristics) {
  // list the services and characteristics found:
  console.log('services: ' + services);
  console.log('characteristics: ' + characteristics);

  // check if each characteristic's UUID matches the shutter UUID:
  for (c in characteristics) {
    if (characteristics[c].uuid === shutterUuid) {
      // this is the characteristic you want to control:
      shutter = characteristics[c];
    }
  }
}
```

Finally, you need a callback for the HTTP GET request. In this function, you'll check to see that you're still connected to the peripheral, then you'll write to the shutter characteristic. After that you'll increment the click counter. Finally, you'll send a response to the HTTP client about what happened.

That's the whole script. Now you're ready to run it.

```
// the function for the HTTP response to GET /click:
function click(request, response) {
  var result = "no peripheral present";  // what you'll send to the client
  // if you're connected and have a valid characteristic:
  if (device.state === 'connected' && shutter != null ) {
    // write to the characteristic:
    var output =  new Buffer([0x01]);
    shutter.write(output, true);
    clickCount++;                               // increment click counter
    result = "Click count: " + clickCount;  // change the response
  }
  response.end(result);    // respond to the HTTP client
}
```

When you run the server script you just wrote by typing `node server.js`, it will start by searching for Bluetooth LE peripherals. If the one that you built earlier is on and in range, you should see output like this:

```
Started scanning
discovered irRemote
Checking services: irRemote
services: {"uuid":"f01a","name":null,"type":null,"includedServic
eUuids":null}
characteristics: {"uuid":"f01b","name":null,"type":null,"properti
es":["read","write"]}
```

Don't be confused that when the program discovered your device it had one name, and when it checked services, it had another. The default device name is set by the library, and sometimes your central app will display that name until it's read the advertised local name. For the ArduinoBLE library, the default device name is "Arduino" and for the CurieBLE library it's "ARDUINO 101 - XXX" where XXX is the last four hex digits of the device's MAC address.

The Web Interface

You won't see anything in the browser if you try to access this server yet, because you haven't written your web pages. You'll need a directory from which to serve those pages, like you did in the webSocket project in Chapter 5. Make a new p5.js project in a subdirectory of the **NobleCameraControl** directory called **public**. In it, copy the p5.js files and libraries from the p5.js example project as you did in Chapter 5. If you're using the command-line p5-manager tool, you can add the p5.js project like so:

```
$ p5 generate --bundle public
```

Then change directories into the **public** directory and get the latest versions of the p5.js libraries like so:

```
$ p5 update
```

Your HTML page will have two elements: a button and a text div to display messages from the server. Both will be generated by some JavaScript you'll write using p5.js. The button will make a GET request back to the server, and the text div will display the server's response.Your **index.html** page is just the default page generated by the p5.js editor. You'll write all your code in the **sketch.js** file.

Try It Your p5.js sketch.js script starts with two global variables for the button and the text div, as you might expect:

```
/*
  p5.js IR Camera Control page
  context: p5.js
*/
var shutterButton; // the shutter button
var messageDiv;    // text div for responses from the server
```

In the `setup()` function you'll create the button and the text div. Position them where you want them to be. You'll need to set a callback function for the `touchEnded` event as well, called `getShutter()`.

```
function setup() {
  // create the shutter button, position it, and give it a callback:
  shutterButton = createButton('shutter');
  shutterButton.position(windowWidth/2, 20);
  shutterButton.touchEnded(getShutter);

  // create the message div and position it:
  messageDiv = createDiv("waiting for messages");
  messageDiv.position(20, 50);
}
```

The getShutter() function makes an HTTP GET request using p5.js's httpGet() function. You give it the path, the data type to expect in return, and the callback for when the server replies. That callback function, called clickDone() here, puts the server's response into the text div.

```
// callback function for the shutter button:
function getShutter() {
  httpGet('/click', 'text', clickDone);
}

// callback function for httpGet() request:
function clickDone(data) {
  messageDiv.html("server response: " + data);
}
```

That's the end of the script.
X.

Once you've saved this, make sure your peripheral is powered, run the **server.js** script again, then open a browser to http://localhost:8080 as you've done for servers you've written in previous chapters. On the command line, you should see the scanning messages. In the browser, you should see the web page shown in Figure 6-18. Initially the text will say `Waiting for messages` and when you click, it will change to Server response: `Client clicked X times`. X will change every time you click, and the IR signal to the camera will fire on the peripheral each time you click.

Figure 6-18
The web page interface for the Bluetooth central server.

This system combines both forms of wireless that you've seen in this chapter, infrared and radio, and merges them with the web protocols you've learned in earlier chapters. Your peripheral translates from the Bluetooth LE protocol to the IR camera control protocol. The node.js server and web browser interface means you can control the camera not just from your local computer, but from any browser-equipped device that can reach your computer's IP address. This offers lots of possibilities for future projects.
X

Conclusion

Wireless communication involves some significant differences from wired communication. Because of the complications, you can't count on the message getting through like you can with a wired connection, so you have to decide what you want to do about it.

If you opt for the least-expensive solutions, you can just implement a one-way wireless link with transmitter-receiver pairs and send the message again and again, hoping that it's eventually received. If you spend a little more money, you can implement a duplex connection, so that each side can query and acknowledge the other. Nowadays, a duplex connection is the standard, as the cost difference is minimal and the availability of transmitter-receiver pairs is waning. Regardless of which method you choose, you have to prepare for the inevitable noise that comes with a wireless connection. If you're using infrared, incandescent light and heat act as noise; if you're using radio, all kinds of electromagnetic sources act as noise, from microwave ovens to generators to cordless phones. You can write your own error-checking routines but, increasingly, wireless protocols like WiFi and Bluetooth are making it possible for you to forget about that, because the modules that implement these protocols include their own error correction.

Just as you started learning about networks by working with the simplest one-to-one network in Chapter 2, you began with wireless connections by looking at simple pairs in this chapter. You also got a glimpse at peer-to-peer networks, in which there is no central controller, and each object on the network can talk to any other object, and you saw two examples of message-based communication, as opposed to session-based, like you've seen with TCP sockets and webSockets. In the next chapter you'll explore message-based communication more.

X

▸▸ Urban Sonar by Kate Hartman, Kati London, and Sai Sriskandarajah
The jacket contains four ultrasonic sensors and two pulse sensors. A microcontroller in the jacket communicates via Bluetooth to your (very old) mobile phone. The personal space bubble, as measured by the sensors, and your changing heart rate, as a result of your changing personal space, paint a portrait of you that is sent over the phone to a visualizer on the internet.

FRONT
LEFT
RIGHT
BACK
START
RESET
Oct 23, 2006 8:54:55 PM
1. FRONT: 214
2. RIGHT: 101
3 .BACK 122
4. LEFT 122
HEART RATE: 83

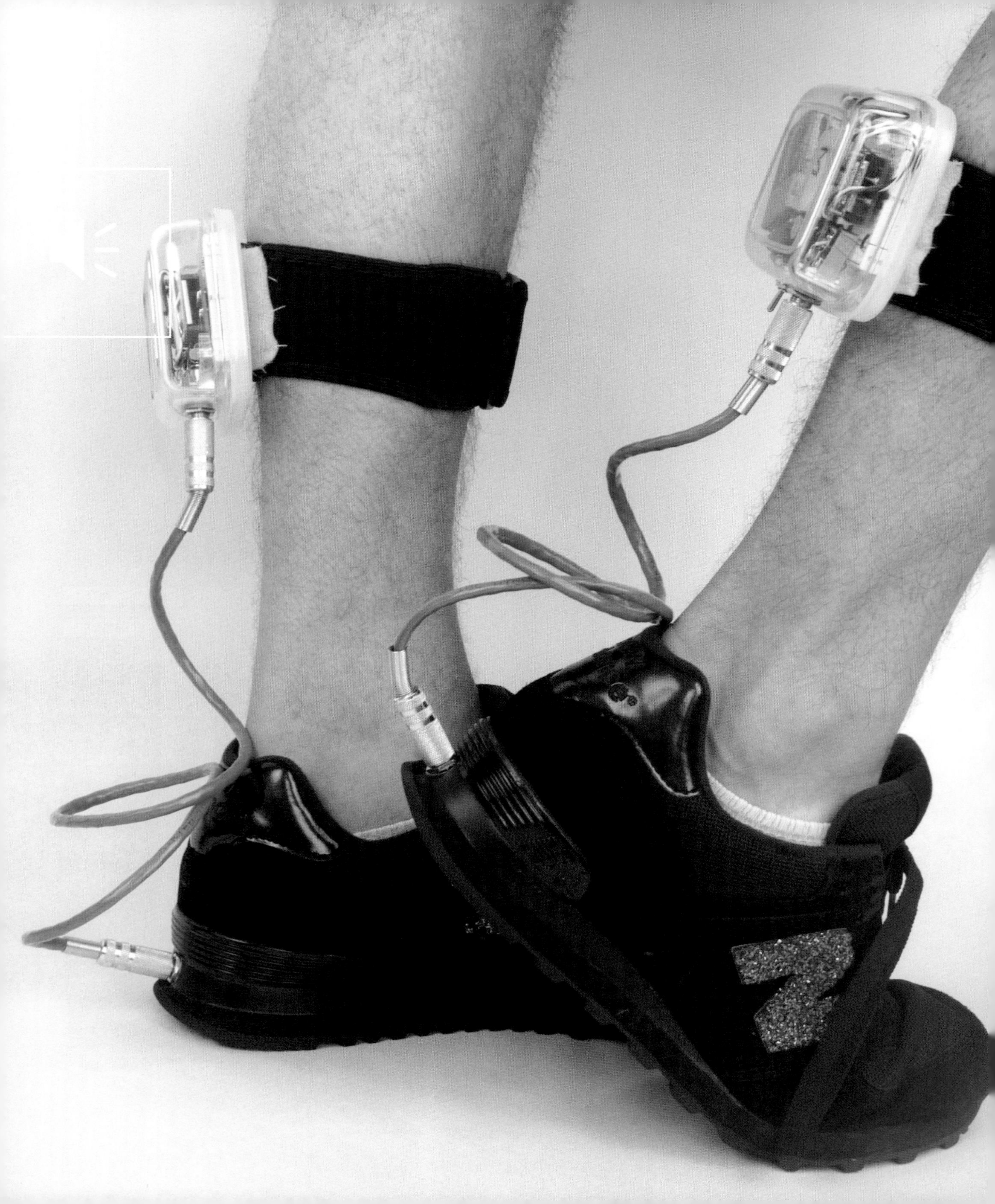

7

MAKE: PROJECTS

Sessionless Networks and Binary Protocols

So far, the network connections you've seen in this book have mostly been dedicated connections between two objects. Serial communications involve the control of a serial port; mail, web, TCP, and webSocket connections involve a network port. In all these cases, there's a device that makes the port available (generally a server), and something that requests access to the port (a client). The projects in Chapter 5 were classic examples of this. In this chapter, you'll learn how to make multiple devices on a network talk to each other directly, or all talk to the other devices in their network at once.

All of the communications protocols you encountered earlier were text-based. Some protocols require you to know how to interpret digital data numerically or bit-wise. You'll also learn a bit more about that in this chapter.

« Perform-o-shoes by Andrew Schneider
The shoes exchange messages with a multimedia computer via Digi radio. When you moonwalk in the shoes, your pace and rhythm controls the playback of music from the computer.

Supplies for Chapter 7

DISTRIBUTOR KEY
A Arduino Store (store.arduino.cc)
AF Adafruit (adafruit.com)
D Digi-Key (www.digikey.com)
F Farnell (www.farnell.com)
J Jameco (jameco.com)
P Pololu (www.pololu.com)
PX Parallax (www.parallax.com)
RS RS (www.rs-online.com)
SF SparkFun (www.sparkfun.com)
SS Seeed Studio (www.seeedstudio.com)

PROJECT 13: Networked Candles
Components listed are for each candle. It's more impressive to have many (between 6 to 30 total),
but you can get an understanding of the tools with just two.

» **1 Arduino MKR1000** **AF** 3156, **RS** 124-0657, **A** ABX00004, GBX00011 (Africa and EU), **D** 1659-1005-ND
Alternatively, an ESP8266-based board will work. **SF** WRL-13231, **AF** 2471
» **1 Sharp GP2Y0A21YK infrared distance ranger** **JJ** 2150256, **D** 425-2063-ND, **AF** 164, **F** 1243869, **RS** 666-6564
» **3-7 WS2812 programmable LEDs (NeoPixels)** **AF** 2226, 2858, or 2859 **D** 1528-1610-ND, **J** 2247947, **SF** BOB-13282, **SS** 104990139
» **1 solderless breadboard** **D** 438-1045-ND, **J** 20723 or 20601, **SF** PRT-12615 or PRT-12002, **F** 4692810, **AF** 64, **SS** 319030002 or 319030001
» **1 3.3V–5V Battery Supply** **AF** 771, **D** BC4AAW-ND, **SS** 320180002
» **1 frosted plexi tube, 4" diameter**
» **1 candy tin, 4" diameter** or some sort of small project enclosure

PROJECT 14: Reporting Toxic Chemicals in the shop
There are three separate circuits for this project. The parts for each are listed separately, though you'll use one adapter for programming the radios.

» **1 USB-XBee adapter** **J** 32400, **SF** WRL-11812, **AF** 247, **PX** 32400. The Digi XStick 802.15.4 is useful as an alternative if you plan to do a lot of XBee work. **D** 602-1200-ND

Sensor Circuit

» **1 solderless breadboard** **D** 438-1045-ND, **J** 20723 or 20601, **SF** PRT-12615 or PRT-12002, **F** 4692810, **AF** 64, **SS** 319030002 or 319030001
» **XBee or XBee Pro S2C 802.15.4 RF module** **AF** 128, **D** 602-1892-ND, **SF** WRL-08665, **J** 2253722, **PX** 32416
» **1 5V regulator** **J** 51262, **D** LM7805CT-ND, **F** 9756078, **RS** 918-1971
» **1 3.3V regulator** **D** 497-1491-5-ND, **J** 242115, **F** 1703357, **RS** 438-4885
» **9–12V DC power supply** Either a 9V battery or plug-in supply will do. **SF** TOL-00298, **AF** 798, **J** 170245, **F** 1176248
» **1 XBee breakout board** **SS** 113100001 **SF** BOB-08276
» **2 rows of 0.1-inch header pins** **D** A26509-20-ND, **J** 103377, **SF** PRT-00116, **F** 1593411
» **2 rows of 2mm female headers** **SF** PRT-08272, **D** 3M9406-ND
» **2 1µF capacitors** **D** 1189-1324-ND, **J** 94161, **F** 8126933, **RS** 475-9009
» **2 10µF capacitors** **D** P11212-ND, **J** 29891, **F** 1144605, **RS** 762-1736
» **1 Hanwei MQ-6 gas sensor** **SF** SEN-09405, **P** 1481, **PX** 605-00009
» **1 gas sensor breakout board** The gas sensor has a pin layout that's not friendly to a breadboard, so these boards correct that. **SF** BOB-08891, **P** 1479 or 1639
» **1 LED** **D** 160-1144-ND or 160-1665-ND, **J** 34761 or 94511, **F** 1855510, **RS** 228-5972 or 826-830, **SF** COM-09592 or COM-09590
» **1 220-ohm resistor** **D** 220QBK-ND, **J** 690700, **F** 9339299, **R** 707-7612
» **1 47-kilohm trimmer potentiometer** **D** A105657-ND, **J** 254028, **RS** 186-205

Server Circuit

» **1 Raspberry Pi** A BeagleBone Green or other embedded Linux processor will work too. **SF** DEV-13825, **AF** 3055 or 3400, **SS** 102010048 or 114990584, **RS** 896-8660, **F** 2525225
» **XBee or XBee Pro S2C 802.15.4 RF module** **AF** 128, **D** 602-1892-ND, **SF** WRL-08665, **J** 2253722, **PX** 32416
» **1 USB-XBee adapter** You can reuse the one you used

for configuration of the radios. **J** 32400, **SF** WRL-11812, **AF** 247, **PX** 32400

Cymbal Monkey Circuit

- 1 solderless breadboard **D** 438-1045-ND, **J** 20723 or 20601, **SF** PRT-12615 or PRT-12002, **F** 4692810, **AF** 64, **SS** 319030002 or 319030001
- XBee or XBee Pro S2C 802.15.4 RF module **AF** 128, **D** 602-1892-ND, **SF** WRL-08665, **J** 2253722, **PX** 32416
- 1 Cymbal Monkey NOTE: If your monkey uses a 3V power supply (such as two D batteries), you won't need the 3.3V regulator. Make sure that there's adequate amperage supplied for the radios. If you connect the circuit as shown and the radios behave erratically, the monkey's motor may be drawing all the power. If so, use a separate power supply for the radio circuit.
- 1 3.3V regulator **D** 497-1491-5-ND, **J** 242115, **F** 1703357, **RS** 438-4885
- 1 XBee breakout board **SS** 113100001 **SF** BOB-08276
- 2 rows of 0.1-inch header pins **D** A26509-20-ND, **J** 103377, **SF** PRT-00116, **F** 1593411
- 2 rows of 2mm female headers **SF** PRT-08272, **D** 3M9406-ND
- 1 LED **D** 160-1144-ND or 160-1665-ND, **J** 34761 or 94511, **F** 1855510, **RS** 228-5972 or 826-830, **SF** COM-09592 or COM-09590
- 1 220-ohm resistor **D** 220QBK-ND, **J** 690700, **F** 9339299, **R** 707-7612
- 1 TIP120 Darlington NPN transistor **D** TIP120-ND, **J** 32993, **F** 9804005 **RS** 808-0502
- 1 1-kilohm resistor **D** 1.0KQBK-ND, **J** 690865, **F** 9339051, **R** 707-7666
- 1 100µF capacitor **D** P10269-ND, **J** 158394, **F** 1144642, **RS** 762-1746

Figure 7-1. New parts for this chapter: **1.** Hanwei gas sensor on breakout board **2.** 3xAAA battery holder **3.** 47-kilohm trimmer pot **4.** Digi XBee S2C module **5.** XBee-to-serial breakout board **6.** Raspberry Pi **7.** Capacitors **8.** 5V and 3.3V voltage regulators **9.** TIP120 transistor **10.** Sharp GP2Y0A21YK infrared distance ranger **11.** 2mm female header pins **12.** WS2812 programmable LEDs **13.** Frosted plexi tube, 4" diameter **14.** candy tin, 4" diameter **15.** Charley Chimp

Sessions vs. Messages

So far, most of the communication in this book has involved opening a dedicated connection between two points for the duration of the conversation. This is session-based communication. Sometimes you want to make a network in which objects can talk to each other more freely, switching conversational partners on the fly, or even addressing the whole network if the occasion warrants. For this, you need a message-based protocol.

Sessions vs. Messages

In Chapter 5, you learned about the Transmission Control Protocol, TCP, which is used for much of the communication on the internet. To use TCP, your device has to request a connection to another device. The other device opens a network port, and the connection is established. Once the connection is made, information is exchanged; then the connection is closed. The whole request-connect-converse-disconnect sequence constitutes a *session*. If you want to talk to multiple devices, you have to open and maintain multiple sessions. Sessions characterize TCP communications.

Not all network communication is session-based. Another common protocol used on the internet, called the *User Datagram Protocol*, or *UDP*, is message-based. With UDP, you compose a message, give it an address, send it, and forget about it.

Unlike the session-based TCP, UDP communication is all about messages. UDP messages are called *datagrams*. Each datagram is given a destination address and is sent on its merry way. There is no two-way socket connection between the sender and the receiver. It's the receiver's responsibility to decide what to do if some of the datagram packets don't arrive, or if they arrive in the wrong order.

UDP doesn't rely on a dedicated one-to-one connection between sender and receiver, so it's also possible to send a broadcast UDP message that goes to every other object on a given subnet. For example, if your address is 192.168.1.45, and you send a UDP message to 192.168.1.255, everybody on your subnet receives the message. Because this is such a handy thing to do, a special address is reserved for this purpose: x.x.x.255, where x is replaced by the first three numbers of your subnet address. This is the *broadcast address*—it goes only to addresses on the same LAN, and does not require you to know your subnet address. This address is useful for tasks such as finding out who's on the subnet.

The advantage of UDP is that data moves faster because there's no error-checking. It's also easier to switch the end device that you're addressing on the fly. The disadvantage is that it's less reliable packet-for-packet, as dropped packets aren't resent. UDP is useful for streams of data where there's a lot of redundant information, like video or audio. If a packet is dropped in a video or audio stream, you may notice a blip, but you can still make sense of the image or sound.

The relationship between TCP and UDP is similar to the relationship between the Bluetooth Serial Port Protocol that you saw in Chapter 2 and the LoRa radio protocol you saw in Chapter 6. The latter radios communicate simply by sending addressed messages out to the network without waiting for a result. Like TCP, Bluetooth is reliable for information-critical applications, but it's less flexible in its pairings than the LoRa radios were. In this chapter, you'll see a new message-based protocol using 802.15.4 radios from Digi.

X

The next section introduces broadcast UDP messages. Some institutional networks won't let you send broadcast UDP messages for security reasons. If you're working on your home router, you should have no trouble. If you're working at a school or business, though, check with your network administrator before experimenting with broadcast UDP messages. UDP is used on the wider internet but generally is used in a directed manner rather than as broadcast messages.

Broadcast vs. Direct Messages

The first advantage to sessionless protocols like UDP is that they allow for broadcasting messages to everyone on the network at once. Although you don't want to do this all the time—because you'd flood the network with messages that not every device needs—it's a handy ability to have when you want to find out who else is on your network. You simply send out a broadcast message asking "Who's there?" and wait for replies.

Querying for Other Devices Using UDP

Node.js includes a library called `dgram` that lets you send and receive datagrams. The script that follows is a simple UDP server. You can communicate to this server using any device that can connect to the internet and send UDP packets.

Arduino's WiFi101 library and others like it include the ability to send and receive UDP packets, so you can write a simple sketch that listens for broadcast messages and responds to them. It's useful when you want to make a large number of networked devices all respond at once. You can use node.js to send your broadcast query.

You'll see code for an MKR1000 or an ESP8266-enabled board here, along with a command-line tool that will work on POSIX-based operating systems as well. The latter should work on any of the embedded processors like the Raspberry Pi or BeagleBone, not to mention your desktop computer.

X

Try It This node.js script sends out a broadcast UDP message and listens for incoming UDP response messages. It uses port 8888. This is an arbitrary number that's high enough that it's probably not used by other common applications.

To broadcast, the script uses the special subnet broadcast address: x.x.x.255. You should keep the broadcast messages in your own subnet, so change the first three bytes to match your network. For example, if your router is 192.168.0.1, then you'll send to 192.168.0.255 and the message will be broadcast to every machine on the subnet.

Unlike session-based messaging, where you wrote and read bytes from a stream like serial communication, UDP messages are all sent with one discrete message, `udp.send()`, which includes the message, the address, and the port for sending.

```
/*
  UDP Server
  context: node.js
*/

var dgram = require('dgram');              // include datagram library
var UDP_PORT = 8888;                       // the port on which to listen
var udpServer = dgram.createSocket('udp4'); // create UDP socket
var broadcastAddress = '192.168.0.255';    // broadcast address

function udpBegin() {
  udpServer.setBroadcast(true);
  console.log('UDP Server started');
}

function readMessage(message, sender) {
  console.log(sender.address + ':' + sender.port +' sent: ' + message);
}

udpServer.bind(UDP_PORT);              // set the UDP port number
udpServer.on('listening', udpBegin);   // start the UDP server
udpServer.on('message', readMessage); // runs when a UDP packet arrives

// send a broadcast message:
var data = "Hello";
udpServer.send(data, 0, data.length, UDP_PORT, broadcastAddress);
```

▸▸ Change the first three numbers to match your own network.

Respond to It Here's an Arduino sketch that listens for UDP messages on the same port and responds in kind. Load it onto an MKR1000, or change to the ESP8266 library to use it on an ESP8266 board. You won't need any external components.

The global configuration and `setup()` function will look familiar from all the previous WiFi-based sketches you've done. You need to connect to a WiFi network just like before. Note that the `ssid` and `password` variables are in the **config.h** file as you've seen them before. Copy that file from previous sketches. The main difference is that you're using the `WiFiUdp` class to set up a UDP object for sending and receiving UDP packets.

```
/*
  UDP Query Responder
  context: Arduino
*/
#include <SPI.h>
#include <WiFi101.h>
// #include <ESP8266WiFi.h>  // use this instead for ESP8266
#include <WiFiUdp.h>
#include "config.h"

WiFiUDP Udp;                 // instance of UDP library
const int port = 8888;       // port on which this client receives

void setup() {
  Serial.begin(9600);
  // while you're not connected to a WiFi AP,
  while ( WiFi.status() != WL_CONNECTED) {
    Serial.print("Attempting to connect to Network named: ");
    Serial.println(ssid);           // print the network name (SSID)
    WiFi.begin(ssid, password);     // try to connect
    delay(2000);
  }

  // When you're connected, print out the device's network status:
  IPAddress ip = WiFi.localIP();
  Serial.print("IP Address: ");
  Serial.println(ip);  Udp.begin(port);
}
```

▸ The `loop()` function listens for incoming UDP packets on the same port on which the node.js sketch broadcasts.

Receiving UDP packets is a bit different than receiving TCP packets. Each packet contains a header that tells you who sent it and from what port, kind of like an envelope. When you receive a packet, you have to parse out that data. The `parsePacket()` method in the `WiFiUdp` class does just that. You can see it in action here. First, you use `parsePacket()` to separate the header elements from the message body. Then, you read the message body using the Stream functions (in this example, `readString()`) just like you've done already with TCP sockets and serial ports.

```
void loop() {
  // if there's data available, read a packet
  if (Udp.parsePacket() > 0) {         // parse incoming packet
    String message = "";
    Serial.print("From: ");            // print the sender
    Serial.print(Udp.remoteIP());
    Serial.print(" on port: ");        // and the port they sent on
    Serial.println(Udp.remotePort());
    while (Udp.available() > 0) {      // parse the body of the message
      message = Udp.readString();
    }
    Serial.print("msg: " + message);   // print it
    sendPacket(message);                // send a reply
  }
}
```

The `sendPacket()` method sends a response packet. In this case, it's sending to the address and port from which the microcontroller received a message. This reply is not a broadcast message, but a *directed message*. It sends the received message as a reply.

```
void sendPacket(String message) {
  // start a new packet:
  Udp.beginPacket(Udp.remoteIP(), Udp.remotePort());
  Udp.print("Received: " + message); // add payload to it
  Udp.endPacket();                  // finish and send packet
}
```

You can see that UDP messaging is different than the socket-based messaging over TCP. Each message packet has to be started and ended. When you `beginPacket()`, the WiFi controller starts saving bytes to send in its memory; when you call `endPacket()`, it sends them out. That's the end of the sketch.

When you've got the microcontroller programmed, open the Serial Monitor. You should see a message like this when the Arduino obtains an IP address:

```
Attempting to connect to Network named: myNetwork
Connected to wifi
SSID: myNetwork
IP Address: 192.168.0.3
signal strength (RSSI):-36 dBm
```

Next, run the node.js script. In the Arduino IDE Serial Monitor, you should see:

```
From: 192.168.0.8 on port: 8888
msg: Hello
```

The address will be your computer's IP address. The `Hello!` is from node.js.

On the command line where you're running the script, you should see:

```
UDP Server started
192.168.0.4:8888 sent: Hello
192.168.0.3:8888 sent: Received: Hello
```

The first line is from the script itself. When you send a broadcast packet, you get it back yourself if you're listening. The second line is from the microcontroller. If you have several microcontrollers on your network all running this sketch, you'll get a response from each one of them. In fact, this exercise is most satisfying when you have multiple units responding.

The microcontrollers are not sending broadcast packets. They're sending directed packets back to whatever address sent to them. That's why they don't see messages from each other, or from themselves when they send, like the node.js script does.

This is a handy routine to add to any networked project. You have to provide a separate UDP instance and port on which to listen, but the `sendPacket()` method can work with other programs to get a response from your device when you send broadcast queries.

X

NetCat

For any device running a POSIX operating system, there's a command-line utility that's good for sending and receiving UDP messages called *netcat*. It's a utility that lets you connect either using TCP or UDP. You can send packets or listen for them. It's a handy tool for debugging network connections.

To use netcat to respond to the previous node.js script, type the following on the command line. Run this on a separate computer from the server script, because the script is already using port 8888:

```
$ nc -u -l 8888
```

If you're using a Raspberry Pi, you'll probably need to run this command using sudo, like so:

```
$ sudo nc -u -l 8888
```

This tells the device to listen for incoming UDP messages on port 8888, just like the previous sketch does for the microcontrollers running it. The `-u` option indicates that it should use UDP, and `-l` indicates that it should listen rather than sending. Once you've got this running on your embedded device (Pi, BeagleBone, etc.), go back to the command line of your personal computer and run the node.js script again. Once the node.js script has sent its broadcast message, you'll see the following on the embedded device's command line:

```
Hello
```

You can type a response to be sent back to the script and it will show up on the node.js script's command line. Press control-C to quit netcat.

You can also use netcat to do more than listen. If you want to send a directed message to the address 192.168.0.4, for example, type:

```
$ nc -u 192.168.0.4 8888
```

After that you can type a message and press Enter. Netcat will wait for a reply and you can send back and forth as much as you want. Again, press control-C to exit the program.

You may notice that if you send to the Arduino sketch you just wrote using netcat as shown here, the Arduino sketch reports a different remote port than 8888. That's because netcat is sending to port 8888 on the destination, but it sends from whatever local port it chooses.

Netcat has some differences between operating systems. For example, on Linux you can send broadcast packets using the `-b` option, but on macOS you can't send broadcast packets. For details on the various options on your system, type:

```
$ man nc
```

Windows doesn't come with netcat, but there are various free and open source versions available for Windows that operate more or less as described here. Cygwin's Net package and the bash shell for Windows 10 that's in beta as of this writing include netcat as well.

Redirecting Output to a UDP Message

There is another way to send UDP packets from a POSIX command-line interface other than netcat. Type:

```
$ echo "my message" > /dev/udp/192.168.0.4/8888
```

You're sending the output of the echo command to /dev/udp/<address>/<port>, which the operating system treats as a file stream. It's handy because you can redirect the output of any program to a UDP packet this way. For example, if you want to send the date as a UDP packet, type:

```
$ date > /dev/udp/192.168.0.4/8888
```

This should work the same on macOS, Linux, and other POSIX systems. It is included with Cygwin's Net package and the beta bash shell for Windows 10, but not on earlier versions of Windows, unfortunately.

Because node.js runs on POSIX systems, you can use node.js to send and receive UDP packets too, of course. However, these command-line utilities for sending and receiving UDP are useful because you can combine them with other programs.

X

 Project 13

Networked Candles

UDP is a handy protocol for networked projects where there are multiple devices all communicating with each other or with a server, and where the exchanges between them are short and sporadic. In this project you'll make a set of networked candles that flicker independently, and signal to each other when you wave your hand over each one.

MATERIALS

For each candle (make 6–30 candles):

- **1 Arduino MKR1000, or WINC1500 or ESP8266-compatible microcontroller**
    - **Features used: Digital I/O, Analog I/O, WiFi**
- **1 Sharp GP2Y0A21YK infrared distance ranger**
- **3–7 WS2812 programmable LEDs (NeoPixels)**
- **1 solderless breadboard or prototyping shield**
- **1 battery supply, 3.3–5V**
- **1 frosted plexi tube, 4" diameter**
- **1 candy tin, 4" diameter**

Inexpensive WiFi radios and programmable LEDs are two of the more exciting pieces of technology to hit the maker market in the past few years. In a recent project, I combined them to make a set of networked electronic candles. You'll duplicate that project here.

The candles each flicker softly like a real candle, fading through different shades of orange, yellow, and red. When you wave your hand over a candle, it suddenly flares to white, then gradually fades back to the warmer colors. When it flares, it also sends a signal to all the other candles to flare as well. That's where the UDP comes in: each candle flare sends a broadcast UDP message, and all the others pick it up and flare to white accordingly. The messages are infrequent and unpredictable. By using UDP, the candles don't need to maintain a connection to each other when there's nothing to say.

Using UDP as the messaging protocol had an interesting effect on the system. Sometimes a given candle would not receive the broadcast message. Remember, UDP doesn't guarantee delivery of every packet. This meant that I never knew which candles would miss the messages. The candles were connected to a router that was not connected to the internet. Because there was no other traffic on the network I used, the missed packets were relatively infrequent, and they added another element of surprise to the system, so I didn't try to correct for this. If it were absolutely essential that every message get through, then TCP would have been the better choice, but it would have been more complex to implement.

Figure 7-2
Networked candles

You can make as many candles as you can afford in time and money. In the original project, I made thirty candles. With less than six, it wasn't much fun.

The Circuit

Each candle in my project had a WiFi-enabled microcontroller in it and a set of programmable LEDs, the WS2812 by Worldsemi. Adafruit sells lots of variations on this LED called NeoPixels. SparkFun and Seeed Studio sell variations as well. These LEDs come in an RGB version and an RGBW version. They're controlled by one wire, they're individually addressable, and they can be daisy-chained so that you need only power, ground, and one output pin from your microcontroller to control lots of them. You can learn all about them at learn.adafruit.com/adafruit-neopixel-uberguide.

Because I was making lots of candles, I needed lots of microcontrollers. The MKR1000 didn't exist yet, so I used lots of ESP8266 radios. You can use either those or the MKR1000. There is an important difference in the circuit between the MKR1000 and the ESP8266 boards. The ESP8266 has only one analog input (labeled ADC on both the SparkFun and Adafruit boards) and it has a 1-volt range. That means your sensor can't output more than 1 volt or you risk damaging the board.

The infrared ranging sensor shown in Figures 7-3 through 7-5 senses distances from 10–80cm, and converts them to an analog output voltage of 0–5V as you move an object closer to or farther from the sensor. Since the ESP8266 analog input can only handle 0–1V input, you need a *voltage divider circuit* to reduce the output voltage. The circuit shown on the right side of Figure 7.4 works well to keep the range close to, but still under, 1 volt. Before you connect your sensor to your microcontroller, test the output from the sensor-voltage divider circuit with a voltmeter to make sure it's in the right range.

The sensor normally operates on 5V, as do the NeoPixels, but I found they would operate successfully on the 4.5V from 3 AA batteries, or even the 3.7V from a Lithium-Polymer (LiPo) battery. Below about 3.3V, both ceased to function, but so did the microcontroller, so they all lasted about the same amount of time on battery power.

Adafruit recommends using a 470-ohm resistor in series between your microcontroller and the NeoPixels' data input pin for protection when you're operating them at 5V. At the lower voltages used here, the resistor reduced the brightness too much for my taste, but you may want to add it for your candles.

In my original candles, I made custom circuit boards, and used three NeoPixels. For this version, I used Adafruit's NeoPixel Jewel board, which has seven NeoPixels on board. You can use more than seven, or fewer, by changing one variable in the code that follows. Each NeoPixel can draw up to 60 milliamps of current, though, so make sure your supply can handle the current that you need for the number of pixels you use.

X

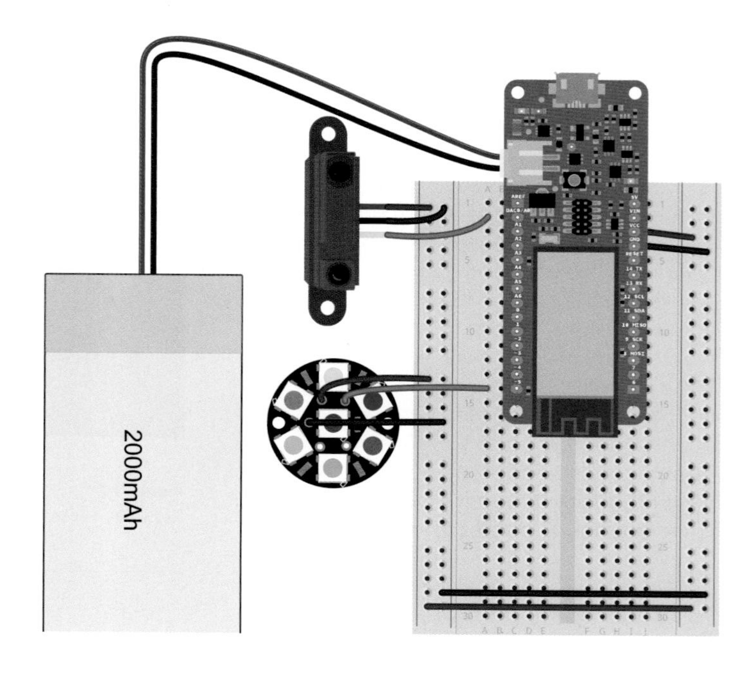

Figure 7-3
Breadboard view for candle project, MKR1000 version. Shown here using a LiPo battery for power. Make sure to use a battery with at least 800mAh of charge. 1200–2000mAh is better. The charging circuit requires it, and the NeoPixels draw a significant amount of current.

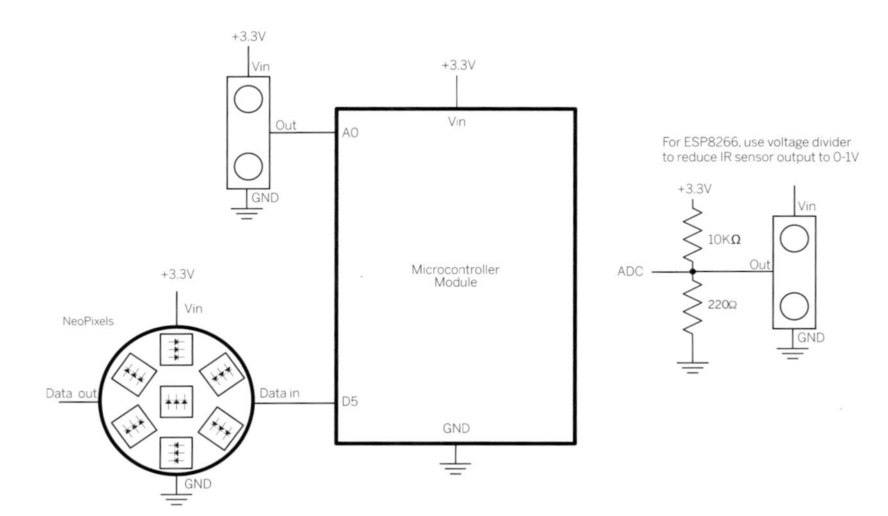

Figure 7-4
Schematic for candle project. The ESP8266 analog input (labeled ADC on SparkFun and Adafruit boards) has a 1-volt input range, so use the voltage divider shown here to reduce the IR ranger's range to 1V.

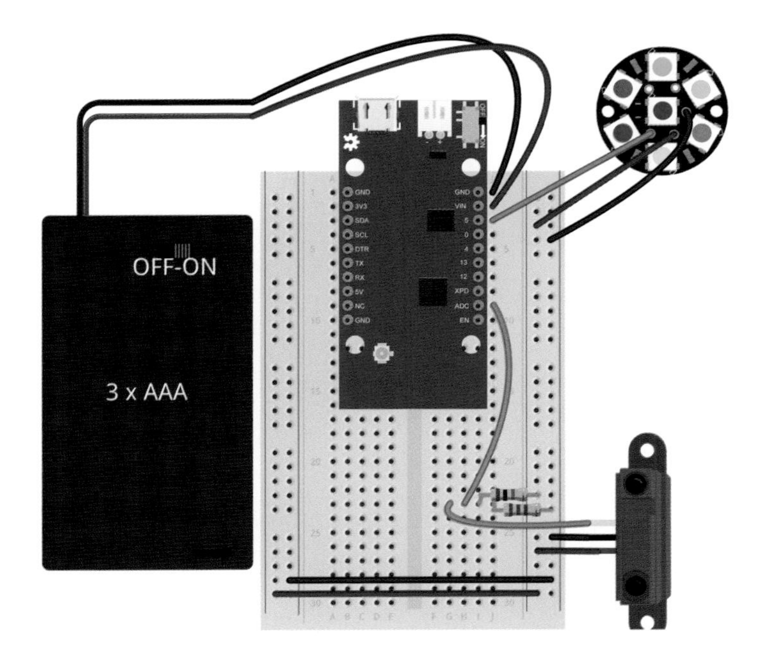

Figure 7-5
Breadboard view for candle project, ESP8266 version. You can also use 3 AAA batteries in series to power this project. Don't go through the charging circuit if you do, though; use the Vin pin instead. The Adafruit Feather Huzzah! ESP8266 has a different pin layout, but you should still use the ADC pin for the sensor and pin 5 for the NeoPixels.

The Code

You'll use a new library to control the pixels. Using the Arduino IDE Library Manager, filter for `NeoPixel` and install the Adafruit NeoPixel library. Once it's installed, you should test your pixels to understand how they work.

The NeoPixel Library

The first thing you'll need to do with the library is to initialize your pixels like so:

```
Adafruit_NeoPixel candle = Adafruit_NeoPixel(
  numPixels, neoPixelPin, NEO_GRB + NEO_KHZ800);
```

The number of pixels and the neoPixelPin are self-explanatory. The final parameter has two parts. The first sets the type of pixel, because there are a few different models of NeoPixel. Some expect the color data as RGB, some as GRB, and some as RGBW, if they contain a fourth white LED. The ones specified here are GRB or GRBW. You should always test your pixels by turning them on to one color at a time at first. The second half of the final parameter is the refresh rate of the pixels; again, there are pixels with different rates. The ones specified here operate at 800KHz.

There are a couple of ways to set the color of a given pixel. You can specify it as three bytes, red, green, and blue, like so:

```
candle.setPixelColor(pixel, red, green, blue);
```

If you're using a four-LED pixel, you can add white like so:

```
candle.setPixelColor(pixel, red, green, blue, white);
```

Or you can set all the colors with one large number, specified in hexadecimal colors like HTML colors are specified:

```
candle.setPixelColor(pixel, color);
```

If you're using a four-LED pixel, the white value is specified first. For example, 0xFFFF0000 turns on white and red at full brightness. If you omit the white value, a four-LED pixel will still work; it will just never turn on the white LED. The code that follows uses the single-number hexadecimal RGB format for colors.

X

Test It Here's a test for the NeoPixel library. This will blink all of your LEDs red. Once you've got it working, change the color to make it blink green, then blue. Then make up some colors of your own.

Just like in HTML color codes, the color is specified like this:

- red: 0xFF0000
- green: 0x00FF00
- blue: 0x0000FF

You can change the value of any of those 3 bytes to any value from 0–255 (0x00–0xFF in hexadecimal). For example, here's a nice teal color: 0x2375FF.

You can change each pixel individually if you want, rather than in a loop as shown here. When you do, the `.show()` command sets all the pixels at once, so you only need to call it after you've set values for all your pixels. The `.clear()` command sets all the pixels to 0x000000.

```
/*
  NeoPixel Test
  context: Arduino
*/
#include <Adafruit_NeoPixel.h>

const int neoPixelPin = 5;          // NeoPixel control pin
const int numPixels = 7;            // number of NeoPixels you're using
Adafruit_NeoPixel candle = Adafruit_NeoPixel(
                              numPixels, numPixels, NEO_GRB + NEO_KHZ800);

void setup() {
  Serial.begin(9600);     // initialize serial communication
  candle.begin();         // initialize pixel strip
  candle.clear();
  candle.show();          // Initialize all pixels to 'off'
}

void loop() {
  for (int pixel = 0; pixel < numPixels; pixel++) {
    candle.setPixelColor(pixel, 0xFF0000);
  }
  candle.show();     // update all the pixels
  delay(1000);
  candle.clear();    // set all the pixels to off
  candle.show();     // update
  delay(1000);
}
```

Creating a Candle Flicker

The actual sketch for this project has three jobs: listen to the sensor, listen for incoming UDP packets, and flicker the candle. You've seen examples of the first two jobs before. Managing the fade of colors in the pixels while still listening for sensor changes or network messages can be tricky. Here's how you do it.

As you know from previous projects, each candle will seem most responsive if its microcontroller is listening more frequently than it's flickering the candles. To achieve this you'll write a flicker routine that changes the candle a tiny amount each time through the main loop. Think of each main loop iteration as one frame of a film. You'll pick a set of target colors, all in the orange and red range. Then you'll set the pixels to one of the colors and pick another target color. Each time through the loop, you'll check to see whether the current color is different than the target color. If it is, you'll change it to be one step closer to the target color. When you reach the target, you'll pick a new target color.

Though it's more efficient in your main code to deal with the colors (red, green, and blue and white if you're using it) as one single 4-byte number, you can't compare two colors numerically to see if they're similar chromatically. When you want to fade a color like this reddish color: `0xCB500F` toward this orange color: `0x853E0B`, you need to separate out the red, green, and blue. To fade it evenly, you'd check to see if the reds are the same, the greens are the same, and the blues are the same. For each color pair that are different, you'll move them closer by one point each time through the loop.

	Red	**Green**	**Blue**
Current	0xCB	0x50	0x0F
Target	0x85	0x3E	0x0B

To separate the colors into red, green, and blue, you can use a technique called *bit shifting*. This technique lets you isolate a byte of a multi-byte value like the colors used in this library. The function that follows demonstrates this. The colors are stored in `unsigned long` integer variables, which are 4 bytes in size. You set up three `byte` variables (which are one byte in size); then you bit-shift the color value until the desired byte is in the lowest byte of the unsigned long variable. Since the `byte` variable can fit only one byte, it takes the lowest byte and disregards the rest. X

Compare It This function takes two colors, a current color and a target color, and compares them. It returns a third color that's closer to the target by one point in each of the red, green, and blue bytes. This is one way to fade between colors that looks fairly pleasant when given two arbitrary colors. You'll use it later in the full candle sketch.

The more you work with NeoPixels, the more color manipulation techniques you'll develop to both change your colors and respond to user input.

```
unsigned long compare(unsigned long thisColor, unsigned long thatColor) {
  // separate the first color:
  byte r = thisColor >> 16;       // shift bits to the right two bytes
  byte g = thisColor >>  8;       // shift bits to the right one byte
  byte b = thisColor;             // get the lowest byte

  // separate the second color:
  byte targetR = thatColor >> 16; // shift bits to the right two bytes
  byte targetG = thatColor >>  8; // shift bits to the right one byte
  byte targetB = thatColor;       // get the lowest byte

  // fade the first color toward the second color:
  if (r > targetR) r--; // if the current is greater, decrement it
  if (g > targetG) g--;
  if (b > targetB) b--;

  if (r < targetR) r++; // if the current is less, increment it
  if (g < targetG) g++;
  if (b < targetB) b++;

  // combine the values to get the new color:
  unsigned long  result = candle.Color(r, g, b);
  return result;
}
```

Bit Shifting and Masking

Often, you need to manipulate the individual bits of a byte, or the bytes of a multi-byte variable. Arduino offers you some commands for reading and writing the bits of a byte:

```
// to read the value of a bit:
myBit = bitRead(someByte, bitNumber);

// to write the value of a bit:
bitWrite(someByte, bitNumber, bitValue);
```

Giving a bit the value 1 is called *setting the bit* in general programming terms, and giving it the value 0 is known as *clearing the bit*. So you also have commands for these:

```
// to make a bit equal to 1:
bitSet(someByte, bitNumber);

// to make a bit equal to 0:
bitClear(someByte, bitNumber);
```

Sometimes it's easier to manipulate several bits at once. This is called *bit shifting*. The shift left and shift right operators in C, Java, and JavaScript allow you to just that. The *shift left operator* (<<) moves bits to the left in a byte, and the *shift right operator* (>>) moves them to the right:

```
0b00001111 << 2; // gives 0b00111100
0b10000000 >> 7; // gives 0b0000001
```

It works for whole bytes as well:

```
0x00FF00 >> 8; // gives 0x0000FF
0x0000CD << 16; // gives 0xCD0000
```

Bit shifting is useful when you need a particular value to be in a specific part of a byte, as you're seeing in this project.

The logical operators AND, OR, and XOR allow you to combine bits in some interesting ways. This is called *bit masking*.

AND (&): if the two bits are equal and not equal to 0, the result is 1. Otherwise, it's 0:

```
1 & 1 = 1
0 & 0 = 0
1 & 0 = 0
0 & 1 = 0
```

OR (|): if either bit is 1, the result is 1. Otherwise, it's 0:

```
1 | 1 = 1
0 | 0 = 0
1 | 0 = 1
0 | 1 = 1
```

XOR (^): if the bits are not equal, the result is 1. Otherwise, it's 0:

```
1 ^ 1 = 0
0 ^ 0 = 0
1 ^ 0 = 1
0 ^ 1 = 1
```

Using the logical, or *bitwise*, operators, you can isolate one bit of a byte, or one byte of a multi-byte variable like so:

```
0b00001111 & 2; // gives 0b00000010, or 2
0b00001111 | 0b10000000; // gives 0b10001111, or 143
```

For example, if you want to isolate the red, green, or blue in one of the combined colors in the NeoPixel colors shown here, you can do it like this:

```
// Check what the green value is:
unsigned long myColor = 0x1A3CFF;  // close to teal
unsigned long green = 0;
// check what the value of green in myColor is:
green = myColor & 0x00FF00;        // gives 0x003C00
```

Combined, these bit-manipulation commands give you tools power to work with binary protocols and multi-byte values. X

Analog Sensor Triggering

The third task for this project's sketch, reading the sensor, will be simple: when the sensor goes above an arbitrary threshold, trigger the candle to flare white and send a UDP message. Once you've done that, set a Boolean variable called `triggered` to true to note that you just triggered the flare action. Variables used in this way are sometimes called *flags*.You'll use this flag to know when you're ready to trigger the flare action again. This prevents you from sending multiple messages every time your sensor goes above the threshold. This technique is similar to watching for a pushbutton to change state, which you saw in earlier projects.

Your candles will be more or less responsive depending on where you reset the `triggered` flag variable to false. In the following code, it's reset when you start fading from white back to your target color. Try this, then see what happens when you reset it after the comparison is complete instead.

Put It All Together Here's the full sketch. Start by including the libraries as usual, and the **config.h** file for your network SSID and password (it's the same as the previous example).

```
/*
  Networked NeoPixel Candle
  context: Arduino
*/
#include <SPI.h>
#include <WiFi101.h>
// #include <ESP8266WiFi.h>  // use this instead for ESP8266
#include <WiFiUdp.h>
#include <Adafruit_NeoPixel.h>
#include "config.h"
```

Next, set up global variables for the UDP library, the broadcast address, and the port on which you'll listen. Following that, set variables for the NeoPixels: the control pin number, the number of pixels, and finally, an instance of the NeoPixel library called `candle`.

```
WiFiUDP Udp;                              // an instance of the UDP library
IPAddress destination(192,168,0,255);  // UDP destination address
const int port = 8888;                    // UDP port

const int neoPixelPin = 5;              // NeoPixel control pin
const int numPixels = 7;                // number of NeoPixels you're using
// an instance of the NeoPixel library:
Adafruit_NeoPixel candle = Adafruit_NeoPixel(
  numPixels, neoPixelPin, NEO_GRB + NEO_KHZ800);
```

After that you'll need an array of target colors for the candle to fade between. You'll call this array `keyColors`. Set a target color variable and a current color variable, and a timestamp for the last time you faded the colors. Set an interval for the time between fade steps; 30 milliseconds works well. Then set the threshold for the sensor, and a trigger flag variable.

```
// colors for the candle, in hex RGB:
unsigned long keyColors[] = {0xCB500F, 0xB4410C, 0x95230C, 0x853E0B};
unsigned long currentColor = keyColors[0];  // current color of the pixels
unsigned long target = keyColors[1];         // the color you're fading to
long lastFadeTime = millis();                // last time the color changed
long interval = 30;                           // fading interval (30 ms)
int threshold = 500;                          // sensor threshold
boolean triggered = false;                    // sensor trigger flag variable
```

The `setup()` function initializes serial communication and the neoPixel instance called `candle`. Then it turns off all the pixels with the `.clear()` function and updates the whole set with the `.show()` command. Any time you want to update the pixels, you have to call `.show()`.

The rest of the `setup()` is familiar: connect to the WiFi SSID, print the IP address, and initialize UDP.

Continued from previous page.

```
void setup() {
  Serial.begin(9600);    // initialize serial communication
  candle.begin();        // initialize NeoPixel control
  candle.clear();        // turn off all pixels
  candle.show();         // refresh the candle

 // while you're not connected to a WiFi AP,
  while ( WiFi.status() != WL_CONNECTED) {
    Serial.print("Attempting to connect to Network named: ");
    Serial.println(ssid);           // print the network name (SSID)
    WiFi.begin(ssid, password);     // try to connect
    delay(2000);
  }
  // When you're connected, print out the device's network status:
  IPAddress ip = WiFi.localIP();
  Serial.print("IP Address: ");
  Serial.println(ip);
  Udp.begin(port);     // initialize UDP communications
}
```

Start the `loop()` function by reading the sensor. Then compare it to the threshold. If it's higher, check the `triggered` flag variable. If it's false, then set the `currentColor` to all white (0xFFFFFF) and send a UDP packet. Then set the `triggered` flag to true.

You'll need to adjust the value of `threshold` once you get the circuit and the candle housing assembled to get a value that works with your particular candle.

```
void loop() {
  int sensorReading = analogRead(A0);   // read the sensor
  if (sensorReading > threshold) {      // it it's above the threshold
    Serial.print(sensorReading);        // print the sensor reading
    if (!triggered) {                   // and it's not already triggered
      currentColor = 0xFFFFFF;          // make the candle all white

      // send the message to the destination UDP address:
      Udp.beginPacket(destination, port);
      Udp.print("ping");
      Udp.endPacket();
      triggered = true;             // note that sensor was triggered
    }
  }
```

Continue the `loop()` function by fading the current pixel color one step closer to the target color. To get a slow fade, use the `millis()` check shown here to do this only once every 30 milliseconds. If the current color and the target are not the same, run the `compare()` function you wrote earlier (insert it after the `loop()` function). If the current color and the target are the same, pick a new target color from the `keyColors` array.

```
  // update candle every 30 ms:
  if (millis() - lastFadeTime >= interval) {
     triggered = false;             // reset sensor trigger

    if (currentColor != target) {
      // fade the current color toward the target color:
      currentColor = compare(currentColor, target);
    } else {
      // pick a new target color randomly from keyColors:
      int next = random(4);
      target = keyColors[next];
    }
```

Next in the `loop()` function, make a `for` loop to update all the pixels' colors; then use the `.show()` command to refresh them all. Save the current `millis()` time in `lastFadeTime` for comparison next time through the loop.

```
  // set the color of all pixels:
    for (int pixel = 0; pixel < numPixels; pixel++) {
      candle.setPixelColor(pixel, currentColor);
    }
    candle.show();            // refresh the candle
    lastFadeTime = millis();  // save the fade time for comparison
  }
```

End the `loop()` function by listening for incoming UDP messages. If there's a new packet, parse it for the remote sender using the `.parsePacket()` function; then read the message if there are bytes available. You always need to call `.parsePacket()` first to get the sender and port information before you can call `.available()` to see the message itself. If the message is `ping`, then set the current color to all white.

The remainder of this sketch is the `compare()` function from earlier. Add it to your sketch and you're ready to upload it to your microcontroller.

X

```
  // if there's data available, read a packet
  if (Udp.parsePacket() > 0) {
    String line = "";
    // if there's a UDP packet, read it:
    while (Udp.available()) {
      line = Udp.readStringUntil('\n'); // read until the newline char
    }
    IPAddress sender = Udp.remoteIP(); // get the sender's IP
    Serial.print("From: ");           // print the sender
    Serial.print(sender);
    Serial.print(", message: ");      // and the message
    Serial.println(line);
    if (line == "ping") {              // if the message is "ping"
      currentColor = 0xFFFFFF;         // set the candle color to white
    }
  }
}      // end of loop() function

// append compare() function here
```

Once you've got the candles programmed, you can test them with each other, or with the node.js script from earlier in the chapter. It will send a broadcast message. Change the message to "ping" to see the candles flare white. Wave your hand over the candle to see it flare white and send a message to other candles as well. The node.js script will receive the messages from the candles as well.

You'll also need to tune the value of the `threshold` variable depending on where you place the NeoPixel and the sensor in the candle. I used a 4-inch candy tin as the base and a 4-inch frosted plexi tube for the candle shade. The sensor and the NeoPixel both need to be at the center of the tube for maximum effect, so you'll have to make a little compromise to fit both. Figure 7-6 shows the prototype layout. Make sure the sensor is far enough away from the edge of the tube so that the tube itself isn't triggering the sensor.

There are a couple of optional features that you can add to make these candles easier to use. First, a switch to disconnect the battery is very helpful in saving battery life. Many AA and AAA battery holders have one built in. Second, adding a function to turn one NeoPixel blue if the WiFi gets disconnected makes it easy to tell when a candle can't reach the network. Add this block of code right before candle`.show()`, both in the `setup()` function and the `loop()`:

```
// set first pixel to blue if network is not connected:
if (WiFi.status() != WL_CONNECTED) {
  candle.setPixelColor(0, 0x0000FF);
}
```

With a half dozen or more candles, this system offers the chance to see lots of interesting dynamics of UDP messaging. Though the broadcast message makes for a nice effect, you can get some fun effects by changing the UDP message to a directed message. Most routers will assign new IP addresses in sequence, so if you have 30 candles, it's likely that the candles will have addresses ranging from 192.168.0.2 to 192.168.0.31. To send to a random candle, you could add these lines right before the `Udp.beginPacket()` line:

```
int addr = random(30) + 2;      // random number from 2 to 31
destination = IPAddress(192, 168, 0, addr);// set IP address
```

You can also create relay effects by having candles retransmit messages they receive. Perhaps each candle transmits to the IP address whose last byte is one higher or lower than its own last byte.

When you're making a system where messages are relayed from one device to another, it's wise to include a number in each message that the receiver can decrement before it sends on, so the message is only passed on a limited number of times. This is a common networking technique, referred to as the *time-to-live (TTL)* of a message. In fact, the header of an Internet Protocol (IP) packet contains a TTL value that tells each router that receives the packet whether to pass it on. Each router decrements the time-to-live before passing the packet on, and a packet with a TTL of 0 is not passed on.

X

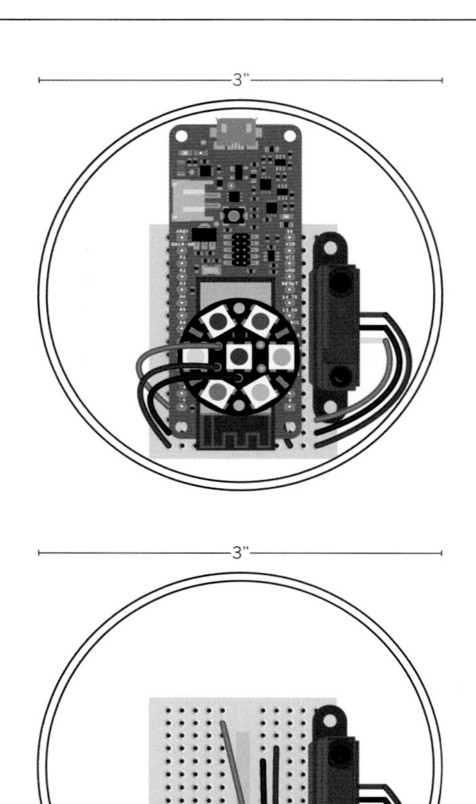

Figure 7-6
The MKR1000 version of the candle circuit shown on a tiny breadboard. Arranged this way, the whole thing fits in a 4" diameter candy tin. The NeoPixel unit, sitting on a layer of insulating rubber to protect it and the microcontroller, is close enough to center to look good, and the sensor is close enough so that the edge of the plexi tube does not interfere. The battery is underneath the circuit. The lower figure shows the wiring without the microcontroller or NeoPixel unit in place.

UDP in Action: Traceroute

Network administrators frequently need to know the route by which data is reaching its destinations from their network. The tool used to do this is called `traceroute`. Traceroute uses UDP and the relay method described on the previous page to learn the most likely route your packets are taking to reach their destination. It shows you the addresses of the routers that your message most likely reached, one after another. Traceroute is handy for network administrators, because when there's a problem, they can find out what network it's in, and contact that network's administrator, or route around it. At the bottom of this page is a typical traceroute output.

What's going on here? The traceroute program sends out a UDP packet to your destination address (in this case, makezine.com, which has the numeric IP address 104.25.44.28). The *time-to-live (TTL)* on the packet it sends is 1, so your router sends it back, and traceroute learns the router's address from the returned packet. Then it sends the packet again with a TTL of 2. Your router passes it on and decrements the TTL, so the next router sends it back, and traceroute gets that router's address. This continues until traceroute gets a reply from the original destination IP address. Through this process, traceroute learns the addresses of all the routers between you and your destination.

Although traceroute defaults to sending UDP messages, you can use TCP or other IP protocols as well. Some routers block particular protocols. The routers' replies are sent using the *Internet Control Messaging Protocol (ICMP)*, which is the same protocol that ping uses.

By default traceroute sends three UDP messages each time, which is why you see three times after the hop address. Those are the times it took to get a reply, in milliseconds.

In the example shown here, you're using the `-a` option. That tells traceroute to list the networks (or *Autonomous Systems*) of each address. You can search any AS number at apps.db.ripe.net/search to learn the network provider.

Note: Windows has its own version of this utility, `tracert`, which does not support the `-a` option shown here.

What's going on in hop 2? The router that received that message didn't reply. There's no definitive way to know why, so traceroute shows us * * * instead of an address.

There's a visual traceroute tool online at www.yougetsignal.com/tools/visual-tracert that shows you a map of your traceroute as it moves around the world. Though visually fascinating, it's not entirely accurate, because there's no definitive database of the geographic location of every device on the internet, fortunately. You'll learn more about this in Chapter 9.

X

```
$ traceroute -a makezine.com
traceroute: Warning: makezine.com has multiple addresses; using 104.25.44.28
traceroute to makezine.com (104.25.44.28), 64 hops max, 52 byte packets
 1  [AS0] 192.168.0.1 (192.168.0.1)  5.359 ms  7.537 ms  10.090 ms
 2  * * *
 3  [AS12271] tge-0-10-0-10.nymanyfo01h.nyc.rr.com (68.173.209.1)  30.503 ms  32.057 ms  34.002 ms
 4  [AS12271] agg115.nyclnyrg01r.nyc.rr.com (68.173.198.64)  30.132 ms  38.111 ms  39.582 ms
 5  [AS19548] bu-ether19.nwrknjmd67w-bcr00.tbone.rr.com (66.109.6.78)  41.176 ms  34.952 ms  24.525 ms
 6  [AS19548] bu-ether12.nycmny837aw-bcr00.tbone.rr.com (66.109.6.27)  34.226 ms  40.067 ms  34.095 ms
 7  [AS7843] 0.ae2.pr0.nyc20.tbone.rr.com (107.14.19.147)  35.052 ms
    [AS19548] 0.ae0.pr0.nyc20.tbone.rr.com (66.109.6.157)  29.482 ms
    [AS19548] 0.ae1.pr0.nyc20.tbone.rr.com (66.109.6.163)  27.959 ms
 8  [AS1299] nyk-b5-link.telia.net (62.115.34.145)  20.895 ms  43.579 ms  47.807 ms
 9  [AS1299] nyk-bb2-link.telia.net (80.91.254.15)  40.781 ms  29.687 ms  29.712 ms
10  [AS1299] nyk-b2-link.telia.net (62.115.134.108)  31.577 ms  29.079 ms  30.539 ms
11  [AS1299] cloudflare-ic-301663-nyk-b2.c.telia.net (213.248.77.162)  30.656 ms  35.082 ms  25.414 ms
12  [AS13335] 104.25.44.28 (104.25.44.28)  29.912 ms  30.884 ms  27.869 ms
```

XBee: Another Message-Based Protocol

Digi's XBee series of radios offer another approach to message-based devices. They can operate as an asynchronous serial radio for a separate device, or they can be configured to read from or write to their own I/O pins independently. They work like modems in that you configure them over an asynchronous serial (UART) connection. They have a message-based binary communications protocol created by Digi that you can use to send them commands over the air, or to read inputs or control outputs.

Digi makes a number of radios that support this protocol, called the XBee API protocol. Each of their products is based on a different link-layer protocol as well: ZigBee, WiFi, and so forth. In the following project you'll use their XBee or XBee Pro S2C 802.15.4 modules. Although Digi has changed the names on these modules as they've updated them, many people still refer to them just as XBees.

You don't write programs for these radios. You just configure their properties and addressing and let them do their business.

NOTE: ZigBee is a trademark of the ZigBee Alliance. XBee is a trademark of Digi, Inc. The two are different protocols. What follows is the XBee protocol.

Digi's XBee protocol is similar to UDP in that it supports both direct messaging and broadcast messaging. It's also similar to the LoRa protocol you saw in Chapter 6 in that you have to set a common *Personal Area Network (PAN) ID* (or sync word, in the former case) for all the radios that you want to network together. The radios have properties you might expect: each radio has an address, a destination address, and a PAN ID. Since the radios also have I/O pins, there are also properties for I/O. Each pin has an I/O mode, and you can set the rate at which the radio reads from the pins (called *sampling*), and the number of readings, or samples, it takes before transmitting. Since the radios have UARTs, you can set the properties of the serial connection as well. Digi has a graphical application for configuring their radios called XCTU that lists all the properties and lets you set them from the GUI. You can download it from www.digi.com/support/product-support. In the section labeled "Select Your Product for Support," enter XCTU and you'll get a download link.

The Digi radios are configured over an asynchronous serial link using a set of commands based on those designed originally for telephone modems, known as the *Hayes AT command protocol*. All commands in the Hayes command protocol (and therefore in the Digi configuration protocol) are sent using ASCII characters. Devices using this protocol have a *command mode* and a *data mode*. To switch from data mode to command mode in the Hayes protocol, you send the string `+++`. There's a common structure to all the commands. Each command sent from the controlling device (like a microcontroller or personal computer) to the modem begins with the ASCII string `AT`, followed by a short string of letters and numbers representing the command, followed by any parameters of the command, separated by commas. The command ends with an ASCII carriage return. The modem then responds to the command with the message `OK`, followed by any information it's expected to return. The Hayes AT command set was ubiquitous in the time of telephone modems, and it's been adapted by makers of many communications products as a configuration interface for their products. You'll see variations on it in Bluetooth radios, telecommunications equipment, and many other places. XCTU's GUI uses this command set to configure the radios, so you don't have to memorize the various settings. When you click Write in the configuration pane, the application puts your radio into command mode, sends the appropriate AT commands, and then puts the radio back into data mode.

To connect a Digi radio to your personal computer, you need a USB serial adapter with an XBee socket. Digi sells a development board for this, and many hobbyist electronics vendors sell less expensive XBee-to-USB tools as well. They're all basically USB-to-serial adapters mounted on a board with pins spaced to fit a Digi radio. Figure 7-9 shows two options: Adafruit's XBee USB adapter board and SparkFun's XBee Explorer.

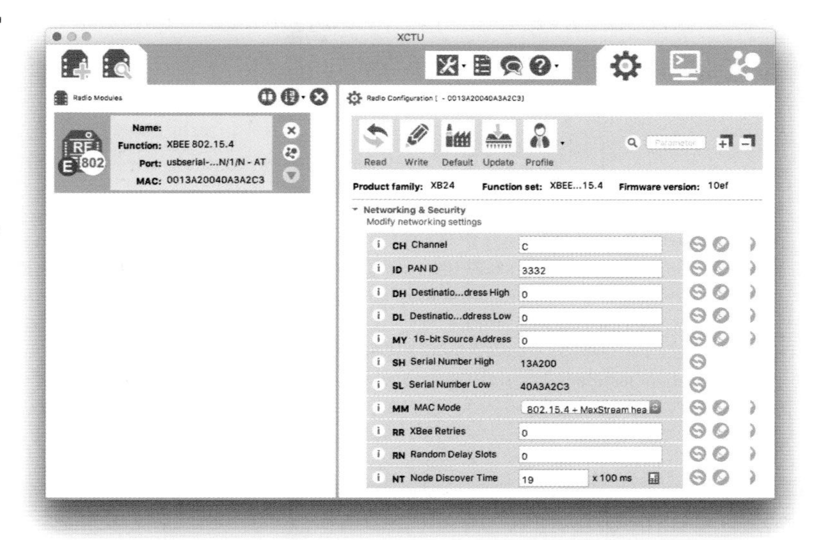

Figure 7-7
XCTU's configuration pane. You can read and write to all of a radio's properties from this pane. When you make a change to a property, click Write to send your change to the radio. XCTU will generate the appropriate AT command to change the property on the radio itself.

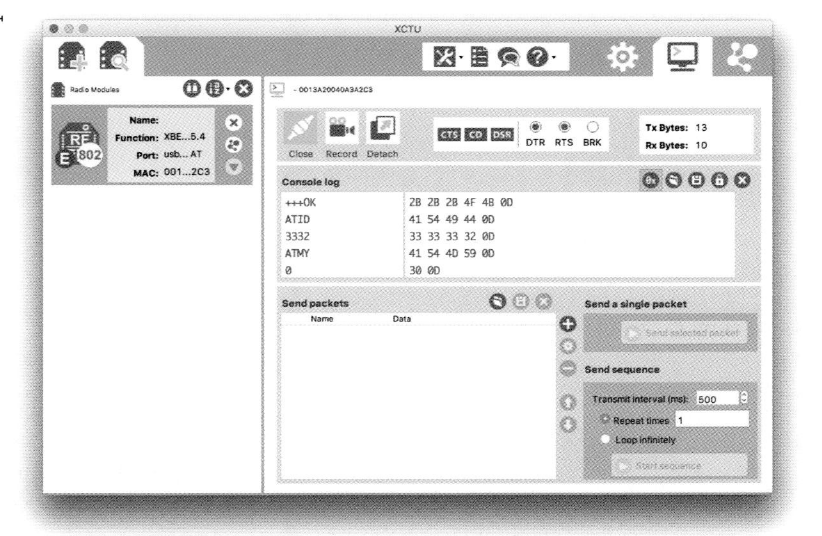

Figure 7-8
XCTU's serial terminal pane. Click the Open/Close button (shown as Close here because the serial port's already open) to connect to a radio. Then you can use this pane to send and receive data directly from the radio's serial port.

Choosing Which XBee Accessories to Buy

There are many Digi-compatible accessories on the market, but there are basically three ways you're likely to use these radios:

XBee to computer via USB-to-XBee adapter. Many of these are available. Figure 7-9 shows two models, the SparkFun XBee Explorer in red, and the XBee USB adapter board—available from Adafruit and Parallax—in blue. You'll need one of these no matter what, to configure your radios. You can use a bare breakout board with an existing USB-to-serial adapter if your adapter can supply 3.3V to the breakout board. Just wire the adapter's TX to the breakout board's RX and vice versa. The Digi XStick 802.15.4 (not shown) is useful as an alternative if you plan to do a lot of XBee work.

XBee to microcontroller. For this, you'll need a breakout board to connect your radio to a microcontroller. The Fio v3 by SparkFun has an Arduino-compatible microcontroller (ATMega32U4) and an XBee mount built in.

XBee standalone. For this, XBee breakout boards work well, as does the XBee LilyPad board. The XBee LilyPad, the XBee Explorer Regulated (SparkFun part no. WRL-11373) all feature built-in voltage regulators, so you can use your XBee with a wider range of power supplies.

The Digi radios operate on 3.3V, so if you're using a bare breakout board (the least expensive option), you'll need a 3.3V supply or a 3.3V regulator.

NOTE: XCTU will ask you to reset your radio if it gets no response from the radio. To do this, you connect the radio's reset pin (pin 5) to ground. Regardless of what mounting solution you use, make sure you're aware of where the connection to pin 5 is, so that you can short it to ground if you need to reset the radio.

Figure 7-9. XBee accessories. Clockwise from top left: XBee USB adapter, XBee Explorer; Fio; XBee Breakout board; XBee LilyPad.

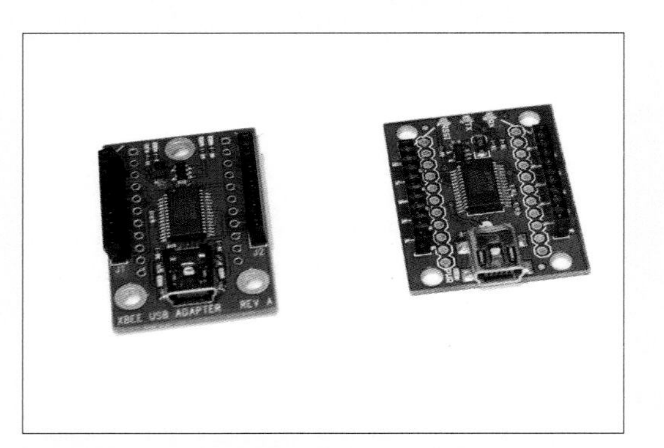

Configuring the Radios

To get started with XCTU, click the Add a Radio button on the top left to choose the serial port to which your radio's attached. When the application finds a radio on the port, it will show you the configuration pane, as shown in Figure 7-7. From there, you can change the properties of your radio, and when you're ready, click Write to send the new settings to the radio. You can set all your radio's properties this way and never have to send AT commands directly.

If you want to get to know the AT command set, however, XCTU also has a serial terminal pane so you can send the commands directly. Click on the second tab on the top-right side of the window to get the serial terminal, as shown in Figure 7-8. You can also use CoolTerm or another serial terminal program if you don't want to use XCTU. You can even send AT commands to configure a radio from a microcontroller sketch, if the controller's connected serially to your radio. The full command protocol for the Digi radios is available on Digi's support site, the same link where you found XCTU. Search for the User Guide for your radio model.

Once you've entered the terminal pane or configured your serial terminal program, open the port and type:

```
+++
```

Don't hit Enter or any other key for at least one second afterward. The XBee should respond like so:

```
OK
```

The `+++` puts it into command mode. The one-second pause after this string is called the *guard time*. If you do nothing, the module will drop out of command mode after 10 seconds. So, if you're reading this while typing, you may need to type `+++` again before the next step.

Once you get the `OK` response, set the radio's address. The XBee protocol uses either 16-bit or 64-bit long addresses, so there are two parts to the address: the high word and the low word (in computer memory, two or more bytes used for a single value are sometimes referred to as a *word*). For the next project, you'll use 16-bit addressing.

To see the radio's source address, type:

```
ATMY\r
```

To change the address, type:

```
ATMY1\r
```

Now you've set the source address to 1. Remember that `\r` indicates you should press Enter. Next, to see the radio's *destination address* (the address to which it will send messages), make sure you're in command mode (`+++`), then type `ATDL\r`.

You'll likely get `0`, because the default destination address on these modules is 0. Don't set a radio's destination address to the same value as its source address, or it will only talk to itself! You can use any 16-bit address for your radios. All values are in hexadecimal, so you can use any value from 0 to FFFF.

To see the PAN ID for your radio, type:

```
ATID\r
```

The default PAN ID is 3332, but you should choose one of your own and set it like you did the source address:

```
ATID1111\r
```

The radio will respond to this command, like all commands, with:

```
OK
```

When you've set all the settings for your radio, add `WR` after your last command, which writes the parameters to the radio's memory. That way, they'll remain the way you want them even after the radio is powered off. For example:

```
ATID1111,WR\r
```

You'll use the same process to set any of a radio's properties if you're not using XCTU's GUI: Put the radio in command mode with `+++`, then send the command, and terminate the last command with WR and a carriage return. To exit command mode, type `ATCN\r`.

X

Project 14

Reporting Toxic Chemicals in the Shop

If you've got a workshop, you'll appreciate this project. By attaching a gas sensor to a Digi 802.15.4 radio, you'll be able to sense the concentration of solvents in the air in your shop. When you're working in the shop by yourself, it's common (yet dangerous) to become insensitive to the fumes of the chemicals with which you're working. This project is an attempt to remedy that issue.

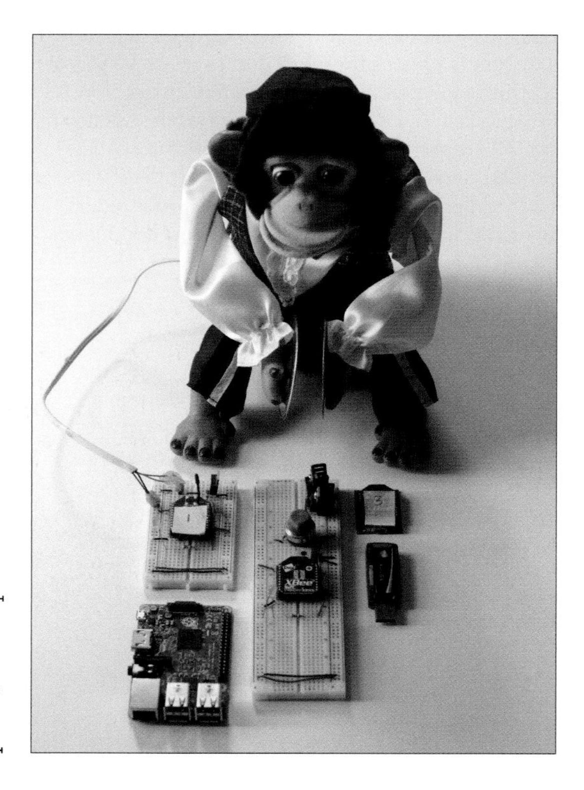

▸▸ Figure 7-10
The completed toxic sensor system: sensor, monkey, and embedded processor server. The Digi XStick (right) is a useful alternative for the server radio.

The gas sensor values will be sent to two other radios. One is attached to an computer or embedded processor like a Raspberry Pi running node.js, which is connected to your local network as a web server. The other radio is attached to a cymbal-playing toy monkey, located elsewhere in the house, that makes an unholy racket when the organic solvent levels in the shop get high. That way, the rest of the family will know immediately if your shop is toxic. If you don't share my love of monkeys, this circuit can control anything that can be switched on from a transistor Figure 7-10 shows the completed elements of the project. Figure 7-11 shows this project's network.

This project is designed for demonstration purposes only. The sensor circuit hasn't been calibrated. It won't save your life; it'll just make you a bit more aware of the solvents in your environment. Don't rely on this circuit if you need an accurate measurement of the concentration of organic compounds. Check with your sensor manufacturer to learn how to build a properly calibrated sensor circuit.

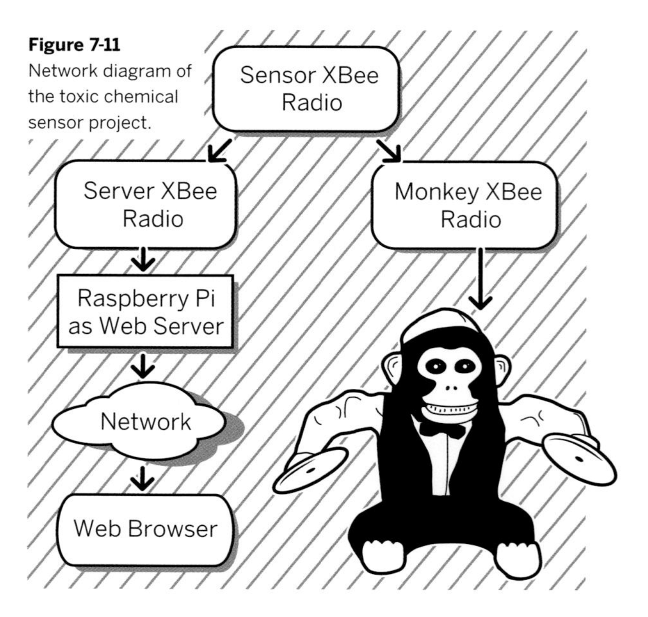

Figure 7-11
Network diagram of the toxic chemical sensor project.

MATERIALS

You'll be building three separate circuits for this project, so the parts list is broken down for each one.

- **1 USB-to-XBee serial adapter** for configuration of the radios

Sensor Circuit

- **1 solderless breadboard**
- **1 XBee or XBee Pro S2C 802.15.4 RF module**
- **1 5V regulator**
- **1 3.3V regulator**
- **1 9–12V DC power supply** Either a 9V battery or plug-in supply will do.
- **1 XBee breakout board**
- **2 rows of 0.1-inch header pins**
- **2 2mm female header rows**
- **2 1µF capacitors**
- **2 10µF capacitors**
- **1 Hanwei MQ-6 gas sensor**
- **1 gas sensor breakout board**
- **1 LED**
- **1 220-ohm resistor**
- **1 47-kilohm trimmer potentiometer**

Server Circuit

- **1 Raspberry Pi, BeagleBone Green, or other embedded Linux processor**
- **1 XBee or XBee Pro S2C 802.15.4 RF module**
- **1 USB-to-XBee serial adapter** You can reuse the one you used for configuration of the radios, or use the Digi XStick, which is an XBee 802.15.4 in a USB stick format.

Cymbal Monkey Circuit

- **1 solderless breadboard**
- **1 XBee or XBee Pro S2C 802.15.4 RF module**
- **1 cymbal monkey**

NOTE: If your monkey uses a 3V power supply (such as two D batteries), you won't need the 3.3V regulator. Make sure there's adequate amperage supplied for the radios. If you connect the circuit as shown and the radios behave erratically, the monkey's motor may be drawing all the power. If so, use a separate power supply for the radio circuit.

- **1 3.3V regulator**
- **1 XBee breakout board**
- **2 rows of 0.1-inch header pins**
- **2 2mm female header rows**
- **1 LED**
- **1 220-ohm resistor**
- **1 TIP120 Darlington NPN transistor**
- **1 1-kilohm resistor**
- **1 100µF capacitor**

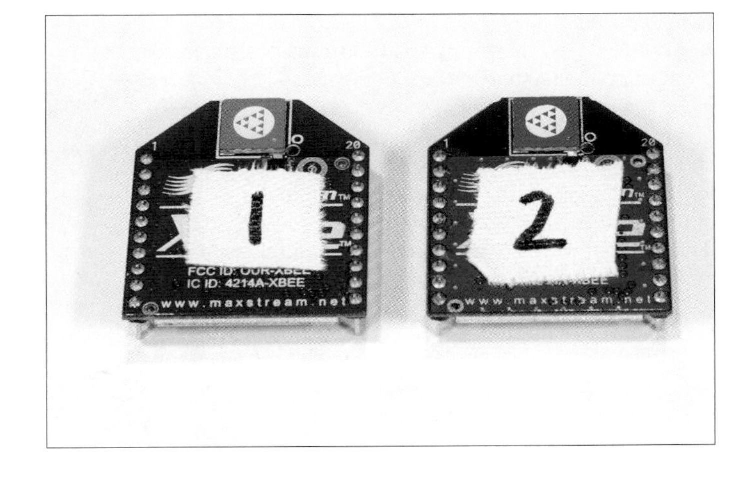

Figure 7-12
It's a good idea to label your radios so you can tell them apart without having to check their configurations serially.

Radio Settings

Connect one of the radios to your USB-to-serial adapter. You'll use this for configuring the radios, then to connect the server radio to the Raspberry Pi or other embedded processor. You've got three radios: the sensor's radio, the monkey's radio, and the server's radio. Earlier, you saw how to configure the radios' source addresses, destination addresses, and Personal Area Network (PAN) IDs. You'll also see how to configure some of their I/O pins' behavior. For example, you can configure the digital and analog I/O pins to operate as inputs, outputs, or to turn off. You can also set them to be digital or analog inputs, or digital or pulse-width modulation (PWM) outputs. You can even link an output pin's behavior to the signals it receives from another radio (this is called *line passing*). You can do this all with XCTU or with the AT commands.

The sensor radio is the center of this project. You'll configure it to read a digital voltage on its first input (D0, pin 20) and broadcast the value that it reads to all other radios on the same PAN. Even though the sensor is analog, you'll be looking for it to cross a threshold voltage, so that's why you're configuring the input pin as digital. Its settings are as follows:

- ATMY01: Sets the sensor radio's source address.
- ATDLFFFF: Sets the destination address to broadcast to the whole PAN.
- ATID1111: Sets the PAN ID.
- ATD03: Sets I/O pin configuration for pin 0 (D0) to act as an analog input.
- ATIR64: Sets the analog input sample rate to 100 milliseconds (0x64 hex). Shorten this if you want more frequent sampling.
- ATIT1: Sets the radio to gather one sample before sending, so it will send every 100 milliseconds (1 samples x 100 milliseconds sample rate = 100 milliseconds).
- ATIAFFFF: This sets the radio to allow line passing to all other devices in the PAN.

The monkey radio will listen for messages on the PAN, and if any radio sends it a message with a sensor reading for pin D0, it will set the state of its pin D0 to the value of the received data. In other words, the monkey radio's digital output will be linked to the sensor radio's digital input. Its settings are as follows:

- ATMY02: Sets the monkey radio's source address.
- ATDL01: Sets the destination address to send only to the sensor radio (address 01). However, this doesn't really matter, as this radio won't be sending.
- ATID1111: Sets the PAN.
- ATD04: Sets pin D0 to be a digital output, defaulting to low.
- ATIU1: Sets the radio to send any I/O data packets out the serial port. This is used for debugging purposes only; you won't actually attach anything to this radio's serial port in the final project.
- ATIA01 or ATIAFFFF: Sets the radio to set its digital outputs using any I/O data packets received from address 01 (the sensor radio's address). If you set this parameter to FFFF, the radio sets its outputs using data received from any radio on the PAN. This effectively links the input of the sensor radio to the output of the monkey radio.

The server's radio listens for messages on the PAN and sends them out its serial port to the embedded processor. This radio's settings are the simplest, as it's doing the least. Its settings are as follows:

- ATMY03: Sets the radio's source address.
- ATDL01: Sets the destination address to send only to the sensor radio (address 01). It doesn't really matter, though,as this radio won't be sending.
- ATID1111: Sets the PAN.
- ATIU1: Sets the radio to send any I/O data packets out the serial port. This data will go to the attached device.

▸▸ **NOTE: If you want to reset your Digi radios to the factory default settings before configuring for this project, send them the command** ATRE\r.

Here's a summary of all the settings:

Sensor radio	Monkey radio	Server radio
MY = 01 DL = FFFF ID = 1111 D0 = 3 IR = 64 IT = 1 IA = FFFF	MY = 02 DL = 01 ID = 1111 D0 = 4 IU = 1 IA = 01 or FFFF	MY = 03 DL = 01 ID = 1111 IU = 1

Make sure to save the configuration to each radio's memory by finishing your commands with WR. XCTU's Write button does this automatically. If you're sending AT commands manually, you can set the whole configuration line by line or all at once. For example, to set the sensor radio, type:

```
+++
```

Then wait for the radio to respond with `OK`. Next, type the following (the 0 in D03 is the number 0):

```
ATMY1, DLFFFF\r
ATID1111, D03, IR64\r
ATIT1, IAFFFF, WR\r
```

For the monkey radio, the configuration is:

```
ATMY2, DL1\r
ATID1111, D04\r
ATIU1, IAFFFF, WR\r
```

And for the server's radio, it's:

```
ATMY3, DL1\r
ATID1111, IU1, WR\r
```

The Circuits

Once you've got the radios configured, set up the circuits for the sensor, the monkey, and the server. In the first two of these circuits, make sure to include the *decoupling capacitors* on either side of the voltage regulators—the Digi radios tend to be unreliable without them. These capacitors smooth out any voltage surges or drops that happen when the radio powers up for transmission.

The Sensor Circuit

The gas sensor takes a 5V supply voltage, so you need a 5V regulator for it, a 3.3V regulator for the radio, and a power supply that's at least 7V to supply voltage to the circuit. A 9V battery will do, or a 9–12V DC power adapter. Figure 7-13 shows the circuit. The gas sensor is an analog sensor, but you're attaching it to a digital input because you just want to know when it crosses a certain threshold. The circuit takes time to warm up because there's a heater element in the sensor. Power up the circuit, and let it heat for a minute or two. Measure the voltage between the sensor's output and ground. The sensor's resistance changes with the concentration of volatile organic compound (VOCs). Actual resistance varies with the particular VOC; see the MQ-6 datasheet for details. You're using a 47-kilohm potentiometer as the base resistor of the circuit so that you can tune the sensor's output. You want the output voltage to rest at about 1.5 to 1.7 volts, going over that only when the concentration of gases is too high. That level will trigger the output transistor on the monkey circuit. I tested with the potentiometer at about halfway (or roughly 22 kilohms), and put hand sanitizer, which contains alcohol, on my hands and placed them over the sensor. I got an output of about 2.3V with the circuit shown here. Be careful not to breathe in the fumes yourself. When I cleared the air, the voltage dropped back to about 1.6V. Your values may vary, so carefully turn the potentiometer until you see the response you want. Once you've got the sensor reading in an acceptable range, connect its output to the radio's D0 input pin, which is pin 20. Make sure to connect the radio's voltage reference pin (pin 14) to 3.3 volts as well.

NOTE: Air out your workspace as soon as you've tested the sensor. You don't want to poison yourself while making a poison sensor!

The Monkey Circuit

To control the monkey, disconnect the monkey's motor from its switch and connect the motor directly to the circuit shown in Figure 7-14. The monkey's battery pack supplies 3V, which is enough for the radio, so you can power the whole radio circuit from the monkey. Connect leads from the battery pack's power and ground to the board. If your monkey runs on a different voltage, make sure to adapt the circuit accordingly so that your radio circuit is getting at least 3V. Figure 7-15 shows the modifications in the monkey's innards. I used an old telephone cord to wire the monkey to the board.

The cymbal monkey circuit takes the variable output that the radio received and turns it into an on-off switch. The PWM output from the radio controls the base of a transistor. The monkey itself has a motor built into it, which is controlled by a TIP120 Darlington transistor in this circuit. When the transistor's base goes high, the motor turns on. When it goes low, the motor turns off.

To test this circuit, make sure that the sensor radio is working, and turn it on. When the sensor's value is low, the motor should be turned off; when the sensor reads a high concentration of VOCs, the motor will turn on and the monkey will warn you by playing his cymbals. Use the sensor potentiometer to affect the motor's activation threshold. Start with the potentiometer set very high, then slowly turn it down until the motor turns off. At that point, expose the sensor to some alcohol—the motor should turn on again. It should go off when the air around the sensor is clear. If you're unsure whether the motor circuit is working correctly, connect an LED and 220-ohm resistor in series from 3V to the collector of the transistor instead of the motor. The LED should turn on when the sensor reading is higher than about 1.5V, and go off below that.

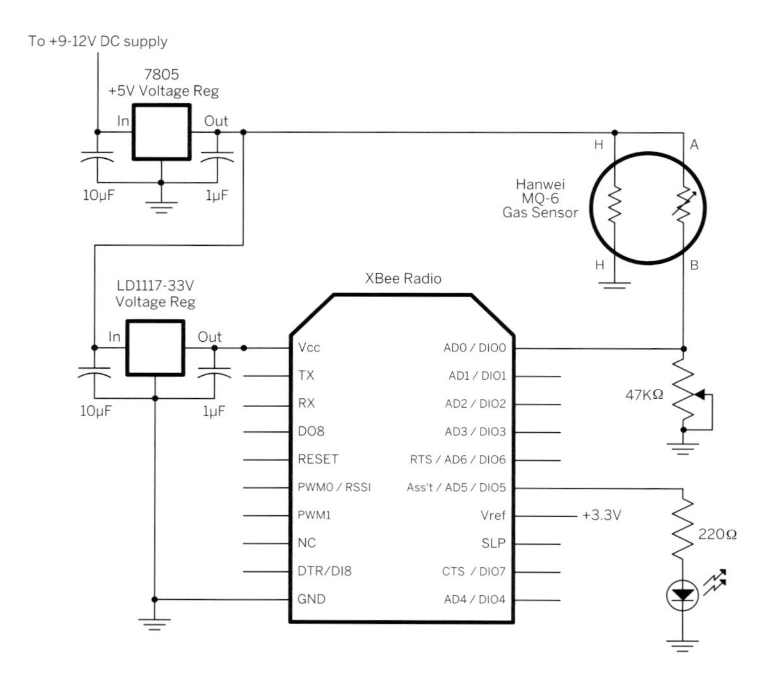

Figure 7-13
Digi radio connected to a gas sensor. The MQ-6 is shown here, but the same circuit will work for many of the Hanwei sensors. The 47K potentiometer lets you set the trigger level for digital input D0.

The radio shown on the breadboard is mounted on a breakout board. The radio's pins will not fit into a breadboard without the breakout board.

A 9V battery will only last about 3 hours on this circuit because the heater consumes considerable current.

The detail breadboard view shows how the decoupling capacitors are connected behind the regulators.

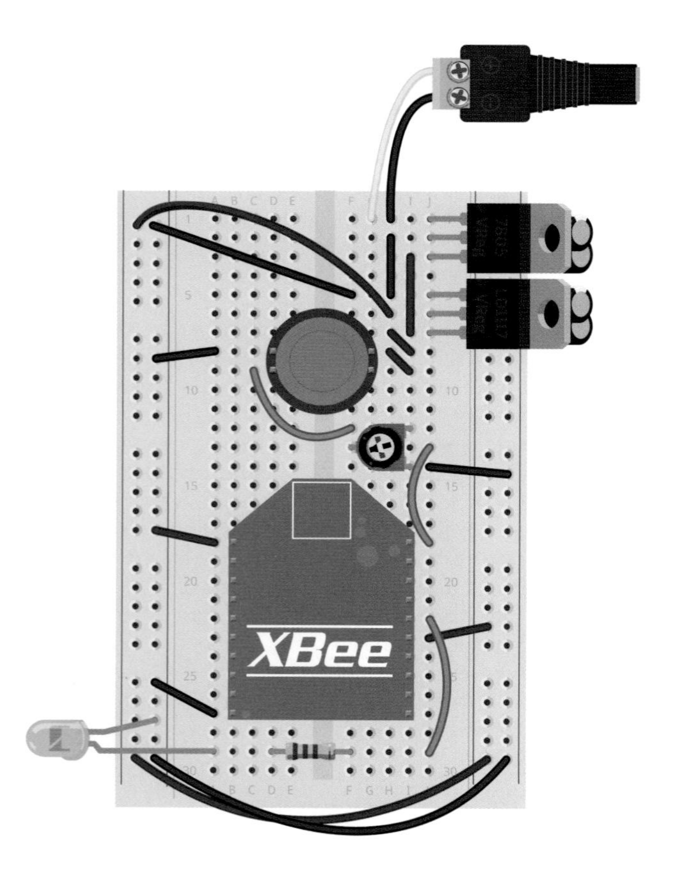

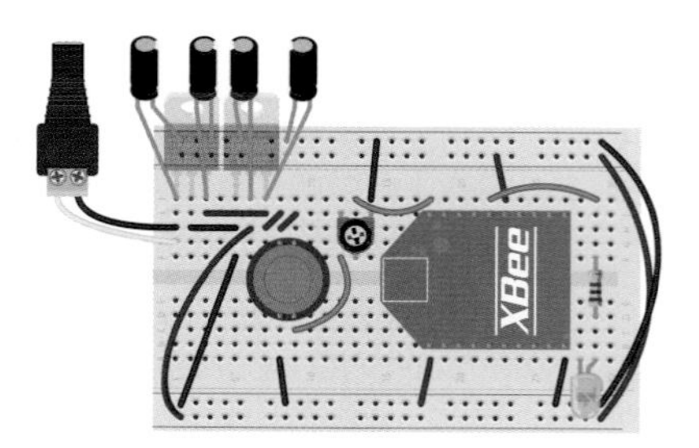

▸▸ **This project is made using the LD1117-33V 3.3V voltage regulator. The pins on this regulator are configured differently from the pins on the 5V regulator. Check the data sheet for your regulator to be sure you have the pins correct.**

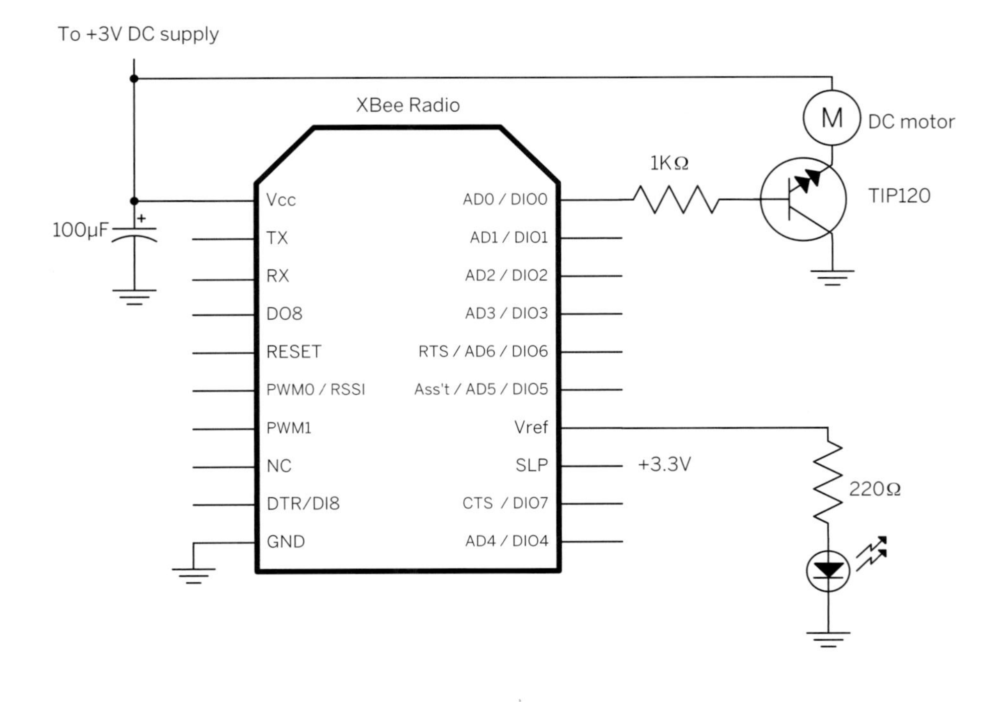

Figure 7-14
Digi radio connected to a cymbal monkey.

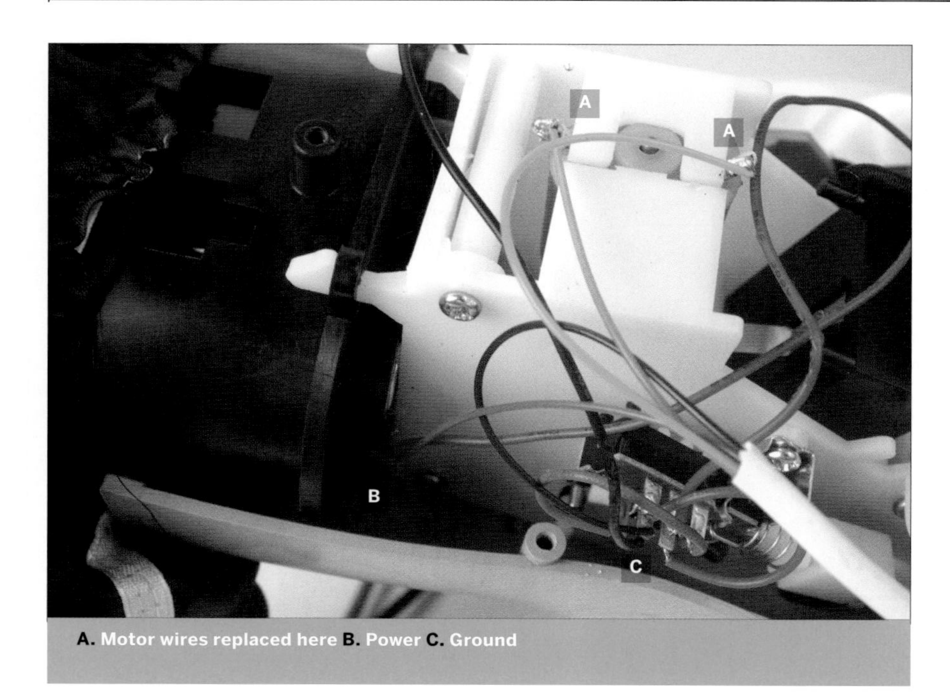

A. Motor wires replaced here B. Power C. Ground

Figure 7-15
The insides of the monkey, showing the wiring modifications. Solder the power to the breadboard to the positive terminal of the battery. Solder the ground wire to the ground terminal. Cut the existing motor leads, and add new ones that connect to the breadboard.

The Server Circuit
To connect the server radio to the Raspberry Pi, all you'll need is your USB-to-serial adapter. You can also use your personal computer instead of a Pi if you prefer. You'll write a script on the embedded processor that reads serial data and runs a web server. Node.js will read the incoming serial data using a library called node-serialport.

When the script receives the message from the gas sensor radio, it will parse the message and extract the sensor value. The script will serve a page with the current state of the sensor. In order to make that happen, it's best to break the task into two parts: first, establish that you can read and parse the radio's messages; then, add the web server component to your code.

The Node-Serialport Library

So far, you've used node.js for lots of tasks, but not serial communication. As with all of those other tasks, there is a library for serial communications in node.js, and you install it with npm like so:

```
$ npm install serialport
```

When you do this, npm will download the source files and compile the library for your platform. If you hit any issues, check github.com/EmergingTechnologyAdvisors/node-serialport#installation-instructions for the most current details of how to install on your platform.

If you're working on a personal computer running macOS or Windows or Linux, you should be ready to start. However, if you're on a Raspberry Pi or other embedded processor, see the sidebar on enabling your serial ports (next page).

The node-serialport library has a number of functions, but the most important ones are what you'd expect. You can open a new serial port like so:

```
var myPort = new SerialPort(portName);
```

Then you can listen for four main events: `open`, `close`, `error`, and `data`. The first three do what they're called. The fourth, `data`, occurs whenever a new byte of data arrives in the serial buffer for the port. There are various options for managing the data flow, but for this project you'll keep it simple and read the data byte by byte.

The structure for a serial script looks like this:

```
// include the serialport library:
var SerialPort = require('serialport');
```

```
var portName = '/dev/ttyUSB0' // change to your port name
var myPort = new SerialPort(portName);

// callback function to read incoming data:
function readData(data) {
  // data is an array of available bytes:
  console.log(data);
}
// Serial listeners:
// called when the serial port opens:
myPort.on('open', portOpen);
// called when there's new incoming serial data:
myPort.on('data', readData);
// called when the serial port closes:
myPort.on('close', portClose);
// called when there's an error with the serial port:
myPort.on('error', portError);
```

Each of the events has a callback function that you define. The important one is the `data` callback. The rest are all for diagnostics.

In the script that follows, you'll parse through the XBee API protocol in `readData()` to extract the useful information about the gas sensor. Then you'll combine the serialport code with an HTTP server using express.js as you've done in previous chapters.
X

Enabling Serial Ports in Raspian

Embedded processors like the Raspberry Pi typically have hardware serial ports that they can use just like the serial ports on your personal computer. However, there are some trade-offs to using them. You can also add serial ports using USB-to-serial adapters if the drivers for your adapter are available for the operating system that you're running.

If you're only using serial devices that you're attaching to the Pi through a USB-to-serial adapter, then you need to make sure the drivers for your USB-to-serial adapter are present. Most of the ones mentioned in this book work automatically in Raspian. To check, plug your device in to one of the USB ports and list the available serial ports like this:

```
$ ls /dev/tty*
```

If you see a port called `/dev/ttyUSB0`, then you're all set. If not, do an internet search to see if there are Raspian or Debian drivers for your particular USB-to-serial adapter.

If you're planning to use the built-in serial port (the one attached to the TX and RX pins on the board's I/O header), then you need to disable the serial console first. As you saw in Chapter 1, the serial terminal is an important way to connect to an embedded processor like the Raspberry Pi. Until you know the board can connect to a network, it can be your only way to operate it, unless you have a keyboard, mouse, and monitor handy. However, once you know your embedded processor can connect to your network, you can use ssh to connect to it like you would any remote computer. This frees your built-in serial port to connect to other devices.

To disable the serial console, log in via ssh (not via a serial console) and run `raspi-config`. Choose the Advanced Options, then choose Serial. When asked `Would you like a login shell to be accessible over serial?`, choose No. Then choose Finish. Then reboot your Pi. You should now be able to use your serial port like a normal port. It's called `/dev/ttyAMA0`.

Reading the XBee Protocol

So far, you've configured the radios, but not looked at the serial output of the ones receiving data from the sensor radio. With the sensor radio operational, connect the server radio to your computer, open XCTU, and click the serial terminal. You'll get output that looks like this:

```
7E 00 0A 83 00 01 1F 00 01 00 01 00 01 59 7E 00 0A 83 00 01
1F 00 01 00 01 00 01 59 7E 00 0A 83 00 01 1F 00 01 00 01 00
01 59
```

The XBee API protocol is a binary protocol, and these are the bytes received from the sensor radio, in hexadecimal notation. This sample represents two messages from the sensor radio. The format of the packet is explained in the Digi XBee 802.15.4 user's manual. Here is the sequence of bytes. Unless stated otherwise, each element is a single-byte value:

- The message starts with a byte of value 0x7E, the start byte value. This never changes.
- Packet size, a 2-byte value (00 and 12 in this example). This depends on your other settings.
- API identifier value, a code that says what type of packet this response is (83 in this example).
- The sender's address, a two-byte value (00 and 01 in this example).
- Received Signal Strength Indicator (RSSI) (24 in this example).
- Broadcast options (not used here) (00 in this example).
- Number of samples in the packet. You set this to 5 using the `IT` command shown earlier (05 in this example).
- Which I/O channels are currently being used, a 2-byte value (00 and 01 in this example). This example assumes only one digital channel, D0, and no analog channels are in use.
- A 2-byte value representing which digital inputs are high (1) or low (0). Each digital input is a bit in the 2-byte value. For example, 00 01 means only pin D0 is high. 00 02 would mean that pin D1 is high (10 in binary).

Because every packet starts with a constant value, 0x7E (that's decimal value 126), you can start by looking for that value. Make a project directory for this project, and install node-serialport in it using npm as you saw previously. Then save this file in the directory as **serialServer.js**.

Read It In this script you'll read bytes and put them in an array. When you get the value `0x7E`, you'll know that's the start of the next message, so you'll parse the array to find the sensor's value.

Start by including the node-serialport library. You'll also need a variable for the port name of your USB-serial adapter. You can get the name from any of the other serial terminal applications you use. Finally, you'll need an array variable to hold incoming serial data. After that, open the serial port.

```
/*
serialServer.js
context: node.js
*/

var SerialPort = require('serialport');  // include the serialport library
var portName = ' /dev/tty.usbserial-xxx'; // replace with your port name
var incoming = [];                        // an array to hold the serial data

// open the serial port:
var myPort = new SerialPort(portName);
```

Next add a function to let you know when the serial port opens successfully, and one to report when there's an error with the serial port. For this script, you can dispense with the function to report when the port closes, because you'll only close the port when you stop the script.

```
function portOpen(portName) {
  console.log('port ' + myPort.path + ' open');
  console.log('baud rate: ' + myPort.options.baudRate);
}

function portError(error) {
  console.log('there was an error with the serial port: ' + error);
  myPort.close();
}
```

Next, add a function to report when new data comes in the serial port. Make a for loop to run through the serial buffer and check if the latest byte's value is `0x7E`, the start of your data message. If so, then print out the array.

```
function readData(data) {
    for (c=0; c < data.length; c++) {   // loop over all the bytes
      var value = Number(data[c]);      // get the byte value
      if (value === 0x7E) {             // 0x7E starts a new message
        console.log(incoming);          // print the previous array
        incoming = [];                  // clear the array for new data
      } else {                          // if the byte's not 0x7E,
      incoming.push(value);             // add it to the incoming array
    }
  }
}
```

Now add the event listeners, and you're ready to run the script. Save it and on the command line type:

```
$ node serialServer.js
```

```
// called when the serial port opens:
myPort.on('open', portOpen);
// called when there's new incoming serial data:
myPort.on('data', readData);
// called when there's an error with the serial port:
myPort.on('error', portError);
```

When you run this script, you'll get a result like what's shown to the right.

Each line that starts with 126 (`0x7E` in hexadecimal) is a single XBee message. Most of the lines have 13 bytes, corresponding to the packet format described earlier. You may wonder why the first line of the output shown here didn't have any bytes. It's simply because there's no way to know what byte the radio is sending when node.js first starts listening. That's OK, because the following code will filter out the incomplete packets.

```
port /dev/tty.usbserial-00001414 open
baud rate: 9600
[]
[[ 0, 10, 131, 0, 1, 35, 0, 1, 0, 1, 0, 1, 85 ]
[ 0, 10, 131, 0, 1, 35, 0, 1, 0, 1, 0, 1, 85 ]
[ 0, 10, 131, 0, 1, 42, 0, 1, 0, 1, 0, 1, 78 ]
[ 0, 10, 131, 0, 1, 41, 0, 1, 0, 1, 0, 1, 79 ]
[ 0, 10, 131, 0, 1, 34, 0, 1, 0, 1, 0, 1, 86 ]
[ 0, 10, 131, 0, 1, 36, 0, 1, 0, 1, 0, 1, 84 ]
```

Declare a new variable at the beginning of the script, after the `incoming` array called `message`. This variable is a JSON object, and it will contain all of the elements of the XBee message as described earlier:

```
var message = {          // the XBee packet as a JSON object:
  packetLength: -1,      // packet length
  apiId: 0,              // message API identifier
  address: -1,           // sender's address
  rssi: 0,               // signal strength
  channels: 0,           // which I/O channels are in use
  sampleData: 0,         // sample data from all pins
  pinStates = []         // array with the state of the digital pins
};
```

Parse It Replace the call to `console.log()` in the `readData()` function with a call to a new function called `parseData()`. Then add this function after the `readData()` function. The new function will parse the message packet byte-by-byte, assembling the relevant values from the various bytes. It will also calculate the average of the samples in the message.

```
if (value === 0x7E) {              // 0x7E starts a new message
  parseData(incoming);             // parse the array
  output = [];                     // clear the array for new data
} else {                           // if the byte's not 0x7E,
incoming.push(value);             // add it to the incoming array
}
```

» Start by checking to see if there are enough bytes in the incoming array. Without the `0x7E` byte, there should be 13 bytes. If that's the case, start reading the bytes into the appropriate elements of the JSON object. For the 2-byte elements, you'll use this formula to combine the bytes:

```
total = highByte * 256 + lowByte
```

Once you've got the sample data, use a for loop to isolate the bits and put them in an array. Since each bit represents one of the digital input pins (D0 through D8), you can use bit masking, which you learned earlier in the chapter, to get each pin's bit value.

```
function parseData(thisPacket) {
  if (thisPacket.length >= 13) {   // if the packet is 13 bytes long
    // read the address. It's a two-byte value, so
    // packetLength = firstByte * 256 + secondByte:
    message.packetLength = (thisPacket[0] * 256) + thisPacket[1];
    // message type is shown in hex in the docs, so convert to a hex
string:
    message.apiId = '0x' + (thisPacket[2]).toString(16);
    // same two-byte formula with address:
    message.address = (thisPacket[3] * 256) + thisPacket[4];
    // read the received signal strength:
    message.rssi = -thisPacket[5];
    // channels is also a two-byte value.
    // It's best read in binary, so convert to a binary string:
    message.channels = ((thisPacket[8] * 256) + thisPacket[9]).toString(2);
    // pin states:
    message.sampleData = (thisPacket[10] * 256) + thisPacket[11];

    // convert the sample data into an array of pin states:
    for (var pin = 0; pin < 9; pin++) {
      // isolate each bit of the sample data:
      var thisPinState = message.sampleData & (1 << pin);
      // push each pin's state to the pin states array:
      message.pinStates.push(thisPinState);
    }
    console.log(message);     // print it all out
  }
}
```

Serve It Now that you can read the messages, add code to make a web server. The rest of this sketch is much like the other node.js web servers you've written. New lines are shown in blue as usual. Make sure to install express with npm in the project's directory. First, include express.js and make an instance of it, and set up static file serving at the beginning of the script.

```
var express = require('express');        // include the express library
var server = express();                  // create a server using express
server.use('/',express.static('public')); // serve static files from /public

// serial port initialization:
var SerialPort = require('serialport'); // include the serialport library
```

At the end of the script, add code to start the server and a listener to listen for GET requests.

```
server.listen(8080);                     // start the server
server.get('/json', respondToClient); // respond to GET requests
```

After the `parseData()` function, add the callback function for the GET request listener, and you'll be done with the script.

```
// define the callback function that's called when
// a client makes a request:
function respondToClient(request, response) {
  // write back to the client:
  response.end(JSON.stringify(message));
}
```

Run the script again. This time, instead of the raw array, you'll get the following JSON object repeatedly in the output:

```
{ packetLength: 10,
  apiId: '0x83',
  address: 1,
  rssi: -33,
  channels: '1',
  sampleData: 1,
  pinStates: [ 1, 0, 0, 0, 0, 0, 0, 0, 0 ] }
```

If you're not getting good values, use the first version of the script that prints out the value of each byte to help diagnose the problem. Seeing every byte that you actually receive before you do anything with those bytes is the best way to solve the problem. There are a few common problems to look for:

- Are you getting anything at all? If not, is the sensor radio transmitting?
- Does each radio have the correct settings? Connect them to a USB-to-serial adapter and check in XCTU or a serial terminal application.
- Is the receiving radio getting adequate power? Check the voltage between pins 1 and 10 of the sensor radio.

Once you know you're receiving properly, open a web browser and enter http://localhost:8080/json in the address bar. You'll get a very bare-bones response with the same values like so:

```
{"packetLength":10,"apiId":"0x83","address":1,"rssi":-37,"channels
":"1","sampleData":1,"pinStates":[1,0,0,0,0,0,0,0,0]}
```

Following is a p5.js sketch that will serve a more informative response.

X

The Web Page

Start as usual by making a p5.js project in a subdirectory of this project called **public**. Either copy the P5.js **empty-example** directory, or use the command line:

```
$ p5 g -b public
```

Then open **sketch.js** file and add some global variables at the beginning, before the `setup()` function.

```
/*
Sensor warning sketch
context: P5.js
*/

var sensorState = 'UNKOWN';                    // state of the sensor
var bgColor = 0;                               // background color
```

The `setup()` function sets the size of the drawing canvas, text size, and fill color. Then it makes an HTTP GET call to get the data from the server script.

```
function setup() {
  createCanvas(windowWidth, windowHeight); // set the canvas size
  textSize(24);                           // set text size
  fill(255);                              // set the text fill color
  httpGet('/json','json',getResponse);    // get a reading from the server
}
```

When the server responds, the `getResponse()` function gets called. It adds the newest reading to the readings array, then makes another HTTP GET call. It will continue to call itself recursively as long as the page is open.

```
// callback for the httpGet() function:
function getResponse(message) {
  // extract pin state 0:
  if (message.pinStates[0] === 1){      // if it's high
    sensorState = 'HIGH';               // change the text
    bgColor = '#FF0000';                // change the fill color
  } else {                              // if it's low
    sensorState = 'low';                // change the text
    bgColor = 0;                        // change the fill color
  }
  httpGet('/json','json',getResponse);  // get another reading
}
```

The `loop()` function just draws the text to the screen. That's the whole sketch. Figure 7-16 shows the possible results in a browser.

```
function draw() {
  // clear the background, and set its color:
  background(bgColor);
  // set the text:
  text('The gas sensor reading is ' + sensorState, 30, 30);
}
```

Now when you're working in your shop, your family or friends can listen for the monkey to clang his cymbals to know whether the air is getting too foul for you to breathe. If you're not near the monkey, you can check the web page as well.

The Digi 802.15.4 radios and the XBee API protocol offer a lot of options for projects once you get to know them. Although they can be used as a radio attachment to another microcontroller, their strength is that they can interface directly to sensors and actuators. In this project, you've used simple line passing to connect the digital input of one radio to the digital output of another. You can also do analog line passing, or can use the API protocol as a way to monitor remote sensors. Many sensor-based projects have simple interaction requirements like this, and don't need a lot of complex programming on the sensor or actuator devices. That's where these radios really excel.

The XBee S2C 802.15.4 RF Module User Guide covers the details of the protocol clearly and in depth, if you want a more advanced introduction than has been covered here. You can find it on the Digi site at www.digi.com/resources/documentation/Digidocs/90001500. Digi uses variations on this protocol in many of its radios, like the WiFi radios it offers, or the cellular modem you'll see in Chapter 10. Because of this, it's easy to work across the whole product line.

X

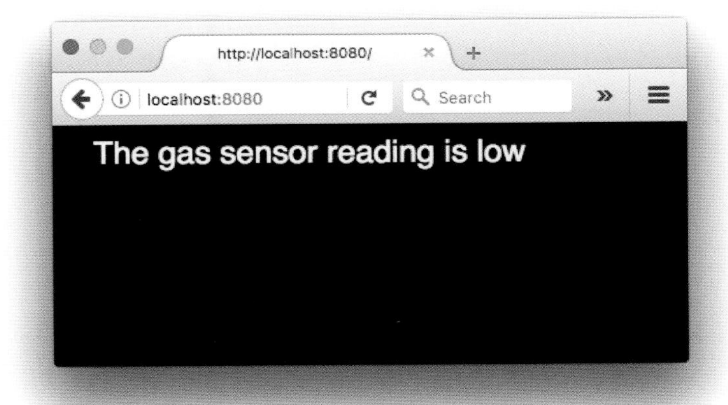

Figure 7-16
The screenshot from the web page of the Digi radio gas server. Both the text and the color give an immediate indication of the state.

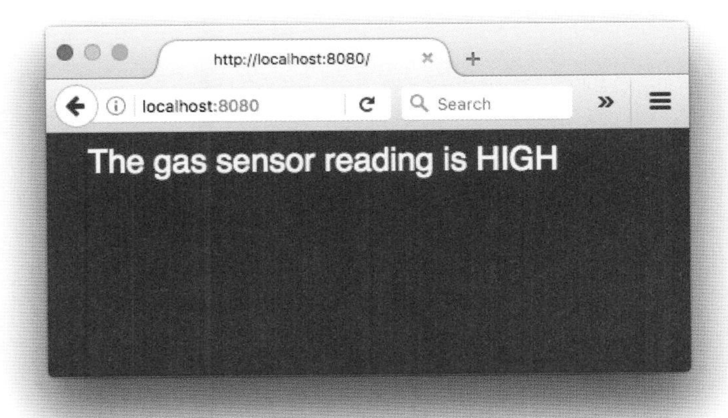

Conclusion

Sessionless networks can be really handy when you're just passing short messages around and don't need a lot of acknowledgment. They involve less complex logic to your communications code because you don't have to maintain the connection. They also give you a lot more freedom to decide how many devices you want to address at once. Binary communications protocols are often handy when using message-based transport protocols, to keep the number of bytes per message to a minimum.

By comparing the two projects in this chapter, you can see that even though the communications protocols are very different (a text-based in the first, XBee API protocol in the second), the network structure is similar. There's not a lot of work to be done to switch from directed messages and broadcast messages in either one. You just use a special address to do it.

It's best to default to directed messages when you can, because it reduces the traffic for those devices that don't need to get every message.

Now that you've got a good grasp of both session-based and sessionless networks, and text and binary communications protocols, the next chapters switch direction slightly, covering two other activities of connecting networks to the physical world: location and identification.

X

▸▸ Wedding Candles by Tom Igoe, Peiqi Su, Deqing Sun, Ben Light, and Andy Sigler
These candles communicate over WiFi using UDP messages and a text-based communications protocol.

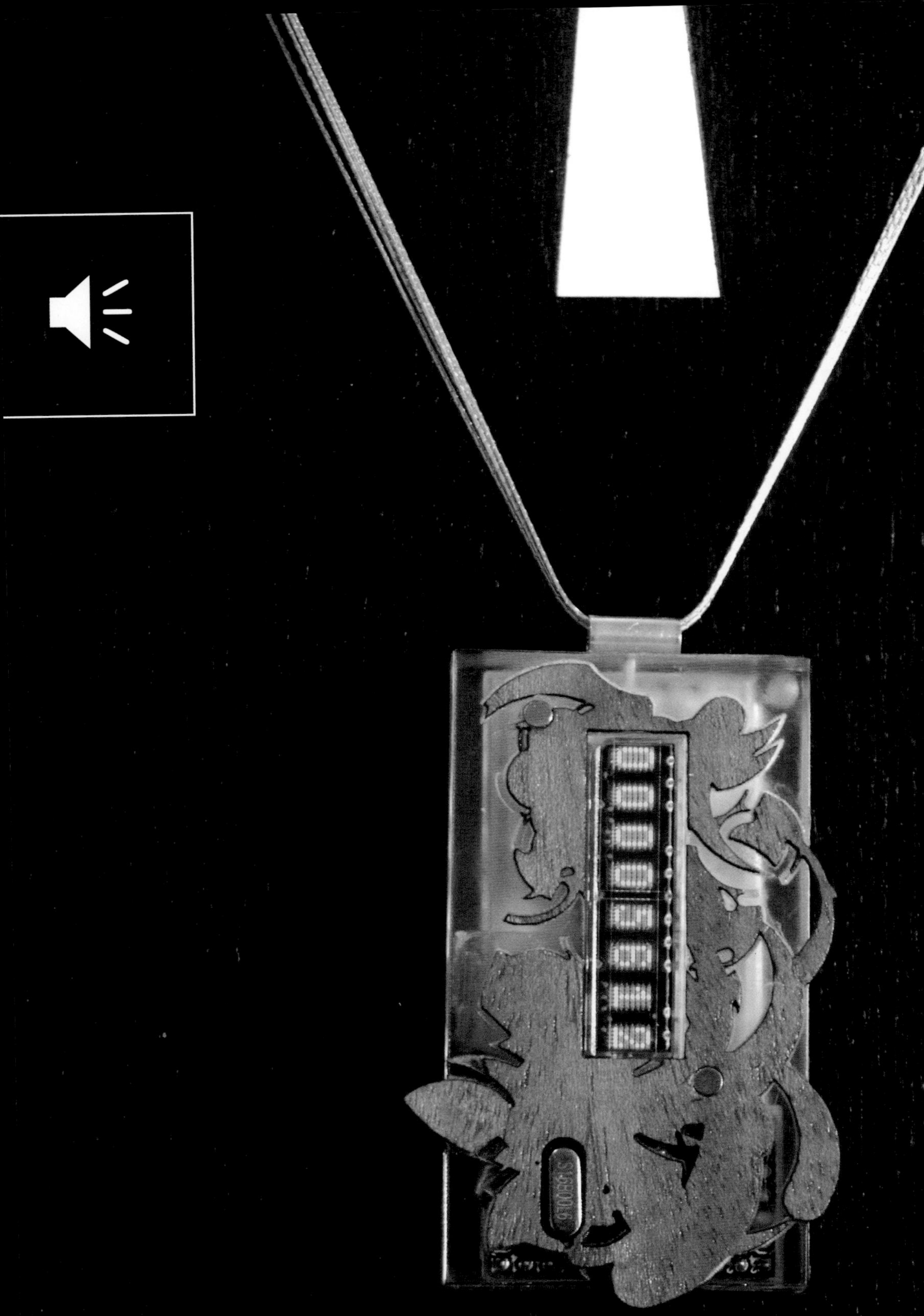

8

MAKE: PROJECTS

How to Locate (Almost) Anything

By now, you've got a pretty good sense of how to make things talk to each other over networks. You've learned about packets, sockets, datagrams, clients, servers, and all sorts of protocols. Now that you know how to talk, this chapter and the next deal with two common questions: "Where am I?" and "Who am I talking to?" Location and identification technologies share some important properties. As a result, it's not uncommon to confuse the two, and to think that a location technology can be used to identify a person or an object, and vice versa. These are two different tasks in the physical world, and often on networks as well. Systems for determining physical location aren't always very good at determining identity, and identification systems don't do a good job of determining precise location. Likewise, knowing who's talking on a network doesn't always help you to know where the speaker is. In the examples that follow, you'll see methods for determining location and identification in both physical and network environments.

◀◀ Address 2007 by Mouna Andraos and Sonali Sridhar
This necklace contains a GPS module. When activated, it displays the distance between the necklace and your home location. *Photo by J. Nordberg.*

Supplies for Chapter 8

This chapter is all about sensing location, so most of the new parts are sensors.

DISTRIBUTOR KEY

A Arduino Store (store.arduino.cc)
AF Adafruit (adafruit.com)
D Digi-Key (www.digikey.com)
F Farnell (www.farnell.com)
J Jameco (jameco.com)
P Pololu (www.pololu.com)
PX Parallax (www.parallax.com)
RS RS (www.rs-online.com)
SF SparkFun (www.sparkfun.com)
SS Seeed Studio (www.seeedstudio.com)

PROJECT 15: INFRARED DISTANCE-RANGER

» **1 microcontroller module** This project will run on just about any Arduino-compatible board.
MKR1000: **AF** 3156, **RS** 124-0657, **A** ABX00004, GBX00011 (Africa and EU), **D** 1659-1005-ND
Arduino 101: **D** 1660-1003-ND, **J** 2239331, **SF** DEV-13787, **AF** 3033, **F** 2520713, **RS** 913-9999, **SS** 114990575, **A** ABX00005, GBX00005 (Africa and EU)
Uno: **D** 1050-1024-ND, **J** 2151486, **SF** DEV-11021, **A** A000099, **AF** 50, **F** 1848687, **RS** 715-4081, **SS** ARD132D2P

» **Infrared distance ranger** **J** 2150256, **D** 425-2063-ND, **AF** 164, **F** 1243869, **RS** 666-6564

» **1 10µF capacitor** **D** P11212-ND, **J** 29891, **F** 1144605, **RS** 762-1736

» **3 0.1-inch header pins** **D** A26509-20-ND, **J** 103377, **SF** PRT-00116, **F** 1593411

» **1 personal computer**

» **All necessary converters to communicate serially from microcontroller to computer** Just like the previous projects.

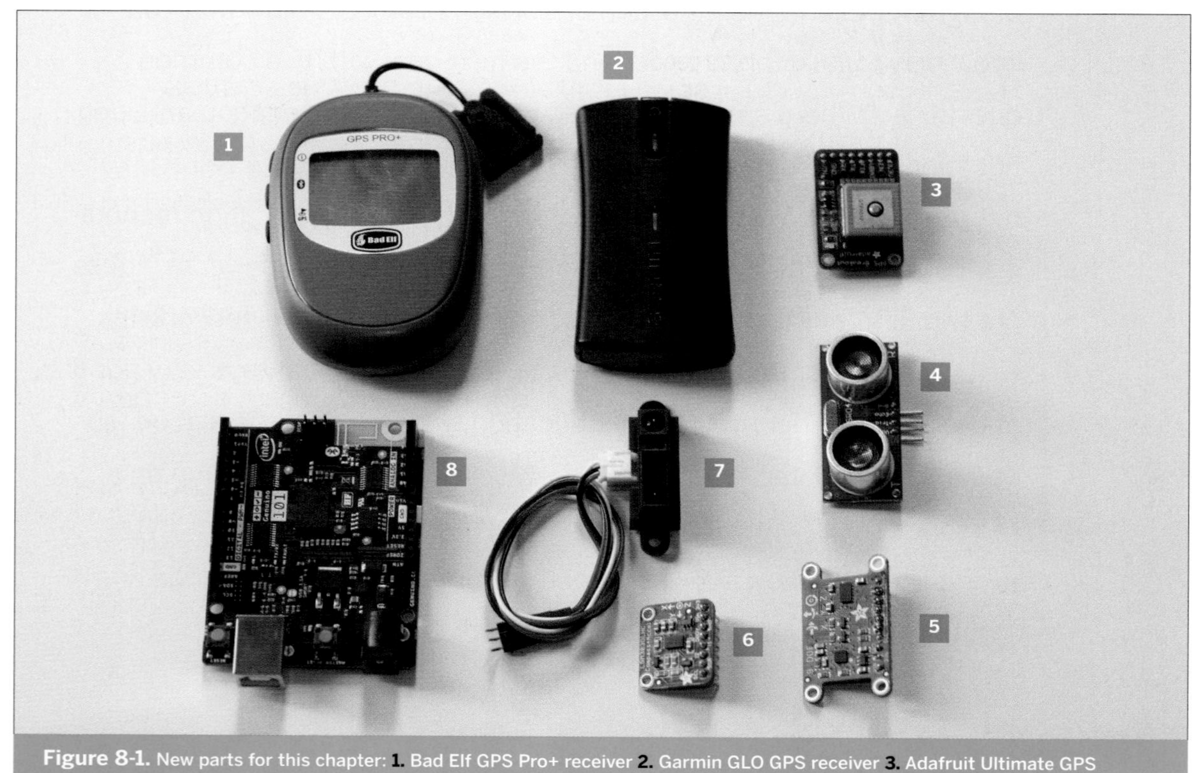

Figure 8-1. New parts for this chapter: **1.** Bad Elf GPS Pro+ receiver **2.** Garmin GLO GPS receiver **3.** Adafruit Ultimate GPS Breakout **4.** HC-SR04 Ultrasonic distance ranger **5.** Adafruit 9DOF IMU breakout board **6.** LMS303DLH digital compass **7.** Sharp GP20Y0A21 IR ranger **8.** Arduino 101

PROJECT 16: ULTRASONIC DISTANCE-RANGER

- 1 microcontroller module This project will run on just about any Arduino-compatible board.
  MKR1000: **AF** 3156, **RS** 124-0657, **A** ABX00004, GBX00011 (Africa and EU), **D** 1659-1005-ND
  Arduino 101: **D** 1660-1003-ND, **J** 2239331, **SF** DEV-13787, **AF** 3033, **F** 2520713, **RS** 913-9999, **SS** 114990575, **A** ABX00005, GBX00005 (Africa and EU)
  Uno: **D** 1050-1024-ND, **J** 2151486, **SF** DEV-11021, **A** A000099, **AF** 50, **F** 1848687, **RS** 715-4081, **SS** ARD132D2P
- 1 ultrasonic rangefinder, SRF04, HC-SR04, or equivalent Another distance-ranging sensor, this one uses sonar and has a different range. **SF** SEN-13959, **D** 1568-1421-ND
- 4 male header pins or 4 jumper wires What you need depends on the interface of your rangefinder.
- 1 personal computer
- All necessary converters to communicate serially from microcontroller to computer Just like the previous projects.

PROJECT 17: READING RECEIVED SIGNAL STRENGTH

- 1 Arduino MKR1000 **AF** 3156, **RS** 124-0657, **A** ABX00004, GBX00011 (Africa and EU), **D** 1659-1005-ND. Alternatively, an ESP8266 board will work. **SF** WRL-13231, **AF** 2471
- 1 WiFi router No internet connection required
- 1 personal computer
- All necessary converters to communicate serially from microcontroller to computer Just like the previous projects.

PROJECT 18: GEOLOCATION SERVICES AND THE NMEA PROTOCOL

- 1 GPS receiver Multiple devices will work in this project. See the text for a description of the differences between models. **AF** 746, **SF** GPS-1275, Garmin GLO GPS available from buy.garmin.com. Bad Elf GPS Pro+ available from bad-elf.com
- 1 Bluetooth Serial adapter **AF** 1588 **SF** WRL-12580 or WRL-12576 or use a wired USB-Serial adapter **SF** DEV-09716 or DEV-14050, **AF** 3309 or 284, **SS** 317990026
- 1 Bluetooth-enabled personal computer If your computer doesn't have a Bluetooth radio, use a USB Bluetooth adapter: **AF** 1327, **RS** 807-7742, **SS** 113990026
- 1 battery power supply check the specifications of your GPS module for power requirements.

PROJECT 19: DETERMINING HEADING USING A DIGITAL COMPASS

- 1 solderless breadboard **D** 438-1045-ND, **J** 20723 or 20601, **SF** PRT-12615 or PRT-12002, **F** 4692810, **AF** 64, **SS** 319030002 or 319030001
- 1 microcontroller module This project will run on just about any Arduino-compatible board.
  MKR1000: **AF** 3156, **RS** 124-0657, **A** ABX00004, GBX00011 (Africa and EU), **D** 1659-1005-ND
  Arduino 101: **D** 1660-1003-ND, **J** 2239331, **SF** DEV-13787, **AF** 3033, **F** 2520713, **RS** 913-9999, **SS** 114990575, **A** ABX00005, GBX00005 (Africa and EU)
  Uno: **D** 1050-1024-ND, **J** 2151486, **SF** DEV-11021, **A** A000099, **AF** 50, **F** 1848687, **RS** 715-4081, **SS** ARD132D2P
- 1 ST Microelectronics model LSM303DLH **AF** 1120, **SF** BOB-13303, **P** 1250, **SS** 101020081
- 4 male header pins To connect the sensor to your breadboard.**D** A26509-20-ND, **J** 103377, **SF** PRT-00116, **F** 1593411
- 1 personal computer
- All necessary converters to communicate serially from microcontroller to computer Just like the previous projects.

PROJECT 20: DETERMINING ATTITUDE

- 1 solderless breadboard **D** 438-1045-ND, **J** 20723 or 20601, **SF** PRT-12615 or PRT-12002, **F** 4692810, **AF** 64, **SS** 319030002 or 319030001
- 1 microcontroller module This project will run on just about any Arduino-compatible board.
  MKR1000: **AF** 3156, **RS** 124-0657, **A** ABX00004, GBX00011 (Africa and EU), **D** 1659-1005-ND
  Arduino 101: **D** 1660-1003-ND, **J** 2239331, **SF** DEV-13787, **AF** 3033, **F** 2520713, **RS** 913-9999, **SS** 114990575, **A** ABX00005, GBX00005 (Africa and EU)
  Uno: **D** 1050-1024-ND, **J** 2151486, **SF** DEV-11021, **A** A000099, **AF** 50, **F** 1848687, **RS** 715-4081, **SS** ARD132D2P
- 1 accelerometer/gyrometer Examples are shown for the ST Microelectronics LSM303 and L3Gxx. If you have an Arduino101, the sensor is built into the board. **AF** 1120, **SF** BOB-13303, **P** 1250, **SS** 101020081
- 6-8 Male header pins **JD** A26509-20-ND, **J** 103377, **SF** PRT-00116, **F** 1593411
- 1 personal computer
- All necessary converters to communicate serially from microcontroller to computer Just like the previous projects.

Network Location and Physical Location

Locating things is one of the most common tasks people want to achieve with sensor systems. Once you understand the wide range of things that sensors can detect, it's natural to get excited about the freedom this affords. All of a sudden, you don't have to be confined to a chair to interact with computers. You're free to dance, run, jump—and still have a computer read your action and respond in some way.

The downside of this freedom is the perception that in a networked world, you can be located by others wherever you are. Ubiquitous surveillance cameras and systems like Wireless E911 (which locates mobile phones on a network) make it seem as though anyone or anything can be located anywhere and at any time—whether they want to be located or not. The reality of location technologies lies somewhere in between these extremes.

Locating things on a network is different than locating things in physical space. As soon as a device is connected to a network, you can get a general idea of its network location using a variety of means—from address lookup to measuring its signal strength—but that doesn't mean that you know its physical location. You just know its relationship to other nodes of the network. You might know that a cell phone is closest to a given cell transmitter tower, or that a computer is connected to a particular WiFi access point. You can use that information along with other data to form a picture of the person using the device. If you know that the cell transmitter tower is less than a mile from you, you'd know that the person with the cell phone is going to reach you soon, and you can act appropriately in response. For many network applications, you don't need to know physical location as much as you need to know relationship to other nodes in the network.

Step 1: Ask a Person

People are really good at locating things. At the physical level, we have a variety of senses to throw at the problem, as well as a brain that's wonderful at matching patterns of shapes and determining distances from different sensory clues. At the behavioral level, we've got thousands of patterns that make it easier to determine why you might be looking for something. Computer systems don't have these same advantages, so when you're designing an interactive system to locate things or people, the best tool you have to work with—and the first one you should consider—is the person for whom you're making your system.

Getting a good location starts with cultural and behavioral cues. If you want to know where you are, ask another person near you. In an instant, she's going to sum up all kinds of things—your appearance, your behavior, the setting you're both in, the things you're carrying, and more—in order to give you a reasonably accurate and contextually relevant answer. No amount of technology can do that, because the connection between where you are and why you want to know is seldom explicit in the question. As a result, the best thing you can do when you're designing a locating system is to harness the connection-making talents of the person who will be using that system. Providing him with cues as to where to position himself when he should take action, and what actions he can take, helps eliminate the need for a lot of technology. Asking him to tell your system where things are, or to position them so that the system can easily find them, makes for a more effective system.

For example, imagine you're making an interactive space that responds to the movements of its viewers. This is popular among interactive artists, who often begin by imagining a "body-as-cursor" project, in which the viewer is imagined as a body moving around in the space of the gallery. Some sort of tracking system is needed to determine his position and report it back in two dimensions, like the position of a cursor on a computer screen.

What's missing here is the reason why the viewer might be moving in the first place. If you start by defining what the viewer's doing, and give him cues as to what you expect him to do at each step, you can narrow down the space in which you need to track him. Perhaps you only need to know when he's approaching one of several sculptures in the space so that you can trigger the sculpture to move in response. If you think of the sculptures as nodes in a network, the task gets easier. Instead of tracking the viewer in an undefined two-dimensional space, now all you have to do is determine his proximity to one of several points in the room. Instead of building a tracking system, you can now just place a proximity sensor near each object, look up which he's near, and read how near he is to it. You're using a combination of spatial organization and technology to simplify the task. You can make your job even easier by giving him visual, auditory, and behavioral cues to interact appropriately. He's no longer passive; he's now an active participant in the work.

Consider a different example: let's say you're designing a mobile phone city-guide application for tourists that relies on knowing the phone's position relative to nearby cell towers to determine the user's position. What do you do when you can't get a reliable signal from the cell towers? Perhaps you ask the tourist to input her current address, or the postal code she's in, or some other nearby cue. Then, your program can combine that data with the location based on the last reliable signal it received, and determine a better result. In these cases, and in all location-based systems, it's important to incorporate human talents in the system to make it better.

Step 2: Know the Environment

Before you can determine where you are, you need to determine your environment. For any location, there are several ways to describe it. For example, you could describe a street corner in terms of its address, its latitude and longitude, its postal code, or the businesses nearby. Which of these coordinates you choose depends in part on the technology you have on hand to determine it. If you're making a mobile city guide, you might use several different ones—the nearest cell transmitter ID, the street address, and the nearby businesses could all work to define the location. In this case, as in many, your job in designing the system is to figure out how to relate one system of coordinates to another in order to give some meaningful information.

Mapping places to coordinate systems is a lot of work, so most map databases are incomplete. *Geocoding* allows you to look up the latitude and longitude of most any U.S. street address. It doesn't work everywhere in the U.S., and it doesn't work in places outside the U.S. unless the data has been gathered and put in the public domain. Geocoding depends on having an accurate database of names mapped to locations. If you don't agree on the names, you're out of luck. The Virtual Terrain Project (www.vterrain.org) has a good list of geocoding resources for the U.S. and international locations at www.vterrain.org/Culture/geocoding.html. There's a free geocoder at geocoder.opencagedata.com that uses data from OpenStreetmaps, and Google's maps API provides geocoding at developers.google.com/maps.

Street addresses are the most common coordinates that are mapped to latitude and longitude, but there are other systems for which it would be useful to have physical coordinates as well. For example, mobile phone cell transmitters all have physical locations. It would be handy to have a database of physical coordinates for those towers. However, cell towers are privately owned by mobile telephone carriers, so detailed data about the tower locations is proprietary, and the data is not in the public domain. Projects such as OpenCellID (www.opencellid.org) attempt to map cell towers by using GPS-equipped mobile phones running custom software.

IP addresses don't map exactly to physical addresses because computers can move. Nevertheless, there are several geocoding databases for IP addresses. These work on the assumption that routers don't move a lot, so if you know the physical location of a router, then the devices gaining access to the internet through that router can't be too far away. The accuracy of IP geocoding is limited, but it can help you determine a general area of the world, and sometimes even a neighborhood or city block, where a device on the internet is located. Of course, IP lookup doesn't work on private IP addresses. In the next chapter, you'll see an example that combines network identity and geocoding.

You can develop your own database relating physical locations to cultural or network locations if the amount of information you need is small, or if you have a large group of people to do the job. It's better to rely on existing infrastructures when you can, though.

Step 3: Acquire and Refine

Once you know where you're going to look, there are two tasks that you have to do continually: acquire a new position, and refine the position's accuracy. *Acquisition* gives a rough position; it usually starts by identifying which device on a network is the center of activity. In the interactive installation example described earlier, you could acquire a new position by determining that the viewer tripped a sensor near one of the objects in the room. Once you know roughly where he is, you can refine the position by measuring his distance with the proximity sensor attached to the object.

Refining doesn't have to mean getting a more accurate physical position. Sometimes you refine what you know about the context or activity, not the position. When you have a rough idea of where something's happening, you need to know about the activity at that location in order to provide a response. In the interactive installation example, you may never need to know the viewer's physical coordinates in feet and inches (or meters and centimeters). When you know which object he's close to in the room—

and whether he's close enough to relate to it—you can make that object respond. You might be changing the graphics on a display when he walks near, or activating an animatronic sculpture as he walks by. In both cases, you don't need to know the precise distance; you just need to know he's close enough to pay attention. In cases like this, sometimes distance-ranging sensors are used as motion detectors to define general zones of activity rather than to measure distance.

Determining proximity doesn't always give you enough information to take action. Refining can also involve determining the orientation of one object relative to another. For example, if you're giving directions, you need to know which way you're oriented. It's also valuable information when two people or objects are close to each other. You don't want to activate the animatronic sculpture if the viewer has his back to the thing!

X

35 Ways to Find Your Location

At the 2004 O'Reilly Emerging Technology Conference (ETech), interaction designer and writer Chris Heathcote gave an excellent presentation on cultural and technological solutions to finding things, entitled 35 Ways to Find Your Location. He outlined a number of important factors to keep in mind before you choose tools to do the job. He pointed out that the best way to locate someone or something involves a combination of technological methods and interpretation of cultural and behavioral cues. His list is a handy tool for inspiring solutions when you need to develop a system to find locations. A few of the more popular techniques that Chris listed are:

- Assume: the Earth. Or a smaller domain, but assume that's the largest space in which you have to look.
- Use the time.
- Ask someone.
- Association: who or what are you near?
- Proximity to phone boxes, public transport stops, and utility markings.
- Use a map.
- Which cell phone operators are available?
- Public phone operators?
- Phone number syntax?
- Newspapers available?
- Language being spoken?
- Post codes/ZIP codes.
- Street names.
- Street corners/intersections.
- Street numbers.
- Business names.
- Mobile phone location, through triangulation or trilateration.
- Triangulation and trilateration on other radio infrastructures, such as TV, radio, and public WiFi.
- GPS, assisted GPS, WAAS, and other GPS enhancements.
- Landmarks and "littlemarks."
- Dead reckoning.

Determining Distance

Electronic locating systems—like GPS, mobile phone location, and sonar—seem magical at first because there's no visible evidence as to how they work. However, when you break the job down into its components, electronic locating becomes relatively straightforward. Most physical location systems are based on one of two methods: measuring the time of a signal's travel from a known location, or measuring the change in its signal strength. Both methods can combine measurements from multiple sources to determine a position in two or three dimensions using trilateration.

Distance ranging techniques use the difference in the strength of a radio, light, or sound signal to determine distance. Sound, light, and radio all decay at known rates over distance. This means that if you compare the signal strength at the source and the strength at a distance, you can calculate the distance. Likewise, the speed of light and of sound are known, so a signal's time delay correlates to the distance between sender and receiver as well.

For example, a GPS receiver determines its position on the surface of the planet by measuring the time delay of received radio signals from several geosynchronous satellites. Mobile phone location systems use the received signal strength from nearby cell towers to determine the phone's distance from the tower, then determine position using trilateration from several towers. Commercial location systems like Skyhook (www.skyhookwireless.com) use several different systems (WiFi, GPS, and cell tower location) to refine their positional accuracy.

Distance ranging can be classified as *active* or *passive*. In active systems, the target has a radio, light, or acoustic source on it and a source of energy such as a battery. The receiver listens for the signal generated directly by the target. By comparing the received signal strength to the sending device's known transmitting signal strength, you can determine the distance. For example, mobile phone location is active because it relies on a two-way transmission between the phone and the cell tower. GPS is also active. In active ranging, there's a powered device on either end of the ranging.

In passive systems, the target doesn't need to generate a signal. The receiver emits a signal of a known strength, then listens for the signal reflected back from the target. The difference in signal strength between the original signal and the reflected one allows you to calculate the distance. The infrared and ultrasonic distance rangers you'll see shortly are examples of this.

Sonar ranging experts sometimes use these terms differently, and this can be confusing. *Active sonar* refers to sending out a sonar signal of a known strength and listening for its reflection off a target. *Passive sonar* refers to listening for sounds generated by a natural object, like a whale, or an undersea landslide. However, passive sonar cannot determine distance unless the loudness of the sound at the source is known. Passive sonar can determine the presence of an object that makes sound, but not its distance from the microphone, unless the loudness at the source is known.

Sometimes ranging is used for acquiring a position; other times, it's used for refining it. In the following examples, the passive distance rangers deliver a measurement of physical distance.

Passive Distance Ranging

Ultrasonic rangers like the SRF04, and infrared rangers like the Sharp GP2Y0A21YK, shown in Figure 8-2, are examples of *distance rangers*. The ultrasonic sensor sends out an ultrasonic signal and listens for an echo. The Sharp sensor sends out an infrared light beam, and senses the reflection of that beam. These sensors work in a short range only. The Sharp sensor can read about 10cm to 80cm, and the ultrasonic sensor reads from about 0 to 3–4m. Passive sensors like these are handy when you want to measure the distance of a person in a limited space, and you don't want to put any hardware on the person. They're also useful when you're building moving objects that need to know their proximity to other objects in the same space as they move.

X

Infrared Distance Ranger

The Sharp GP2Y*x* series of infrared-ranging sensors give a decent measurement of short-range distance by bouncing an infrared light signal off the target, and then measuring the returned brightness. Figure 8-2 shows a circuit for a Sharp GP2Y0A21 IR ranger, which can detect an object in front of it within about 10cm to 80cm.

The Sharp sensors' outputs are not linear, so if you want to get a linear range, you need to make a graph of the voltage over distance, and do some math. The sensor's datasheet at www.sharp-world.com/products/device/lineup/data/pdf/datasheet/gp2y0a21yk_e.pdf includes a graph of voltage over the inverse of the distance. It shows a pretty linear relationship between the two from about 10 to 80cm, and the slope of that line is about 21.7V*cm. The sensor's response time is given as 39ms. This is dependent on your actual voltage input to the sensor, however.

MATERIALS

- **» 1 Arduino module**
  - **» Features used: analog input at +5V**
- **» 1 Sharp GP2Y0A21K IR ranger**
- **» 1 10µF capacitor**
- **» 3 male header pins**

These sensors modulate the IR light on a carrier wave, and filter out all but that carrier wave when they receive the reflected signal, so they're pretty good at filtering out other IR sources, but not perfect. They work well indoors, but in direct sunlight, you might have problems.

Although this sensor's rated operation range is 4.5–5.5V, it will operate on lower voltages. It will output a voltage that's dependent on distance when powered by the 3.3V output of an MKR1000 or 101, for example. However, the distance calculation for the sensor will not hold in that case, and you'll need to work our your own formula by measuring the distance to your target and recording the voltage at that distance, for multiple distances.

This sketch reads the sensor and converts the results to a voltage. Then, it uses the result explained earlier to convert the voltage to a distance measured in centimeters.

The conversion formula gives only an approximation. For many applications, though, you don't need the absolute distance, but the relative distance. Is the person nearer or farther away? Has she passed a threshold that should trigger some other interaction? For such applications, you won't need this conversion. You can just use the output of the `analogRead()` command and choose a value for your threshold by experimentation.

```
/*
  Sharp GP2xx IR ranger reader
  Context: Arduino
*/
void setup() {
  Serial.begin(9600); // initialize serial communications
}

void loop() {
  int sensorValue = analogRead(A0); // read the sensor value
  // convert to a voltage:
  float voltage = sensorValue * (5.0 / 1024.0);
  // calculate the distance:
  float distance = 21.7 / voltage;
  Serial.print(voltage);
  Serial.print(" V\t");
  Serial.print(distance);
  Serial.println(" cm");
  // wait 39 milliseconds before the next reading
  delay(39);
}
```

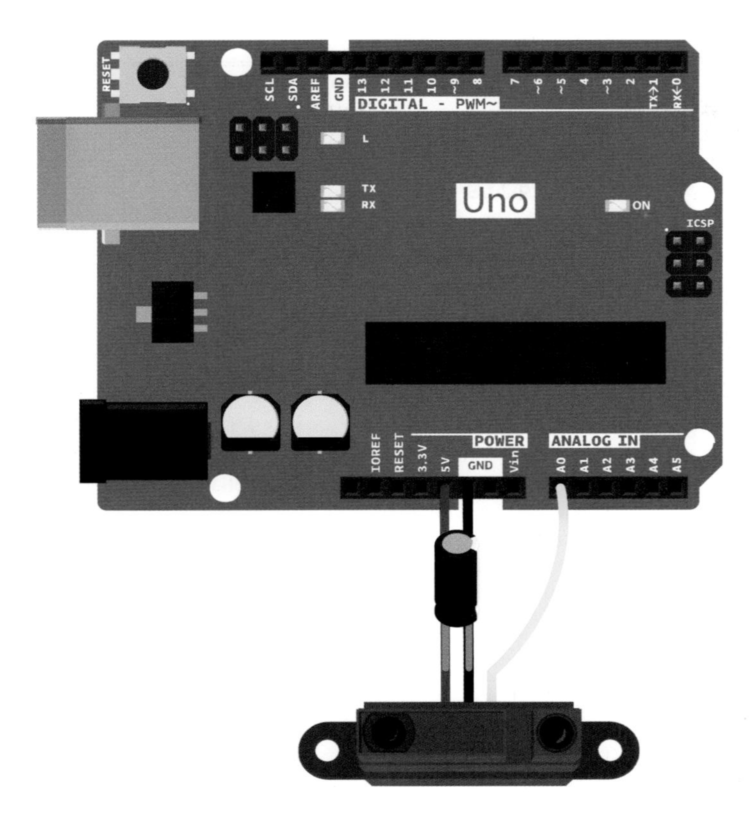

Figure 8-2
The Sharp GP2Y0A21YK IR ranger attached to a microcontroller. The capacitor attached to the body of the sensor reduces fluctuations due to the sensor's current load.

> For accurate distance sensing using the datasheet's formula, the sensor and microcontroller's input must operate at 5V. The sensor will operate on 3.3V, however, as you saw in the last chapter.

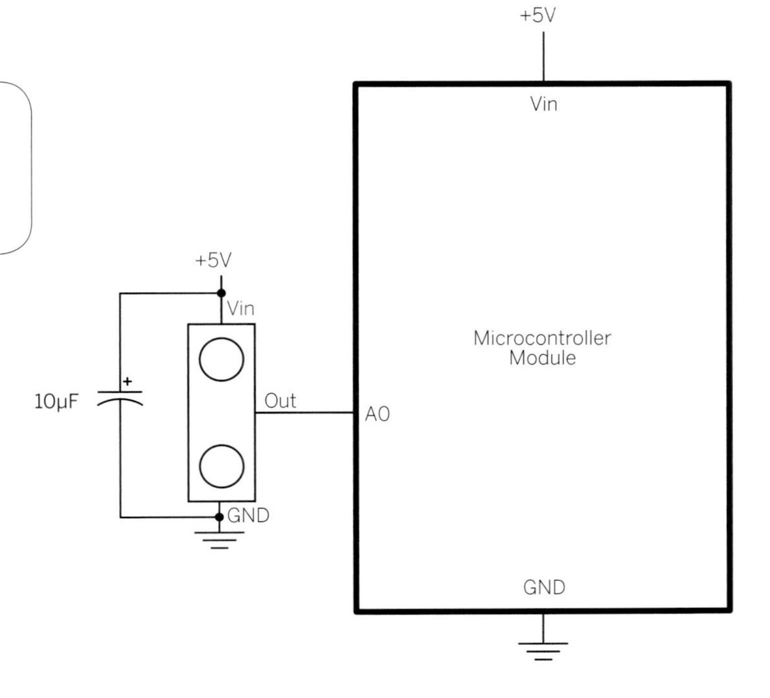

Project 16

Ultrasonic Distance Ranger

Ultrasonic sensors measure distance using a similar method to the Sharp sensors, but theirs have a greater sensing range. Instead of infrared, they send out an ultrasonic signal and wait for the echo. Then they measure the distance based on the time required for the echo to return. There are a variety of ultrasonic sensor modules, most of them based on ProWave's ceramic transducers (www.prowave.com.tw). The earliest modules available on the hobbyist market were from Devantech (www.robot-electronics.co.uk), but now you can find them from nearly every retailer thanks to Devantech's open designs.

MATERIALS

- **1 Arduino module**
  - **Features used: Digital out, Digital in**
- **1 ultrasonic ranger, SRF04, HC-SR04, or equivalent**
- **4 male header pins or 4 jumper wires**

There are many variations on the Devantech SRF04 sensor's open design, and they all share the same interface. They have a trigger pin and an echo pin. When you pulse the trigger pin high for 10 microseconds, it causes one of the sonar transducers to send out an ultrasonic signal. The other transducer listens for the echo. When it's received, the echo pin pulses high. The pulsewidth of the echo pin tells you the distance: the pulsewidth divided by 58 gives the distance in centimeters, and divided by 148 gives it in inches.

Distance rangers are great for measuring linear distance, but they have a limited conical field of sensitivity, so they're not great for determining location over a large two-dimensional area. The SRF04 sensor, for example, has a cone-shaped field of sensitivity that's about

Figure 8-3
HC-SR04 ultrasonic sensor connected to an Arduino module.

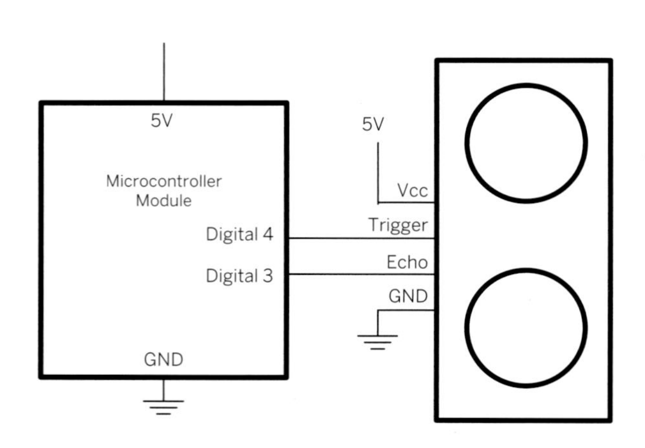

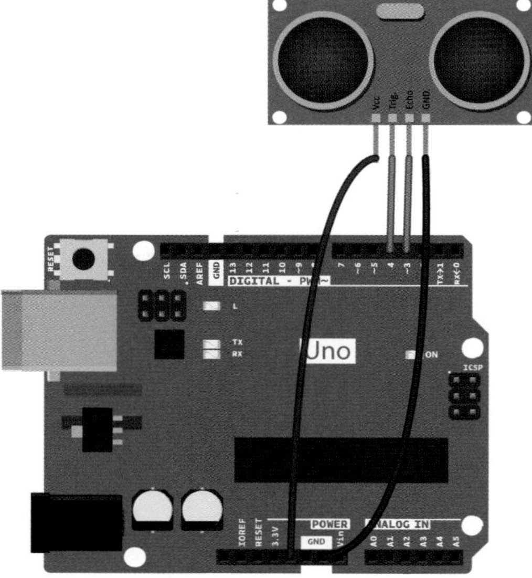

70-degrees wide (though the sensitivity drops off at the edges) and 3–4 meters long. In order to use it to cover a room, you'd need to use several of them and arrange them creatively. Figure 8-4 shows one way to cover a space using five of the rangers. In this case, you'd need to make sure that none of the sensors were operating at the same instant, because their signals would interfere with each other. The sensors would have to be activated one after another in sequence. Because each one takes up to 50 milliseconds to return a result, you'd need up to 250 milliseconds to make a complete scan of the space.

These sensor's I/O pins work with the 3.3V logic I/O pins of a 101 or MKR1000, but the sensor requires a 5V power supply. It will operate from the 5V supplied by the USB connection for your microcontroller.

This reads the sensor and converts the results to a pulsewidth, then converts that to a distance measured in centimeters. The conversion formula again gives only an approximation.

```
/*
  Ultrasonic sensor reader
  context: Arduino
*/
const int triggerPin = 4;
const int echoPin = 3;

void setup() {
  // set pin states and initialize serial communications:
  pinMode(triggerPin, OUTPUT);
  pinMode(echoPin, INPUT);
  Serial.begin(9600);
}

void loop() {
  digitalWrite(triggerPin, HIGH);  // pulse the trigger pin HIGH
  delayMicroseconds(10);           // for 10 microseconds
  digitalWrite(triggerPin, LOW);  // to start the sensor

  // measure a pulse on the echo pin in microseconds:
  long pulsewidth = pulseIn(echoPin, HIGH);
  float distance = pulsewidth / 58.0;  // cm = microseconds/58
  Serial.println(distance);
}
```

Figure 8-4
Measuring distance in two dimensions using ultrasonic distance rangers. The square in each drawing is a 4m × 4m floor plan of a room, approximately. In order to cover the whole of a rectangular space, you need several sensors placed around the sides of the room.

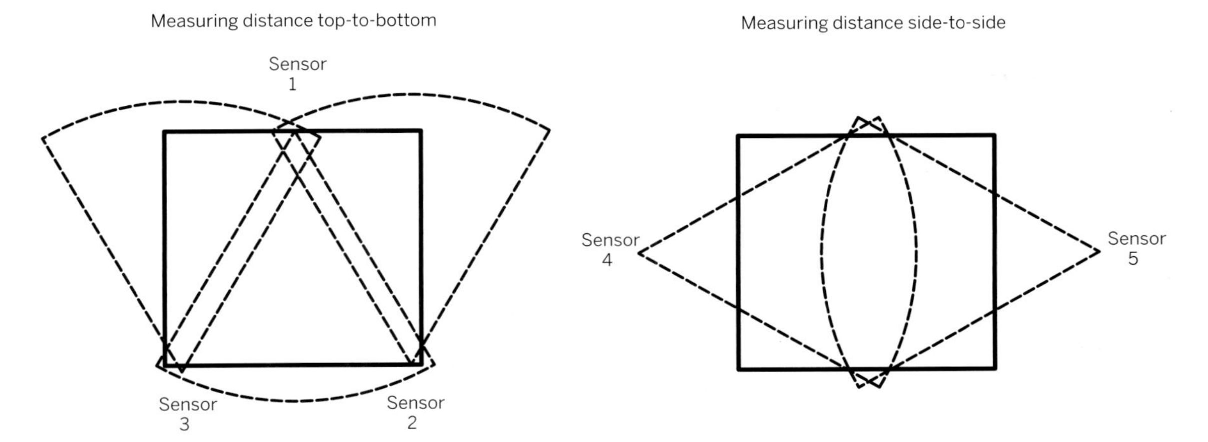

Active Distance Ranging

The ultrasonic and infrared rangers in the preceding sections are passive distance-sensing systems. Mobile phones and the global positioning system (GPS) measure longer distances by using ranging as well. These systems include a radio beacon (the cell tower or GPS satellite) and a radio receiver (the phone or GPS receiver). The receiver determines its distance from the beacon based on the received signal from the beacon. These systems can measure much greater distances on an urban or global scale. The disadvantage of active distance ranging is that you must have a powered device at both ends. You can't measure a person's distance from somewhere using active distance ranging unless you attach a receiver to the person.

Radio receivers don't actually give you the distance from their radio beacons, just the relative signal strength of the radio signal. In order to relate this to distance, you need to be able to calculate that distance as a function of signal strength. The main function of a GPS receiver is to calculate distances to the GPS satellites based on signal strength, timing, and known positions of the satellites, and then determine the receiver's position using that information. WiFi, Bluetooth, 802.15.4, and other short range protocols give you received signal strength, but they don't do the distance calculations for you.

In many applications, though, you don't need to know the distance—you just need to know how relatively near or far one person or object is to another. For example, if you're making a pet door lock that opens in response to the pet, you could imagine a Bluetooth beacon on the pet's collar and a receiver on the door lock. When the signal strength from the pet's collar is strong enough, the door lock opens. In this case, and in others like it, there's no need to know the actual distance. You would only know the pet's identity if you checked the MAC address of the collar radio, though. For more on identity, see the next chapter.

Figure 8-5
Active vs. passive distance ranging

Response signal generated by mobile unit (e.g., cell phone).
Base unit (sensor) sends out signal, reads reflection from mobile object or person
Initial signal generated by base unit (e.g., cell tower)
Active distance ranging
Passive distance ranging

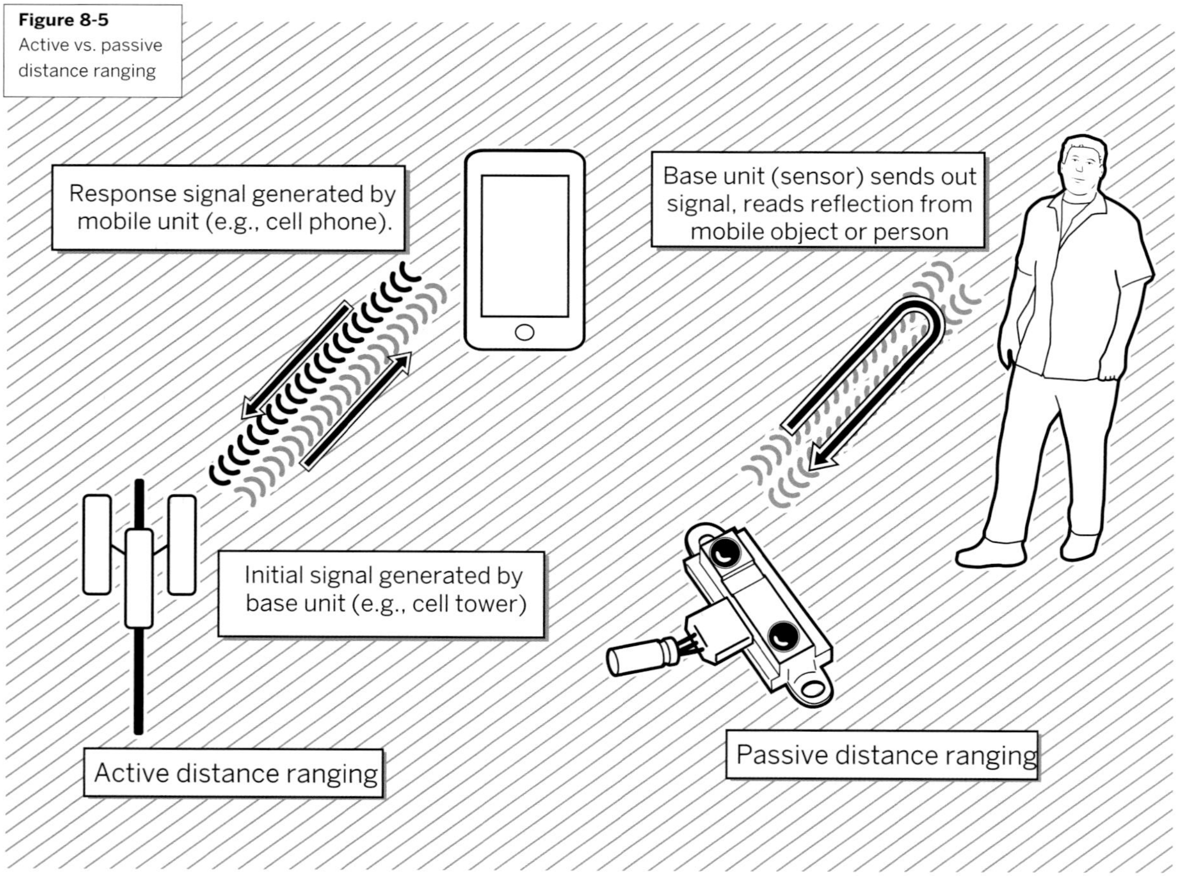

Project 17

Reading Received Signal Strength

In the WiFi projects in Chapter 4, the LoRa radio project and the Bluetooth LE project in Chapters 6 and the Digi 802.15.4 project in Chapter 7, you saw a property of those radios called received signal strength indicator, or RSSI. RSSI tells you how strong the last received radio signal was. With no obstructions, received radio signal strength is inversely proportional to the square of the distance from the transmitter. So if you know the received signal strength, you can approximate the distance between sender and receiver.

Using RSSI to estimate distance is not exact. RSSI is a measure of the strength of the last message received. For one thing, it's not measured in real time. The last message might be quite old, and the receiver could have moved in the meantime. Furthermore, the received radio signal may not have traveled directly from the sender, but might have bounced off a reflecting surface instead. Although you can't really use RSSI to measure a precise distance, you can use it to tell if the receiver is nearer or further from the sender.

Radio signal strength is measured in decibel-milliwatts (dBm). You might wonder why the signal reads -65dBm. How can the signal strength be negative? The relationship between milliwatts of power and dBm is logarithmic. To get the dBm, take the log of the milliwatts. So, for example, if you receive 1 milliwatt of signal strength, you've got log 1 dBm. Log 1 = 0, so 1 mW = 0 dBm. When the power drops below 1 mW, the dBm drops below 0, like so: 0.5 mW = (log 0.0005) dBm or -3.01 dBm. 0.25mW = (log 0.00025) dBm, or -6.02 dBm.

If logarithms confuse you, just remember that 0 dBm is the maximum transmission power, which means that signal strength is going to start at 0 dBm and go down from there. Power is decreased by half for every 3 dBM drop. The more sensitive a receiver is, the longer the range of signals it can detect. In a perfect world, with no obstructions to create errors, the relationship between signal strength and distance would be a logarithmic curve.

All of the radios you've used so far have an interface that lets you read the RSSI. In the WiFi status sketch in Chapter 4, you read the signal strength on the MKR1000 or ESP8266 like so:

```
long rssi = WiFi.RSSI();
```

For the LoRa radios in Chapter 6, you read the RSSI on the microcontroller like so:

```
int rssi = LoRa.packetRssi();
```

In the Bluetooth LE project, you read the signal strength of the message from the peripheral in node.js like so:

```
console.log('signal strength: ' + peripheral.rssi);
```

You can get Bluetooth RSSI without pairing, because the radios use the advertising packets to determine RSSI. In the Digi 802.15.4 project in Chapter 7, you parsed the signal strength out of the Digi API messages with this line of code:

```
message.rssi = -thisPacket[5];
```

You could use the signal strength from any of these devices to make a rough proximity measurement between sender and receiver.

In this project you'll make a node.js server that accepts a report of signal strength from any client and displays the latest report via a p5.js client sketch. The server accepts data from a client reporting an RSSI via POST request like so:

```
POST /rssi/value
```

where `value` is the signal strength. The server reports the rssi and proximity when a client requests it via GET request like so:

```
GET /rssi
```

Serve It Start by making a new node.js project with its own directory. Save the following file there as **server.js**. Install express as usual from the command line like so:

```
$ npm install express
```

You'll need a global variable, `rssi`, to keep track of the latest reported RSSI from any client.

```
/*
  RSSI server
  context: node.js
*/
var express = require('express'); // include the express library
var server = express();            // create a server using express
var rssi = -100;                   // the latest received signal strength
```

Write two response functions, `getRssi()` and `postRssi()`. They'll be the callbacks for two listeners which you'll write at the end of the script. The first, `postRssi()`, responds to a POST request that has the RSSI value appended at the end of the request URL. The second, `getRssi()`, takes the latest reported RSSI and generates a narrative description of the receiver's proximity to the sender based on the signal strength.

```
function postRssi(request, response) {
  rssi = request.params.rssi;    // get the signal strength from the request
  response.send("rssi received: " + rssi);  // respond to the client
  response.end();               // close the connection
}

function getRssi(request, response) {
  // make a string to send to the client:
  // auto-refresh meta tag:
  var message = "<meta http-equiv=\"refresh\" content=\"3\">\n";
  var message = "Last RSSI received: " + rssi + " dBm<br>";
  message += "The receiver is ";
  if (rssi <= -60 ) {          // weakest signal strength
    message += "far away ";
  }
  if (rssi > -60 && rssi <=-40 ) {// a little stronger
    message += "within a few meters ";
  }
  if (rssi > -40 && rssi <=-20 ) {
    message += "a few steps away ";
  }
  if (rssi > -20 ) {           // really strong signal
    message += "arm's reach";
  }
  message += " from the sender."  // add an end to the message
  response.send(message);         // send message to the client
  response.end();                 // close the connection
}
```

Finally, start the server and establish the GET and POST listeners. That's the whole script.

```
// start the server:
server.listen(8080);

// define what to do when the client requests something:
server.get('/rssi', getRssi);          // GET /rssi
server.post('/rssi/:rssi', postRssi);  // POST /rssi/value
```

This server doesn't care what kind of RSSI you're reporting. You could use it with any of the radio projects you've seen previously. Furthermore, the client making the GET request doesn't have to be the client that's reporting the signal strength. You may need to adjust thresholds based on your device's sensitivity, though.

The sketch that follows will make the HTTP POST request from an ESP8266 or a MKR1000 client, reporting the microcontroller's RSSI from its WiFi access point. No external hardware is needed for this sketch. Figure 8-6 shows the output in a browser.

X

Report It Add the **config.h** file containing the username and password variables as you did with the examples in previous chapters, and change the server variable to match your node.js server's IP address. The `setup()` function connects to the WiFi router as usual.

```
/*
  RSSI HTTP Client
  Context: Arduino, with WINC1500 module
*/
// include required libraries and config files
#include <SPI.h>
#include <WiFi101.h>
//#include <ESP8266WiFi.h>     // use this instead for ESP8266 modules
#include <ArduinoHttpClient.h>
#include "config.h"

WiFiClient netSocket;                  // network socket to server
const char server[] = "192.168.0.8"; // server address; fill in yours
String route = "/rssi/";               // API route

void setup() {
  Serial.begin(9600);                 // initialize serial communication
  // while you're not connected to a WiFi AP,
  while ( WiFi.status() != WL_CONNECTED) {
    Serial.print("Attempting to connect to Network named: ");
    Serial.println(ssid);           // print the network name (SSID)
    WiFi.begin(ssid, password);     // try to connect
    delay(2000);
  }
  Serial.print("Connected to: "); // Now that you're connected,
  Serial.println(ssid);          // print out the network name
}
```

In the loop, you're going to make the HTTP POST request every three seconds. That's the whole sketch.

Run the server sketch on the previous page, and upload this sketch to your board. Then open a browser and open http://localhost:8080/rssi to see the results of the GET request. Then move the microcontroller around your home while watching the result a browser. You should see the RSSI report change as you move. This is your microcontroller's signal strength relative to your router.

```
void loop() {
  HttpClient http(netSocket, server, 8080); // make an HTTP client
  int rssi = WiFi.RSSI();                   // get the RSSI
  http.post(route + rssi);                  // make a POST request

  while (http.connected()) {                // while connected,
    if (http.available()) {                 // if there is a response,
      String result = http.readString();    // read it
      Serial.println(result);               // and print it
    }
  }
  // when there's nothing left to the response,
  http.stop();                              // close the request
  delay(3000);                              // wait 3 seconds
}
```

The Multipath Effect

When you moved your microcontroller around your space, you probably saw that you didn't always get RSSI values that would correspond precisely to measurable distances. The narrative descriptions were accurate, however, because they told you approximately where your device was.

The biggest source of error in RSSI distance ranging is what's called the *multipath* effect (see Figure 8-7). When electromagnetic waves radiate, they bounce off things. If there's a wall between your microcontroller and your router, for example, the signal may be bouncing off the wall rather than coming directly from your router. This is more likely if your walls have galvanized metal studs or aluminum-backed insulation, which can block some radio signals.

Similarly, if you're outdoors with a mobile phone and you're seeing wildly changing signal strength, the problem may be because of architecture. Your phone may receive multiple signals from a nearby cell tower if, for example, you're positioned near a large obstacle, such as a building. The reflected waves off the building create "phantom" signals that look as real to the receiver as the original signal. This issue makes it impossible for the receiver to calculate the distance from the beacon accurately, that causes degradation in the signal quality of mobile phone reception, as well as errors in locating the phones. For GPS receivers, multipath results in a much wider range of possible locations, as the error means that you can't calculate the position as accurately. It is possible to filter for the reflected signals, but not all radios incorporate such filtering.

X

Figure 8-6
Output of the RSSI server script. The server you just wrote doesn't have to report RSSI values from a WiFi device. You could write a client that reports RSSI from a LoRa radio, or a Bluetooth LE radio, or any radio that returns its received signal strength. As long as the client can make the HTTP POST call, it will work.

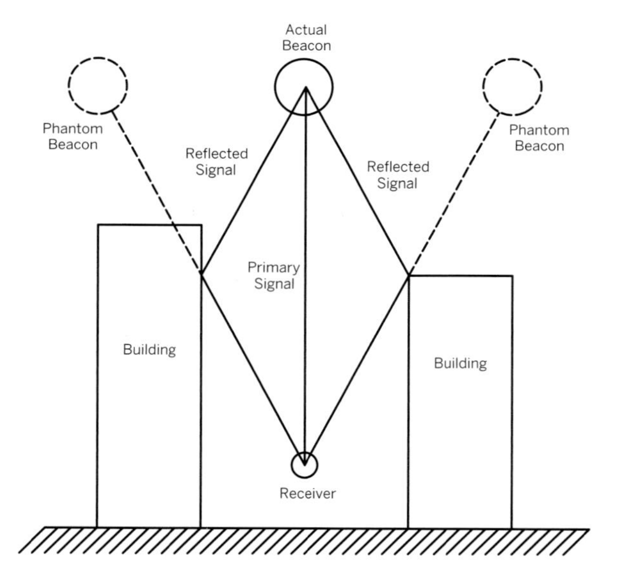

Figure 8-7
The multipath effect. Reflected radio waves create phantom beacons that the receiver can't tell from the real beacon, causing errors in calculating the distance based on signal strength.

Determining Position Through Trilateration

Distance ranging tells you how far away an object is from your measuring point in one dimension, but it doesn't define the whole position. The distance between your position and the target object determines a circle around your position (or a sphere, if you're measuring in three dimensions). Your object could be anywhere on that circle.

In order to locate an object within a two- or three-dimensional space, though, you need to know more than distance. The most common way to do this is by measuring the distance from at least three points. This method is called *trilateration*. If you measure the object's distance from two points, you get two possible places it could be on a plane, as shown in the middle of Figure 8-8. When you add a third circle, you have one distinct point on the plane where your object could be. A similar method, *triangulation*, uses two known points and calculates the position using the distance between these points; it then uses the angles of the triangle formed by those points and the position you want to know.

The global positioning system uses trilateration to determine an object's position. GPS uses a network of satellites circling the globe. The position of each satellite can be determined from its flight path and the current time. Each one is broadcasting its clock signal, and GPS receivers pick up that broadcast. When a receiver has signals from at least three satellites, it can determine a rough position using the time difference between transmission and reception. Most receivers use at least six satellite signals to calculate their position, in order to correct any errors. Mobile phone location systems like Wireless E911 calculate a phone's approximate position in a similar fashion, by measuring the distance from multiple cell towers based on the *time difference of arrival* (TDOA) of signals from those towers.

Figure 8-8
Trilateration on a two-dimensional plane. Knowing the distance from one point defines a circle of possible locations. Knowing the distance from two points narrows it to two possible points on the plane. Knowing the distance from three points determines a single point on the plane.

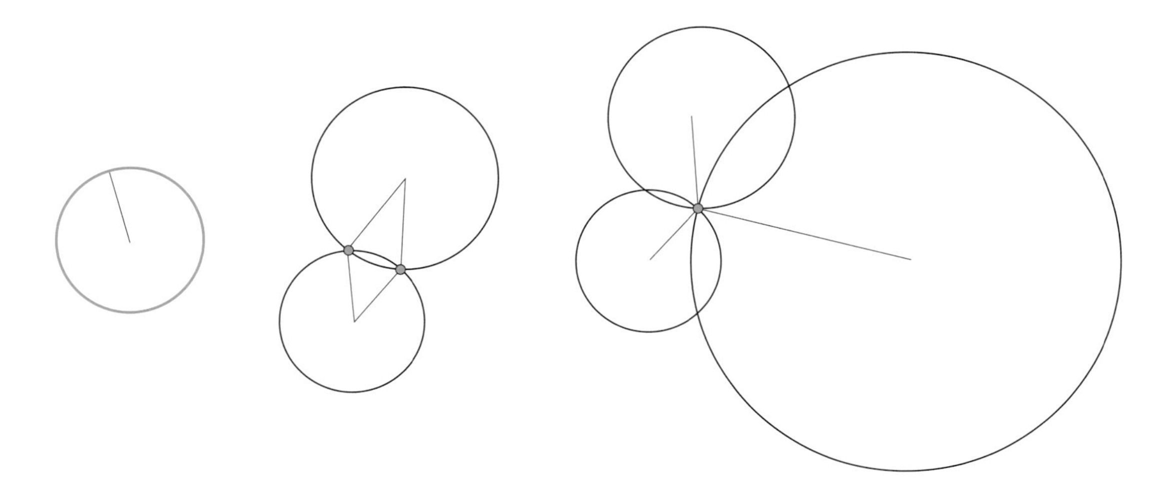

 **Project 18**

Geolocation Services and the NMEA Protocol

If you're using GPS, you never have to do trilateration or triangulation calculations—GPS receivers do the work for you. They then give you the position in terms of latitude and longitude. There are several data protocols for GPS receivers, but the most common is the NMEA 0183 protocol established by the National Marine Electronics Association (NMEA) in the United States. Just about all receivers on the market output NMEA 0183, and usually other protocols as well.

MATERIALS

- **1 GPS receiver; see text for recommendations**
- **1 Bluefruit (depending on GPS unit)**
- **1 battery power supply for Bluetooth and Bluefruit (depending on GPS unit)**

NMEA 0183 is an asynchronous serial protocol. Most receivers send this data using either RS-232 or TTL serial levels. The standard rate is 4800bps, but many devices send at other data rates. It's just one of many protocols used in mapping GPS data. The great thing about NMEA, though, is that it's nearly ubiquitous among GPS receivers. No matter what other protocols they use, they all tend to have NMEA as an option. So it's a good one to know about no matter what GPS tools you're working with.

Nowadays you seldom need an external GPS receiver for everyday use, since most smartphones and many tablets and laptops have built-in GPS radios. The internal GPS receivers on consumer devices don't always have very sensitive antennas, though, so if you're in a remote location, a handheld GPS device with a good antenna is a must. Garmin, Trimble, Bad Elf, and many others make good ones. Garmin's GLO device and Bad Elf's GPS Pro and Pro+ are particularly easy ones to use. Though not all external GPS devices communicate wirelessly, these three all connect via Bluetooth, and appear to your computer as a serial port.

There are also many GPS receivers available that can connect to a microcontroller. Figure 8-9 shows Adafruit's Ultimate GPS Breakout module connected to the Bluefruit serial radio you saw in Chapter 2. Connected in the this way, the unit would operate just like the Garmin GLO or the Bad Elf units.

Parsing NMEA Sentences

If you want to get to know the NMEA 0183 protocol, the best way is to simply connect to a receiver through a serial terminal on your personal computer. If you're using an Adafruit Ultimate GPS Breakout or SparkFun's unit, you can connect using a Bluefruit serial radio as shown in Figure 8-9, or using any USB-to-serial adapter. If you're using a Garmin GLO or a Bad Elf device, you can just pair them with your computer via your Bluetooth control panel; then you'll see a new serial port in the serial terminal program's list after you're paired. When you open the serial port (all of these operate at 9600 bits per second), you'll see a series of strings like those shown here.

```
$GPGGA,180226.000,4040.6559,N,07358.1789,W,1,04,6.6,75.4,M,-34.3,M,,0000*5B
$GPGSA,A,3,12,25,09,18,,,,,,,,,6.7,6.6,1.0*36
$GPGSV,3,1,10,22,72,171,,14,67,338,,25,39,126,39,18,39,146,35*70
$GPGSV,3,2,10,31,35,228,20,12,35,073,37,09,15,047,29,11,09,302,20*7D
$GPGSV,3,3,10,32,04,314,17,27,02,049,15*73
$GPRMC,180226.000,A,4040.6559,N,07358.1789,W,0.29,290.90,220411,,*12
$GPGGA,180227.000,4040.6559,N,07358.1789,W,1,04,6.6,75.4,M,-34.3,M,,0000*5A
$GPGSA,A,3,12,25,09,18,,,,,,,,,6.7,6.6,1.0*36
$GPRMC,180227.000,A,4040.6559,N,07358.1789,W,0.30,289.06,220411,,*1C
```

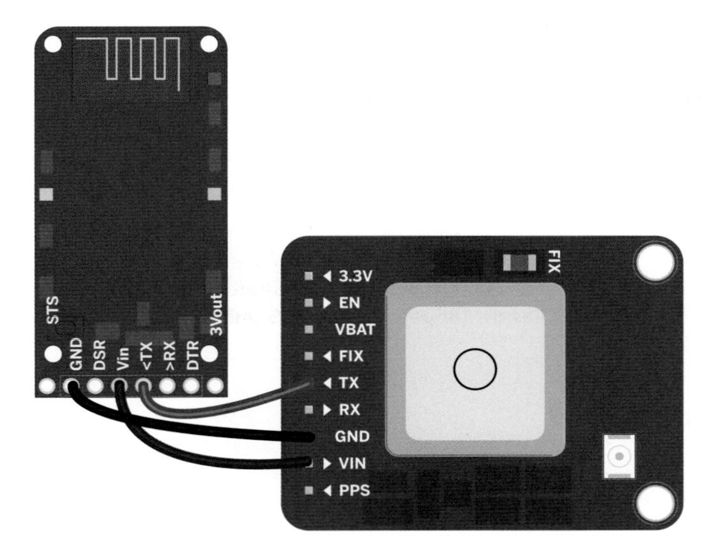

Figure 8-9
Adafruit Ultimate GPS Breakout attached to a Bluetooth serial radio. In order to get a real GPS signal, you'll have to go outside, so wireless data and a battery power source are handy. If you soldered the battery connector that comes with the Bluefruit on the back of the radio, then you can power the GPS unit and the Bluefruit from a rechargeable LiPo battery.

There are several types of *sentences* within the NMEA protocol, and each serves a different function. Some tell you your position, some tell you about the satellites in view of the receiver, some deliver information about your course heading, and so on. Each sentence begins with a dollar sign (`$`), followed by five letters that identify the type of sentence. After that come each of the parameters of the sentence, separated by commas. An asterisk comes after the parameters, then a checksum, then a carriage return and a linefeed.

Take a look at the `$GPRMC` sentence shown in blue as an example: RMC stands for Recommended Minimum specifiC global navigation system satellite data. It gives the basic information almost any application might need. This sentence contains the information shown in the table.

There is another NMEA sentence that's useful for diagnostics: the `$GPGSV` sentence (GPS Satellites in View). The third number in that sentence tells you the number of satellites in view. If that number is less than four, your receiver's not going to acquire a position too well. Satellites move, though, so if you don't get anything, wait 10 to 15 minutes and try again.

Message identifier	`$GPRMC`
Time	`180226.000` or `18:02:26 GMT`
Status of the data (valid or not valid)	`A` = valid data (`V` = not valid)
Latitude	`4040.6559` or `40°40.6559'`
North/South indicator	`N` = North (`S` = South)
Longitude	`07358.1789` or `73°58.1789'`
East/West indicator	`W` = West (`E` = East)
Speed over ground	`0.29` knots
Course over ground (heading)	`290.90`° from north
Date	`220411` or `April 22, 2011`
Magnetic variation	none
Mode	none
Checksum (there is no comma before the checksum; magnetic variation would appear to the left of that final comma, and mode would appear to the right)	`*12`

Parsing these sentences is a matter of separating the sentences from each other at the linefeed, then separating the elements of each sentence at the commas, as you've done in projects like the Monski Pong project in Chapter 2.

Before you start to write an NMEA-parsing program, though, give some thought to the device or application that you're building. Much of the time, the most important use for GPS is to show the user where she is, and for that, a mobile device with a browser is a better tool than a microcontroller. There's a geolocation standard in HTML5 for reading from a device's geolocation service, whether it's being provided via internal or external GPS, cellular location, WiFi tracking, or some combination of all of them. If you're looking to display GPS using a browser-based environment like p5.js on a mobile device. there's no need to parse the NMEA sentences, The HTML5 geolocation API will do the work for you.

You can even use the geolocation API to read some external devices on iOS and Android via Bluetooth. On iOS, if you pair the Garmin GLO or the Bad Elf units mentioned here, the operating system will simply use the data from the external GPS device for location instead of its internal receiver. Google's Android OS won't do this automatically, but you can enable it with a free app called Bluetooth GPS, available at play.google.com/store/apps/details?id=googoo.android.btgps.

Choosing Which GPS Accessories to Buy

There are many GPS receiver modules on the market, and it can get confusing choosing the right tool for this job. Here are a few things to consider.

Most common GPS receivers communicate with the microcontroller via TTL serial, using the NMEA 0183 protocol. So when it comes to connecting them to your microcontroller, and reading their data, they're largely interchangeable. The big difference between them is how well they receive a GPS signal, which is influenced by how many channels they receive (generally more is better), what kind of antenna they have (generally larger is better), and how much power they consume. SparkFun has a nice selection of GPS receivers and a good buying guide on their site at www.sparkfun.com/pages/GPS_Guide. To connect a GPS receiver, all you need is a serial transmit connection, power, and ground. Some receivers are serially configurable, in which case you need a serial receive connection as well.

In more recent projects, I've used the Garmin GLO GPS as a separate unit, because it's compact, has no user interface, but has a Bluetooth serial connection. For many projects, it removes the need for a separate microcontroller to read the GPS signal because it can speak directly to your mobile device.

X

EM-506 GPS receiver. 48 channels, good reception, less expensive than other alternatives.

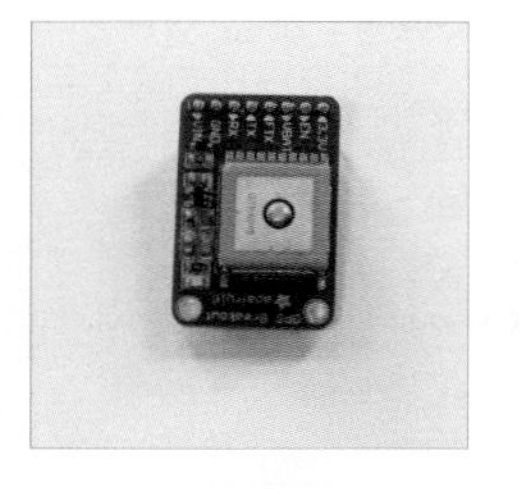

Adafruit Ultimate GPS receiver. 22 channels (66 searching), Has memory to log data onboard as well.

Garmin GLO. Built-in Bluetooth serial interface, and already in a case!

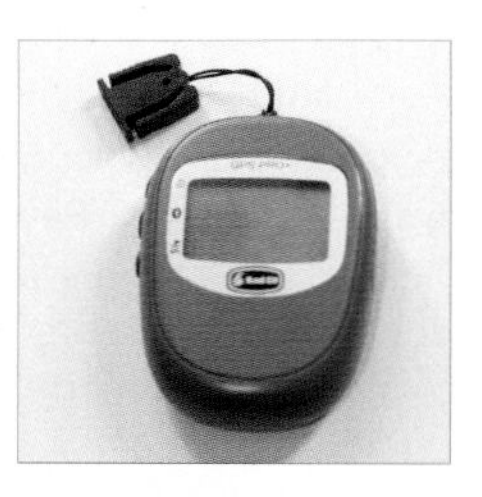

Bad Elf GPS Pro+. Compact form, user interface screen, Bluetooth serial interface.

Geolocation in the Browser

The HTML5 geolocation API starts with the `navigator` interface. It's supported by most modern browsers. To use it, you can use the `getCurrentPosition()` function to get the current geolocation of the client, or the `watchPosition()` function to watch for position updates. Both functions can have a success callback function, a failure callback function, and a position options parameter. Following is a p5.js sketch to demonstrate it.

Using the HTML5 geolocation API is pretty simple, but serving the HTML page that contains it is another matter. Many browsers don't allow you to access geolocation services from a page that was served using HTTP. They insist that the page be served using HTTPS, the version of HTTP which supports authentication and encryption. You'll see how to do that shortly.

Create a new project directory wherever you store your projects called **GeolocationServer**. Inside it, make a **public** directory and copy the files needed for a p5.js sketch: **index.html**, **sketch.js**, and the p5.js libraries, as you've done for previous projects. If you're using the command-line p5-manager tool, create it as follows:

```
$ p5 g -b public
$ cd public
$ p5 u
```

The **index.html** page is just the standard page for p5.js, so you don't need to change the one you copied from the example project. The **sketch.js** file is where all the action happens, as usual. Open it and change it as shown here.

Locate It Start the **sketch.js** with a text label to put on the page. Then set up JSON object called `options` to hold the settings for the geolocation watcher.

In the `setup()` function, create the canvas, choose a fill color for the text and a size, align the text to the center. The text size is large so that it's easily readable on a phone or tablet. Finally, start the geolocation watcher with the `watchPosition()` function. It has two callback functions, one for success and one for failure (you'll write these shortly), and takes the options you set at the beginning of the sketch.

```
/*
Geolocation
Context: p5.js
*/
 var label = "Checking to see if your browser supports geolocation...";
 var options = {
   enableHighAccuracy: true,
   timeout: 10000,
   maximumAge: 0
 };

function setup() {
  // get the canvas width and height from the window:
  createCanvas(windowWidth, windowHeight);
  fill('#A3B5CF');        // pale slightly blue text
  textSize(36);           // text size
  textAlign(CENTER);      // text alignment
  // start the geolocation watch:
  navigator.geolocation.watchPosition(success, failure, options);
}
```

The `draw()` function just sets the background color and draws the text on top of it.

```
function draw () {
  background('#0D1133');
  text(label, width/2, 100);
}
```

The `success()` callback function reads the position, formats it nicely, and puts it in the `label` variable. You enabled high accuracy in the options, but this function limits the latitude and longitude to five significant digits using `.toFixed()`. You can change this if you want more or less precision.

The `failure()` callback function writes any error message to the label variable.

That's the whole sketch. Save it, then move on to the next section to write the server.

Continued from previous page.

```
function success(position) {
  var coordinates = position.coords;
  var now = new Date(position.timestamp);
  label = 'Your position is:'
    + '\nLatitude : ' + coordinates.latitude.toFixed(5)
    + '\nLongitude: ' + coordinates.longitude.toFixed(5)
    + '\nWithin ' + coordinates.accuracy + ' meters, at'
    + '\n' + new Date(position.timestamp);
}

function failure(error) {
  label = 'Error code ' + error.code + ': \n' + error.message;
}
```

Serving Files via HTTPS

Since most browsers expect you to serve any document containing the geolocation API via HTTPS, you'll need to serve this page with an HTTPS server. It's a good excuse to learn a little more about secure connections.

When a client makes an HTTPS request, it assumes that the server is going to send an encrypted *certificate* that the client can then check with a known *certificate authority*. Once the certificate is verified, the request proceeds like HTTP, except the bytes are encrypted before they are sent. If you're only serving a page for your own purposes, you can create your own certificate from the command line, as shown next. If you're serving a wider audience, however, you should obtain a certificate from a known certificate authority. For more on HTTPS and certificates, see the "HTTPS and Certificates" sidebar following.

In your **GeolocationServer** project directory, make a new text file for the node server called **https-server.js**. Finally, make a new subdirectory called **keys**. This is where your certificate will live. To create the certificate, type the following (this assumes you're in the project directory):

```
$ openssl req -newkey rsa:2048 -nodes -keyout keys/domain.
key -x509 -days 365 -out keys/domain.crt
```

Windows Users: **Cygwin includes openSSL, as does the bash shell on Windows 10. You can also generate these files on a Linux or macOS machine, then copy them to your Windows machine. Use OpenSSL on a Mac or your web host if it runs Linux, or a Raspberry Pi to do this.**

You'll be asked a series of questions about your location and identity. These are for the certificate. Fill them in as you please. This will create two files in the keys directory, **domain.crt** and **domain.key**. The first is the certificate for your server, and the second is the *private key* for your server. The server will use this key to encrypt all responses. Included in the certificate contains the server's *public key*, which the client will use to encrypt its requests. Now that you've created these, you're ready to write the server. Install express.js as usual using npm:

```
$ npm install express
```

Then open **https-server.js** and write it as follows.

Figure 8-10
The Geolocation example page in a browser on an Android device. Note the crossed out https in the address bar. This is how the browser tells you that the server is using an untrusted certificate (in this case self-signed).

Serve It This server will only serve static files and redirect HTTP requests to HTTPS. You'll need the express.js library along with a few core node.js libraries, `http` and `https` and `fs`. The latter gives you functions to read and write from your local filesystem.

```
/*
HTTP/HTTPS server
context: node.js
*/
// include libraries and declare global variables:
var express = require('express');  // include the express library
var https = require('https');      // require the HTTPS library
var http = require('http');        // require the HTTP library
var fs = require('fs');            // require the filesystem library
var server = express();            // create a server using express
```

Next set up some options for the HTTPS server, to tell it where to read the certificate and private key, using the `fs.readFileSync()` function.

You'll use the `express.static()` middleware to serve files as you've done before, but first you'll need a middleware function to redirect all HTTP requests to HTTPS.

```
var options = {                              // options for the HTTPS
server
  key: fs.readFileSync('./keys/domain.key'), // the key
  cert: fs.readFileSync('./keys/domain.crt') // the certificate
};

server.use('/', httpRedirect);          // set a redirect function for http
server.use('/',express.static('public'));    // set a static file directory
```

The redirect function comes next. It checks if the incoming request is secure (i.e., an HTTPS request), and if not, it redirects it to an HTTPS request. If the request is already secure, this function passes it on to the next middleware function (i.e. `express.static()`).

```
function httpRedirect(request,response, next) {
  if (!request.secure) {
    console.log("redirecting http request to https");
    response.redirect('https://' + request.hostname + request.url);
  } else {
    next();      // pass the request on to the express.static middleware
  }
}
```

Finally, start the server. You need to start both an HTTP server and an HTTPS server. HTTPS is normally served on port 443.

```
// start the server:
http.createServer(server).listen(8080);           // listen for HTTP
https.createServer(options, server).listen(443);  // listen for HTTPS
```

In order to run this server on macOS or Linux, you'll need to use the superuser do command, sudo. Run the server like so:

```
$ sudo node https-server.js
```

Then open a browser and go to http://localhost:8080 in the address bar. The server should redirect your request to an HTTPS request, and then the browser will complain that it can't verify the server's certificate. Give it permission to proceed anyway (in Chrome, this option is under the Advanced link). Then the index page will load.

When you load the page, your browser will ask you if you want to allow location services. Allow it, and you'll get your location. Figure 8-10 shows what the page looks like.

It may take several minutes before you get a position with this sketch if you're using an external GPS receiver. Different receivers take varying times to acquire an initial position when first started. Pick a spot with a clear view of the skies and be patient.

Many mobile phones on the market these days feature a GPS receiver, and seem to have a signal most of the time, which may make you wonder why an external receiver can't get a signal as fast. Mobile phones tend to use both GPS and cellular location tracking together. Even if they don't have enough satellite signals, they can generally determine their position relative to the nearest cell towers, and—using the known locations of those towers—approximate a fix.

X

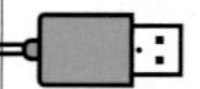

HTTPS and Certificates

As you've seen already, the Hypertext Transport Protocol, HTTP, is a text-based protocol that runs over TCP sockets. When a client makes a basic HTTP request to a server, the packets are sent from the client to its router and through the ensuing chain of routers to the server unencrypted. Any other connected device can read the packets flowing across the network. It's as if you're sending a letter without an envelope. Just as you wouldn't send a check to pay a bill without an envelope, you shouldn't send payment information or other sensitive data over an unencrypted HTTP connection. *HTTPS* is the protocol for securing HTTP connections using *Transport Layer Security (TLS)*. Here's how it works: Some forms of device data, like geolocation, are deemed sensitive enough that they should be carried over HTTPS as well.

When a client contacts a server that's using HTTPS, the server offers a digital certificate to verify itself. The client then verifies that certificate with a *certificate authority (CA)*. Certificate authorities are companies whose business is to verify the identity of certificate holders; they confirm that the given domain is hosted at a given IP address, registered to a given individual or company. The certificate contains the server's *public encryption key* that the client can use to encrypt its messages to the server. When the client has verified the server's certificate, it uses the server's public key to encrypt its requests to the server. Messages encrypted with the public key can only be decrypted with the server's *private key*, which only the server holds.

As you've seen in this project, it's possible to generate your own certificate, but your browser won't recognize it unless it's registered with a known certificate authority. If the browser doesn't recognize your certificate, it will generate a warning like the one you saw when you first loaded index page in this project. CAs voluntarily undergo regular auditing to verify their security in order to maintain their status with the popular web browsers. If your certificate authority is trusted by your browser, the certificate check will happen in the background, invisible to the user. If the certificate's not valid, you'll get a warning message like the one you saw with your self-signed certificate.

Though there are many CAs, most certificates are issued by a small handful of known companies: Symantec, Comodo, GoDaddy, and GlobalSign among them. Many web hosting companies will register your certificate through one of these for a nominal fee.

Recently, the Internet Security Research Group set up a not-for-profit CA, Let's Encrypt, that will register certificates for free. The Electronic Frontier Foundation offers a nice interface for generating a free certificate through Let's Encrypt at certbot.eff.org. If you're new to HTTPS and certificates, this is probably the easiest way to get one.

Determining Orientation and Attitude

Once you know where you are, it's also useful to know which way you're oriented relative to the world around you. People have an innate ability to determine their orientation relative to the world around them, but objects don't. Orientation sensors are typically used for determining orientation and attitude. There are three common sensors for this: magnetometers, accelerometers, and gyrometers. Using these sensors, you can determine which way you're heading and which way is up.

Orientation, or compass heading, is how you determine your direction if you're level with the earth. If you've ever used an analog compass, you know how important it is to keep the compass level in order to get an accurate reading. If you're not level, you need to know your tilt relative to the earth as well. In navigational terms, your tilt is called your *attitude*, and there are two major aspects to it: roll and pitch. *Roll* refers to how you're tilted side-to-side. *Pitch* refers to how you're tilted front-to-back.

Pitch and roll are only two of six navigational terms used to refer to movement. Pitch, roll, and *yaw* refer to angular motion around the X, Y, and Z axes. These are called *rotations*. *Surge*, *sway*, and *heave* refer to linear motion along those same axes. These are called *translations*. Figure 8-11 illustrates these six motions. These are often referred to as six degrees of freedom. *Degrees of freedom* refer to how many different parameters the sensor is tracking in order to determine your orientation.

Magnetometers read magnetic field strength, and can therefore be used to determine orientation relative to Earth's magnetic field. *Accelerometers* measure linear acceleration, and can therefore be used to measure the effect of the earth's gravitational field, since acceleration due to the earth's gravity is constant. *Gyrometers* measure rate of rotation, and are used to determine rotation independent of heading or attitude. In other words, magnetometers help determine heading; accelerometers help determine pitch, roll, surge, and sway; and gyrometers help determine yaw, pitch, roll. These three sensors are increasingly common in all kinds of devices. You'll find an accelerometer and magnetometer in every mobile phone or tablet. The Arduino 101 has a built-in accelerometer and gyrometer, as do some other microcontrollers.

Figure 8-11
Rotations and translations of a body in three dimensions.

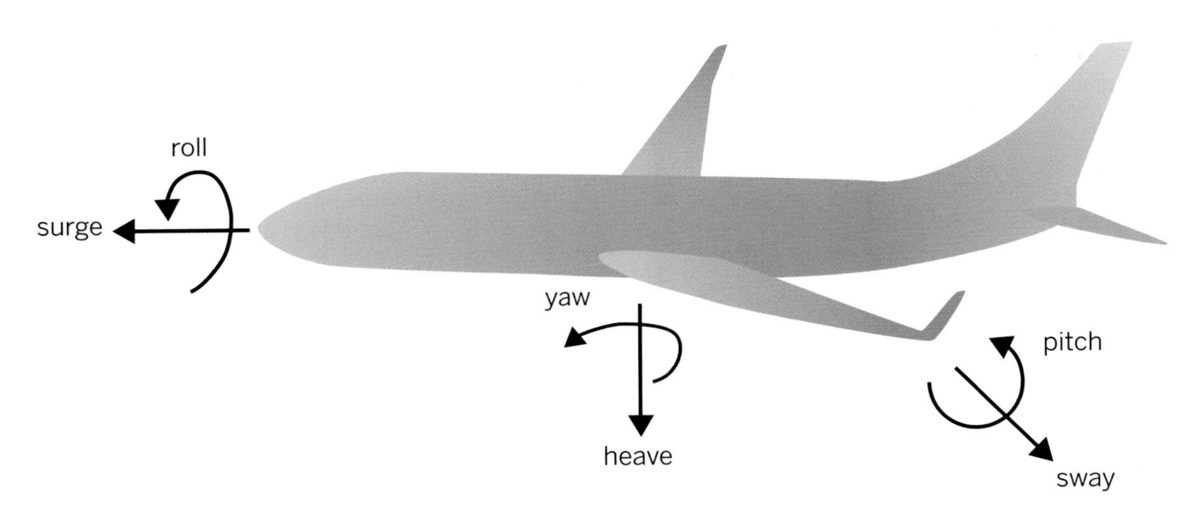

Orientation Sensor Terminology

You'll hear a number of different terms for these sensors. The combination of an accelerometer and gyrometer is sometimes referred to as an *inertial measurement unit*, or *IMU*, and the library for the Arduino 101 is called the CurieIMU library. Many companies make versions of these sensors. Some of the most popular come from Analog Devices, Freescale, ST Microelectronics, and Bosch. When an IMU is combined with a magnetometer, the combination is referred to as an *attitude and heading reference system*, or *AHRS*. Sometimes they're also called *magnetic, angular rate, and gravity*, or *MARG*, sensors. You'll also hear them referred to as *6-degree of freedom*, or *6-DOF*, sensors. There are also *9-DOF sensors* that incorporate all three types of sensors. Each axis of measurement is another degree of freedom. Adafruit even has a 10-DOF sensor that adds a barometric pressure sensor for determining altitude.

Orientation sensors come in a variety of interfaces. There are some that provide an analog output voltage corresponding to each axis, some that use the SPI synchronous serial protocol, and others that use the I^2C synchronous serial protocol. All of these sensors usually read changes on three axes, X, Y, and Z. The Z axis is usually perpendicular to the earth, so when you've got the sensor flat, that axis won't change much.

Magnetic field strength is measured in *gauss*. Linear acceleration is measured in *meters per second squared* (m/s^2), or *g*'s, which are units of Earth's gravity ($1g = 9.8\ m/s^2$). Rotational rate is measured in *degrees per second (dps)*.

These sensors come in a variety of ranges. For example, the ST Microelectronics LSM303 is a combined accelerometer and magnetometer. The accelerometer can be set to measure in ranges from +/-2g to +/-16g. The magnetometer on the LSM303 can be set to a range from +/-1.8 to +/-8.1 gauss. The L3GD20 gyrometer can be set to measure from 250 to 2000 degrees per second. The sensitivity you need depends on your application. If you're measuring attitude of a stationary object, a 2g accelerometer will work fine, but if you're measuring the force of a ball hitting a bat, the force can be an order or magnitude or two higher. Measuring the rotation of a person's hand may be fine at 250 degrees per second, but measuring the rotational change of a high-speed motor might be several thousand degrees per second.

Regardless of their scale, all of these sensors will output the middle value of their range when they're at rest. Movement in one direction on an axis will lower the value, and movement in the other direction will raise the value.

When you're using these sensors to determine which way an object is oriented, you may not need to convert their output to their physical properties. You're measuring the relative magnitudes of the three axes; what units you use to measure them are not relevant in this case.

For example, if you want to know which way an accelerometer is facing, all you need to know is which axis is perpendicular to the ground. That will be the axis with the greatest magnitude relative to the other two. Figure 8-12 illustrates this. When the sensor is facing up, the Z axis would have the highest reading, and the X and Y axes would be at rest, in the middle of their range. If you flipped it upside down, the Z axis would read the same magnitude, but a negative value. If you rotate it on the Y axis, then the X axis will have the greatest magnitude, and the other two axes would be at rest in the middle of their range. If you turn it on the X axis, then the Y axis would have the greatest magnitude.

If you do need to convert the sensor output to its physical property, you calculate it from the sensor's data sheet. You need to know the measurement range and the resolution of its readings. For example, a 2g accelerometer that has a 16-bit resolution would output 2^{16} possible values, or a range from -32,768 to 32,767. You know the measurement range is from -2g to +2g. From that, you can do the following math:

```
sensorReading/32,768 = acceleration/2.0
```

or

```
acceleration = sensorReading * 2.0 / 32,768
```

The sign of the sensor reading will give you the direction of acceleration.

Make sure to check the data sheet of your sensor, because sometimes other factors need to be considered in the calculation, like how the reading is formatted, or whether the sensor has a nonlinear output profile.

X

Project 19

Determining Heading Using a Digital Compass

Magnetometers, also called digital compasses, measure the change in Earth's magnetic field, just as an analog compass does. Like analog compasses, they are subject to interference from other magnetic fields, including those generated by strong electrical induction. You can calculate heading using a digital compass if you are in a space that doesn't have a lot of magnetic interference.

MATERIALS

- **1 solderless breadboard**
- **1 Arduino module**
    - **Features used: I²C**
- **1 digital compass, ST Microelectronics model LSM303DLH. See chapter opening for options.**
- **4 male header pins**

This example uses a digital compass from ST Microelectronics, model LSM303. Adafruit, SparkFun, and Pololu all carry this sensor on one or more of their boards. Figure 8-12 shows the Adafruit 9-DOF board, but the connections are the same for the others. It communicates via synchronous serial data sent over an I²C connection. This sensor measures magnetic field strength along three axes, and has a three-axis accelerometer built in as well.

This compass offers 16 bits of resolution per axis, or -32,768 to 32,767 for each axis. That maps to +/- 1.3 gauss at the default settings.

Figure 8-13 shows the compass connected to an MKR1000, but you can use any Arduino-compatible board that supports I²C communication.

The compass operates on 3.3–5V. The pins used are as follows:

- Vin: voltage input. Connect to the microcontroller's voltage output.
- GND: ground. Connect to the microcontroller's ground.
- SCL: Serial clock. Connect to microcontroller's SCL pin.
- SDA: Serial data. Connect to microcontroller's SDA pin.

You won't use the sensor's other pins in this application.

There are a few different libraries for this sensor, but I like Pololu's one best, as it's the simplest. It relies on the Wire library to communicate via I²C with the compass. To install it, open the Libraries Manager in the Sketch menu under Include Library...→Manage Libraries..., then search for LSM303 by Pololu. Once it's installed, you're ready to start.

X

Figure 8-12
The axes of a 3D sensor. If you're using the accelerometer in this sensor, its readings would be as shown here.

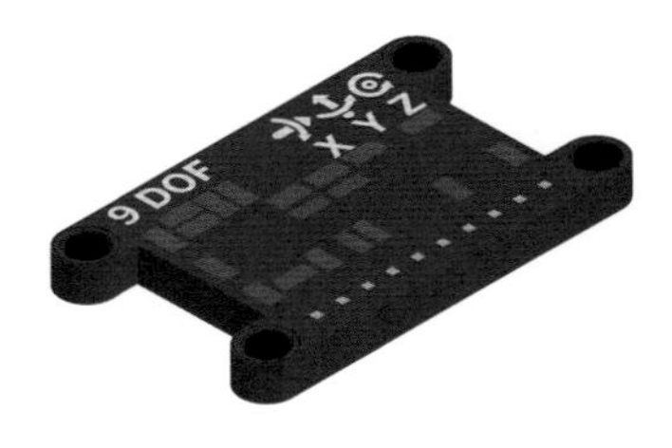

Z at greatest magnitude, X and Y at rest

X at greatest magnitude, Z and Y at rest

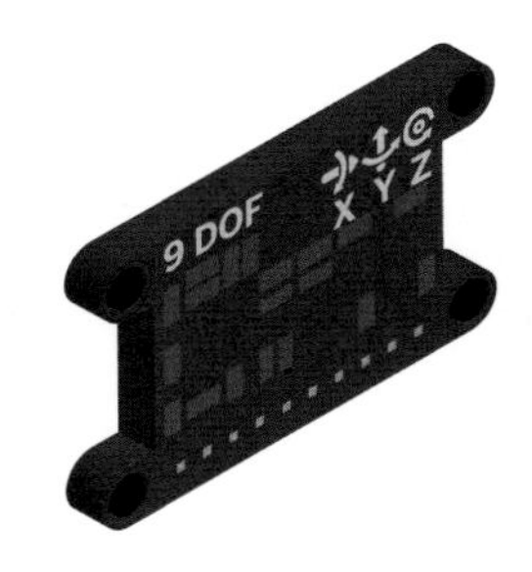

Y at greatest magnitude, Z and X at rest

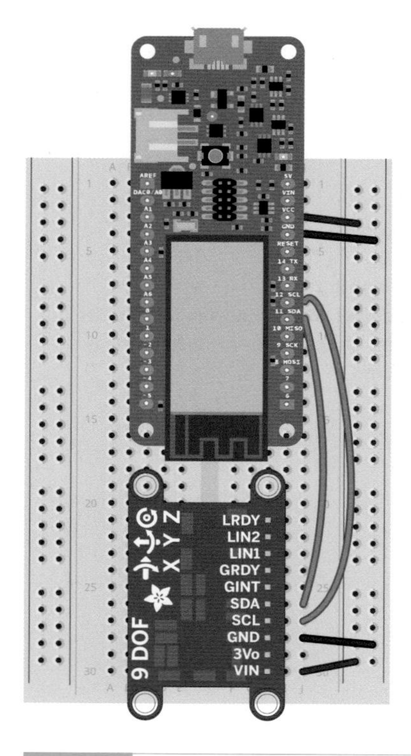

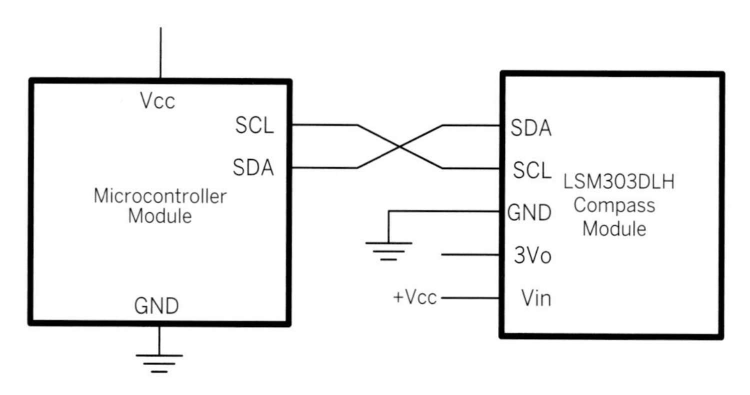

Figure 8-13
Adafruit's 9 DOF board, which features ST Microelectronics LSM303 compass, connected to an MKR1000. This will work with any Arduino-compatible module that supports I2C, however. The compass is on the Adafruit 9 DOF sensor board in this drawing. Only the pins used are shown in the schematic.

Try It Include the library and make an instance of it in a variable called `compass`. In the `setup()` function, initialize serial communication, then the Wire library to initialize I2C communication, then the compass itself. Finally, the `enableDefault()` function sets the compass's default settings. In this state, the compass will read a +/-1.3 gauss range, which is fine for sensing the planet's magnetic field.

```
/*
  LSM303 compass example
  context: Arduino
*/
#include <Wire.h>
#include <LSM303.h>

LSM303 compass;              // an instance of the sensor library

void setup() {
  Serial.begin(9600);        // start serial communication
  Wire.begin();              // start I2C communication
  compass.init();            // initialize the compass
  compass.enableDefault();   // enable defaults on the compass
}
```

The `loop()` method just reads the compass using the `read()` function, then calculates the heading using the `heading()` function.

```
void loop() {
  compass.read();                         // read the compass
  float myHeading = compass.heading();   // get the heading
  Serial.println(myHeading);             // print it
  delay(100);                             // wait 100ms
}
```

Calibrating the Compass

The compass reads magnetic field strength, so anything that generates a magnetic field will influence it: electrical wiring, magnets, motors, and other electrical equipment can sometimes affect your reading. If possible, test it somewhere away from electrical equipment. When testing the compass, you need to know the cardinal directions precisely, so make sure you know which way is north in your area. Get a magnetic needle compass and check properly.

The LSM303 compasses are auto-calibrated at the factory, but the library also comes with a calibration sketch that you can use to increase the accuracy. There's a variation on the calibration sketch included in the github repository for this book. You can also do it yourself as described next.

To calibrate the compass, you need to rotate the sensor through all three axes slowly in order to read the maximum and minimum values read on all three axes. To see the actual compass readings, add the following lines to the `loop()` function:

```
Serial.print(compass.m.x);
Serial.print("\t");
Serial.print(compass.m.y);
Serial.print("\t");
Serial.println(compass.m.z);
```

The defaults for the expected maxima and minima are -32,768 (minimum) and 32,767, since this is a 16-bit sensor. When you know the actual maximum and minimum values for each axis in your location, you can set the expected maxima and minima in your `setup()` function like so:

```
compass.m_min.x = -680;
compass.m_min.y = -623;
compass.m_min.z = -616;
compass.m_max.x = 422;
compass.m_max.y = 451;
compass.m_max.z = 424;
```

These values were taken at my location. The values for your location will probably be different.

Setting these expected maxima and minima should help increase the accuracy of your compass. For more on the compass's API, see the library repository at github.com/pololu/lsm303-arduino.

X

Introducing the I²C Interface

The LSM303 compass uses a form of synchronous serial communication called Inter-Integrated Circuit, or I²C. Sometimes called Two-Wire Interface, or TWI, it's the other common synchronous serial protocol besides SPI, which you learned about in Chapter 2.

I²C is comparable to SPI in that it uses a single clock on the master device to coordinate the devices that are communicating. Every I²C device uses two wires to send and receive data: a serial clock pin, called the *SCL* pin, that the microcontroller pulses at a regular interval; and a serial data pin, called the *SDA* pin, over which data is transmitted. For each serial clock pulse, a bit of data is sent or received. When the clock changes from low to high (known as the rising edge of the clock), a bit of data is transferred from the microcontroller to the I²C device. When the clock changes from high to low (known as the falling edge of the clock), a bit of data is transferred from the I²C device to the microcontroller. Unlike SPI, I²C devices don't need a chip select connection. Each has a unique address, and the master device starts each exchange by sending the address of the device with which it wants to communicate.

I²C requires just two connections between the controlling device (or master device) and the peripheral device (or slave), as follows:

- **Clock (SCK): The pin that the master pulses regularly.**
- **Data (SDA): The pin that data is sent on, in both directions.**

All devices on an I²C bus can share the same two lines.

The Wire library is the standard interface for I²C on Arduino-compatible devices. Most higher-level libraries that use I²C, like the one shown here, rely on the Wire library to handle the communication. The I²C pins are different from board to board, but they are labeled SCL and SDA on all the Arduino and most Arduino-compatible boards.

X

 Project 20

Determining Attitude

Compass heading is an excellent way to determine orientation if you're level with the earth, but heading is only one aspect of determining your attitude relative to your surroundings. In order to know which way is up, you also need to know your pitch and your roll. You can do this using accelerometers, which measure linear acceleration, and gyrometers, which measure angular rotational rate. In this project, you'll learn more about determining attitude using accelerometers and gyrometers.

MATERIALS

» **1 solderless breadboard or prototyping shield**
» **1 Arduino module**
» **1 accelerometer/gyrometer** You can use any accelerometer/gyrometer for this. Examples are shown for the ST Microelectronics LSM303 and L3Gxx, and the Arduino 101's built-in IMU.
» **6–8 male header pins** depending on your sensor's needs.

Accelerometers measure changing linear acceleration. At the center of an accelerometer is a tiny mass that's free to swing in three dimensions. As the accelerometer tilts relative to the earth, the gravitational force exerted on the mass changes. The force, and therefore the acceleration measured, will be greatest on the axis that's closest to perpendicular. From this, you can determine which face of your sensor is up. In order to read the actual angle of the roll, pitch, and yaw of an object, though, you need both data from an accelerometer and from a gyrometer. The accelerometer can give you the roll and pitch at rest, but you need a gyrometer to give you the yaw, or heading, without a compass.

Accelerometers can be used for other purposes as well, and therefore come in a variety of resolutions. For determining attitude, you only need about 2g. Normal human movement doesn't generate much more than that. Most commercial products using accelerometers, such as gaming controllers or mobile phones, use an accelerometer in the range 3–6g range. More exertive activities can require a much higher range, though. For example, a boxer's fist can decelerate at up to 100g when he or she hits!

This project will show you how to read the raw values from accelerometer/gyrometer combinations, and how to convert them to their physical quantities. You'll also see how to calculate yaw, pitch, and roll using a library based on algorithms developed by Sebastian Madgwick, and you'll use that data in a Processing sketch to control the orientation of a 3D object onscreen.

The following examples show two different options: the LSM303 that you used in the previous project (it has an accelerometer in addition to a compass), used with ST Microelectronics' L3G gyro, and the Arduino 101's Bosch BM160 accelerometer/gyrometer. If you're using the LSM303 for this project, use the circuit as shown in Figure 8-13. For the Arduino 101, the accelerometer and gyrometer are built into the board, so no circuit diagram is necessary.

X

Listen to It (101)

This sketch reads the inputs from the Arduino 101's accelerometer and gyrometer. The CurieIMU library has a function that will return the sensor reading scaled as accelerations in g's, and read the gyro as degrees per second (dps). You'll need to include the CurieIMU library using the Library Manager and make sure you've selected the Arduino 101 in the Tools→Board menu before this will work. Since the accelerometer is built into the board, no additional circuit is necessary.

NOTE: Use this sketch with the Arduino 101 accelerometer/gyrometer.

```
/*
  CurieIMU reader
  Context: Arduino
  Reads the Arduino 101 IMU's scaled values
*/
#include "CurieIMU.h"

void setup() {
  Serial.begin(9600); // begin serial communication
  CurieIMU.begin();  // initialize the IMU

  // set the accelerometer range to +/-2g:
  CurieIMU.setAccelerometerRange(2);
  // set the gyro range to +/-250 dps:
  CurieIMU.setGyroRange(250);
}

void loop() {
  float x, y, z;    // variables for the scaled accelerometer values
  float gx, gy, gz; // variables for the scaled gyro values
  // read and scale:
  CurieIMU.readAccelerometerScaled(x, y, z);
  CurieIMU.readGyroScaled(gx, gy, gz);
  Serial.print("x: ");      // print x reading
  Serial.print(x);
  Serial.print(",\ty: ");  // print comma, tab, y reading
  Serial.print(y);
  Serial.print(",\tz: ");  // print comma, tab, z reading
  Serial.print(z);
  Serial.print("\tgx: ");  // print comma, tab,  gx reading
  Serial.print(gx);
  Serial.print(",\tgy: "); // print comma, tab, gy reading
  Serial.print(gy);
  Serial.print(",\tgz: "); // print comma, tab, gz reading
  Serial.println(gz);
}
```

The next sketch reads the inputs from two sensors: ST Microelectronics' LSM303 accelerometer, which you saw in the compass project, and their L3Gxx gyro. You'll find this combination of sensors on Adafruit's 9DOF board, and some of Pololu's boards as well. SparkFun sells these sensors separately as the L3G4200D gyro board and the LSM303 board.

You'll need to include the same library that you used for the compass project previously, as well as Pololu's L3G library, which you can find using the Library Manager.

If you're using the Adafruit board, you can use the same circuit as shown in Project 19. Both sensors connect via I²C. If you're using separate modules for each sensor, connect the SDA pin of both sensors to the SDA of the microcontroller and the SCL pins to the microcontroller's SCL pin. Multiple I²C sensors can share the same connections.

Listen to It (I^2C)

At the top of the sketch, include the Wire library for I^2C communication, and the libraries for the two sensors, which you installed using the Library manager. Then make instances of the libraries in variables called `accelerometer` and `gyro`, respectively.

NOTE: Use this sketch with the LSM303 accelerometer and the L3Gxx gyrometer.

```
/*
  LSM303 accelerometer reader/ L3G gyro reader
  Context: Arduino
  Reads an LSM303 accelerometer's values
  and an L3G gyro's values
*/
#include <Wire.h>
#include <LSM303.h>
#include <L3G.h>

LSM303 accelerometer;
L3G gyro;
```

In the `setup()` function, initialize serial and I^2C communication and the sensors. Enable their default settings as well.

```
void setup() {
  Serial.begin(9600);    // begin serial communication
  Wire.begin();          // initialize I2C
  accelerometer.init(); // initialize accelerometer
  accelerometer.enableDefault(); // enable default range (+/- 2g)
  gyro.init();           // initialize gyro
  gyro.enableDefault(); // enable default range (+/- 250dps)
}
```

In the `loop()` function, read the sensors, then use functions called `readAccelerometer()` and `readGyro()` to convert them to their physical quantities. You'll write those functions next.

```
void loop() {
  accelerometer.read(); // read accelerometer
  gyro.read();           // read gyro

  // convert accelerometer readings to g's:
  float x = readAcceleration(accelerometer.a.x);
  float y = readAcceleration(accelerometer.a.y);
  float z = readAcceleration(accelerometer.a.z);

  // convert gyro readings to degrees per second:
  float gx = readGyro(gyro.g.x);
  float gy = readGyro(gyro.g.y);
  float gz = readGyro(gyro.g.z);
}
```

The `readAcceleration()` function converts the raw analog readings to accelerations as you saw in the CurieIMU sketch, and the `readGyro()` function does the same for the gyrometer. For the LSM303, the sensor's 16-bit resolution (from -32,768 to +32,767) maps to a -2g to +2g range. For the gyro, the resolution maps to a +/- 250dps range.

```
float readAcceleration(int rawValue) {
  // LSM303 accelerometer has a +/- 2g range
  // and a 16-bit resolution:
  float result = (rawValue * 2.0) / 32768.0;
  return result;
}

float readGyro(int rawValue) {
  // L3G gyro has a +/- 250 dps range
  // a 16-bit resolution:
  float result = (rawValue * 250.0) / 32768.0;
  return result;
}
```

Determining Which Way Is Up from an Accelerometer

Force equals mass times acceleration, and the mass of the accelerometer is constant. Likewise, the acceleration due to gravity is constant, at 9.8 m/s^2. With this information, you can treat an accelerometer as shown in Figure 8-12 and determine which face is up like so:

```
if x's magnitude is greatest,
  if x > 0
    face 1 is up
  else
    face 2 is up
if y's magnitude is greatest,
  if y > 0
    face 3 is up
  else
    face 4 is up
if z's magnitude is greatest,
  if z > 0
    face 5 is up
  else
    face 6 is up
```

Orient It (Basic)

Using this algorithm, you can determine the sensor's facing with any of accelerometer. Using either of the examples on the previous pages, add the following lines to your `loop()` function. New lines are shown in blue:

```
/*
Accelerometer orientation
  context: Arduino

  Determines which face is up,
  using an accelerometer.
*/
// Include any library includes, global variables,
// and the setup() function from your previous sketch here.

void loop() {
  // the rest of your previous loop code goes here

  int orientation = readOrientation(x, y, z);
  Serial.println(orientation);
}
```

Now add the following function to determine facing. You're passing it the accelerometer x, y, and z values that you determined in the previous sketches, but you could use the raw sensor values if you prefer. The algorithm does not depend on getting the sensor data in g's.

```
int readOrientation(float x, float y, float z) {
  int result = -1;     // result to return orientation
  int absX = abs(x);   // absolute values of sensor readings
  int absY = abs(y);
  int absZ = abs(z);

  // determine which axis had the greatest magnitude:
  int bigger = max(absX, absY);
  int biggest = max(bigger, absZ);

  if (biggest == absX) {  // if x axis is greatest
    if (x > 0) {          // positive x is up
      result = 1;
    } else {              // negative x is up
      result = 2;
    }
  }
```

»

When you run this sketch, turn your board so that it faces up, down, left, right, forward, and backward. You'll see that each facing outputs a different orientation value.

This function works well for determining attitude roughly when you've only got an accelerometer, and often that's all you need. It won't tell you which way the sensor is rotated around its vertical axis, however.

X

Continued from previous page.

```
  if (biggest == absY) {  // if y axis is greatest
    if (y > 0) {          // positive y is up
      result = 3;
    } else {
      result = 4;         // negative y is up
    }
  }
  if (biggest == absZ) {  // if z axis is greatest
    if (z > 0) {
      result = 5;         // positive z is up
    } else {
      result = 6;         // negative z is up
    }
  }
  return result;
}
```

Determining Yaw, Pitch, and Roll

Three-axis accelerometers measure the linear acceleration of a body on each axis—in other words, the surge, sway, or heave of a body. They don't give you the roll, pitch, or yaw. However, you can calculate the roll and pitch when you know the acceleration along each axis. That calculation takes some trigonometry. For a full explanation, see Freescale Semiconductor's application note on accelerometers at cache.freescale.com/files/sensors/doc/app_note/AN3461.pdf. Here are the highlights.

The force of gravity always acts perpendicular to the earth's surface. So when an object is tilted at an angle (called theta, or θ), part of that force acts along the X axis of the object, and part acts along the Y axis (see Figure 8-14). The X-axis acceleration and the Y-axis acceleration add up to the total force of gravity using the Pythagorean Theorem: $x^2 + y^2 = z^2$.

Since you know that, you can calculate the portions of the acceleration using sines and cosines. The X-axis portion of the acceleration is gravity * sinθ, and the Y-axis portion is gravity * cosθ (remember, sine = opposite/hypotenuse, and cosine = adjacent/hypotenuse).

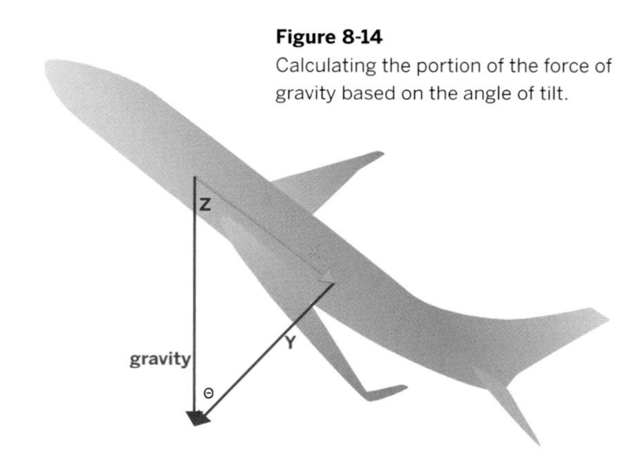

Figure 8-14
Calculating the portion of the force of gravity based on the angle of tilt.

From there, you have to calculate the portions along all three axes at once. It turns out that:

$$roll = \arctan\left(x\text{-axis} / \sqrt{y\text{-axis}^2 + z\text{-axis}^2}\right)$$

and:

$$pitch = \arctan\left(y\text{-axis} / \sqrt{x\text{-axis}^2 + z\text{-axis}^2}\right)$$

In order to know the acceleration on each axis, though, you have to convert the raw accelerometer into g's, as you did earlier. Once you've got the acceleration for each axis, you can apply these trigonometric formulas to get the pitch.

The Madgwick Method

In 2010, Sebastian Madgwick made a detailed study of how to sense orientation using the kinds of sensors described here. He developed a more efficient set of algorithms for determining yaw, pitch, and roll using the data from IMU sensors, available at x-io.co.uk/open-source-imu-and-ahrs-algorithms/. The mathematics of his work is more detailed than space allows here, but you can find it at x-io.co.uk/res/doc/madgwick_internal_report.pdf.

Helena Bisby converted Madgwick's algorithms into a library for Arduino, improved upon by Paul Stoffregen and members of the Arduino staff, so now it's relatively simple to determine attitude on a microcontroller using this method. First you need to install the Madgwick library. Open the Libraries Manager in the Sketch menu under Include Library...→Manage Libraries..., then search for Madgwick by Arduino. Once it's installed, make the following modifications to your previous sketch.

Orient It (Advanced)

To use the Madgwick library, create an instance called `filter`. You'll update the filter at a regular rate, in this case 25Hz. You'll need a time stamp and an interval of 40 milliseconds to make sure the filter's updated at the right time. New lines are shown in blue as usual.

```
/*
  Accelerometer Madgwick orientation
  context: Arduino
  Can be used with any accelerometer/gyrometer combo
*/
// Include any library includes, global variables
// from your previous sketch here.
#include <MadgwickAHRS.h>
long lastReading = 0;             // time stamp of last reading
long readingInterval = 40;        // 40 ms = 25 Hz sample rate
Madgwick filter;                  // instance of the Madgwick library
```

In your `setup()` function, initialize the filter at 25Hz.

```
void setup() {
  // the rest of your previous setup code goes here
   filter.begin(25);              // update filter at 25Hz
}
```

In the `loop()`, read your sensors and convert to their physical properties as you did before, then set up an if statement to check if 40 milliseconds have passed. If so, update the filter with the sensor readings.

Remove the `readOrientation()` function and the call to print the resulting orientation from your previous sketch.

Once you've updated the filter with the current sensor readings, call `getRoll()`, `getPitch()` and `getYaw()` to get your sensor's attitude. Print them out separated by commas, with a newline at the end. Finally, update the time stamp. That's the end of the sketch.

```
void loop() {
  // the rest of your previous loop code goes here

  if (millis() - lastReading >= readingInterval) {
    // update the Madgwick filter:
    filter.updateIMU(gx, gy, gz, x, y, z);

    // get the heading, pitch and roll and print it:
    float roll = filter.getRoll();
    float pitch = filter.getPitch();
    float heading = filter.getYaw();
    Serial.print(heading);
    Serial.print(",");
    Serial.print(pitch);
    Serial.print(",");
    Serial.println(roll);
    lastReading = millis();      // update time stamp
  }
}
```

Connect It This **Processing sketch** reads the incoming data from the microcontroller and uses it to change the attitude of a disc onscreen in three dimensions. It will work with either of the previous accelerometer sketches, because they both output the same data in the same format. Make sure the serial port opened by the sketch matches the one to which your microcontroller is connected.

```
/*
 Madgwick display
 Context: Processing

 Takes the values in serially from an accelerometer
 attached to a microcontroller and uses them to set the
 attitude of a disk on the screen.
 */
import processing.serial.*;      // import the serial lib

float heading, pitch, roll;      // pitch and roll for rotation
float position;                  // position for translation

Serial myPort;                   // the serial port
```

The `setup()` method initializes the window and the serial connection, and sets the text formatting.

```
void setup() {
  // draw the window:
  size(400, 400, P3D);
  // calculate translate position for disc:
  position = width/2;

  // List all the available serial ports
  printArray(Serial.list());

  // Open whatever port is the one you're
using.
  myPort = new Serial(this, Serial.list()[0], 9600);
  // only generate a serial event when you get a newline:
  myPort.bufferUntil('\n');
  // set up text formatting:
  textSize(12);
  textAlign(CENTER, CENTER);
}
```

You will probably need to look at the output of Serial.list() and change this number to match the serial port that corresponds to your microcontroller.

The `draw()` method just refreshes the screen in the window, as usual. It calls a method, `tilt()`, to actually tilt the plane.

```
void draw () {
  // colors inspired by the Amazon rainforest:
  background(#20542E);
  fill(#79BF3D);
  // draw the disc:
  tilt();
}
```

The 3D system in Processing works on rotations from zero to 2*pi. `tilt()` maps the yaw, pitch, and roll into that range. It uses Processing's `translate()` and `rotate()` methods to move and rotate the plane of the disc to correspond with the sensors' movement.

```
void tilt() {
  // translate from origin to center:
  translate(position, position, position);

  // X is front-to-back:
  rotateX(radians(roll+90));
  // Y is left-to-right:
  rotateY(radians(pitch));
  // Z is heading:
  rotateZ(radians(heading));
  // set the disc fill color:
  fill(#79BF3D);
  // draw the disc:
  ellipse(0, 0, width/3, width/3);
  // set the text fill color:
  fill(#20542E);
  // Draw some text so you can tell front from back:
  text(heading + "," + pitch + "," + roll, 0,0,1);
}
```

The `serialEvent()` method reads all the incoming serial bytes and parses them as comma-separated ASCII values, just as you did in the Monski Pong project in Chapter 2.

Run this sketch with the accelerometer/gyrometer attached to your microcontroller, and your microcontroller programmed to output the yaw, pitch and roll, and you'll see something like Figure 8-15 as you move the sensor.

```
void serialEvent(Serial myPort) {
  // read the serial buffer:
  String myString = myPort.readStringUntil('\n');

  // if you got any bytes other than the linefeed:
  if (myString != null) {
    myString = trim(myString);
    // split the string at the commas
    String items[] = split(myString, ',');
    if (items.length > 2) {
      heading = float(items[0]);
      pitch = float(items[1]);
      roll = float(items[2]);
      println(heading, pitch, roll);
    }
  }
}
```

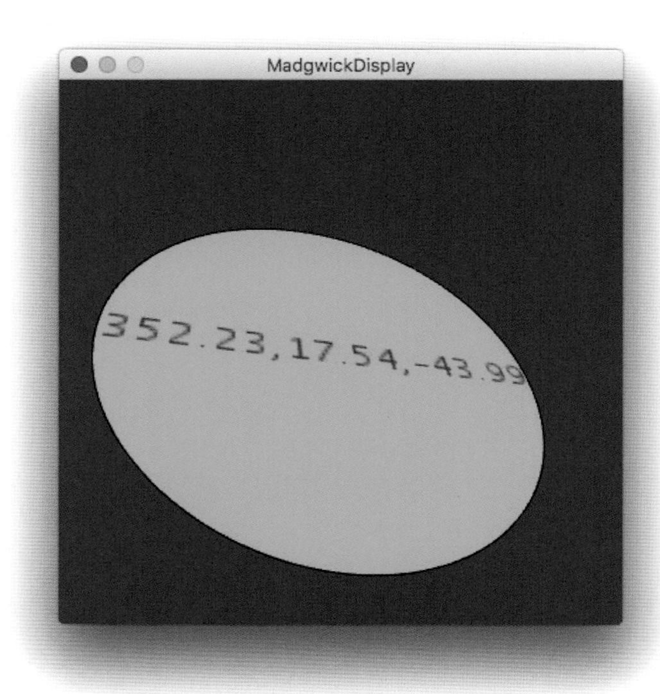

Figure 8-15
The output of the Processing Madgwick display sketch.

Though it may seem like a lot of work to go from the raw output of an accelerometer to the visualization shown in Figure 8-15, it's useful to understand the process. You read accelerations and rotations along three axes as raw sensor values, then converted those to acceleration and rotation measurements. Then, you used the math in the Madgwick library to convert the results to angles in degrees.

The advantage of having the results in degrees is that it's a known standard measurement, so you didn't have to do a lot of mapping when you sent the values to Processing. Instead, Processing could take the output from an accelerometer that gave it pitch and roll in degrees.

You don't always need this level of standardization. For many applications, all you care about is that the accelerometer readings are changing, as you saw in the first orientation function. However, if you want to convert those readings into a measurement of attitude relative to the ground, the latter method is more useful.

You may find that the Madgwick method drifts a bit on the Z axis. If so, and if you're also using a compass like the one in the LSM303 sensor, you can replace the yaw reading with the compass's heading to get a more reliable reading for the heading.

X

▸▸ Address 2007 by Mouna Andraos and Sonali Sridhar
Address shows that location technologies don't have to be purely utilitarian. *Photo by J. Nordberg.*

Conclusion

When you start to develop projects that use location systems, you usually find that less is more. It's not unusual to start a project thinking you need to know position, distance, and orientation, then pare away systems as you develop the project. The physical limitations of the things you build and the spaces you build them in solve many problems for you.

This effect, combined with your users' innate ability to locate and orient themselves, makes your job much easier. Before you start to solve all problems in code or electronics, put yourself physically in the place for which you're building, and do what you intend for your users to do. You'll learn a lot about your project, and save yourself time, aggravation, and money.

The examples in this chapter are all focused on a solitary person or object. As soon as you introduce multiple participants, location and identification become more tightly connected. This is because you need to know whose signal is coming from a given location, or what location a given speaker is at. In the next chapter, you'll see methods crossing the line from physical identity to network identity. X

9

MAKE: PROJECTS 

Identification

In the previous chapters, you assumed that identity equals address. Once you knew a device's address on the network, you started talking. Think about how disastrous this would be if you used this formula in everyday life: you pick up the phone, dial a number, and just start talking. What if you dialed the wrong number? What if someone other than the person you expected answers the phone?

Networked objects mark the boundaries of networks, but not of the communications that travel across them. We use these devices to send messages to other people. The network identity of the device and the physical identity of the person are two different things. Physical identity generally equates to presence (is it near me?) or address (where is it?), but network identity also takes into consideration network capabilities of the device and the state it's in when you contact it. In this chapter, you'll learn some methods for giving physical objects network identities. You'll also learn ways that devices on a network can learn each other's capabilities through the messages they send and the protocols they use.

◂◂ Sniff, a toy for sight-impaired children, by Sara Johansson
The dog's nose contains an RFID reader. When it detects RFID-tagged objects, it gives sound and tactile feedback—a unique response for each object. Designed by Sara Johansson, a student in the Tangible Interaction course at the Oslo School of Architecture and Design, under the instruction of tutors Timo Arnall and Mosse Sjaastad.
Photo courtesy of Sara Johansson.

Supplies for Chapter 9

Cameras and RFID readers are your main new components in this chapter. You'll be using them to identify colors, faces, tags, and tokens.

Distributor KEY

A Arduino Store (store.arduino.cc)
AF Adafruit (adafruit.com)
AMZ (www.amazon.com)
B Belkin (www.belkin.com)
D Digi-Key (www.digikey.com)
F Farnell (www.farnell.com)
ID Identive (www.identiveusa.com)
J Jameco (jameco.com)
P Pololu (www.pololu.com)
PX Parallax (www.parallax.com)
RS RS (www.rs-online.com)
SF SparkFun (www.sparkfun.com)
SS Seeed Studio (www.seeedstudio.com)

PROJECT 21: Color Recognition Using a Webcam

» **Personal computer with webcam**
» **Colored objects**

PROJECT 22: Face Detection Using a Webcam

» **Personal computer with webcam**
» **Face**

PROJECT 23: 2D Barcode Recognition Using Webcam

» **Personal computer with webcam**
» **Printer**

PROJECT 24: Reading RFID Tags

» **RFID reader** Identiv SCL3711 USB Smart Card Reader **ID** SCL3711
» **Mifare Classic RFID tags** The retailers listed sell tags in a variety of physical packages, so choose the ones you like best. **AF** 359 or 360**, SF** SEN-10128 or SEN-11319, **SS** 113990013
» **Raspberry Pi SF** DEV-13825, **AF** 3055 or 3400, **SS** 102010048 or 114990584, **RS** 896-8660, **F** 2525225

Figure 9-1. New parts for this chapter: **1.** Seeed Studio PN532 NFC shield **2.** Adafruit PN532 breakout board **3.** Identiv SCL3711 USB NFC reader **4.** USB camera (if your computer doesn't have one) **5.** Belkin WeMo switch **6.** Brightly colored objects **7.** Solenoid lock **8.** QR Codes **9.** Mifare Classic RFID tags

PROJECT 25: Reading and Writing NDEF Messages

- **RFID reader** Identiv SCL3711 USB Smart Card Reader **ID** SCL3711
- **Mifare Classic RFID tags** The retailers listed sell tags in a variety of physical packages, so choose the ones you like best. **AF** 359 or 360, **SF** SEN-10128 or SEN-11319, **SS** 113990013
- **Raspberry Pi SF** DEV-13825, **AF** 3055 or 3400, **SS** 102010048 or 114990584, **RS** 896-8660, **F** 2525225

PROJECT 26: NFC Meets Home Automation

- **1 solderless breadboard D** 438-1045-ND, **J** 20723 or 20601, **SF** PRT-12615 or PRT-12002, **F** 4692810, **AF** 64, **SS** 319030002 or 319030001
- **Perma-Proto Half-sized Breadboard PCB** If you plan to change to a soldered perf board. **A** 1609
- **4 1-1/2" hex standoffs, 4-40 threads female-female** For mounting in the enclosure **D** 36-2206-ND
- **2 1/2" hex standoffs 4-40 threads female-female** For mounting in the enclosure **D** 36-2203-ND
- **12 1/4" 4-40 thread pan head machine screws** For mounting in the enclosure **D** 36-9300-ND
- **1 project enclosure** using mat board as in Chapter 5
- **1 Arduino MKR1000 AF** 3156, **RS** 124-0657, **A** ABX00004, GBX00011 (Africa and EU), **D** 1659-1005-ND.
  Alternatively, an ESP8266-based board will work. **SF** WRL-13231, **AF** 2471
- **1 PN532 NFC reader module** These devices offer SPI, I2C, and UART connections. There's a DFRobot reader module that only exposes the UART that wasn't compatible with non-AVR-based boards at the time of writing **AF** 364, **D** 1528-1781-ND, **SS** 113030001
- **2 Mifare Classic tags AF** 359 or 360, **SF** SEN-10128 or SEN-11319, **SS** 113990013
- **2 Belkin WeMo switch modules B** P-F7C027
- **1 personal computer**
- **All necessary converters to communicate serially from microcontroller to computer** Just like the previous projects.

PROJECT 27: Two-Factor Authentication Using NFC

- **RFID reader** Identiv SCL3711 USB Smart Card Reader **ID** SCL3711
- **Mifare Classic RFID tags** The retailers listed sell tags in a variety of physical packages, so choose the ones you like best. **AF** 359 or 360, **SF** SEN-10128 or SEN-11319, **SS** 113990013
- **Raspberry Pi SF** DEV-13825, **AF** 3055 or 3400, **SS** 102010048 or 114990584, **RS** 896-8660, **F** 2525225
- **TIP120 transistor D** TIP120-ND, **J** 32993, **F** 9804005 **RS** 808-0502
- **1-kilohm resistor D** 1.0KQBK-ND, **J** 690865, **F** 9339051, **RS** 707-7666
- **Solenoid Lock AF** 1512 **AMZ** "uxcell DC 12V Open Frame Type Solenoid for Electric Door Lock"
- **12V power supply J** 170245, **F** 1176248
- **5.5mm outside diameter, 2.1mm inside diameter power jack D** CP3-1000-ND , **J** 28760, **SF** PRT-10287, **A** 369 **F** 1737256
- **Small breadboard SF** PRT-12044, **AF** 65, **D** 923273-ND
- **Perma-Proto Half-sized Breadboard PCB** If you plan to change to a soldered perf board. **A** 1609
- **4 3/4" hex standoffs, 4-40 threads female-female D** 36-2204-ND
- **8 1/4" 4-40 thread pan head machine screws** For mounting in the enclosure **D** 36-9300-ND
- **1 project enclosure** using mat board like Chapter 5.

PROJECT 28: IP Geocoding

- **Personal computer with an internet connection**

Physical Identification

The process of identifying physical objects is such a fundamental part of our experience that we seldom think about how we do it. We use our senses, of course: we look at, feel, pick up, shake and listen to, smell, and taste objects until we have a reference—then we give them a label. The whole process relies on some pretty sophisticated work by our brains and bodies, and anyone who's ever dabbled in computer vision or machine learning in general can tell you that teaching a computer to recognize physical objects is no small feat. Just as it's easier to determine location by having a human narrow it down for you, it's easier to distinguish objects computationally if you can limit the field—and if you can label the important objects.

We identify things using information from our senses, and so do computers. They can identify physical objects only by using information from their sensors. Two of the best-known digital identification techniques are *optical recognition* and *radio frequency identification* (*RFID*). Optical recognition can take many forms, from video color tracking and shape recognition to the ubiquitous barcode. Once an object has been recognized by a computer, the computer can give it an address on the network.

The network identity of a physical object can be centrally assigned and universally available, or it can be provisional. It can be used only by a small subset of devices on a larger network or used only for a short time. RFID is an interesting case in point. The RFID tag pasted on the side of a book may seem like a universal marker, but what it means depends on who reads it. The owner of a store may assign that tag's number a place in his inventory, but to the consumer who buys it, it means nothing unless she has a tool to read it and a database in which to categorize it. She has no way of knowing what the number meant to the store owner unless she has access to his database. Perhaps he linked that ID tag number to the book's title, or to the date on which it arrived in the store. Once it leaves the store, he may delete it from his database, so it loses all meaning to him. The consumer, on the other hand, may link it to entirely different data in her own database, or she may choose to ignore it, relying on other means to identify it. In other words, there is no central database linking RFID tags and the things to which they're attached or the people who possessed them.

Like locations, identities become more uniquely descriptive as the context they describe becomes larger. For example, knowing that my name is Tom doesn't give you much to go on. Knowing my last name narrows it down some more, but how effective that is depends on where you're looking. In the United States, there are dozens of Tom Igoes. In New York City, there are at least three. When you need a unique identifier, you might choose a universal label, like using my Social Security number, or you might choose a provisional label, like calling me "Frank's son, Tom." Which you choose depends on your needs in a given situation. Likewise, you may choose to identify physical objects on a network using universal identifiers, or you might choose to use provisional labels in a given temporary situation.

The capabilities assigned to an identifier can be fluid as well. Considering the RFID example again: in the store, a given tag's number might be enough to set off alarms at the entrance gates, or to cause a cash register to add a price to your total purchase. In another store, that same tag might be assigned no capabilities at all, even if it's using the same protocol as other tags in the store. Confusion can set in when different contexts use similar identifiers. Have you ever left a store with a purchase and tripped the alarm, only to be waved on by the clerk who forgot to deactivate the tag on your purchase? When two companies use the same type of RFID tags but don't differentiate by the tags' unique IDs, you're likely to trip the alarms when you bring a product from one company into the other's store. It's not enough just to read the tag. You have to associate its ID with a record in your database to give it meaning.

Figure 9-2
Video color recognition. Simple color tracking works best with colors that are vibrantly different than their backgrounds.

Video Identification

All video identification relies on the same basic method: the computer reads a camera's image and stores it as a two-dimensional array of pixels. Each pixel has a characteristic brightness and color that can be measured using any one of a number of palettes: red-green-blue is a common scheme for video- and screen-based applications, as is hue-saturation-value. Cyan-magenta-yellow-black is common in print applications. The properties of the pixels, taken as a group, form patterns of color, brightness, and shape. When those patterns resemble other patterns in the computer's memory, the computer can identify those patterns as objects. Figure 9-2 shows an example, using a bright pink monkey. The application in question recognizes a shade of pink, and is indicating the pixel closest to that color pink.

In the three projects that follow, you'll use computer vision to read an image from your personal computer's camera, and then analyze the image. The first, and simplest, will look for a color. The second will look for something resembling a face. The third will look for a 2D barcode called a *QR (Quick Response) code*.

Computer vision (CV) has gotten considerably more robust and common since the last edition of this book, as have cameras in computers. It's difficult to find a personal computer or mobile device that doesn't have a camera anymore. Camera drivers are no longer a problem, since video is now standard USB device type. Image capture from a camera is a standard feature built into HTML5 now, so you can access the camera on a computer from any browser using JavaScript. If you don't like the JavaScript libraries being used in these projects, there are several others you can choose from. Do a little web searching to find ones you like better.

Color Recognition

Recognizing objects by color is a relatively simple process, if you know that the color you're looking for is unique in the camera's image. This technique is used in film and television production to make superheroes fly. The actor is filmed against a screen of a unique color, usually green, which isn't a natural color for human skin. Then, the pixels of that color are removed, and the image is combined with a background image.

Color identification can be an effective way to track physical objects in a controlled environment. Assuming you've got a limited number of objects in the camera's view, and each object's color is unique and doesn't change with the lighting conditions, you can identify each object reasonably well. Even slight changes in lighting can change the color of a pixel, however, so lighting conditions need to be tightly controlled, as the following project illustrates.

X

Project 21

Color Recognition Using a Webcam

In the next three projects, you'll get a firsthand look at how computer vision works. The p5.js sketch shown here uses your computer's video camera to generate a digital image, looks for pixels of a specific color, and then marks them on the copy of the image that it displays.

MATERIALS

- **» Personal computer with working camera**
- **» Colored objects**

Color tracking is the most basic of computer vision applications, and a good place to start. The following p5.js sketch uses the HTML5 video library to get the pixels from a camera image and scan through them for a color you choose. To use this, you'll need to have a working camera attached to your computer. It will work on a mobile device, but the small screen might make clicking on a color difficult. You'll also need some small colored objects—stickers or toy balls work well.

Since all three of the next projects are camera projects, and therefore require an HTTPS server, make a copy of the geolocation project from the last chapter to start this section. You can use the same server for all three projects.

The next three projects are all video tracking applications built in p5.js. Because they use the camera, they must be served using HTTPS, just like the geolocation sketch in Chapter 8. You'll use that same server, https-server.js, for all three. Don't forget that it requires your private key and certificate. When you copy the geolocation project to start this project, make sure to copy the keys subdirectory as well.

Name this new project **CameraProjects**. Delete the contents of the **public** directory. Make a new p5.js project in **public** called **colorTracking**. Include all the files you need for a p5.js project in the **colorTracking** directory: the **index.html** file, the **sketch.js** file, and the **libraries** directory, which contains the p5.js libraries. In this project you'll only edit the **sketch.js** file as usual. When you get to the next two projects, you'll also make subdirectories for them in the **public** directory.

Open the **sketch.js** file and add the following global variables at the top of the file, before the `setup()` function. The `video` variable will make an instance of the video API. `trackColor` will hold the color values for the color you plan to track, and `differenceThreshold` will set the threshold of difference for matching colors. `dSlider` is a slider to set the difference threshold.

The `setup()` function sets the initial conditions, initializes the video capture instance, sizes it, and positions it. Then you'll add a canvas on top of it, so you can draw on top of the video.

```
/*
 ColorTracking with HTML5 video
 Context: p5.js
 Based on an example by Daniel Shiffman
 */
 var video;                          // the video capture object
 var trackColor = [255,0,0];         // the color you're searching for
 var differenceThreshold = 10;       // the threshold of color difference
 var dSlider;                        // slider for the threshold

 function setup() {
   video = createCapture(VIDEO);     // take control of the camera
   video.size(400, 300);             // set the capture resolution
   video.position(0, 0);          // set the position of the camera image
   var canvas = createCanvas(400, 300); // draw the canvas over the image
   canvas.position(0,0);             // set the canvas position
   dSlider = createSlider(0, 100, 10, 1); // initialize the slider
   dSlider.position(10, height-30);   // position the slider
   dSlider.touchEnded(setDifference);  // set a touchEnded callback
 }
```

The `draw()` method begins by reading the camera, drawing it to the window, and saving the resulting image in a pixel array.

Next comes a pair of nested for loops that iterates over the rows and columns of pixels. This is a standard method for examining all the pixels of an image. The pixel array contains four items for each pixel: red, green, blue, and alpha, or transparency. With each pixel, you determine its position in the array with the following formula (which you'll see frequently):

```
arrayLocation = x + (y * width) * 4;
```

Once you have the pixel's position, you extract the red, green, and blue values, as well as the values for the color you wish to track.

```
function draw() {
  image(video,0,0);    // draw the video
  video.loadPixels(); // get the video pixels in an array

// Loop through every pixel:
  for (var x = 0; x < video.width; x++ ) {       // rows
    for (var y = 0; y < video.height; y++ ) {    // columns
      // calculate the pixel's position in the array
      // based on its width and height:
      var loc = (x + y * video.width) * 4;
      // get the values of the current pixel:
      var r1 = video.pixels[loc];               // red value of the pixel
      var g1 = video.pixels[loc + 1];           // green value
      var b1 = video.pixels[loc + 2];           // blue value
      // get the values of the tracked color:
      var r2 = trackColor[0];                   // red value
      var g2 = trackColor[1];                   // green value
      var b2 = trackColor[2];                   // blue value
```

Next, calculate the difference between the two colors. By treating the red, green, and blue values of each color as positions in 3-dimensional space, you can find the difference by calculating the Euclidean distance between them. p5.js's `dist()` function is handy for this.

Once you know the difference, compare that to the `differenceThreshold` variable. If it's less, then this pixel is a match. Draw a black point over it, so you can tell which pixels match the color being tracked.

```
      // use the dist() function to get the difference between the colors:
      var difference = dist(r1, g1, b1, r2, g2, b2);

      // If current pixel's color is within the difference threshold:
      if (difference < differenceThreshold) {
        stroke(0);          // set stroke (and point) color to black
        point(x, y);        // draw a point at the current pixel position
      }
    }
  }
```

Now draw text at the bottom of the screen so you know what color's being tracked and what the color threshold is. That's the end of the `draw()` function.

You'd like to be able to pick a color to track by clicking on it, so use the `mousePressed()` function to make that possible. When you click on a pixel, this function will put that pixel's color values into `trackColor`.

```
  // put some text at the bottom of the screen:
  fill(255);
  text("search color: " + trackColor, 10, height-60);
  text("difference threshold: " + differenceThreshold, 10, height-40);
} // end of draw() function


// Save color where the mouse is clicked in trackColor variable:
function mousePressed() {
  if (mouseY < dSlider.y) {            // if you clicked above the slider,
    trackColor = video.get(mouseX,mouseY);  // get the color you clicked
  }
}
```

»

Finally, you need a way to change `differenceThreshold`. Use the slider for this. You set a callback handler for it in the `setup()` function, so when the user stops touching the slider handle, this function will get called. It will put the slider's value into `differenceThreshold`.

Continued from previous page.

```
// if the slider moves, change the differenceThreshold:
function setDifference() {
    differenceThreshold = dSlider.value();
}
```

Running Video Sketches

To run this sketch, start **http-server.js** from the command line as you did in the last chapter:

```
$ sudo node http-server.js
```

You'll be asked for your password again. Then open a browser and go to http://localhost:8080/colorTracking. You'll be warned that this is an insecure page again, and asked if this page is allowed to use the camera. Click yes, and the video will load. You'll follow the same process for the video sketches in the next two projects as well.

Lighting for Color Tracking

As you can see when you run this sketch, it's not the most robust color tracker! Changing the `color-Threshold` helps, but eventually it bleeds in similar colors, as you can see in Figure 9-3. You can get it to be more precise by controlling the setting very carefully. There are some lighting tricks you can use as well:

- DayGlo colors under ultraviolet fluorescent lighting tend to be the easiest to track, but they lock you into a very specific visual aesthetic.
- Objects that produce their own light are easier to track, especially if you put a filter on the camera to block out stray light. A black piece of 35mm film negative (if you can still find 35mm film!) works well as a visible light filter, blocking almost everything but infrared light. Two polarizers, placed so that their polarizing axes are nearly perpendicular, are also effective. Infrared LEDs track very well through this kind of filter, as do incandescent flashlight lamps.
- Regular LEDs don't work well as color-tracking objects unless they're relatively dim. Brighter LEDs tend to show up as white in a camera image because their brightness overwhelms the camera's sensor.

Color recognition doesn't have to be done with a camera; color sensors can do the same job. Texas Advanced Optoelectronic Solutions (www.taosinc.com) makes a few different color sensors, including the TAOS TCS34725, which you can get from Adafruit. This sensor contains four photodiodes, three of which are covered with color filters; the fourth is not, so it can read red, green, blue, and white light.

X

Figure 9-3
Video color recognition in p5.js. Threshold set to 10, 32, and 47, left to right. The higher the difference threshold, the more you match pixels outside the desired object.

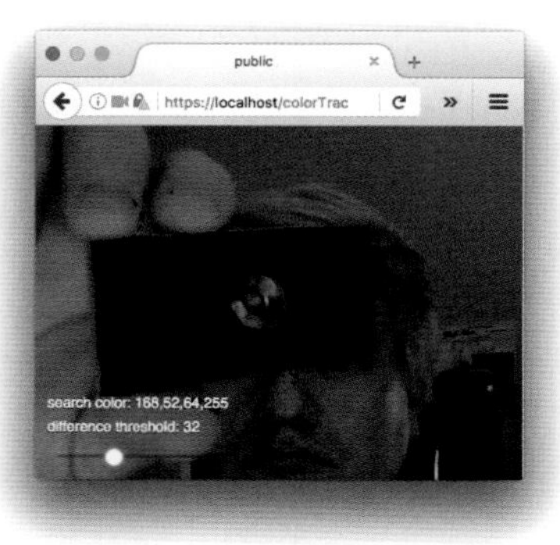

Challenges of Identifying Physical Tokens

Designer Durrell Bishop's marble telephone answering machine from 1992 (vimeo.com/19930744) is an excellent example of the challenges of identifying physical tokens. With every new message the machine receives, it drops a marble into a tray on the front of the machine. The listener hears the messages played back by placing a marble on the machine's "play" tray. Messages are erased and the marbles are recycled when they are dropped back into the machine's hopper. Marbles become physical tokens representing the messages, making it very easy to tell at a glance how many messages there are.

Bishop tried many different methods to reliably identify and categorize physical tokens representing the messages:

> I first made a working version with a motor and large screw (like a vending machine delivery mechanism), with pieces of paper tickets hung on the screw, and had different color gray levels on the back. When it got a new message, the machine read the next gray before it rotated once and dropped the ticket. It was a bit painful, so I bought beads and stuffed resistors into the hole which was capped (soldered) with sticky-backed copper tape. When I went to Apple and worked with Jonathan Cohen, we built a properly hacked version for the Mac with networked barcodes. Later, again with Jonathan but this time at Interval Research, we used the Dallas ID chips.
>
> —*Durrell Bishop, personal communication*

Color by itself isn't enough to give you identity in most cases, but there are ways in which you can design a system to use color as a marker of physical identity. However, it has its limitations. In order to tell the marbles apart, Bishop could have used color recognition to read the marbles, but that would limit the design in at least two ways. First, there would be no way to tell the difference between multiple marbles of the same color. If, for example, he wanted to use color to identify the different people who received messages on the same answering machine, there would then be no way to tell the difference between multiple messages for each person. Second, the system would be limited by the number of colors between which the color recognition can reliably differentiate.

Shape and Pattern Recognition

Recognizing a color is relatively simple computationally, but recognizing a physical object is more challenging. To do this, you need to know the two-dimensional geometry of the object from every angle so that you can compare any view you get of the object.

A computer can't actually "see" in three dimensions using a single camera. The view it has of any object is just a two-dimensional shadow. Furthermore, it has no way of distinguishing one object from another in the camera view without some visual reference. The computer has no concept of a physical object. It can only compare and match patterns in the camera's pixel array. It can rotate the view, stretch it, and do all kinds of mathematical transformations on the pixel array, but the computer doesn't understand an object as a discrete entity the same way a human does.

Face Detection

If you've used a digital camera developed in the past 10 years or so, chances are it's got a face-detection algorithm built in. It'll put a rectangle around each human face and attempt to focus on it. Face detection is a good example of visual pattern recognition. The camera looks for a predefined pattern that describes, generically, a face. It has a particular proportion of height to width: there are two darker spots about one-third of the way from the top, a second darker spot about two-thirds of the way down, and so forth. Facial detection is not facial identification—a face-detection algorithm generally isn't looking specifically enough at an image to tell you who the person is, just that they have something that more or less resembles a face. The following project shows you how to detect faces.

X

Project 22

Face Detection Using a Webcam

Now that you've got a basic understanding of how optical detection works from the color-tracking project, it's time to try some simple pattern detection. In this project, you'll use facial-detection methods to look for faces in a camera image.

MATERIALS

- » **Personal computer with working camera**
- » **Faces**

A number of facial recognition libraries are available for JavaScript, and each uses different algorithms for finding faces. All of them use pattern description files to describe the characteristics of a particular pattern. This describes the subregions of a given pattern, including their relative sizes, shapes, and contrast ratios. The patterns are designed to be general enough to allow for some variation, but specific enough to tell it from other patterns. The tracking.js library used here comes with patterns for the following human features: frontal face view, eyes, and mouth. The sketch attempts basic facial recognition; see if you can fool it.

To get started, make a new p5.js project called **faceRecognition** in the **public** directory of **CameraProjects**. Copy all the files necessary for a p5.js project into this new directory. Then download tracking.js from trackingjs.com. Unzip it, and copy the contents of the **build** subdirectory into the **libraries** subdirectory of the **faceRecognition** p5.js project. This should include the files **tracking.js**, and **tracking-min.js**, and a directory called **data**, which contains the pattern description files for eyes, mouth, and faces.

Now open the **index.html** file for the p5.js project, and add the following lines after the other libraries' `<script>` tags:

```
<script src="libraries/tracking-min.js"></script>
<script src="libraries/data/face-min.js"></script>
```

If you want to try tracking eyes and mouths, add lines like the last one for **eye-min.js** and **mouth-min.js** as well. Save the file, then open **sketch.js** and write it as follows.

Before the `setup()` function, initialize the global variables as usual. You only need the canvas and an instance if the video library. The beginning of the `setup()` function is similar to the previous project. Once you've set the video and the canvas, make a new ObjectTracker using the tracker library, and set its parameters using the functions `setInitialScale()`, `setStepSize()`, and `setEdgesDensity()`. Changing these will affect the accuracy of your tracking. Try changing them to see how.

When you start tracking, you pass it the video element. Then you set a callback function for when you get a successful track.

Finally, the `setup()` function sets the drawing conditions for ellipses, so you can draw circles over the faces.

```
/*
  Face detection using tracking.js
  Context: P5.js
*/
var canvas;      // the drawing createCanvas
var tracker;     // instance of the tracker library

function setup() {
  // setup camera capture
  var video = createCapture(VIDEO); // take control of the camera
  video.size(400, 300);             // set the capture resolution
  video.position(0, 0);             // set the position of the camera image
  canvas = createCanvas(400, 300);  // draw the canvas over the image
  canvas.position(0,0);             // set the canvas position
  // set up the tracker:
  tracker = new tracking.ObjectTracker('face');
  tracker.setInitialScale(4);       // set the tracking parameters
  tracker.setStepSize(2);
  tracker.setEdgesDensity(0.1);
  tracking.track(video.elt, tracker); // track on the video image
  tracker.on('track', showTracks);    // callback for the tracker
  ellipseMode(CORNER);                // draw circles from the corner
}
```

The `draw()` function does nothing, since all the action is driven by the callback function `showTracks()`. When the tracker finds faces, it generates an array of coordinates for rectangles in `event.data`. Make a for loop to iterate over that array and draw an ellipse at each rectangle.

```
function draw() {
  // nothing to do here.
}  // end of draw() function

function showTracks(event) {
    clear();
    var faces = event.data;
    for (f in faces) {
      fill(0xFF, 0x00, 0x84, 0x3F);   // a nice shade of fuchsia
      noStroke();                      // no border
      ellipse(faces[f].x, faces[f].y, faces[f].width, faces[f].height);
    }
}
```

When you run this, point your face at it, and you'll get a nice fuchsia mask, as shown in Figure 9-4.

Try the program on different versions of faces and different conditions.

There's much more that tracking.js can do, as documented on their website.

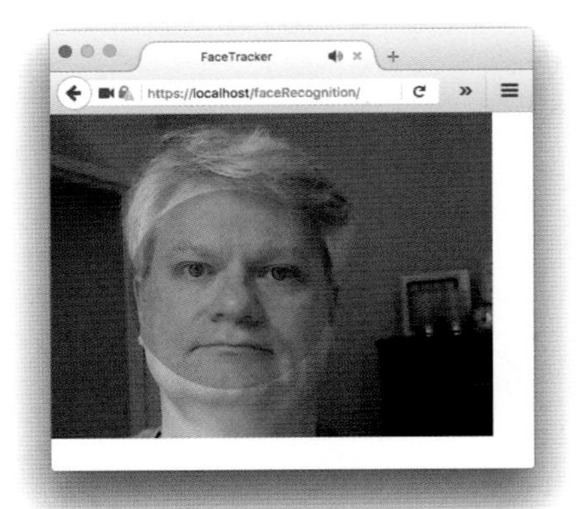

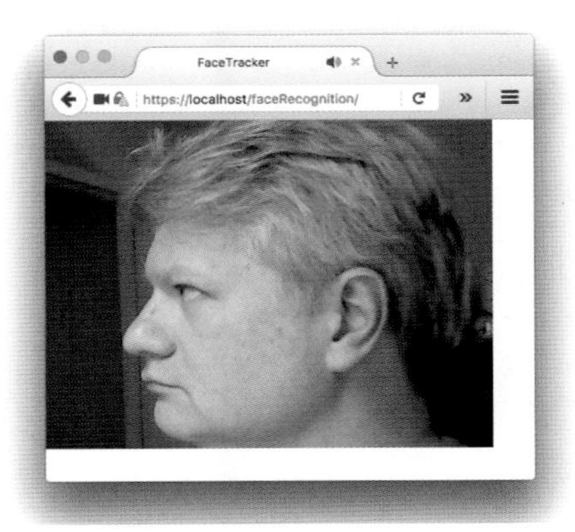

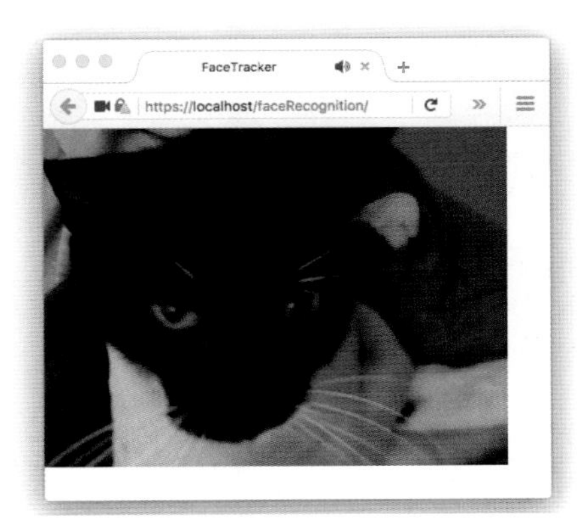

Figure 9-4
The face detection finds me fairly well, but turning sideways, I disappear.

It does reasonably well with a photo of people. Facial hair and other markings make a person harder to detect.

Animals are not detected by this algorithm. Noodles is not detected (and not happy either).

Barcode Recognition

A barcode is simply a pattern of dark and light lines or cells used to encode an alphanumeric string. A computer reads a barcode by scanning the image and interpreting the widths of the light and dark bands as zeroes or ones. This scanning can be done using a camera or a single photodiode, if the barcode can be passed over the photodiode at a constant speed. Many handheld barcode scanners work by having the user run a wand with an LED and a photodiode in the tip over the barcode, and reading the pattern of light and dark that the photodiode detects.

The best known barcode application is the *Universal Product Code*, or *UPC*, used by nearly every major manufacturer on the planet to label goods. There are many dozen different barcode symbologies, which are used for a wide range of applications. For example, the U.S. Postal Service uses POSTNET to automate mail sorting. European Article Numbering (EAN) and Japanese Article Numbering (JAN) are supersets of the UPC system developed to facilitate the international exchange of goods. Each symbology represents a different mapping of bars to characters. The symbologies are not interchangeable, so you can't properly interpret a POSTNET barcode if you're using an EAN interpreter. This means

Figure 9-5
A one-dimensional barcode. This is the ISBN barcode for this book.

that either you have to write a more comprehensive piece of software that can interpret several symbologies, or you have to know which one you're reading in advance. There are numerous software libraries for generating barcodes and several barcode fonts for printing the more popular symbologies.

Barcodes such as the one shown in Figure 9-5 are called *one-dimensional barcodes* because the scanner or camera needs to read the image only along one axis. There are also *two-dimensional barcodes* that encode data in a two-dimensional matrix for more information density. As with one-dimensional barcodes, there are a variety of symbologies. Figure 9-6 on page 336 shows a typical two-dimensional barcode. This type of code, the *QR (Quick Response) code*, was created in Japan and originally used for tracking vehicle parts, but it's since become popular for all kinds of product labeling. It's also shown up in many mobile apps, to allow a quick exchange of data, like WeChat's contact exchange. The following project uses a JavaScript library to read QR codes in p5.js.

Backslash, by Pedro Oliveira and Xuedi Chen
Created as a critical design project, Backslash aims to help protestors maintain their safety and connection to remote supporters through use of digital technologies. The kit includes a QR code-patterned bandana for embedding messages and public keys, which can be read from a photo of the protestor More details available at www.backslash.cc.

Photo courtesy of Roy Rochlin

Project 23

2D Barcode Recognition Using a Webcam

In this project, you'll generate some two-dimensional barcodes from text using an online QR code generator. Then you'll decode your tags using a camera and a browser. Once this works, try decoding the QR code illustrations in this book.

MATERIALS

- » **Personal computer with working camera**
- » **QR codes**

This sketch reads QR codes using a camera attached to a computer or mobile device. The video component is very similar to the previous examples. You'll recognize the video setup code. Before you start on the sketch, though, you'll need some QR codes to read. Fortunately, a number of QR code generators are available online. Just type the term into a search engine and see how many pop up. There's a good one at webqr.com/create.html, from which you can generate URLs, phone numbers, or plain text. The more text you enter, the larger the symbol. Generate a few codes and print them out for use with this sketch.

To get started, make a new directory called **qrCodeReader** in the **public** directory of **CameraProjects**. Copy all the files necessary for a p5.js project into this new directory. Then download the qrCodeJS library .zip file from github.com/lagoLast/qrcodejs. Unzip it, and copy the **qrcode.js** file from the of the **dist** subdirectory into the libraries subdirectory of the **qrCodeReader** p5.js project. Open the **index.html** file for the **qrCodeReader** project, and add the following line after the other libraries' `<script>` tags:

```
<script src="libraries/qrcode.js"></script>
```

Now open the sketch.js and write it as follows:

The global variables for this sketch include an instance of the qrReader library and a configuration JSON object for the reader. You need to set the callback functions for the QR code reader's success and error events here, and the name of the video canvas element.

```
/*
QR Code reading with HTML5 video
Context: p5.js
*/
var video;                          // the video capture object
var message;                        // the QR code message
var qrReader;                       // the QR code reader instance
var config = {                      // config for the QR code reader
  sucessCallback: getMessage,       // callback for when a code is read
  errorCallback: onError,           // callback for errors
  videoSelector: 'video',           // name of the canvas
  stopOnRead: false                 // don't stop when you get a code
}
```

The `setup()` for this sketch is similar to the last two. You'll create a video capture canvas and positon it as you did previously. You'll also give it an ID so that the qrReader can access it. Then you'll make a new instance of the qrReader using the configuration you set earlier.

```
function setup() {
  video = createCapture(VIDEO);         // take control of the camera
  video.size(400, 300);                 // set the capture resolution
  video.position(0, 0);                 // position the camera image
  var canvas = createCanvas(400, 300); // draw the canvas over the image
  canvas.position(0,0);                 // set the canvas position
  canvas.elt.id = "video";              // name the canvas element
  qrReader = new QrReader(config);      // make a new instance of qrReader
}
```

The `draw()` method draws the camera image and prints a status message to the screen. If the decoder detects a QR code, it changes the status message.

```
function draw() {
  image(video,0,0);                      // draw the video
  video.loadPixels();                    // get the video pixels in an array
  fill(255);
  text("read: " + message, 10, height - 20);
} // end of draw() function

function getMessage(result) {           // if the reader reads a QR code,
  message = result;                      // put it in the message variable
}

function onError(err) {                 // if there's a reader error,
  console.error(err);                    // print it to the console
}
```

When you run this, notice how the read quality is more reliable when the image is better. The distortion from the analog-to-digital conversion through the camera can cause errors. This error is made worse by poor optics or low-end camera imaging chips. As mobile phone cameras have improved, so has QR code scanning. Even with a good lens, if the code to be scanned isn't centered, the distortion at the edge of an image can throw off the pattern-recognition routine. You can improve the reliability of the scan by guiding the user to center the tag before taking an image. Even simple graphic hints like putting crop marks around the tag, as shown in Figure 9-6, can help. When you do this, users framing the image tend to frame to the crop marks, which ensures more space around the code and a better scan. Such methods help with any optical pattern recognition through a camera, whether it's one- or two-dimensional barcodes, or another type of pattern altogether.

Figure 9-6
A two-dimensional barcode (a QR code, to be specific) with crop marks around it. The image parsers won't read the crop marks, but they help users center the tag for image capture.

Optical recognition forces one additional limitation besides those mentioned earlier: you have to be able to see the barcode. By now, most of the world is familiar with barcodes, because they decorate everything we buy or ship. This limitation is not only aesthetic. If you've ever turned a box over and over trying to get the barcode to scan, you know that it's also a functional limitation. A system that allowed for machine recognition of physical objects—but didn't rely on a line of sight to the identifying tag—would be an improvement. For this, RFID and NFC are often viable alternatives to QR codes or other visual barcodes.

X

Radio Frequency Identification (RFID) and Near Field Communication (NFC)

Like barcode recognition, RFID relies on tagging objects in order to identify them. Unlike barcodes, however, RFID tags don't need to be visible to be read. An RFID reader sends out a short-range radio signal, which is picked up by an RFID tag. The tag then transmits back a short string of data, a unique ID number. Depending on the size and sensitivity of the reader's antenna and the strength of the transmission, the tag can be several feet away from the reader, enclosed in a book, box, or item of clothing, and still be read.

There are two types of RFID system: passive and active, just like distance-ranging systems. Passive RFID tags contain an integrated circuit that has a basic radio transceiver and a small amount of nonvolatile memory. The tags have no battery. They are powered by the current that the reader's signal induces in their antennas. The received energy is just enough to power the tag to transmit its data once, and the signal is relatively weak. Most readers of passive tags can only read tags a few inches to a few feet away. Figure 9-7 shows the range of a typical passive RFID reader. More expensive, and larger, readers can read for a few meters, but those are generally beyond the scope of the hobbyist in cost.

In an active RFID system, the tag has its own power supply and radio transceiver, and it transmits a signal in response

Figure 9-7
The field of an RFID reader, by Timo Arnall. This stop-motion photo shows the effective range and shape of the RFID reader's field. You can see it's just a few centimeters.

Figure 9-8
RFID tags in all shapes and sizes, photo by Timo Arnall. All of these items have RFID tags in them. For more information on RFID design research by Arnall and his colleagues, see www.elasticspace.com.

to a received message from a reader. Active systems can transmit for a much longer range than passive systems, and they are less error-prone. They are also much more expensive. If you're a regular automobile commuter, and you pass through a toll gate during your commute, you're probably an active RFID user. Systems like E-ZPass use active RFID tags so that the reader can be placed several meters away from the tag.

You might think that because RFID is radio-based, you could use it to do radio distance ranging as well, but that's not the case. Neither passive nor active RFID systems are typically designed to report the signal strength from the tag's reply. Without this information, it's impossible to use RFID systems to determine the actual location of a tag. All the reader can tell you is that the tag is within reading range.

As shown in Figure 9-8, RFID tags come in a number of different forms: sticker tags, coin discs, key fobs, credit cards, playing cards, even capsules designed for injection under the skin. The latter are used for pet tracking and are not designed for human use, though there are some adventurous hackers who have inserted these tags under their own skin. Like any radio signal, RFID can be read through a number of materials, but it is blocked by any kind of RF shielding, such as wire mesh, conductive fabric lamé, metal foil, or adamantium skeletons. This feature means that you can embed it in all kinds of projects, as long as your reader has the signal strength to penetrate the materials.

There are many different standard RFID frequencies. Short-range passive readers come in at least three common frequencies: two low-frequency bands at 125 and 134.2kHz, and high-frequency readers at 13.56MHz. The higher-frequency readers allow for faster read rates and longer-range reading distances. In addition to different frequencies, there are different protocols. The International Standards Organization, ISO, maintains the protocols for RFID, among other technologies. There are multiple ISO standards for different aspects of the technology. Some standards define how the radio communication works. Others define how data is structured on the tags. Different manufacturers implement these standards in their own brands—Philips' I-Code, Texas Instruments' Tag-IT HF, Picotag, Philips' Mifare and Mifare UL, ST's SR176, and many others. You can't expect one reader to read every tag. You can't even count on one reader to read all the tags in a given frequency range. You have to match the tag's protocol to the reader's. Within a given protocol, though, manufacturers' products are generally interoperable. Figure 9-9 shows the Near Field Communication protocol stack, which uses many related protocols.

⚠ **Most RFID capsules are not sterilized for internal use in animals (humans included), and they're definitely not designed to be inserted without qualified medical supervision. Besides, insertion hurts. Don't RFID-enable yourself or your friends. Don't even do it to your pets—let your vet do it. If you're really gung-ho to be RFID-tagged, make yourself a nice set of RFID-tag earrings, or wear an RFID ring.**

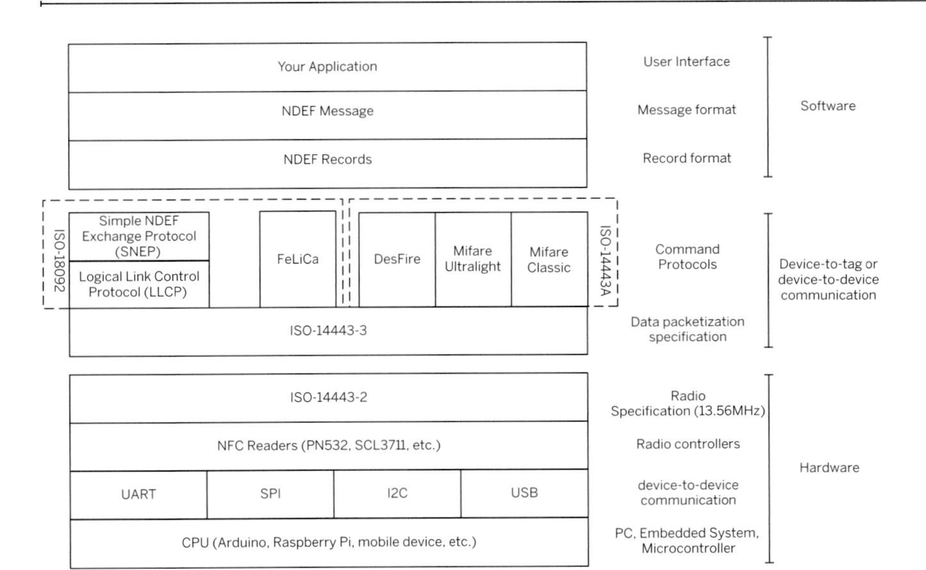

Figure 9-9
The NFC Protocol Stack. This gives you an idea how many different protocols underlie a typical RFID or NFC application. Protocol stacks make it possible for manufacturers to make their products somewhat interoperable, and for you to only worry about how your application code works with the upper layers of the stack. Image adapted from **Beginning NFC**, by Tom Igoe, Don Coleman, and Brian Jepson (O'Reilly, 2014)

RFID systems vary widely in cost. Active systems can cost tens of thousands of dollars to purchase and install. Commercial passive systems can also be expensive. A typical passive reader that can read a tag a meter away from the antenna typically costs a few thousand dollars. At the low end, short-range passive readers can come as cheap as $30 or less. Longer range readers go up significantly in price. There are a number of inexpensive and simple readers on the market. For example, ID Innovations makes a range of 125kHz readers with a serial output, capable of reading EM Microelectronics' EM4001 protocol tags. SparkFun sells these readers and matching tags. Parallax also sells a 125kHz reader that can read these tags. These were featured in earlier editions of this book. Though these readers are very simple to use—they usually have an asynchronous serial interface—they don't offer the range of functions afforded by the Near Field Communications devices that follow.

Before picking a reader, think about the environment in which you plan to deploy it, and how that affects both the tags and the reading. Will the environment have a lot of RF noise? In what frequency range? Consider a reader outside that range. Will you need a relatively long-range read? If so, look at the high-frequency readers or active tags. If you're planning to read existing tags rather than tags you purchase yourself, research carefully in advance, because not all readers will read all tags. Pet tags can be some of the trickiest—many of them operate in the 134.2kHz range, in which there are fewer readers from which to choose.

Near Field Communication

The ISO 14443 family of protocols defines standards used in contactless payment systems and smart cards, which can store information on the card, encrypt the transaction for security, and verify card data. These protocols define both radio specifications and data packetization standards. For example, many mass transit systems use fare cards and readers based on these standards. They form the basis of *Near Field Communication*, or *NFC* protocol, which expands RFID's capabilities. NFC is designed to provide the basic functionality of RFID, and adds the capability for peer-to-peer communications between devices in addition to the device-to-tag communication afforded by RFID.

NFC tags can carry more than just a unique ID. The *NFC Data Exchange Format (NDEF)* supports several data types, including text records, URLs, and digital signatures. For example, you can store a URL on an NFC tag, and any device that has an NFC reader and a web browser is expected to be able to open that URL and show you what's there.

The following projects all use NFC-compatible tags and readers. There are a number of readers on the market, from Identiv, Adafruit, and Seeed Studio, among others. These projects only show a fraction of what NFC can do, however. For a more in-depth discussion of NFC's capabilities, see **Beginning NFC**, by Tom Igoe, Don Coleman, and Brian Jepson (O'Reilly, 2014).

X

 Project 24

Reading RFID Tags

In this project, you'll install software tools to control NFC readers, and read some tags to get a sense for how the readers behave. You'll read the unique ID on the tag and see how far away from your reader a tag can be read. This is a good place to start any time you're adding RFID or NFC to a project.

Although we're talking about Near Field Communication (NFC), remember that it's based on RFID. In this project you'll install the software tools to use an NFC reader, but you'll only read the unique ID of the tags. In other words, you'll use the RFID capabilities that underlie NFC.

Philips Mifare Classic and Mifare Ultralight RFID protocols, common in transit systems, are ISO 14443–compatible. Philips was one of the companies in the consortium that developed NFC, so its RFID standard, Mifare, became one of the default types of tags that could work with NFC.

The Identiv SCL3711 NFC reader can read both Mifare Classic RFID tags and other NFC-compatible tags, and connects to a computer via USB. It works well on a Raspberry Pi as well as other computers with USB host ports, like a personal computer. To use it, you'll need to install a few extra software tools using the command-line interface of the Pi. Much of this project is about installing the tools for using RFID and NFC on the Pi. Once installed, however, you'll be able to use these same tools for future projects.

This project assumes you're running a current version of the Raspian distribution of the Linux operating system, as explained in Chapter 1. The software shown here will work on any POSIX-style operating system, including Apple's OSX, Ubuntu, the Beagle Bone boards running Debian, and probably other embedded boards as well, though you'll have to check the websites of the various tools used to learn what to do differently for your system if you're not using a Pi.

MATERIALS

» **Identiv SCL3711 USB Smartcard Reader**
» **Mifare Classic RFID tags**
» **Raspberry Pi**

Power up your Pi and log into it via a serial terminal or ssh as you did in Chapter 1. The serial terminal is faster if you're not sure of your Pi's address on your network, or if it's not networked. Once you're logged in, you're going to download the necessary libraries and install them. You'll need:

- libnfc-dev—a library for controlling NFC readers
- libfreefare—a library for reading and formatting Mifare RFID tags

To install the libraries needed, you'll use *apt-get*, a software package manager that's built into Raspbian. You've already been using `npm`, a package manager to install node.js packages. The `apt-get` tool performs a similar function, but for a wide variety of software packages, not just for node.js.

First you need to make sure `apt-get`'s list of packages is up to date so it can find the libraries you need. Make sure your Pi can connect to the internet, then type:

```
$ sudo apt-get update
```

This will check with the Raspbian remote repository to get a list of the latest packages available. It will take several minutes, and you'll see a lot of text scroll by. When it's done, you should see this:

```
Reading package lists... Done
```

Now you're ready to install new packages. Install the libnfc packages first, all on one line, like so:

```
$ sudo apt-get install libnfc-dev libnfc-bin libnfc-examples
```

Again you'll see a lot of text go by. The `apt-get` tool will download all the files, compile them if needed, and store them in the correct directories in your system. When you get a command prompt again, install the libfreefare libraries like so:

```
$ sudo apt-get install libfreefare-dev libfreefare-bin
```

Now reboot your system and log back in. Assuming everything worked out, you've now got working NFC tools. Plug in your NFC reader and type:

```
$ nfc-list
```

You should get the following response:

```
nfc-list uses libnfc 1.7.1
NFC device: SCM Micro / SCL3711-NFC&RW opened
```

Now you know the NFC reader is recognizable by the system.

Libnfc and libfreefare installed a number of useful utilities and examples that you can use to read, format, and write tags, as follows:

- nfc-anticol
- nfc-dep-initiator
- nfc-dep-target
- nfc-emulate-forum-tag2
- nfc-emulate-forum-tag4
- nfc-emulate-tag
- nfc-emulate-uid
- nfc-list
- nfc-mfclassic
- nfc-mfsetuid
- nfc-mfultralight
- nfc-poll
- nfc-read-forum-tag3
- nfc-relay
- nfc-relay-picc
- nfc-scan-device
- mifare-classic-format
- mifare-classic-read-ndef
- mifare-classic-write-ndef
- mifare-desfire-access
- mifare-desfire-create-ndef
- mifare-desfire-ev1-configure-ats
- mifare-desfire-ev1-configure-default-key
- mifare-desfire-ev1-configure-random-uid
- mifare-desfire-format
- mifare-desfire-info
- mifare-desfire-read-ndef
- mifare-desfire-write-ndef
- mifare-ultralight-info

To get help on any of the tools, type its name with the `-h` flag, like so:

```
$ nfc-mfclassic -h
```

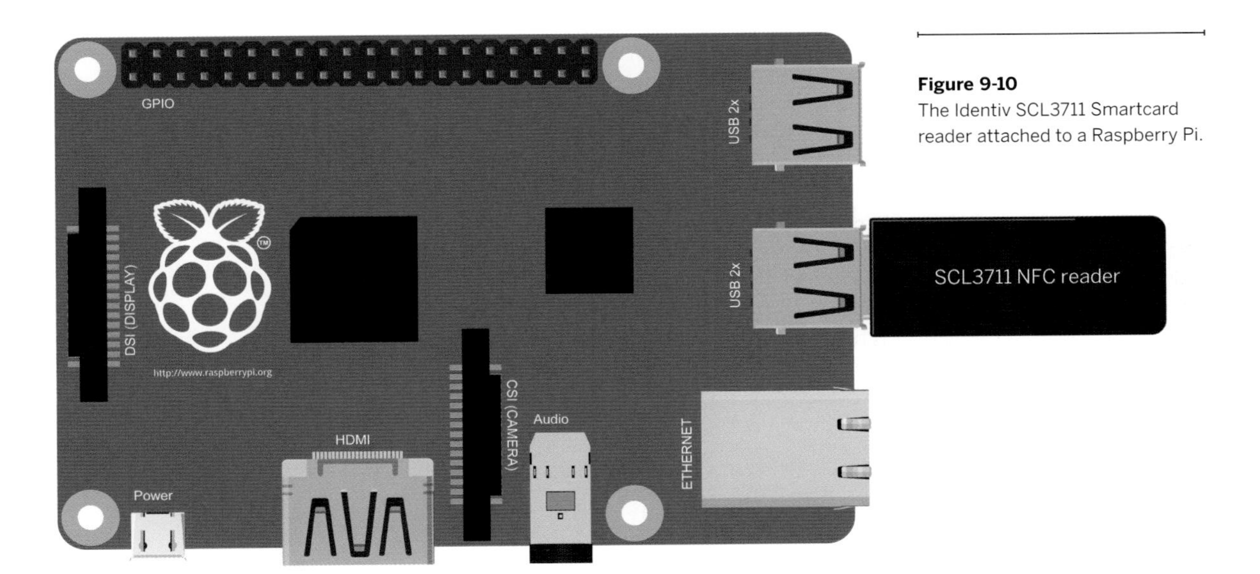

Figure 9-10
The Identiv SCL3711 Smartcard reader attached to a Raspberry Pi.

Reading Tags

The two libraries libnfc and libfreefare installed some useful command-line tools that you can use to read and write Mifare RFID tags and NFC tags. To start with, try reading a tag. Start with the `nfc-poll` tool:

```
$ nfc-poll
```

You should get a response like this:

```
nfc-poll uses libnfc libnfc-1.7.1-150-gbf31594
NFC reader: SCM Micro / SCL3711-NFC&RW opened
NFC device will poll during 30000 ms (20 pollings of 300 ms
for 5 modulations)
```

The reader will wait for you to bring a tag in range. When you do, it will respond like so:

```
ISO/IEC 14443A (106 kbps) target:
    ATQA (SENS_RES): 00  44
       UID (NFCID1): 04  8e  41  62  b7  20  80
      SAK (SEL_RES): 08
nfc_initiator_target_is_present: Target Released
Waiting for card removing...done.
```

That UID is the unique ID number of the RFID tag, written in hexadecimal notation. For simple RFID applications, this is all you'd get, and all you'd need. In a typical RFID application, you would associate the UIDs of a set of tags with the items to which you attach them, so that you could identify the items with an RFID reader.

When you have this working, test the range of your reader to see from how far away you can read a tag. It won't be a long distance, a couple of centimeters at most. You should also see that the reader can't read more than one tag at a time—this is common among passive RFID and NFC readers. Although industrial RFID readers can read multiple tags at once, the less expensive ones cannot.

X

No matter what RFID protocol you're using, you'll always be able to do what you did in the last project: read a unique ID number from a tag. NFC gives you a bit more than just a UID, though. The most powerful feature of NFC is the NFC Data Exchange Format, or NDEF. Think of it this way:

- RFID: Tags typically have only a unique ID
- NFC: Tags can read and write a limited amount of data
- NDEF: Provides a common data format for NFC across different tag types

NDEF is a bit like the common storage drive formats for hard drives, SD cards, and other mass storage devices. Even though there are different types of storage devices made by different vendors, they all use a couple of standard drive formats. This is why you can store files to an SD card, a thumb drive, or a hard drive, and not have to think about the difference. Like mass storage devices, NFC tags store data in sectors and blocks on the tag. Each tag technology has a different format for reading from and writing to its memory, but NDEF provides a common abstraction on top of these tag technologies. Using NDEF, you can exchange data while ignoring the actual storage implementation.

Each NDEF-formatted tag contains an NDEF *message*, and each message is made up of one or more NDEF *records*. Records can be one of several different record types. The most common record types are text messages; URI records, which contain web URIs; and MIME media records, which can contain various multimedia formats. There are a few specialty record types as well, which are not covered here. For these projects, you'll just use text and URI records.

There are four *NFC tag types* officially supported by the NFC specification:

- *Type 1*: Based on ISO-14443A specification, e.g., Innovision Topaz, Broadcom BCM20203.
- *Type 2*: Based on NXP/Philips Mifare Ultralight tag (ISO-14443A) specification, e.g., NXP Mifare Ultralight.
- *Type 3*: Based on the Sony FeliCa tag (ISO-18092 and JIS-X-6319-4) specification, e.g., Sony FeliCa.
- *Type 4*: Based on NXP DESFire tag (ISO-14443A) specification, e.g., NXP DESFire, SmartMX-JCOP.

In the projects you're doing here, you're using Mifare Classic tags. Mifare Classic is not an official NFC tag type, but it can but read and write NDEF as long as you're using a reader that's compatible with Mifare Classic. More recent NFC readers ignore these tags, so they may not work with some Android NFC-enabled devices.

X

 Project 25

Reading and Writing NDEF Messages

In this project, you'll get an introduction to Near Field Communication (NFC) and the NFC Data Exchange Format (NDEF) by reading from and writing to tags using node.js. You'll write a command-line reader and a writer.

MATERIALS

- » **Identiv SCL3711 USB Smartcard Reader**
- » **Mifare Classic RFID tags**
- » **Raspberry Pi**

As you might expect, there are node.js libraries for working with NDEF and Mifare Classic tags, and they work with the libnfc and libfreefare libraries that you just installed. These libraries offer simple JavaScript wrappers around libnfc and libfreefare, so you can use the utilities in your node scripts.

Make a new project directory on your Pi called **ndefReadWrite**. Change directories to this new directory and make a new file called **writeNdef.js**. Finally, install the ndef and mifare-classic libraries using npm as usual:

```
$ npm install ndef mifare-classic
```

In this script, you'll write an NDEF message with a couple of records to a tag, a text record and a URI record.

Write it Start the script by importing the ndef and mifare-classic libraries, and setting up an array to hold the NDEF message. The message array will contain records as its elements.

```
/*
  NDEF message writer
  context: node.js
*/
var ndef = require('ndef');               // import ndef library
var mifare = require('mifare-classic');   // import mifare classic library
var ndefMsg = new Array();                // array for NDEF message
```

» Next, create a text record and a URI record and push the records to the message array. Then encode the array as a byte stream in preparation for writing it to the tag using `ndef.encodeMessage()`.

```
var textRecord = ndef.textRecord("Here's a string");
var uriRecord = ndef.uriRecord("http://www.example.com");

ndefMsg.push(textRecord);                 // add a text record
ndefMsg.push(uriRecord);                  // add  a URI record
var bytes = ndef.encodeMessage(ndefMsg);// encode message as a byte stream
```

» The `mifare.write()` function, coming up shortly, needs a callback function, so write one called `writeResponse()`. This will just report whether the tag was successfully written to or not.

Finally, write to the tag using `mifare.write()`.

```
function writeResponse(error){            // write function
    if (error) {                          // if there's an error,
      console.log("Error: " + error);     // report it
    } else {                              // otherwise, report success
      console.log("Tag written successfully");
    }
}

// write to the tag:
mifare.write(bytes, writeResponse);
```

That's the end of the NDEF writer script. Save it, then place a tag near the reader and run it like so:

```
$ node writeNdef.js
```

You should get the following reply:

```
NDEF file is 38 bytes long.
Found Mifare Classic 1k with UID 0a64ef28.
Tag written successfully
```

Now you need a reader script. Create a new script in the **ndefReadWrite** directory called **ndefRead.js**.

Read It Just like you did the last one, by importing the libraries.

```
/*
  NDEF message reader
  context: node.js
*/
var ndef = require('ndef');               // import ndef library
var mifare = require('mifare-classic'); // import mifare classic library
```

As you might expect, the `mifare.read()` function, coming shortly, will require a callback. Call it `listTag()`. The read function will return a buffer, which you need to format to JSON, then look for a property called `data`. This is the tag's payload.

Once you've got the payload, decode the message and you'll get an array of records. Use a for loop to iterate over the array and print each record.

```
// callback function for when you successfully read a tag:
function listTag(error, buffer) {
  if (error) {                              // if there's an error
    console.log("Read failed:  " + error);  // report it
  } else {                                  // otherwise
    var bytes = buffer.toJSON();            // convert the tag data to JSON
    if (bytes.hasOwnProperty('data')) {     // if it's got a data property
      bytes = bytes.data;                   // then get the data
    }
    var message = ndef.decodeMessage(bytes);// decode the message
    for (record in message) {               // loop over the message array
      console.log(message[record].value);  // get each record's value
    }
  }
}
```

Finally, call the `mifare.read()` function to read the tag. A successful read will call `listTag()`, of course.

```
mifare.read(listTag);   // read for tag
```

Save the file, then run it the usual way:

```
$ node readNdef.js
```

You should get the following output when you bring the tag you just wrote in range:

```
Here's a string
http://www.example.com
```

Since you can write multiple records, you can use the different records to do different things. In the project that follows, you'll see separate records used to store the name of a user and an IP address associated with that user. You can store most anything you want to in an NDEF record (up to 2^{32}-1 bytes). Depending on the application reading the tag, you can use the different tag types to trigger different actions. For example, if you bring this tag in range of an Android device that can read Mifare Classic tags, you'll see that the device prompts you to open the address in the URI record in a browser. Tags can be tokens to indicate locations on the web. This opens a range of interesting possibilities.

Now that you can read and write NDEF messages to tags in node.js, the next project shows you how you can also do so on an Arduino-compatible microcontroller.

X

Project 26

NFC Meets Home Automation

Between my officemate and me, we have dozens of devices drawing power in our office: several computers, two monitors, four or five lamps, a few hard drives, a soldering iron, Ethernet hubs, speakers, and so forth. Even when we're not in the office, the room is drawing a lot of power. What devices we have turned on at any given time depends largely on which of us is here, and what we're doing. This project is a system to reduce our power consumption, particularly when we're not there. When we come into the office, all we have to do is touch our keys on a plate by the door, and the room turns on or off the devices we normally use. Each of us has a key ring with an NFC-tag key fob. The module behind the plate has an NFC reader in it, which reads the tags.

MATERIALS

- **1 solderless breadboard**
- **1 Arduino MKR1000, or WINC1500 or ESP8266-compatible microcontroller**
  - **Features used: SPI or I²C, WiFi**
- **1 PN532 NFC reader module**
- **2 Mifare Classic tags**
- **1 solderless breadboard**
- **2 Belkin WeMo switch modules**

The NFC reader is connected to a microcontroller module that communicates over the WiFi. Each of the various power strips is plugged into a Belkin WeMo switch, which affords wireless network control of AC power. Depending on which tag is read, the microcontroller knows which modules to turn on or off. Figure 9-11 shows the system. You can duplicate this with as many WeMo switches as you want; just plan which tags turn on which switches.

The Circuit

Since the microcontroller for this project communicates with the WeMo switches over WiFi, you'll need to use a WiFi-enabled controller. The MKR1000 or ESP8266 or any variant of the ESP8266 or WINC1500 boards should work for this project.

The NFC reader in this project is based on a PN532 chip from NXP Semiconductors. Several companies make modules with this chip: Adafruit's PN532 NFC/RFID controller breakout board (shown), their NFC shield, and Seeed Studio's NFC Shield for Arduino have been tested with the following code; Seeed's Grove NFC module should work as well. Any board based on the PN532 chip and that uses SPI or I²C should work.

The PN532 has I²C, SPI, and high-speed UART interfaces, and most modules offer jumpers that you can set to choose which interface you want to use. In the schematic in Figure 9-12, you can see the jumpers on the Adafruit board set for SPI. There are also two LEDs attached to the microcontroller, one to indicate when a tag is present and one to indicate when the network is connected.

The Communication Protocols

You'll write an NDEF message containing three NDEF text records to each tag:

- Username
- Number of the WeMo switch associated with that user
- IP address of the switch

When the microcontroller reads an NFC tag containing this message, it will make the HTTP call to the appropriate WeMo switch to toggle it on or off.

Although Belkin doesn't publish the API for their WeMo products, they encourage customers to customize them using If This Then That (ifttt.com), a web service for linking connected devices to web services via HTTP. Therefore, it's well known that the WeMo products respond to HTTP requests, and there are dozens of online tutorials on how to do it. Belkin treats these unofficial applications with benign neglect, which is good for hobbyists.

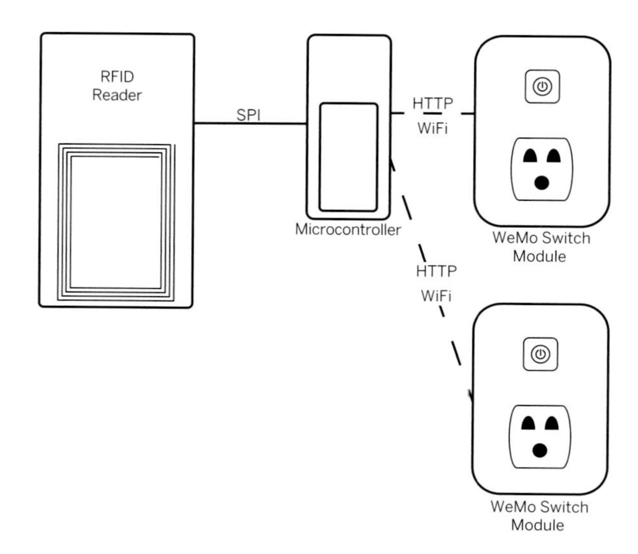

Figure 9-11
An NFC-controlled home (or office) automation system using WeMo switch modules.

To get started, follow the WeMo's installation instructions that come with your WeMo switch. You'll need to install the WeMo app on your Android or iOS smartphone. Once you've done that, you'll be able to get the switches' IP addresses from the Settings menu, under Settings & About→Hardware Info→`Your Switch Name`→IP Address. Write them down.

The WeMo communication protocol implements *Universal Plug & Play*, or *uPnP*. This is a protocol designed to make networked devices "just work." It also uses the *Simple Object Access Protocol (SOAP)*, which is an XML-based protocol for device control that works over HTTP or SMTP. Despite its name, SOAP requires a lot of text to send a simple message. Since it's running over HTTP, however, you can use the ArduinoHttpClient library to send SOAP messages, nice and clean. The code will be similar to the Air Quality project in Chapter 4.

The message you'll send will look like this:

```
<?xml version="1.0" encoding="utf-8"?>
  <s:Envelope xmlns:s="http://schemas.xmlsoap.org/soap/envelope/"
    s:encodingStyle="http://schemas.xmlsoap.org/soap/encoding/">
    <s:Body>
      <u:SetBinaryState xmlns:u="urn:Belkin:service:basicevent:1">
        <BinaryState>1</BinaryState></u:SetBinaryState>
     </s:Body>
  </s:Envelope>
```

You can see that XML, like HTML, is made up of nested tags, and that this XML document has an envelope and a body. In the body, it's going to set the state of a tag called `BinaryState` to 1 or 0. That turns the switch on or off.

The message is sent using a POST request, and the HTTP header that you'll send will look like this:

```
Content-type: text/xml; charset=utf-8
SOAPACTION: "urn:Belkin:service:basicevent:1#SetBinaryState"
Connection: keep-alive
Content-Length: 230
```

To do this, you'll use the ArduinoHttpClient's ability to set custom headers.

The URL path for the WeMo devices is the same on all of them: /upnp/control/basicevent1. You'll use this for the HTTP path in your code. The WeMo devices listen on port 49153.

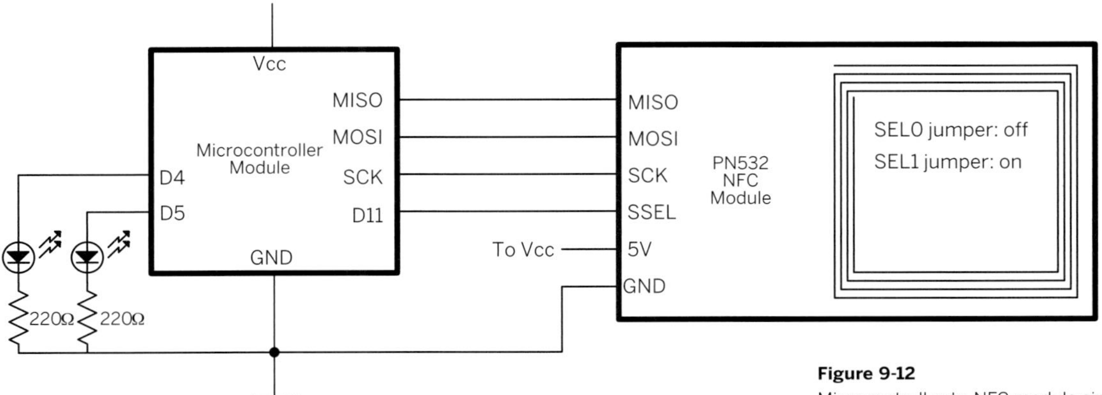

Figure 9-12
Microcontroller-to-NFC module circuit. An MKR1000 is shown here. Change connections for your controller, but make sure that SPI connections are matched up. Make sure also that the left-side ground row on the breadboard is connected to ground under the NFC reader.

> ⚠ **This project uses SPI to interface with the NFC reader. The MKR1000 uses SPI to communicate with its WiFi radio as well. It is possible to mix devices on the same SPI bus, but each device needs its own chip select pin. The MKR1000 uses pin 7 as the chip select for the WiFi radio, so you'll use pin 11 as the chip select for the NFC module. Note that this is different than the Uno's standard chip select, which is pin 10. You can also change the code in this project to use the I²C interface if you choose. See the library examples for how to make the code switch if needed. Change the jumpers accordingly.**

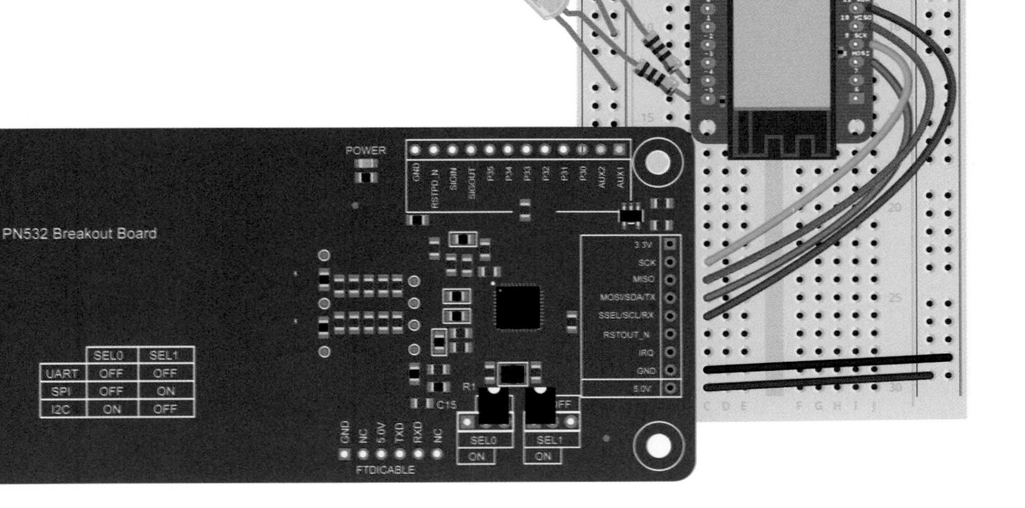

Install the Libraries

You'll need two new libraries for this project, Don Coleman's NDEF library for Arduino available at github.com/don/NDEF, and Yihui Xiong's PN532 library, available at github.com/Seeed-Studio/PN532. The NDEF library formats NDEF messages and records, and depends on the PN532 library to communicate with the reader. Neither of these is available in the Library Manager as of this writing, so download the .zip files and install them locally. Unzip both files, then open the Arduino IDE and click on the Sketch menu→Include Library→Add .ZIP library..., then choose the unzipped library folder in your Downloads directory. For the NDEF library, you can use the unzipped directory or the .zip file itself. For the PN532 library, look inside the main folder for the PN532_I2C, PN532_SPI, and PN532_HSU folders. Install all three, even though you're only using the I²C one. When you've installed them successfully, you should see examples for all the libraries in the File→Examples menu. Only the NDEF ones are relevant to this project.

The upcoming code is for the NDEF writer, followed by the code for the NDEF reader for an Arduino. Try the writer first, and when you get a successful write, try the reader.

X

Write the Tags

The first thing you need to do is to write the NDEF messages to the tags. You could use the node.js scripts from the last project, or you could write to the tags from your Arduino directly. First include the libraries for SPI and the NFC reader.

```
/*
  NDEF card writer
  context: Arduino
*/
#include <SPI.h>           // include SPI library
#include <PN532_SPI.h>     // include SPI library for PN532
#include <PN532.h>         // include PN532 library
#include <NfcAdapter.h>    // include NFC library

PN532_SPI pn532spi(SPI, 11);         // initialize adapter
NfcAdapter nfc = NfcAdapter(pn532spi);
```

In the setup, initialize serial communications and communication with the reader.

```
void setup() {
  Serial.begin(9600); // open serial communications
  nfc.begin();        // open NFC communications
}
```

In the loop, wait for a tag to be present. When one is there, create an NDEF message; then add the NDEF records to the message. Then try to write to the tag and report the results.

```
void loop() {
  Serial.println("Put a formatted Mifare Classic tag on the reader.");
  if (nfc.tagPresent()) {
    NdefMessage message = NdefMessage();
    message.addTextRecord("Tom Igoe");
    message.addTextRecord("1");
    message.addTextRecord("192.168.0.17");
    if (nfc.write(message)) {
      Serial.println("Success. Tag written.");
    } else {
      Serial.println("Write failed");
    }
  }
```

▸▸ Replace with your own username, WeMo switch number, and WeMo switch IP address

Add a 3-second delay after the tag is written so that the user can remove the tag before you try to write again. That's the end of the writer sketch.

```
  Serial.println();     // blank line at the end
  delay(3000);          // give the user time to remove the tag
}
```

Read the Tags

The reader is a bit more complex than the writer, because you have to parse the NDEF message. Set up the libraries at the beginning as you did for the writer. This time you'll also add an LED to turn on when there is a card present.

```
/*
  NDEF card reader
  Context: Arduino
*/
#include <SPI.h>          // include SPI library
#include <PN532_SPI.h>    // include SPI library for PN532
#include <PN532.h>        // include PN532 library
#include <NfcAdapter.h>   // include NFC library

PN532_SPI pn532spi(SPI, 11);        // initialize adapter
NfcAdapter nfc = NfcAdapter(pn532spi);
const int tagLed = 5;               // LED to indicate tags' presence
```

The `setup()` is also the same as with the writer, except for the addition of the LED.

```
void setup() {
  Serial.begin(9600);              // initialize serial communication
  nfc.begin();                     // initialize NFC communication
  pinMode(tagLed, OUTPUT);         // make tagLed an output
}
```

In the `loop()`, wait for the tag, then turn on the LED. Then read the tag. If there is an NDEF message in the tag, then get the number of records.

```
void loop() {
   if (nfc.tagPresent()) {                              // read NFC tag
    digitalWrite(tagLed, HIGH);                        // indicate tag presence
    NfcTag tag = nfc.read();                           // get data
    if (tag.hasNdefMessage()) {                        // if tag has a message
      NdefMessage message = tag.getNdefMessage();  // get message
      int recordCount = message.getRecordCount();  // get records count
```

Use a for loop to iterate over the records and get each one. The routine here shows how to create a byte array of the right length, then put the record into the array, then convert the byte array to a String. Later on, you'll need to manipulate these strings to talk to the WeMo.

```
      for (int r = 0; r < recordCount; r++) {
        NdefRecord record = message.getRecord(r);  // get record
        int payloadLength = record.getPayloadLength();// get payload length
        byte payload[payloadLength];               // array for payload
        record.getPayload(payload);                // get payload
        String payloadString;                      // clear payloadString
        for (int c = 3; c < payloadLength; c++) {  // get bytes from byte 3
          payloadString += (char)payload[c];       // copy to payloadString
        }
        Serial.println(payloadString);            // print payloadString
      }     // end of record for-loop
    }       // end of if tag.hasMessage()
  }         // end of if tagPresent()
```

Finally, turn the LED off, and delay so the user can remove the tag.

```
  digitalWrite(tagLed, LOW);                          // indicate tag is gone
  delay(3000);                                         // let the user remove
tag
}
```

You should have a working NFC writing and reading system now. You have to reprogram the Arduino to switch from one to the other, but you can see now that it's possible to do both.

Since there will be no other physical user interface, the LED is important to let the user know when the tag is being read. Likewise, the delay at the end is a quick way to ensure that the tag isn't read twice before the reader removes it.

Once you know the reader is reading reliably, it's time to add the HTTP client code to talk to the WeMo switches.

X

Refine It These additions to your reader sketch add WiFi functionality, and make an HTTP POST request to the WeMo with the data from the NDEF records. Additions and changes are shown in blue as usual. Start by adding the WiFi101 headers that you added for all the other WiFi projects. You'll also need to add **config.h** as you did for those projects, to store the network SSID and password.

You'll also need some new global variables for a WiFi connection indicator LED pin number, the WiFi Client instance and port number, the request route, an array for the states of the two WeMo switches, Strings for the username and IP address, and an integer for the WeMo switch number. You'll also need to add the SOAP request as a long string. It's broken up for readability here, but you could remove the line continuation characters (\) and put it all on one line if you prefer.

```
/*
  WeMo NFC HTTP Client
  Context: Arduino, with WINC1500 module
*/
#include <SPI.h>          // include SPI library
#include <WiFi101.h>      // include WiFi101 library
//#include <ESP8266WiFi.h>// use this instead of WiFi101 for ESP8266 modules
#include <ArduinoHttpClient.h>
#include "config.h"
#include <PN532_SPI.h>    // include SPI library for PN532
#include <PN532.h>        // include PN532 library
#include <NfcAdapter.h>   // include NFC library

PN532_SPI pn532spi(SPI, 11);          // initialize adapter
NfcAdapter nfc = NfcAdapter(pn532spi);

const int tagLed = 5;                // LED to indicate tags' presence
const int wifiLed = 4;               // LED to indicate WiFi connection
WiFiClient netSocket;                // network socket to device
const int port = 49153;              // port number
String route = "/upnp/control/basicevent1";  // API route
boolean wemoStates[] = {0, 0};       // state of the wemo switches
String username = "";                // the username
String wemoAddress = "";             // address of the current wemo
int wemoNumber = -1;                 // the current wemo number

// string for the SOAP request:
String soap = "<?xml version=\"1.0\" encoding=\"utf-8\"?> \
<s:Envelope xmlns:s=\"http://schemas.xmlsoap.org/soap/envelope/\" \
s:encodingStyle=\"http://schemas.xmlsoap.org/soap/encoding/\"> \
<s:Body> \
<u:SetBinaryState xmlns:u=\"urn:Belkin:service:basicevent:1\"> \
<BinaryState>1</BinaryState></u:SetBinaryState></s:Body> \
</s:Envelope>";
```

In the `setup()` add a line to set the WiFi LED to be an output, and then add the WiFi connection routine that you've used with other WiFi projects.

```
void setup() {
  Serial.begin(9600);              // initialize serial communication
  nfc.begin();                     // initialize NFC communication
  pinMode(tagLed, OUTPUT);         // make tagLed an output
  pinMode(wifiLed, OUTPUT);        // make wifiLed an output

  // while you're not connected to a WiFi AP,
  while ( WiFi.status() != WL_CONNECTED) {
    Serial.print("Attempting to connect to Network named: ");
    Serial.println(ssid);           // print the network name (SSID)
    WiFi.begin(ssid, password);     // try to connect
    delay(2000);
  }
```

Once you're connected, print out a serial message to that effect.

```
  // When you're connected, print out the device's network status:
  IPAddress ip = WiFi.localIP();
  Serial.print("IP Address: ");
  Serial.println(ip);
}
```

The loop() is largely unchanged. Most if it has been cut here for brevity. You'll add a call to a new function, copyRecords(), after you print the payloadString of each record to copy the records into the global variables you added earlier. You'll also add a conditional statement at the end to check the WiFi connection status and set the WiFi status LED accordingly.

```
void loop() {
       // ... unchanged lines above here cut for brevity
        Serial.println(payloadString);             // print payloadString
        copyRecords(r, payloadString);
      }      // end of record for-loop
    }        // end of if tag.hasMessage()
  }          // end of if tagPresent()
  digitalWrite(tagLed, LOW);                      // indicate tag is gone
  if (WiFi.status() == WL_CONNECTED) {          // indicate  WiFi connection
    digitalWrite(wifiLed, HIGH);
  } else {
    digitalWrite(wifiLed, HIGH);
  }
  delay(3000);                                   // let the user remove tag
}
```

Next comes the copyRecords() function. This takes the record number and record payload string and copies each to the appropriate global variable. When it gets the third record, which is the IP address, it calls another new function, wemoRequest(), to make the POST request.

```
void copyRecords(int recordNum, String recordString) {
  switch (recordNum) {  // do something different for each record
    case 0:      // the username record; nothing to do
      break;
    case 1:      // the wemo number record; convert to integer
      wemoNumber = recordString.toInt();
      break;
    case 2:      // the IP address record; make request
      wemoAddress = recordString;            // if you have an IP address,
      wemoRequest(wemoNumber, wemoAddress); // then make a request
      break;
  }   // end of case statement
}
```

Finally, write the wemoRequest() function. The function starts by checking the state of the WeMo switch in question and changing the SOAP request string using the String.replace() function. Methods like this are handy when you need to send a long message where only one or two characters change.

Next comes an HTTP POST request using the ArduinoHttpClient library's commands as you did in Chapters 4 and 5. Since there are custom headers, you need to build the request in parts, using beginRequest() to start it and endRequest() to end the header. The body of the POST request is sent after the endRequest() using http.println(). You can use any of the Stream functions to send the body in a POST request with custom headers like this.

Once the WeMo switch responds, this function prints out the response.

Continued from previous page.

```
void wemoRequest( int thisWemo, String wemo) {
  if (wemoStates[thisWemo] == 0) { // if the wemo's off
    soap.replace(">0<", ">1<");      // turn it on
  } else {                           // otherwise
    soap.replace(">1<", ">0<");     // turn it off
  }
  wemoStates[thisWemo] = !wemoStates[thisWemo];   // toggle wemoState

  HttpClient http(netSocket, wemo.c_str(), port); // make an HTTP client
  http.connectionKeepAlive();              // keep the connection alive
  http.beginRequest();                     // start assembling the request
  http.post(route);                        // set the route
  // add the headers:
  http.sendHeader("Content-type", "text/xml; charset=utf-8");
  String soapAction = "\"urn:Belkin:service:basicevent:1#SetBinaryState\"";
  http.sendHeader("SOAPACTION", soapAction);
  http.sendHeader("Connection: keep-alive");
  http.sendHeader("Content-Length", soap.length());
  http.endRequest();                       // end the request
  http.println(soap);                      // add the body
  Serial.println("request sent");

  while (http.connected()) {               // while connected to the server,
    if (http.available()) {                // if there's a server response,
      String result = http.readString();   // read it
      Serial.println(result);              // and print it
    }
  }
}
```

As you saw with the previous reader sketch, and with this one, the NFC reader only reads when a new tag enters its field. It doesn't have the ability to read multiple tags if more than one tag is in the field. That's an important limitation. It means that you have to design the interaction so that the person using the system places only one tag at a time, and then removes it before placing a second one. In effect, it means that two people can't place their key tags on the reader at the same time. Users of the system need to take explicit action to make something happen. Presence isn't enough.

The simplest solution is to design the project physically so that the user has to remove the tag once he passes it by the reader, as shown in Figure 9-15. This is the most common commercial solution as well.

You'll also notice that since the HTTP request and the switches' responses take a few seconds. These devices have a simple relay inside to turn on or off the power, and they can't be switched more than about once every second, or they'll get damaged. The delay at the end of your loop() ensures that they won't get triggered too frequently. As you design systems with multiple devices, all with their own timing limitations, you have to consider factors like these in order to make your system reliable. X

Figure 9-13
The NFC reader box, under construction. The reader antenna needs to be at the top of the box, requiring a 2-level construction, with the controller underneath. Different height standoffs make this possible.

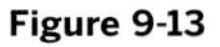

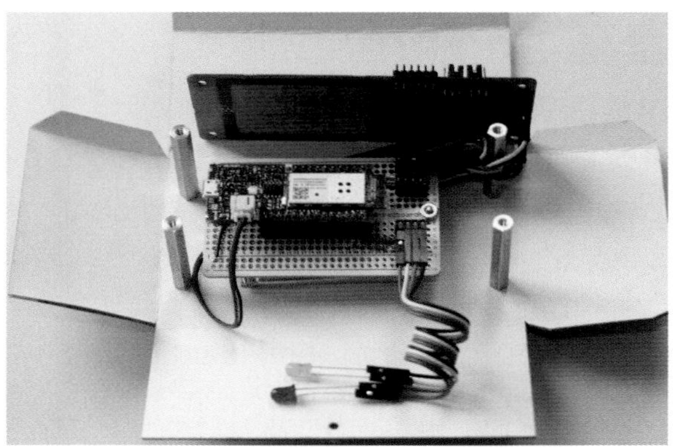

Figure 9-14
The NFC reader box exploded. The reader and the LEDs are attached with hook-up wire and right-angle headers.

Construction

The box for this project is similar to the ones for the video controller in Chapter 5. In fact, it's made from the same template, with the dimensions changed slightly. There's only two LED holes, since there are no user controls other than the reader. The Adafruit perma-protoshield, used for the microcontroller board, has the same layout as a solderless breadboard, so the circuit layout is as shown in Figure 9-12. Figures 9-13 and 9-14 show two views of the box construction. Figure 9-15 shows the finished reader. X

Figure 9-15
The finished NFC reader box, Mounted by the door, it ensures that the user has to remove the tag. This works around the reader's limitation of being able to read only one tag at a time.

“ Security of Networked Devices

Physical identity is increasingly linked to networked data, and networked data is increasingly vulnerable to attack. Everyone should know some basic techniques to protect their identity and their devices.

Identity thieves can damage your financial profile when they access your credit cards and bank accounts online, but when they can find your physical address and affect the devices therein, they can affect your physical safety as well. Here are a few basic tactics you can use to keep your networked devices secure.

Keep It in the Local Network

If all your communications with your devices are local, they are less likely to be compromised. A device with only a private IP address on a local area network is invisible to the rest of the internet unless the LAN itself is compromised. Consider this when you're building devices. Do you need to control your lights from outside the house, or can they be automated locally while you're gone? The latter is safer. Consider it also when you're shopping for devices, and give preference to those that don't share data back to the vendor without your explicit permission. If you don't need regular updates from your thermostat, why does the company who built it need it? If they want your usage data, they can ask you.

Use Two-Factor Authentication

Two-factor authentication means that you have to provide something unique that you know and something unique that you have in order to gain access to a device. Sometimes simple methods like a physical pushbutton are used as the physical factor. This is how you authenticate your mobile device with a Philips Hue Hub. You have to press the button, then connect to the hub with your mobile device within a few seconds in order to authenticate the device. Sometimes the system texts you a separate code to your phone after you enter your password, as Google's two-factor authentication does. In this case, the phone is the something you have, and your password is the something you know. Still others, like iOS's two-factor authentication, use biometric data like a fingerprint. In the next project, you'll use an NFC tag and a challenge phrase in a two-factor authentication system.

Use HTTPS and Public Key Cryptography

In Chapter 8, you learned about HTTPS and certificates, a form of encryption that uses public keys and private keys to encrypt and verify messages between a server and a client. This kind of encryption is called *public key* or *asymmetric encryption*. It works like this.

First you generate a pair of keys, a private key and a public key. For the HTTPS server, you did this when you generated the certificate (which contained the public key) and the private key. The *private key* can encrypt messages. The *public key* can only decrypt messages. If you have someone's public key, you can verify that an encrypted message came from them. Any message you can decrypt with their public key could only have been encrypted with their private key. You can't pretend to be the sender, since you can't encrypt messages the same way they can with the private key. You can give your public key out to lots of people, safe in the knowledge that they can read your encrypted messages with it but cannot copy your encryption scheme.

For secure two-way communication, both parties to the conversation generate their own key pairs and share their public key with the other. Then when you send a message, you encrypt it first with the other's public key, then with your own private key. This second step is called *signing*, because the receiver needs to decrypt using your public key, thus verifying that it came from you, before they decrypt a second time, verifying that you have their public key. When a website is using HTTPS, its certificate contains the site's public key so that you can verify that all the content that you can unlock with that key came authentically from the site.

This is different from *symmetric encryption*, in which both parties have the same encryption key and can encrypt and decrypt messages as they wish. You saw symmetric encryption in Chapter 3, when you used the aes-256-cbc encryption algorithm to both encrypt and decrypt your password, in two different scripts. You can't verify the sender of a message using symmetric encryption; you can only verify that they know the encryption scheme.

X

 Project 27

Two-Factor Authentication Using NFC

In this project, you'll create a two-factor authentication system using NFC tags and a Raspberry Pi. This project will give you a system that you can use whenever you want to protect a system with two-factor authentication.

MATERIALS

- » **Identiv SCL3711 USB Smartcard Reader**
- » **Mifare Classic RFID tags**
- » **Raspberry Pi**
- » **TIP120 transistor**
- » **1-kilohm resistor**
- » **Solenoid lock**
- » **12V power supply**
- » **5.5mm OD, 2.1mm ID power jack**
- » **Small breadboard**

In this project, you'll have to show something you have, an NFC tag, and reveal something you know, a challenge phrase, in order to open a lock. It will work like this.

You'll open a web page served by node.js and tap an NFC tag to a reader that the server is controlling. The server will read the tag and extract an encrypted challenge phrase. You'll enter the challenge phrase in the browser. If the challenge phrase you enter (that's the something you know) matches the phrase that's on the tag (that's the something you have), then the server will grant you access, opening a solenoid lock when you're successfully authenticated.

The challenge phrase on the tag will be encrypted, so that you can't just read it out with any NFC reader. You'll write a script that encrypts a phrase that you choose and writes it to a tag using a private key. Then you'll write a separate script to verify the phrase using the corresponding public key. This adds to the security of the system, because it means you can have a verification server that doesn't know how to write tags, only how to read and verify them, thus reducing the chance that someone might make fake authentication tags.

Since you want this to be secure, you'll start with the same HTTPS server that you used in the Geolocation project in Chapter 8 and the camera projects earlier in this chapter, and add some modifications. You'll generate a second pair of keys for the tag encryption as well.

The Code

This will all run on the Raspberry Pi, using the NFC and NDEF tools you used in Project 26. You'll write this application in several modules, using node.js's capability for combining modules to put it all together. The JavaScript interfaces for libraries like express or ndef are modules, too. Now you'll get a sense of how they are written.

Create a directory for this project called **ndefSignature**. It's actually two applications, one that writes and formats tags and another that reads them. They're almost identical, and technically they could be combined in the same server script. However, to make the application more secure, they should be separate. The reader script won't even have functions to write or format tags, so that it can't be exploited. Make two copies of the **https-server.js** script from the video projects, and call them **https-server-writer.js** and **https-server-reader.js**. Copy the **keys** directory as well, since you'll need an SSL certificate. The writer will need an NDEF writer module, and the reader will need an NDEF reader module and a GPIO module to turn on and off the solenoid lock. Both will need web pages written in p5.js. Make a **public** directory for the static pages as usual, and make two p5.js projects in **public**, called **reader** and **writer**. Then install the following libraries using npm:

```
$ npm install express body-parser ndef mifare-classic onoff
```

Make three other files: **writeNdefSignature.js**, **readNdefSignature.js**, and **gpioControl.js**.

You need a pair of keys for encryption as well. You could use the SSL certificate and its public key, but it's better to use a separate key. Generate a new key pair in the **keys** directory, using openSSL as follows:

```
$ openssl genrsa -out keys/private.key 2048
$ openssl rsa -pubout -in keys/private.key -out keys/public.key
```

Now you're ready to get started.

X

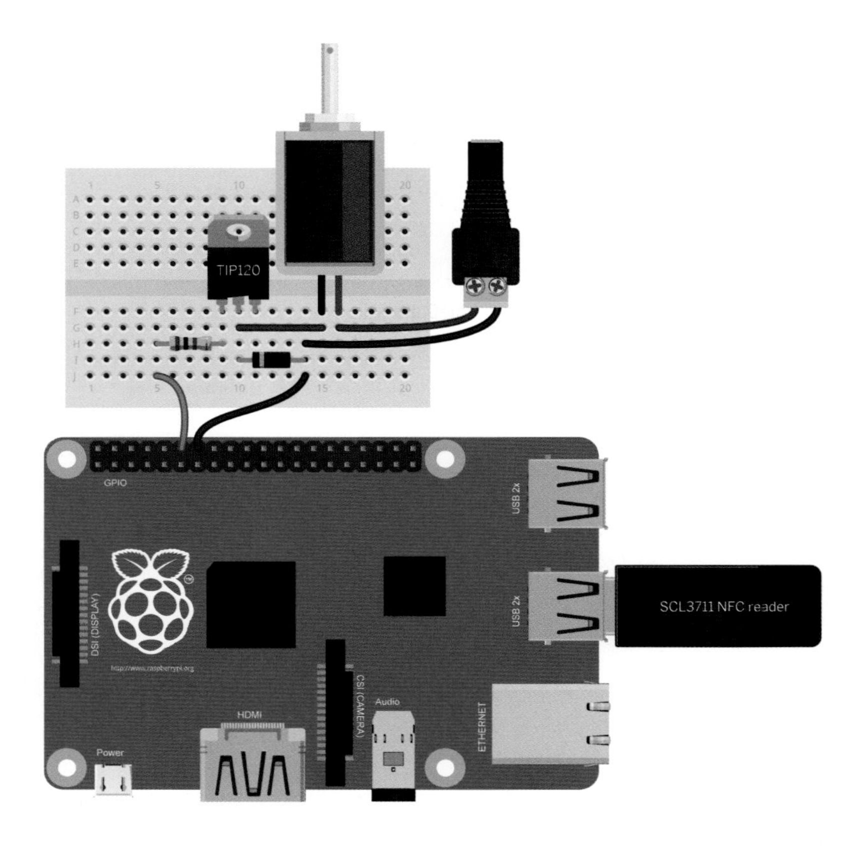

Figure 9-16
Identiv SCL3711 NFC reader, TIP120 transistor, and solenoid attached to a Raspberry Pi. The solenoid runs on 12V and 1A, so a separate 12V DC power supply is needed. Attach the ground of the 12V DC supply to the Pi's ground, but not the voltage side. 12V DC will destroy your Pi.

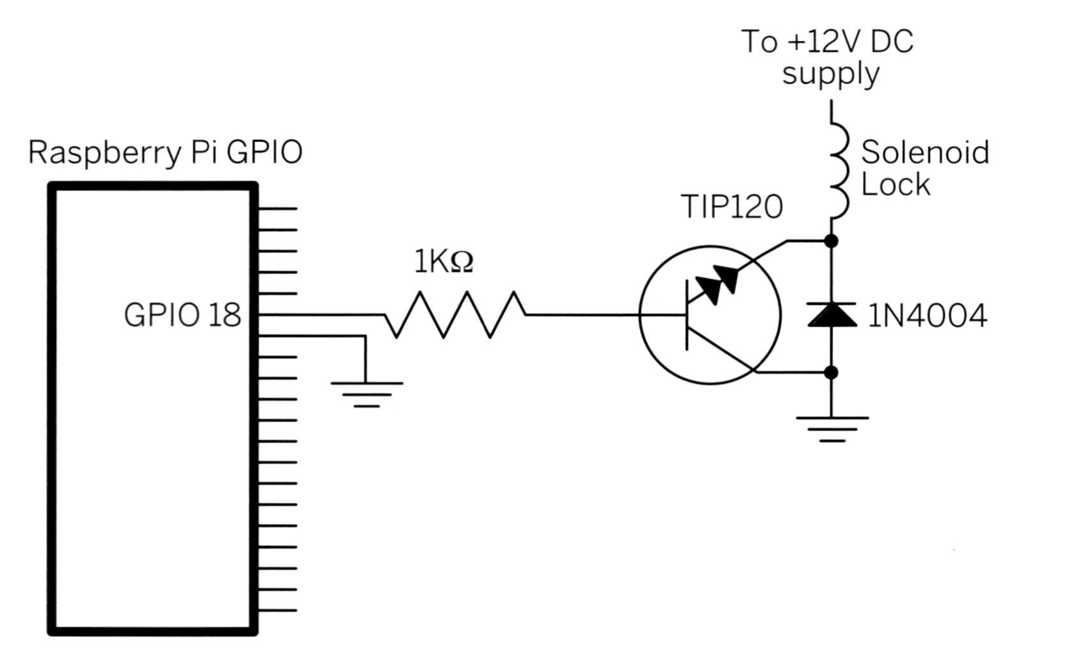

GPIO Control on the Pi

The circuit for this project is mostly the same as the one you used in Projects 25 and 26. The only addition will be a transistor and a solenoid on one of the Raspberry Pi's I/O pins. Figure 9-16 shows the circuit. Make sure to connect the ground of the 12V supply to the ground of the Pi, however.

Before you work with the NFC reader, you'll write a script to get to know the physical output of the Pi, **gpioControl.js**. In this script, you'll use the onoff library for node.js to turn on or off the general purpose I/O pins of the Pi.

The Pi's GPIO pins are similar to those of a microcontroller like the Arduino in that there are a variety of functions they can do: asynchronous serial (UART), which you've already used; SPI; I²C; and PWM. The Pi has no analog inputs, unfortunately. Figure 9-17 shows the pin diagram. All of its pins operate on 3.3V, except for the two 5V power pins. The onoff library (github.com/fivdi/onoff), which you installed already, gives you the ability to set them as inputs or outputs and to read, write, or watch them for changes (when they are inputs).

Like the pins of a microcontroller, the Pi's GPIO pins can't supply or receive large amounts of current, but they can control LEDs or turn on transistors. That's what you'll do in this script. The TIP120 turns on the solenoid, but only needs a small amount of current from the GPIO pin to turn on.

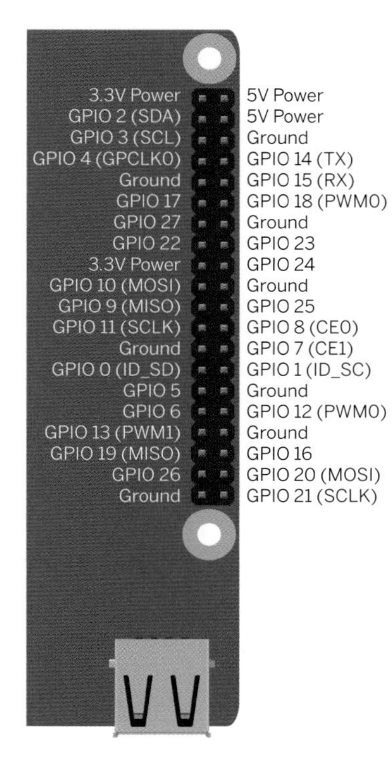

Figure 9-17
The GPIO pin diagram of the Raspberry Pi model 2, 3, and Zero.

Move It The **gpioControl.js** script starts like any other, by including the library it uses. Then it sets GPIO pin 18 to be an output, and stores it in a variable called `lock`, which you'll use to change the pin.

```
/*
 GPIO control
 context: node.js
*/

var Gpio = require('onoff').Gpio; // include onoff library
var lock = new Gpio(18, 'out');   // set solenoid I/O pin as output
```

» To turn a pin on, write 1 to it. To turn it off, write 0. The `open()` and `close()` functions reflect this. In `open()`, you'll also add a timeout to call `close()` after two seconds, so that the solenoid closes automatically.

```
function close() {
  lock.writeSync(0);        // turn the lock pin off
}

function open() {
  lock.writeSync(1);        // turn the lock pin on
  setTimeout(close, 2000);  // set a 2 second timeout
}
```

»

This conditional statement checks the command-line arguments for a `-o` flag, and calls `open()` if it's there. This allows you to test the script from the command line.

Continued from previous page.

```
// If there are command-line arguments, process them:
if (process.argv[2] === '-o') {
  open();
}
```

The final section is all new code. This is how you export some of your script's functions for use by other scripts. You'll see how it's used in the HTTPS server later on.

```
// this section exports some functions for use by other scripts:
module.exports = {
  close: close,
  open: open
};
```

Testing the GPIO

You can test the **gpioControl.js** script function from the command line like so:

```
$ node gpioControl.js -o
```

When you do this, GPIO pin 18 will turn on for 2 seconds, then turn off. Assuming you have the solenoid connected as shown in Figure 9-14, the solenoid should fire when you do this. If you want to test with something simpler, connect the anode of an LED to pin 18, and the cathode to ground. You should see the LED turn on when you run the script.

Once you know it works, you may want to disconnect the solenoid so that it doesn't interfere with the NFC read and write functionality as you're writing the next scripts. You'll combine this with them, and with the server scripts, later on.

Once the **gpioControl.js** script works, you're ready to write the NDEF reader and writer scripts.
X

Node.js Modules

Node.js is designed so that you can write code in a modular way, and you've been using modules all through this book. Every library is a module. By writing your code in discrete modules as you're doing here, you can have a team of programmers working on an application without having to coordinate on the same file. You can also reuse code easier.

In making a module, you can decide which functions or variables you want to be private to the module, and which ones you want to be used by other scripts that import your module just by using the `module.exports` object as you see in the GPIO example. Anything that you include in `module.exports` is available as a public function or variable. Anything that you don't include is private to the module and can only be used by other functions in the module.

When you want to use a module in another script, first you include it using `require()` as you've done with all the libraries you've seen. Then you can call its functions from the variable that you assigned using `require`. For example, to use the gpio module that you just made, you'd do this:

```
var lockControl = require('./gpioControl.js');
```

When a module is not in the node_modules directory, you need to give a relative path. Then you can call its functions like so:

```
lockControl.open();
```

You'll see more of this in action in the pages that follow.

Write It Now that you can control the lock, you need to be able to format and read tags. Open **writeNdef-Signature.js** and write it as shown to the right.

Include the ndef-js and mifare-classic libraries, and the crypto and filesystem (fs) libraries that you used in Chapter 3. You'll need them for encrypting and decrypting your NDEF messages.

Add variables to hold the private key and the secret to be encrypted.

Next, add a function to set the value of the secret. Although you don't need this for the command-line version of the script, it will be needed when you combine this script with the server script.

```
/*
  NDEF signed message writer
  Uses public-key cryptography to sign a record,
  then writes it to an NDEF message and then to a tag
  context: node.js
*/
var ndef = require('ndef');              // include ndef library
var mifare = require('mifare-classic'); // include mifare classic library
var crypto = require('crypto');          // include crypto library
var fs = require('fs');                  // include filesystem library

var privKey = fs.readFileSync('keys/private.key'); // read the private key
var privateKey = privKey.toString();               // convert to a string
var secret = null;                                 // secret to be encrypted

// function to set the secret:
function setSecret(data) {
  secret = data;
}
```

The main function of this script is `setMessage()`, which converts your secret into an encrypted array of bytes in an NDEF record to be written to the NFC tag. This function is similar to the NDEF writer you saw in Project 25, with the addition of the encryption. It returns the byte array so that you can use it in the `mifare.write()` function when you write to the tag.

```
// this function makes an NDEF message, signed or unsigned:
function setMessage(record, signed) {
  var ndefRecord;                  // string for the NDEF record
  var ndefMsg = new Array();       // array for the NDEF message

  if (signed === true) {  // if the record is to be signed, then sign it
    var signer = crypto.createSign('RSA-SHA256');    // make a signer
    signer.update(record);                           // add the record
    var signature = signer.sign(privateKey, 'hex'); // make the signature
    ndefRecord = ndef.textRecord(signature);         // make NDEF record
  } else {                        // if the record shouldn't be signed:
    ndefRecord = ndef.textRecord(record);            // make unsigned record
  }

  ndefMsg.push(ndefRecord);                   // add the record to the message
  var bytes = ndef.encodeMessage(ndefMsg); // encode record as byte array
  return bytes;                               // return the byte array
}
```

The main action of this script is a call to `setMessage()` followed by a call to `mifare.write()`, which will come at the end. You'll need a callback function for `mifare.write()`, so write it here. Call it `showResponse()`. It just shows the error messages or a success message.

```
// callback function for the write() and format() functions:
function showResponse(error){
    if (error) {
      console.log('error ' + error);
    } else {
      console.log('success');
    }
}
```

»

▸ The next section of the script reads what you type on the command line after node **writeNdefSignature.js**. `process.argv[]` is an array holding all the arguments from the command line. If you add `-f`, the script will format a tag. If you add a challenge phrase, the script will set the secret using the phrase, then create an NDEF message with the encrypted secret, then write it to the tag.

Continued from previous page.

```
// If there are command-line arguments, process them:
if (process.argv[2] != null) {
  if (process.argv[2] === '-f') {           // if the -f flag is included,
    mifare.format(showResponse);            // format the tag
  } else {                                  // otherwise,
    setSecret(process.argv[2]);            // get secret from command line
    var response = setMessage(secret, true);  // Add the secret, encrypted
    mifare.write(response, showResponse);    // attempt to write the tag
  }
}
```

The final section exports the public functions for other scripts to use, like the gpioControl script did. This is the end of the **writeNdefSignature.js** script.

```
// this section exports some functions for use by other scripts:
module.exports = {
  format: mifare.format,
  setMessage: setMessage,
  write: mifare.write
};
```

Now you're ready to test this script. Place a Mifare classic tag on the reader and type:

```
$ node writeNdefSignature.js -f
```

You should get a response like this:

```
Found Mifare Classic 1k with UID 1e6f5ba5.
Formatting 16 sectors [...4...8...12...16] done.
success
```

The script just erased and formatted your tag. Now to write a phrase to it, type:

```
$ node writeNdefSignature.js 'this is my secret phrase'
```

You'll get a response like so:

```
NDEF file is 522 bytes long.
Found Mifare Classic 1k with UID 1e6f5ba5.
success
```

The script will write any phrase (without the quotes) in an encrypted string. Remember the string you used, as you'll need to verify the tag with the next script.

Read It Now open up **readNdefSignature.js** and write it as follows. This script decrypts messages written to the NFC tags by the previous script.

You'll notice the same included libraries as the last script. However, instead of importing the private key, this script uses the public key. The two scripts could exist on separate devices and still function. This script never needs the private key to decrypt messages, and the other script never needs the public key.

The `setSecret()` function is the same as in the previous script.

```
/*
NDEF signed message reader
Reads an NDEF tag for a message, then
uses public-key cryptography to verify the records.
context: node.js
*/
var ndef = require('ndef');               // include ndef library
var mifare = require('mifare-classic'); // include mifare classic library
var crypto = require('crypto');          // include the crypto library
var fs = require('fs');                  // include filesystem library

var pubKey = fs.readFileSync('keys/public.key');  // read the public key
var publicKey = pubKey.toString();                // convert it to a string
var secret = null;                                 // set the secret message

// function to set the secret:
function setSecret(data) {
  secret = data;
}
```

You'll read tags using `mifare.read()` a bit later on. First, write a callback function for the read function. It prints out errors if there are any, or calls another function, `parseMessage()`, to parse the NDEF message read from the tag.

```
// callback function for when you successfully read a tag:
function getMessage(error, buffer) {
  if (error) {                              // if there's an error
    console.log('error ' + error);         // print it
  } else {                                  // otherwise
    console.log(parseMessage(buffer));     // check the message
  }
}
```

`parseMessage()` is like the tag reader script from Project 25. It checks to see that the NDEF message has data, and if so, it decodes it into an array of NDEF records. It iterates over the array and uses another function, `verifyRecord()`, to see if the contents of any record match the secret challenge phrase.

```
function parseMessage(buffer) {
  var result = null;                          // result to return
  var bytes = buffer.toJSON();               // convert the tag data to JSON
  if (bytes.hasOwnProperty('data')) {        // if it's got a data property
    bytes = bytes.data;                       // then get the data
    var message = ndef.decodeMessage(bytes);// decode the message
    for (r in message) {                      // loop over the message array
      var record  = message[r].value;        // get each record's value
      result = verifyRecord(record, secret);// verify it
    }
  } else {
    result = false;                   // nothing to verify, so return false
  }
  return result;
}
```

The `verifyRecord()` function uses the crypto library's signature verification functions to check that the record you give it was signed using the private that corresponds to the public key that this script is using. It doesn't actually show you the decoded phrase; it just verifies that it was authentically signed.

```
function verifyRecord(signature, secret) {
  var verifier = crypto.createVerify('RSA-SHA256');// make a verifier
  verifier.update(secret);                          // update with the secret
  // verify the incoming encrypted signature:
  var result = verifier.verify(publicKey, signature, 'hex');
  return result;
}
```

The next section takes action by reading from the command line. Like the previous script, it takes a phrase from the command line, and uses it to set the secret challenge. Then it reads the NDEF message from the tag using `mifare.read()`, which also calls the verification functions you just wrote.

Finally, write a section to export the public functions for other scripts. That's the end of the **readNdefSignature.js** script.

```
// If there are command-line arguments, process them:
if (process.argv[2] != null) {
 setSecret(process.argv[2]);  // get the secret from the command line
 mifare.read(getMessage);     // attempt to read the tag
}

// this section exports some functions for use by other scripts:
module.exports = {
  setSecret: setSecret,
  read: mifare.read,
  parseMessage: parseMessage
}
```

To test this script, put the same tag that you just wrote with the previous script on your NFC reader. Then type:

```
$ node readNdefSignature.js 'this is my secret phrase'
```

You should get a response like this:

```
Found Mifare Classic 1k with UID 1e6f5ba5.
NFC Forum application contains a "NDEF Message TLV".
true
```

The script just read your tag and verified that the phrase you typed is on the tag, and is signed using the private key associated with this script's public key.

Try formatting, reading, and writing with a few different tags and different phrases until you're comfortable with how it works. When you're ready, you'll combine these scripts with the HTTPS server to make a human interface for the whole process. Once you have the server working, you'll finish with the GPIO module to control the lock.

You'll write two versions of the HTTP server script as well, one to write and format tags, and one to read tags. They're almost identical, and technically, they could be combined in the same script. However, to make the application more secure, they should be separate. The reader script won't even have functions to write or format tags, so it can't be exploited.

Serve It Open up **https-server-writer.js** and modify it as follows. New lines are shown in blue, as usual. First you need to add a few libraries: You'll use the body-parser middleware that you used in Chapter 3 to parse the body of POST requests. You'll include the modules you just wrote using `require()` as well. Note that when they're not in the **node_modules** directory; you need to give a relative path to them.

The instantiation of the server, the options for reading the SSL certificate, the inclusion of the middleware functions using `server.use()`, and the HTTP redirect are mostly the same as before. The only addition is the inclusion of the body-parser middleware. This enables your functions to parse the body of an HTTP POST request.

```
/*
HTTP/HTTPS 2-Factor Authentication server
Tag formatter/writer version
context: node.js
*/
// include libraries and declare global variables:
var express = require('express');// include the express library
var bodyParser = require('body-parser'); // include body parser middleware
var https = require('https');      // require the HTTPS library
var http = require('http');        // require the HTTP library
var fs = require('fs');            // require the filesystem library
// include the external modules:
var tagWriter = require('./writeNdefSignature.js');

var server = express();                       // instantiate server
var options = {                               // options for the HTTPS
server
  key: fs.readFileSync('./keys/domain.key'), // the key
  cert: fs.readFileSync('./keys/domain.crt') // the certificate
};

server.use('*', httpRedirect);              // redirect function for http
server.use('/',express.static('public'));   // set a static file directory
server.use(bodyParser.urlencoded({extended: true})); // enable body parsing

function httpRedirect(request,response, next) {
  if (!request.secure) {
    console.log('redirecting http request to https');
    response.redirect('https://' + request.hostname + request.url);
  } else {
    next();      // pass the request on to the express.static middleware
  }
}
```

At the end of the script you'll write the server listener and route listeners as usual, but first, write the function to respond to requests. The process will be more or less the same whether formatting a tag or writing a tag: first you'll get the body of the request, then you'll write the beginning of the response to the client so they know that the server's acknowledged their request. Reading takes a second or two, so an immediate response is helpful.

```
function processTag(request, response) {
  var command = request.path;              // get the command
  command = command.slice(1);              // slice off the leading slash
  var data = JSON.stringify(request.body);// convert the body to a string

  // write the beginning of the response:
  response.writeHead(200, {"Content-Type": "text/html"});
  response.write('accessing tag...<br>');
```

You're going to format or write the tag, depending on the request. You'll determine which by looking at the `request.path`. You'll do this using the `.format()` and `.write()` functions that you exposed as public functions of your **writeNdefSignature.js** module. Both functions require a callback function to report success or error. You'll write this function, `finishResponse()`, next. It's inside the request handler function (`processTag()`) so that it's got access to the request response and can finish the response.

```
  // function to finish the response:
  function finishResponse(error, buffer) {
    if (error) {
      response.write('Failure: ' + error + '<br>');
    } else {
      response.write(command + ' success.<br>');
    }
    response.end();
  }
```

Finish the `processTag()` function by calling `.format()` or `.write()`, depending on the request. As usual, you're putting your callback function, `finishResponse()`, before the commands that call it.

The write request will include the data to be written to an NDEF message in the body of the request, so you'll call `.setMessage()` to create the NDEF message first.

```
  // take action depending on the command:
  if (command === 'formatTag') {
    tagWriter.format(finishResponse);    // attempt tag formatting
  }
  if (command === 'writeTag') {
    var message = tagWriter.setMessage(data, true);
    tagWriter.write(message, finishResponse); // attempt tag writing
  }
}
```

Finally, you'll create the HTTP and HTTPS servers as you did in the earlier projects, and add the route listeners. Both the `writeTag` and `formatTag` listeners will respond with the same function, `processTag()`, that you just wrote. This is the end of the script.

```
// start the server:
http.createServer(server).listen(8080);           // listen for HTTP
https.createServer(options, server).listen(443);  // listen for HTTPS
server.post('/writeTag', processTag);              // listener for writeTag
server.post('/formatTag', processTag);             // listener for formatTag
```

Add Web Page Now that you've got the server, you need a web interface for it. From the **writer** p5.js project that you created in the **public** directory, open the **sketch.js** file. Write it as follows.

The only global variables are `userField`, `challengeField`, and `responseDiv`, which you'll use in a few functions.

The `setup()` function creates and positions the input fields and their labels and a div for responses from the server. It also creates `writeButton` and `formatButton` buttons and gives them callback functions. Both buttons use a function called `submit()`, which you'll write next, as their callback.

This whole script is driven by user input on the buttons, so there is no `draw()` function.

```
/*
  Writer/formatter page script
  context: p5.js
*/
var userField, challengeField;  // fields for username and challenge phrase
var responseDiv;                // div for responses from the server

function setup() {
  noCanvas();    // no drawing, so no canvas needed
  // create the response div:
  responseDiv = createDiv('Enter your username and challenge phrase \
   to write to the tag.');
  responseDiv.position(10, 130);

  var userLabel = createSpan('Username');      // create user field label
  userLabel.position(10, 10);
  userField = createInput('','text');          // create user field
  userField.position(100, 10);
  var challengeLabel = createSpan('Phrase');  // create challenge field label
  challengeLabel.position(10, 40);
  challengeField = createInput('','password'); // create challenge field
  challengeField.position(100, 40);
  var writeButton = createButton('Write to tag');// create write button
  writeButton.position(10, 70);
  writeButton.id('writeTag');
  writeButton.touchEnded(submit);
  var formatButton = createButton('Format tag')  // create format button
  formatButton.position(10, 100);
  formatButton.id('formatTag');
  formatButton.touchEnded(submit);
}
```

The `submit()` callback gets the ID of the button that calls it, to use as the route for an HTTP POST call. Then it assembles a JSON object using the contents of the user field and response field to use as the body of the POST request. Then it makes the POST request.

The `getResponse()` function is the callback for the POST request. It just gets the server's response and puts it in `responseDiv`.

That's the whole sketch. Save it, then you're ready to test the server.

```
function submit(event) {
  var route = '/' + event.target.id;    // get the ID of the button pressed
  var data = {                          // create the body of the POST request
    'user' : userField.value(),
    'challenge' : challengeField.value()
  };
  // change the response div in case the user hasn't touched a tag:
  responseDiv.html('Touch a tag to the reader.<br>')
  httpPost(route, data, 'text', getResponse);  // make the POST request
}

function getResponse(data) {      // callback from the POST request
  responseDiv.html(data);         // put the response in the div
}
```

Testing the Writer Application

You've got a lot of pieces in this project, only a few of which you've edited so far. Before testing the writer, here's a checklist of what you should have. The ones you've edited are marked in blue:

```
ndefSignature                 your project directory
  public                      your HTML directory
    writer                    p5.js project for writer
          index.html
          libraries
          sketch.js
    reader                    p5.js project for reader
          index.html
          libraries
          sketch.js
  node_modules                npm-added libraries
  https-server-reader.js      reader server
  https-server-writer.js      writer server
  keys                        certificates and keys
    domain.key                SSL private key
    domain.crt                SSL certificate
    private.key               NDEF writer private key
    public.key                NDEF reader public key
  writeNdefSignature.js       NDEF writer module
  readNdefSignature.js        NDEF reader module
  gpioControl.js              GPIO module
```

To test the writer server, change directories to the root of the project on your Pi and run it using sudo, like so:

```
$ sudo https-server-writer.js
```

Then open a browser on your personal computer and go to `https://your.raspberry.pi.address/writer`. Fill in the IP address of your Pi for `your.raspberry.pi.address`, and note the `https://`. You should get the page shown in Figure 9-18.

Place a tag on the reader and tap the Format button. You should get the following response:

```
Touch a tag to the reader.
accessing tag...
formatTag success.
```

When the tag's formatted, enter a username and a challenge phrase (notice that the phrase is obfuscated like a password) and tap the Write button. You should get a response like this:

```
Touch a tag to the reader.
accessing tag...
writeTag success.
```

If all of that happened as described, congratulations! You've got a working writer and formatter application.

If it didn't work, there are a few things to check:

- Are all the files in the right place?
- Is your NFC reader working properly? Use the **writeNdefSignature.js** and **readNdefSignature.js** modules on the command line to test it.
- Did you generate the key pair for the writer and reader?

Once you've got it all working, you'll write the reader.

X

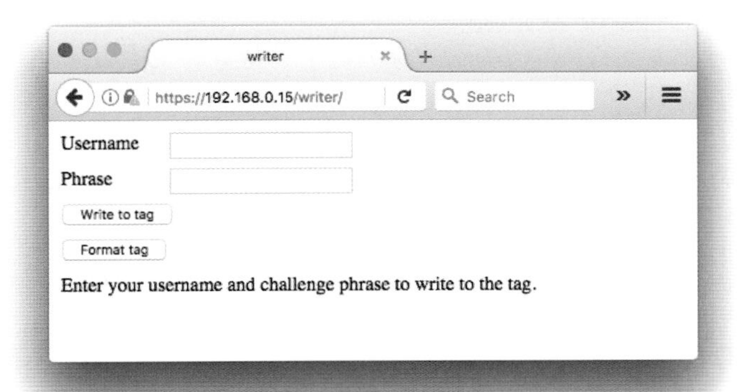

Figure 9-18
The web page of the NDEF writer application. Additional messages from the server will appear in the bottom div.

Serve It (Reader)

The **https-server-reader.js** script is very similar to the **https-server-writer.js** script. Copy the latter and modify it for this script. Changes are shown in blue. Chunks of identical code are removed for brevity. They're noted in comments.

The opening library includes are almost identical. The only change is that instead of including the **writeNdefSignature.js** module, you'll include **readNdefSignature.js** and **gpioControl.js.**

Nothing else changes until you get to the `processTag()` function. You'll write the same header to the response, but instead of calling `.write()` or `.format()` at the end of the function, you'll call `.setSecret()` and `.read()` from the **readNdefSignature.js** module. The `finishResponse()` function will be different too. Once it's got the data from the tag, it will call `.parseMessage()` from the **readNdefSignature.js** module to verify the tag.

```
/*
HTTP/HTTPS 2-Factor Authentication server
Tag reader/verifier version
context: node.js
*/
// include libraries and declare global variables:
// same library requires as https-server-write.js go here
// include the external modules:
var tagReader = require('./readNdefSignature.js');
var lockControl = require('./gpioControl.js');

// ... same server instantiation, middleware, and HTTPS redirect
// as https-server-write.js go here

function processTag(request, response) {
  var command = request.path;              // get the command
  command = command.slice(1);             // slice off the leading slash
  var data = JSON.stringify(request.body);  // convert the body to a string

  // write the beginning of the response:
  response.writeHead(200, {"Content-Type": "text/html"});
  response.write('accessing tag...<br>');

  // function to finish the response:
  function finishResponse(error, buffer) {
    if (error) {
      response.write('Failure: ' + error + '<br>');
    } else {
      response.write(command + ' success.<br>');
      var verified = tagReader.parseMessage(buffer);
      if (verified) {
        response.write('You entered the correct response.<br>');
        lockControl.open();
      } else {
        response.write('However, your response does not \
          match this tag.<br>');
      }
    }
    response.end();
  }

  // take action depending on the command:
  tagReader.setSecret(data);
  tagReader.read(finishResponse);    // attempt tag reading
}
```

The only change at the end of the script is that you're listening for a POST call to `/readTag` instead of `/writeTag` or `/formatTag`. That's the end of the script.

```
// start the server:
http.createServer(server).listen(8080);            // listen for HTTP
https.createServer(options, server).listen(443);  // listen for HTTPS
server.post('/readTag', processTag);              // listener for readTag
```

Add Web Page (Reader)

Like the server, the p5.js script for the web page of the reader is very similar to that of the writer. Changes are shown in blue. The global variables are the same.

```
/*
  Reader page script
  context: p5.js
*/
var userField, challengeField;  // fields for username and challenge phrase
var responseDiv;                // div for responses from the server
```

The setup is virtually the same, except that the initial text of the `responseDiv` is different, and `readButton` replaces `writeButton` and `formatButton`.

```
function setup() {
  noCanvas();    // no drawing, so no canvas needed
  // create the response div:
  responseDiv = createDiv('Enter your username and challenge phrase \
   to verify the tag.');
  responseDiv.position(10, 130);
  // same setup from writer sketch.js goes here,
  // to create userfield and challengeField
  var readButton = createButton('Verify tag');// create read button
  readButton.position(120, 70);
  readButton.id("readTag");
  readButton.touchEnded(submit);
}
```

The rest of the script is identical to the writer sketch. It has the same `submit()` and `getResponse()` functions. That's the end of the script.

```
// ... same submit() and getResponse() functions from
// writer sketch.js go here.
```

Testing the Reader Application

Now you can test the whole reader application. You should run one last test of the **gpioControl.js** module, just to make sure you plugged the solenoid back in. When you know it works, run the reader server like so:

```
$ sudo https-server-reader.js
```

Then open a browser on your personal computer and go to https://`your.raspberry.pi.address`/reader. Fill in the IP address of your Pi for `your.raspberry.pi.address`, and note the `https://`. You should get the reader page, shown in Figure 9-19.

Place the tag you formatted and wrote to earlier, enter the same username and a challenge phrase that you used before, and tap the Verify button. You should get a response like this:

```
Touch a tag to the reader.
accessing tag...
readTag success.
You entered the correct response.
```

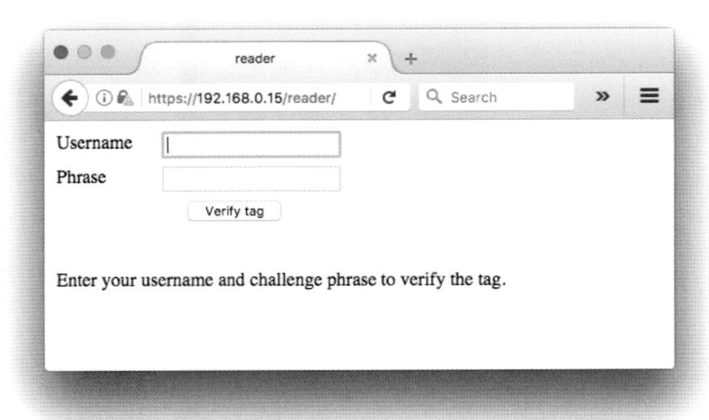

Figure 9-19
The web page of the NDEF reader application. Additional messages from the server will appear in the bottom div.

The solenoid should fire as well. When it works, try using the wrong phrase or username. You'll get this:

```
Touch a tag to the reader.
accessing tag...
readTag success.
However, your response does not match this tag.
```

You probably noticed when you were writing the servers that neither the writer nor the reader server breaks up the body of the POST request. The whole JSON object, including the username and phrase, is turned into a string encrypted, and written to a single NDEF record. The verifier script just checks that what's on the tag matches; it doesn't bother to parse them out.

It might seem more secure to separate the username and challenge phrase into separate NDEF records and encrypt each. There can be value in doing that, in order to verify the two pieces of data separately. However, the single blob of data is potentially more secure. If the username were unencrypted, someone with a tag reader might guess your phrase if they know your username. If they read the tag and get no clue as to what it is, though, they have a much harder time divining your username or challenge phrase.

Although it is easy to combine the reader and writer into one server script, the advantage of not doing this, as mentioned before, is that you can physically separate the reader from the writer by using two devices. The reader will not need the private key, so if you only put the public key and reader scripts on it, no one can exploit that device to write fake tags.

This application is secure because you need both an RFID tag and a device with a browser, as well as your username and challenge phrase, to open the lock. It's not terribly convenient, but if you were building a safe or a vault, you might sacrifice the convenience for the security. Even if you're not using this application for physical access, it can still work as a two-factor authentication interface for access to restricted websites.

X

Figure 9-20
The solenoid lock will replace the normal door hasp. The wire goes to the Raspberry Pi.

Network Identification

So far, you've identified network devices computationally by their address. For devices on the internet, you've seen both IP and MAC addresses. Bluetooth and 802.15.4 devices have standard addresses as well. However, the address of a device doesn't tell you much about what the device is or what it does.

Recall the networked air-quality project in Chapter 4. The microcontroller made a request via HTTP and the server sent back a response. Because the server had an API for programs other than a browser, you could get a response that was short enough for the microcontroller to process efficiently, formatted in a way that made it easy to read. You didn't have to read through the whole web page to get the info. But what if you wanted to customize the server's response depending on the capabilities of the device that's making the request?

Most internet communications protocols include a basic exchange of information—as part of the initial header messages—about the sender's and receiver's identity and capabilities. As a client, you should be aware of what information you're giving up. As a system designer, you can use these to your advantage when designing network systems like the ones you've seen here. Following is an example of how you can use HTTP header information to customize the content that a server delivers.

When a server-side program like the node.js scripts you've been writing receives an HTTP request, it has access to a lot more information about the client than you've seen thus far.

To see some of it, create a new node.js project directory called **clientHeaders**. Add a script and name it **server.js**. Install express using npm, run the script, and then open it in a browser.

```
/*
server, showing client headers
context: node.js
*/

var express = require('express');     // include express
var server = new express();            // make an instance of it

// GET response listener callback function:
function respond(request, response) {
  console.log(request.headers);
  console.log("server hostname: " + request.hostname);
  console.log("client IP address: " + request.ip);
  response.end("hello client!");
}

server.listen(8080);        // start the server on port 8080
server.get('/', respond);   // listener for GET requests
```

You should get something like this in your browser:

```
{ host: '114.226.112.201:8080',
connection: 'keep-alive',
'cache-control': 'max-age=0',
'upgrade-insecure-requests': '1',
'user-agent': 'Mozilla/5.0 (Macintosh; Intel Mac OS X 10_12_3)
AppleWebKit/537.36 (KHTML, like Gecko) Chrome/56.0.2924.87
Safari/537.36',
accept: 'text/html,application/xhtml+xml,application/
xml;q=0.9,image/webp,*/*;q=0.8',
'accept-encoding': 'gzip, deflate, sdch',
'accept-language': 'en-US,en;q=0.8' }
server hostname: 114.226.112.201
client IP address: ::ffff:206.175.85.178
```

As you can see, there's plenty of information: the web host's IP address, the client's IP address, the browser type, the operating system version, and the language of the client device, among other things. You probably never knew you were giving up so much information by making a simple HTTP request! This is very useful when you want to write server-side scripts that can respond to different clients in different ways.

The field `user-agent` tells you the name of the software browser and the operating system with which the client connected. From that, you can determine whether it's a mobile phone, desktop, or something else, and serve appropriate content for each. `accept-language` tells you the language in which the client would like the response. When you combine `client IP address` with the IP geocoding example to follow, you can even make an estimation as to where the client is.

All of this information comes at a cost: you're telling the servers that you contact a lot about yourself. You may not want to give away so much information every time, but if you want the information at the website you're contacting, you may not have much of a choice.

If you run this script on a server that's off your local network and make a request from your browser, you'll notice that the IP address for the request is probably not your device's address. Unless your device has a public IP address, as explained in Chapter 3, then the server will see your router's address and not your device's address. That's because your router is your device's public face to the network.

You're in public when you're on the network—as public as you are when you go out of your house to go shopping. The specific device from which you're connecting may not be immediately visible, but the local network through which you connect is visible, and it means that data about your browsing habits is available to your network provider and to whomever they choose to share it with.

Anonymous Browsing

While the server script is running, try connecting to it through an anonymous browser window (in Chrome, it's called Incognito Window; in Firefox and Safari it's Private Window; in Microsoft Edge, it's inPrivate Window). You'll see that the server sees your client as the exact same IP address as when you connected through a regular browser window. Surprise! Not so private, is it?

While there's no such thing as truly anonymous browsing, it is possible to further obfuscate the address and network from which you're connected to the internet. The Tor Project (www.torproject.org) is an open source software network that makes software to help clients anonymize their browsing somewhat. The Tor network is a network of volunteer servers that collectively agree to pass network traffic on without revealing the original source of any given request. When you use the Tor browser, for example, your request is sent to a server in the Tor network first, and then bounces around inside that network a few times. Eventually, another server (called the *exit node*) in the network makes the HTTP request that you made, and sends the result back to you through the Tor network. Instead of appearing as yourself to the end server, you appear as one of the nodes in the Tor network. For example, in testing the clientHeaders.js script, I connected to the server through the Tor browser. The connection went from my service provider in New York to a Tor node in the Netherlands to another node in France to another node in Norway. The node in Norway made the HTTP request to my server. When I used `nslookup` and `whois` (see Chapter 4) to look up the hostname of the IP address from which the request came, I got the name and address of a Swedish company.

Browsing through Tor is not a perfect solution. It's slower, because your request bounces around the network more, in order to make your connection more secure. However, seeing Tor in action gives you a greater understanding of how network traffic operates, and to what degree you can (or can't) protect your identity while online.

X

Project 28

IP Geocoding

The next example uses the client's IP address to get its latitude and longitude. It gets this information from www.freegeoip.net, a community-based IP geocoding project. Seeing the results from this site will give you an idea of the accuracies and inaccuracies of geocoding IP addresses. This script also uses the HTTP user agent to format its response appropriately for each client.

Freegeoip.net has an API that allows you to request the geolocation of a given IP address like so: http://freegeoip.net/<format>/<ip address>. The format can be CSV, JSON, or XML. You can send any IP address, and if you leave it off, the site uses the IP address from which you're connecting. I like to use it to look up the locations of IP addresses that add comments to blogs. Here's a typical result:


```
{"ip":"91.200.12.34","country_code":"UA","country_
name":"Ukraine","region_code":"","region_name":"","city":"","zip_
code":"","time_zone":"","latitude":50.45,"longitude":30.5233,"met
ro_code":0}
```


In the script that follows, you'll wait for client requests and use the client's IP address to get a geolocation. If the client is a web browser capable of displaying images, you'll redirect the request to Google Maps to display the client's location on a map.

If you connect to the previous script from most browsers, you'll see they reference "Mozilla/5.0" in their user agent declaration. For example, here is the user agent string for Chrome on macOS:

```
Mozilla/5.0 (Macintosh; U; Intel Mac OS X 10_5_8; en-US)
AppleWebKit/532.5 (KHTML, like Gecko) Chrome/4.0.249.0
Safari/532.5
```

Most modern web browsers claim to be "Mozilla-compatible," meaning that, at minimum, they support the features supported by the Mozilla Foundation's specifications. The foundation has its roots in Netscape, one of the pioneering companies of the Worldwide Web. Mozilla, its flagship product, is an open source browser derived from NCSA Mosaic, the earliest graphical browser. Mozilla is the basis for Firefox, and is often used as the reference among web browsers because the foundation plays an active role in web and internet standards bodies.

This means that if you're testing to see if your client can support the features of a modern browser—images, CSS, possibly JavaScript—then seeing the Mozilla/5.0 string in the user agent is a good sign. You'll use that in the following script.

For the most interesting results, run this script on a remote web host, not locally.

Locate It Create a new project directory called **geoIP**. In that, create a file called **server.js**. Install express using npm as you've done before. Start the script by including express and node.js's built-in http library, which you'll use to make a client request to freegeoip.net. Make an instance of express in the variable `server`; then set up global variables that you'll use later to make the request to freegeoip.net, and to redirect browsers to Google Maps when appropriate.

```
/*
server that looks up client's location by IP address
context: node.js
*/
var http = require('http');          // include http library
var express = require('express');    // include express
var server = new express();           // make an instance of it
// address for Google Maps server:
var mapsAddress = 'https://www.google.com/maps/place/';
// freegeoip server options:
var geoOptions = {
  host: 'freegeoip.net',
  path: '/json/'
};
```

»

At the end of your script you'll write a listener for GET requests just like in the last script. It will have a callback function called `respond()`. In `respond()`, you're going to use the client's IP address to make an HTTP request to freegeoip.net in order to determine the client's location.

Continued from previous page.

```
// GET response listener callback function:
function respond(request, response) {
  console.log("client IP address: " + request.ip);
```

The HTTP client command, `http.request()`, to make a request is at the end of `respond()`. It has a callback function, `getIPAddress()`. In this function, you'll listen for events from the freegeoip server. When new data arrives, you'll add it to the result.

The geolocation request and all its associated functions have to happen inside of `respond()` so that you can write a response to the original client when you're done.

```
  function getIPAddress(geoResponse) {
    var result = '';                   // string to hold the response
    // listen for events from  the freegeoip server:
    geoResponse.on('data', collectData);  // response data listener
    geoResponse.on('end', finishResponse);  // response close listener

    function collectData(data) {    // response data may arrive in chunks;
      result += data;               // add chunks together as they arrive.
    }
```

When the response ends, you'll finish your response to the original client in a function called `finishResponse()`. You'll send different content depending on the user-agent field.

```
    // when response closes, process it:
    function finishResponse() {
      var location = JSON.parse(result);
      var latLong =  location.latitude + "," + location.longitude;
      console.log(latLong);
      // if the user-agent is a browser (most modern browsers
      // identify as Mozilla/5.0), then redirect to Google Maps.
      // if not, supply the location as a JSON string:
      if (request.headers['user-agent'].includes('Mozilla/5.0')) {
        response.redirect(mapsAddress + latLong);
      } else {
        response.end(JSON.stringify(location));
      }
    }
  }

  // start the geoIp request:
  geoOptions.path = '/json/' + request.ip;    // add client IP to geoIP path
  var geoRequest = http.request(geoOptions, getIPAddress);  // start request
  geoRequest.end();                    // end it
}
```

The server start command and the route listener come at the end of the script as usual.

```
server.listen(8080);         // start the server on port 8080
server.get('/', respond);    // listener for GET requests
```

If you call this script from a browser, you'll get redirected to Google Maps and see your location on a map. If you call it from another client, like the command-line program `curl` that you saw in Chapter 3, you'll get a JSON string instead. Try this from the command line. Fill in your server's address in place of example.com.

```
$ curl example.com:8080
```

You should get a response like this:


```
{"ip":"208.113.160.6","country_code":"US","country_name":"United States","region_code":"CA","region_name":"California","city":"Brea","zip_code":"92821","time_zone":"America/Los_Angeles","latitude":33.9269,"longitude":-117.8612,"metro_code":803}
```


If you run it from a browser instead, you'll get a Google Maps location like the one shown in Figure 9-21. Try it from different browsers and you'll get the same result, but if you try it from Tor, you'll get the location of the exit node that made your request. Try it from your mobile phone as well, to see where your carrier's servers might be.

As you can see from this, IP geocoding is not always accurate. The geocode of an IP address depends on how the coding database gathered its data, and is only possible for public IP addresses. Even then, the location is more likely to be where your internet service provider's nearest routers are located rather than where you actually are. Furthermore, it's not too hard to obfuscate your location by going through a remote proxy. In that case, the proxy's IP address will show up as your location. So the next time you see an action movie or TV show where the police claim "We're tracking their IP address! They're at Smith and Main!" look at the screen and in a loud voice, shout, "No they're not! Their ISP is!"

The other notable aspect of this script is the value of the user agent field. To use it, all you need to do is add a line to your HTTP requests from your microcontroller or your client program, and add a conditional statement to your server to listen for the user agent you specify. For example, if you wanted to let the server know that your client is an Arduino-compatible microcontroller, your request would look like this:

```
// make a new HTTP client:
HttpClient http(netSocket, server, 8080);
http.beginRequest();   // begin the request
http.get(route);       // make it a GET request
/// add the user-agent header:
http.sendHeader("user-agent","arduino");
http.endRequest();       // send the request to the server
```

If you want to send custom user-agent strings from `curl`, you can that too. Here's how you'd send the Mozilla string needed for the previous sketch:

```
$ curl -L -A 'Mozilla/5.0' example.com:8080
```

The `-L` flag tells curl to follow redirect links, and the `-A` flag followed by a string sets the string as the user agent.

Using the user agent variable like this can simplify your development a great deal. It means that you can customize the server's response depending on the client, and you can easily use a browser or the command line to debug programs that you're writing for any type of client.
X

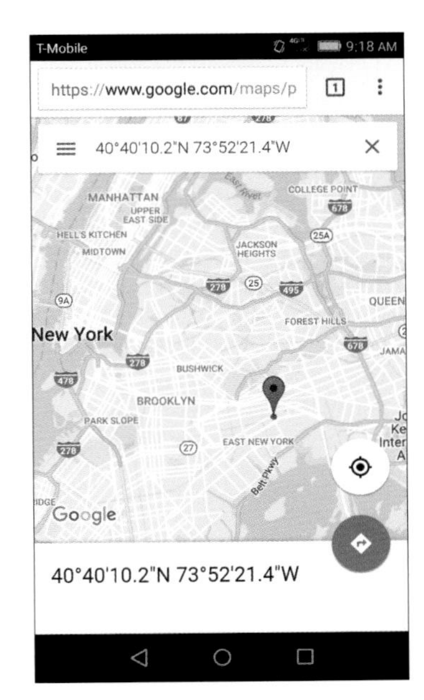

Figure 9-21
The geocoding of your IP address usually corresponds with your service provider's location. T-Mobile landed me in Cypress Hill, New York, this time, when I was actually in lower Manhattan. The same request from the same location through the Tor browser located me in west Texas.

Conclusion

In this chapter, you've seen several examples of how physical identity and network identity can be linked. The boundary between the two always introduces the possibility for confusion and miscommunication. No system for moving information across that boundary is foolproof. Establishing identity, capability, and activity are all complex tasks, prone to misinterpretation. The more human input you can incorporate into the situation, and the more transparency and agency you give the people involved, the better your results will be.

Security is essential when you're transmitting identifying characteristics because it maintains the trust of the people using what you make and keeps them safe. Once you're connected to the internet, nothing's truly private and nothing's truly closed, so learning to work with the openness makes your life easier. In the end, keep in mind that clear, simple ways of marking identity are the most effective, whether they're universal or not. Consistency is more important than universality. Both beginners and experienced network professionals often get caught on this point, because they feel that identity has to be absolute and clear to the whole world. Don't get caught up in how comprehensively you can identify things at first. It doesn't matter if you can identify someone or something to the whole world—it only matters that you can identify them for the context in which you're working. Once that's established, you've got a foundation on which to build.

When you start to develop projects that use identification systems, you usually find that less is more. Use only the identifying information you need, and let the users have their privacy about things you don't need to know.
X

10

MAKE: PROJECTS 

Mobile Phone Networks and the Physical World

Ethernet and WiFi are handy ways to talk to people and things on the internet, but there's a great big chunk missing: the mobile telephone network. Nowadays, telephony and the internet are so intertwined that it doesn't make sense to talk about them separately. It's getting increasingly easy to connect physical devices other than mobile phones through mobile phone networks. In this chapter, you'll learn how to connect these two networks, and when it's useful to do so.

◂◂ ohai lion, how r u??? txt me l8r!!!
This lion can send you an SMS. Groundlab, in conjunction with Living with Lions and Lion Guardians, developed this tracking collar that utilizes a GPS/GSM module to transmit lions' locations via SMS to researchers and Maasai herders. This open source system aims to help conservationists protect the last 2,000 lions living in the wild in Southern Kenya, and safeguard the Maasai herders' cattle, restoring Maasai land to a working ecosystem. *Photo courtesy of Groundlabs.*

Supplies for Chapter 10

Many of the basic hardware parts in this chapter will seem familiar. You will get the chance to work with 120V or 220V AC, though, and to work with conductive fabrics and threads.

DISTRIBUTOR KEY
A Arduino Store (store.arduino.cc)
AF Adafruit (adafruit.com)
D Digi-Key (www.digikey.com)
DL D-Link (www.dlink.com)
F Farnell (www.farnell.com)
J Jameco (jameco.com)
L Less EMF (www.lessemf.com)
MS Maker SHED (www.makershed.com)
RS RS (www.rs-online.com)
SF SparkFun (www.SparkFun.com)
SS Seeed Studio (www.seeedstudio.com)

PROJECT 29: CatCam Redux

» **1 solderless breadboard D** 438-1045-ND, **J** 20723 or 20601, **SF** PRT-12615 or PRT-12002, **F** 4692810, **AF** 64, **SS** 319030002 or 319030001 or prototyping shield **AF** 2077, **A** TSX00083, **SF** DEV-07914 with mini breadboard **SF** PRT-12044, **AF** 65, **D** 923273-ND
» **1 Arduino MKR1000 AF** 3156, **RS** 124-0657, **A** ABX00004, GBX00011 (Africa and EU), **D** 1659-1005-ND
Alternatively, an ESP8266-based board will work. **SF** WRL-13231, **AF** 2471
» **IP-based camera** The project uses a D-Link DCS-930L, DCS-5222L, or DCS-960L but you can also make one with a Raspberry Pi and Pi Cam as described in the text. **D-link IP Cameras** if you opt to not use a Raspberry Pi

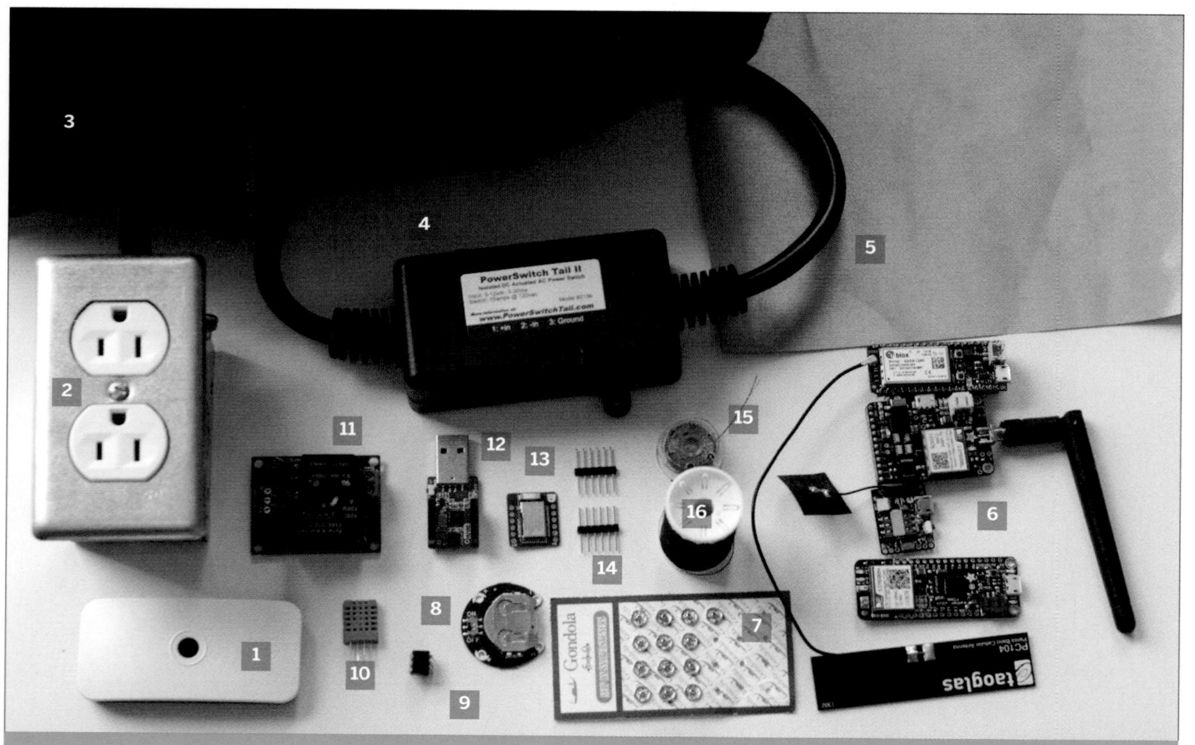

Figure 10-1. New parts for this chapter: **1.** Raspberry Pi Zero with Pi camera in Pi case **2.** AC junction box, outlet, and extension cord **3.** Hoodie **4.** PowerSwitch Tail **5.** Conductive fabric **6.** Cellular modems **7.** Sewable snaps **8.** Battery holder **9.** Optoisolator **10.** DHT11 temperature and humidity sensor **11.** AC relay **12.** MK20 USB adapter **13.** BLE Nano **14.** Long header pins **15.** Conductive thread **16.** Embroidery thread

DL DCS-930L, DCS-5222L, or DCS-960L
Raspberry Pi **SF** DEV-13825, **AF** 3055, **SS** 114990584, **RS** 896-8660, **F** 2525225
Pi Cam **SF** 14028, **AF** 3099 **SS** 113990214
Pi Zero W Camera Kit **A** 3414

» **DTH11 Temperature and Humidity Sensor** **AF** 386, **J** 2245415, **SS** 101020011
» **Relay Control PCB** SparkFun sells a relay kit you'll need to solder together. Alternately, you can use the PowerSwitch Tail.
SF KIT-13815
» **1 PowerSwitch Tail** This is an alternative device to the relay board above. A 240V version is available from www.powerswitchtail.com. **SF** COM-10747 **AF** 2935
» **1 1-kilohm resistor** **D** 1.0KQBK-ND, **J** 690865, **F** 9339051, **RS** 77-7666
» **1 100-ohm resistor** **D** 100QBK-ND, **J** 690620, **F** 9337660, **RS** 755-0707
» **1 4N35 optoisolator** **J** 41056, **F** 1244500, **D** 160-1304-5-ND, **RS** 597-302
Alternatively, SparkFun makes a breakout board **SF** BOB-09118
» **Cat**
» **Air conditioner**
» **Country estate (optional)**

PROJECT 30: Phoning the Thermostat

» **Completed Project 29**
» **Twilio account**
» **Telephone**

PROJECT 31: Personal Mobile Datalogger

» **Android or iOS device**
» **RedBear BLE Nano and MK20 USB board** Sold together **MS** MKRBL5, **SF** WRL-14071
You could also use an Arduino 101 board, though it would be larger to wear. **D** 1660-1003-ND, **J** 2239331, **SF** DEV-13787, **AF** 3033, **F** 2520713, **RS** 913-9999, **SS** 114990575 , **A** ABX00005, GBX00005 (Africa and EU)
» **Extra-long male header pins, 0.1" spacing** **AF** 400, **SF** PRT-12693 Alternatively, you can use IC hooks with pigtails **SF** CAB-09741 and CAB-00501
» **Battery Holder and Battery** **SF** DEV-10730 or DEV-13883 with battery PRT-00338, **AF** 1870 or 1871 with 654. If you prefer a solution with more capacity **SF** DEV-11893 and PRT-13851
» **1 270-kilohm resistor** **J** 691446, **D** CF14JT270KCT-ND, **RS** 845-7577
» **Thick conductive thread** **AF** 603, **SF** DEV-11791 **L** 304
» **Conductive fabric** **AF** 1168 **L** 1220
» **Sewable snaps, 5mm diameter** **AF** 1126 **SF** DEV-11347
» **Hoodie**
» **Embroidery thread** Available at most fabric or yarn shops.

Cellular modems

» **Adafruit Fona 800** USA **AF** 3147, Europe **AF** 2691
» **Adafruit Fona Feather** **AF** 3027
» **Seeed Studio Xadow GSM+BLE** **SS** 102040005
» **XBee Cellular** D 602-1976-ND,
» **Particle Electron Kit** Americas/Australia **SF** WRL-14211, Africa/Asia/Eurpoe **SF** WRL-14212

One Big Network

Before the internet, there was the telephone network. All connections were analog electrical circuits, and all phone calls were circuit-switched, meaning that there had to be a dedicated circuit between callers. Then modems came along, which allowed computers to send bits over those same analog circuits. Gradually, switchboards were replaced with routers, and now telephone networks are mostly digital as well. Circuits are virtual, and what takes place behind the scenes of your phone calls is not that different from what occurs behind the scenes of your email or chat conversation: a session is established, bits are exchanged, and communication happens. The difference between a phone call and an email is now a matter of network protocols, not electrical circuits.

There are plenty of IP-based telephony tools that blur the line between phone call and internet connection, including the open source telephony server Asterisk (www.asterisk.org), as well as telephony services from companies like Twilio (www.twilio.com), Google Voice (voice.google.com), and Skype (www.skype.com). These voice services are compatible with those offered by your phone company. What the phone company is giving you on top of the software service is a network of wires and routers that prioritizes voice services, so you are guaranteed a quality of service that you don't always get on IP-only telephony services.

Telephony services and internet services meet on gateway servers and routers that run software to translate between protocols. For example, mobile phone carriers all offer SMS gateway services that allow you to send an email that becomes an SMS, or to send an SMS that emails the person you want to reach. Google Voice and many of the other online telephony services offer voicemail-to-text, in which an incoming call is recorded as a digital audio file, then run through voice-recognition software and turned into text, and finally emailed to you. The major task when building projects that use the telephony network is learning how to convert from one protocol to another.

A Computer in Your Pocket

Mobile phones are far more than just phones now. The typical smartphone—such as an Android phone or iPhone—is a computer capable of running a full operating system. The processing power is well beyond that of older desktop machines, and smartphones run operating systems that are slightly stripped-down versions of what you find on a laptop, desktop, or tablet computer. Most smartphones also incorporate the sensors you've already used, such as a camera, accelerometer, light sensor, and GPS. And, of course, they are networked all the time. What they lack, however, is the capability for adding sensors, motors, and other actuators—the stuff for which microcontrollers are made. When you treat the phone as a multimedia computer and mobile network gateway, you open up a whole host of possibilities for interesting projects.

Interfacing phones and microcontrollers can be done in a number of ways, depending on the phone's capabilities. For the purposes of this chapter, I'll be talking about smartphones, so you can assume most or all of the following capabilities, many of which you'll use in the projects to come:

- Programmability
- Touchscreen or keyboard
- Mobile network access
- Bluetooth connection
- USB connection
- Microphone, speaker
- Onboard accelerometer
- Onboard GPS

There are a few possible system arrangements afforded by these capabilities and the kinds of things you might want to make: WiFi or mobile service for remote connections, and Bluetooth for local connections. Once you're connected via an IP connection, you can use all the internet protocols we've discussed, like HTTP, email, SSH, or FTP.

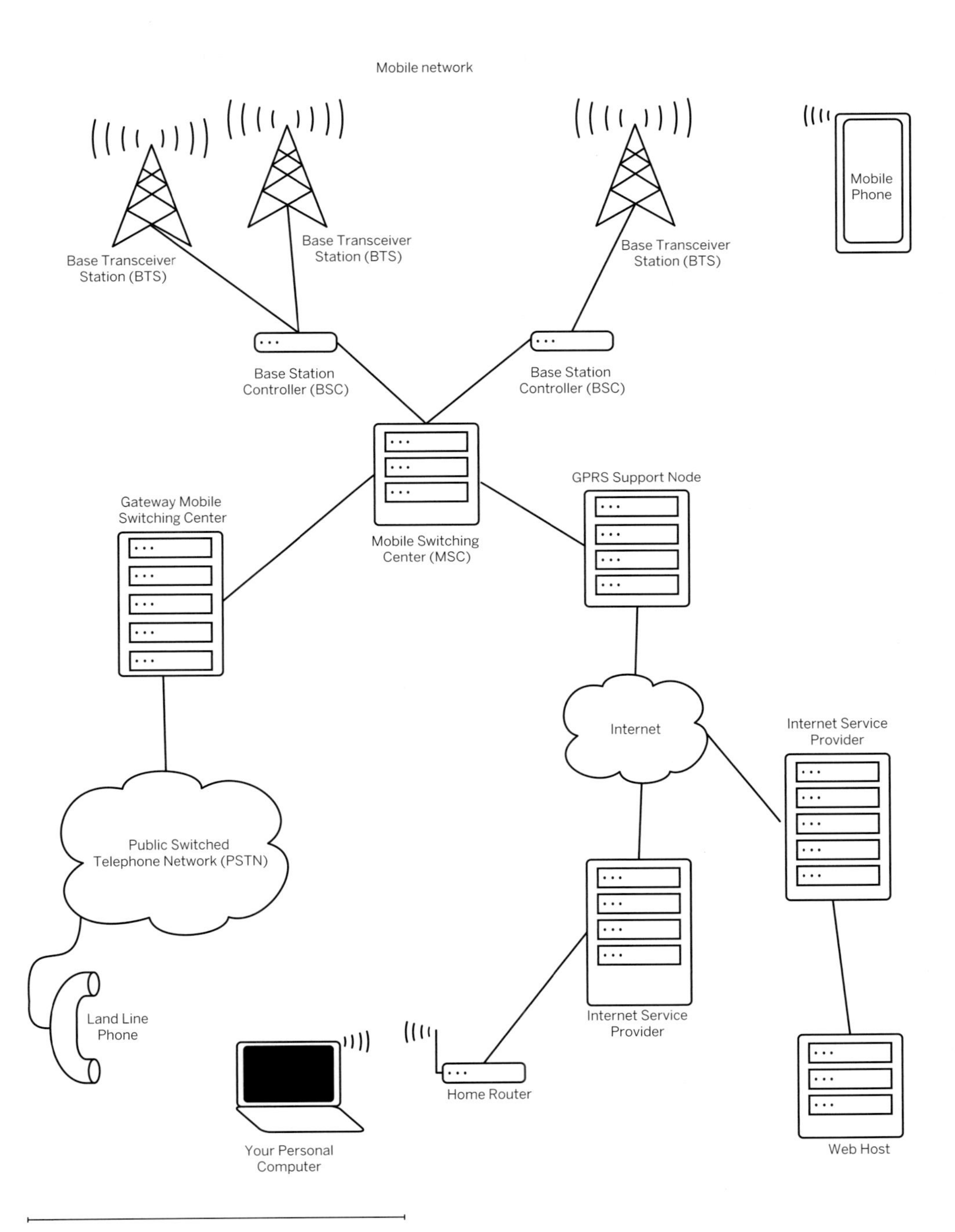

Figure 10-2
How the mobile phone network meets the internet

Figure 10-2 shows how the mobile phone network connects to the collection of networks we think of as the internet. Mobile gateways provide the link between the phone network and the internet.

Start With What Happens

By now this should be a common theme to you: start with what you want people to do with the thing you're building, and design whatever makes that easiest to do. Are you making a system where your users have real-time interactive control over a nearby model helicopter or boat? Are they controlling a remote device like their home thermostat while on vacation? Consider how tight an interactive coupling you need. Do you need real-time interaction like you saw in Chapter 5, or will updates every few seconds or minutes be enough? And what about the distance between the devices—do you need to be able to communicate at long range, or will a local connection like Bluetooth suffice? If it's a remote connection, how will your users know what happened on the other side? Can they see the result? Do they get an email or SMS message telling them about it? Evaluate whether you need a wired or wireless connection, based on how the people need to move and manipulate the devices involved. Consider what networks are available where you plan to use your system. Got WiFi? Got mobile networks? Got Ethernet? Can you create a local network between the two devices using Bluetooth or WiFi?

Browser Interfaces

If you want to control an existing networked device from a mobile phone, the easiest thing to do is to give the device an internet connection and control it from your phone's browser. You don't have to learn anything specific to a particular phone operating system to do this—you just need to be able to make an HTTP connection between them. Most of the projects you've seen in this book are accessible from your phone's browser.

Native Application Interfaces

If you don't want to use the browser, you can build your own application on the phone to interact with the microcontroller via Bluetooth, network, or SMS. You'll see an example of this later in the chapter. The advantage of building a dedicated app is that you have complete control over what happens. The disadvantage is the challenge of learning a new programming environment, as well as the limitation of not being able to use your program on a phone with a different operating system. Like the desktop browser, there are still some limits to what hardware you can access from the phone browser, though that is changing rapidly.

SMS and Email Interfaces

Not every project requires real-time interaction between your phone and the thing you're building. If you just want to get occasional updates from remote sensors, or you want to start a remote process that can then run by itself—like turning on a sprinkler or controlling your house lights via a remote timer—SMS or email might be useful. You can make a microcontroller application that can receive SMS messages using a GPRS modem, or you can write a networked server that checks email and sends messages to an internet-connected microcontroller when it gets a particular email message. The advantage of this method is that you can control remote things using applications that are already on your phone, email, and SMS.

Voice Interfaces

Don't forget the phone's original purpose: two-way communication of sound over distance. Voice interfaces for networked devices can be simple and fun. Telephony gateways like Asterisk and Twilio allow you to use voice calls to generate networked actions.

Phone As Gateway

Not all applications in this area involve control of a remote device from the phone. Sometimes you just want to use the mobile phone as a network connection for sensors on your body or near you. In these cases, you'll make a local connection from the microcontroller to the phone, most likely over Bluetooth or USB, and then send the data from those sensors to a remote server. In these cases, you probably don't need to build a complex application—you just need to make the connection between the two devices, and between the phone and the network.

 Project 29

CatCam Redux

Since the phone has a web browser, you don't need to know how to program it to make a mobile application. In this project, you'll make a variation on the CatCam from Chapter 3. But this time, you'll use a new protocol—MQTT—and an IP-based camera that needs no computer to connect to the internet.

MATERIALS

- **1 solderless breadboard**
- **1 Arduino MKR1000, or WINC1500 or ESP8266-compatible microcontroller**
  - **Features used: GPIO, WiFi**
- **IP-based camera**
- **Temperature sensor**
- **Relay Control PCB**
- **1 1-kilohm resistor**
- **1 100 ohm resistor**
- **1 optoisolator**
- **1 LED**
- **Cat**
- **Air conditioner**
- **Country estate (optional)**

My neighbor, let's call her Luba, has a cat named Gospodin Fuzzipantsovich. Luba has a country estate with cherry trees. She likes to summer there, at least on summer weekends, but she can't take Fuzzipantsovich with her. Because the country estate has huge expenses (samovars and the like), Luba can't afford an air conditioner with a thermostat for her city place. So, I made her a web-based temperature monitor. It works as follows.

Luba opens a browser on her phone or laptop and logs into the thermostat's web interface. She sees the current temperature in her city apartment and the air conditioner's state, along with a camera view of Fuzzipantsovich. If it's too hot, she sets the trigger temperature lower. If it's too cool, she sets it higher. The interface updates the thermostat, and shows her the new trigger point.

Before you build this project, you should diagram the devices and protocols involved, and break the action into steps to understand what needs to happen. Figure 10-3 shows the system, and Figure 10-4 details the interaction.

There are several pieces of hardware involved and protocols to communicate between them. Many of them are familiar

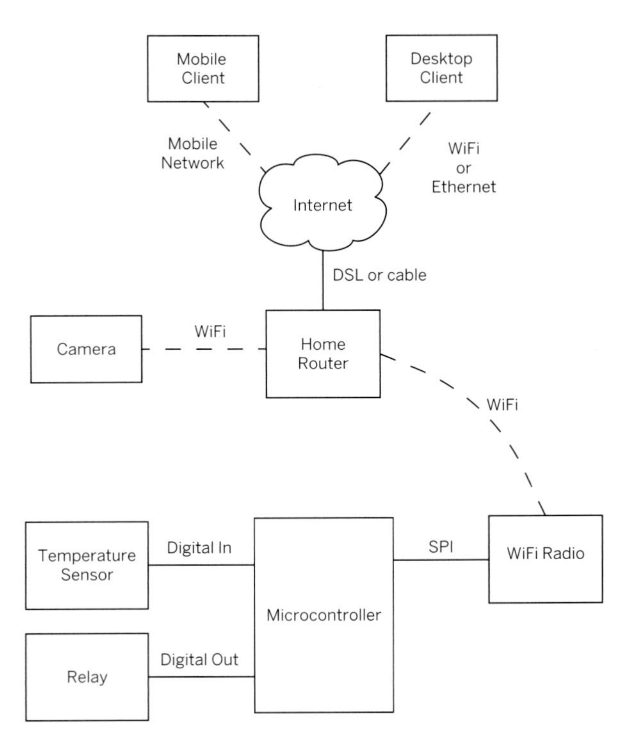

Figure 10-3
The system diagram for the CatCam 2 and air conditioner controller.

to you by now. The WiFi radio on the MKR1000 communicates with the microcontroller via SPI. The temperature sensor is a digital input, and the 120V relay to control the air conditioner is a digital output. The IP camera in this project is a consumer item that is WiFi-connected, so all you have to do is configure it to speak to your router and send images to your web host. The clients are just web browsers on your smartphone, tablet, or personal computer.

For the web application, you'll make a web interface in p5.js, which can run on your phone, and will send the camera image to the web host from the camera via FTP. To manage the air conditioner, you'll need to program the microcontroller, which you'll do with an MKR1000 or one of the ESP8266-based boards. You'll also need to write a server to run on your public web host, both to serve the web pages and to manage the connection with the microcontroller. You'll do that in node.js.

MQTT

Although you could run a web server on the microcontroller and use port forwarding, which you learned about in Chapter 3, to give it and the camera a public IP address, that's not good security practice. Instead, you're going to serve the web page and the image from a public web host, and you'll run a new protocol on the web host, *Message Queueing Telemetry Transfer*, or *MQTT*. MQTT (mqtt.org) is a lightweight messaging protocol designed for communications with remote sensor systems, where the sensors might not always be online. An MQTT server program is called a *broker*. Data is organized in *topics*, and within each topic there are *messages*. Devices subscribe to different topics, and are notified by the broker when new messages arrive. They can also publish new messages to a topic as well. This is why MQTT is known as a *Publish & Subscribe*, or *PubSub* system.

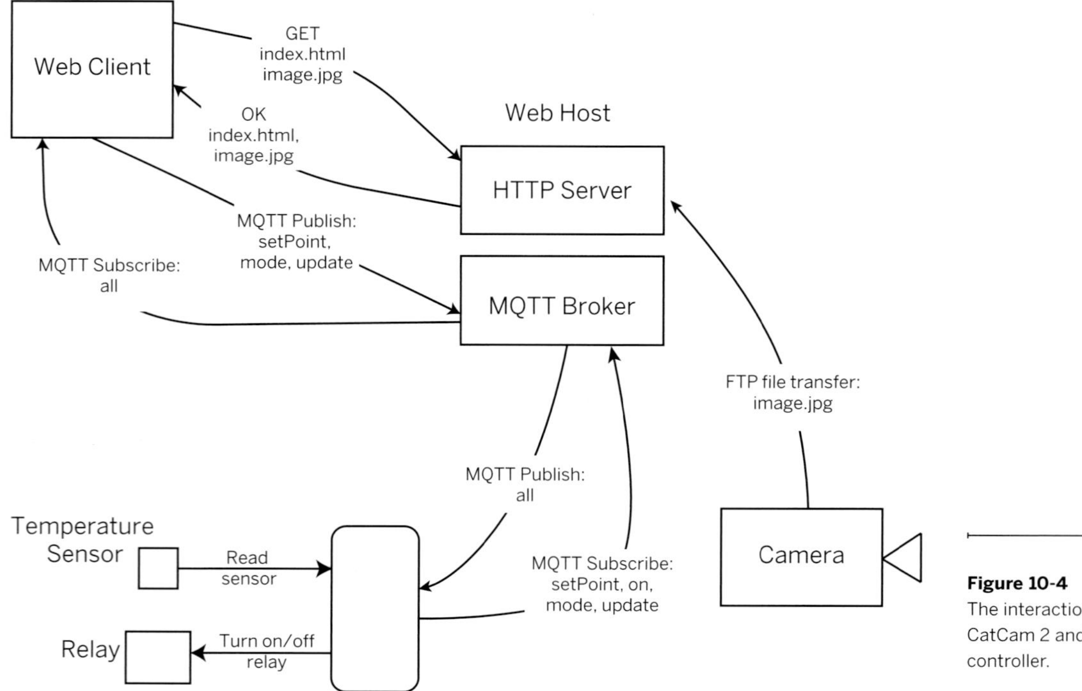

Figure 10-4
The interaction diagram for the CatCam 2 and air conditioner controller.

Figure 10-5
The relay board is connected to an AC cord, which has been cut on one side to attach it to the relay.

Figure 10-6
An AC junction box with the relay inside, and the control lines from the relay connecting to an Arduino. This keeps the AC circuit protected. You could also use a PowerSwitch Tail.

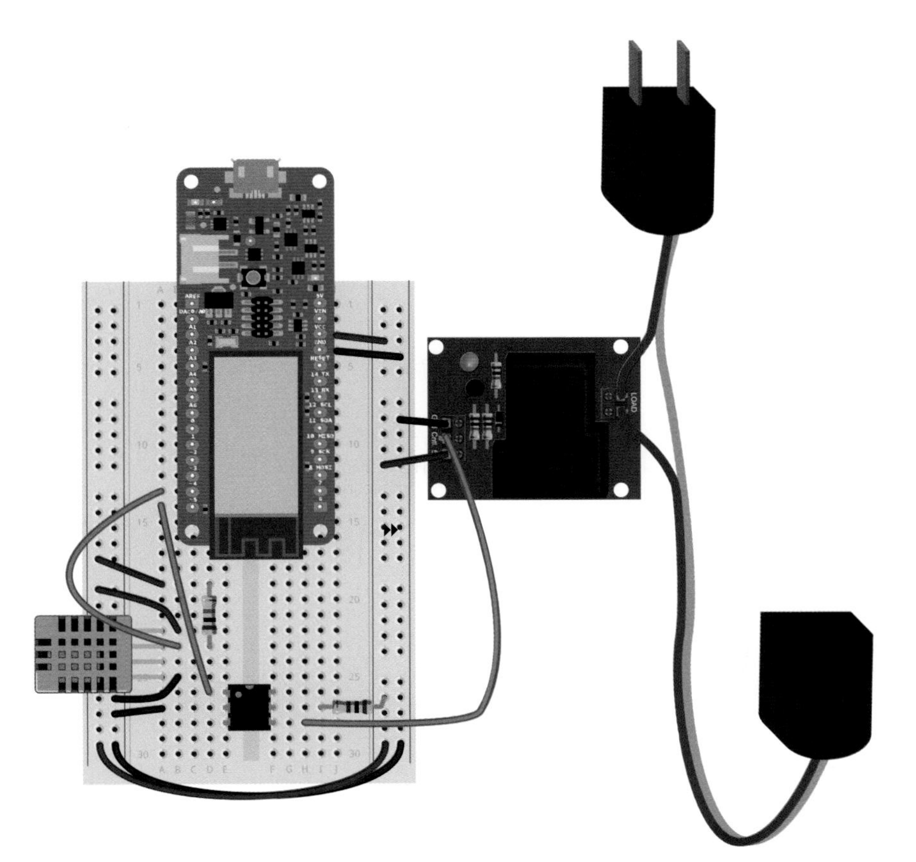

Figure 10-7
The CatCam 2 and air conditioner circuit, shown with an MKR1000 microcontroller.

For example, your device is the air conditioner, and its properties are the topics, like so:

- airConditioner/on—whether it's on or off
- airConditioner/mode—on, off, or auto
- airConditioner/temperature
- airConditioner/setPoint—temperature at which the air conditioner turns on or off, when in auto mode.

When the temperature changes, the microcontroller would publish a new message in the /temperature topic. The web client would send a new message the /setPoint topic to change the thermostat setpoint, or it might publish a message to the /mode topic to turn the air conditioner on or off, or to set it running from the thermostat.

Because the broker updates all connected clients, the microcontroller and the web client would get each other's messages in close to real time. MQTT can be tricky, though; when a given client is offline, it won't get all of the topic updates that it missed unless you design a way for it to do so. Because of that, you'll add two other topics, air-Conditioner/connected, which the microcontroller will use when it logs on or off, and airConditioner/update. When the microcontroller receives a new message in the /update topic, it will publish an update for all its topics.

The Circuit

Once you've got a grasp on the plan and the system, you're ready to build it. The circuit is shown in Figure 10-7 and the schematic in Figure 10-8. You only need to add the temperature sensor, the optoisolator, and the relay.

The relay contains a coil wrapped around a thin switch. When the coil gets voltage, it forms a magnetic field through induction, pulling the two sides of the switch together. To use it, take an alternating current circuit (AC) and break one of the wires, as shown in Figures 10-5 through 10-7.

The optoisolator in the circuit works in place of a transistor to turn the relay on or off. It's got an LED and a phototrans-istor inside it. Your microcontroller turns on the LED and the LED connects the phototransistor's collector and emitter, which turns on the relay. The two sides of the optoisolator are electrically separate. This way, you'll have less chance of destroying your microcontroller circuit if there's a short on the AC side. To be really safe, you could power the right side of the optoisolator from a separate 5V supply.

Working with AC can be dangerous, so make sure every-thing is connected properly before plugging it in. If you're uncomfortable building an AC circuit, the PowerSwitch Tail

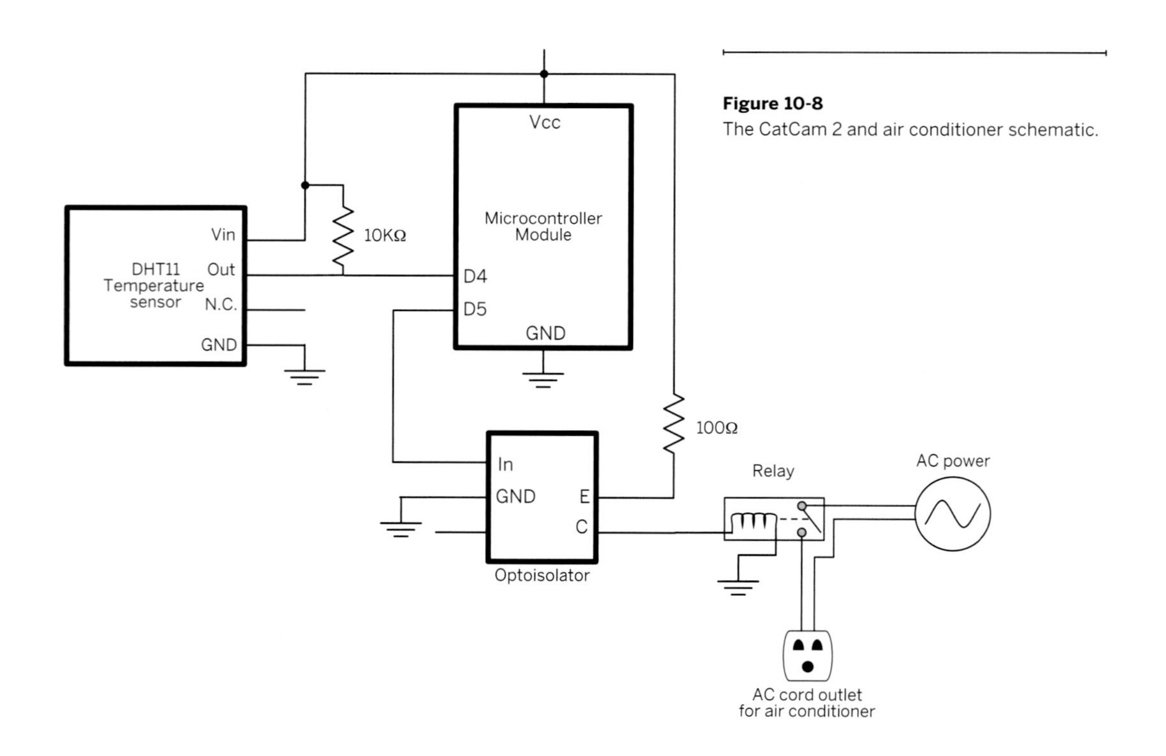

Figure 10-8
The CatCam 2 and air conditioner schematic.

will allow you to control 120V AC or 240VAC, depending on your model. A 240V version was recently announced at www.powerswitchtail.com. It's available from Adafruit or SparkFun. You could also rewrite this project using the WeMo Switch™ you used in the last chapter.

SparkFun's AC relay board is a safe and less expensive option if you want to build your own circuit. Solder the relay to the relay board. Then take an AC cord, cut one of the two wires, and attach it to the relay board's LOAD connections. Figure 10-5 shows a close-up of a household AC cord attached to the relay. The relay's ground, control, and 5V connections connect to your microcontroller. To build a more robust version—such as the one shown in Figure 10-6—with a household electrical outlet junction box, see Nathan Seidle's excellent tutorial at www.SparkFun.com/tutorials/119.

To test the circuit safely *before you plug it in*, set your multimeter to measure continuity, and to measure between one of the plug's pins and the corresponding socket hole on the other end of the cord. One pin/hole pair should have continuity all the time. The other should not, unless you power the 5V and ground contacts of the relay and connect the control line to 5V. Then you'll get continuity. When you connect the control line to ground, you won't get continuity. When you're sure the circuit's working, test it by attaching the control line to a pin of your microcontroller and use the Blink sketch to turn the relay on and off.

Make sure to install your circuit in an insulated box and add plenty of strain relief to the AC lines. If they accidentally disconnect, they can cause a short circuit—and AC shorts are not pleasant. An electronic project case is safest. You can add strain relief to the wires by using hot glue or rubber caulking, though some cases come with wire strain relief glands to hold the wire in place. Once you've done that, give the wires a tug to make sure they won't move. Take your time and get this part very secure before you proceed.

The Code

Since the sketch is complex, it's built up piece-by-piece, as follows. Save this as **AirConditionerArduino**:

Try It The temperature sensor for this project is the DHT11. This sensor shows up in many companies' sensor module kits because it's inexpensive, robust, and stable. It uses Dallas Semiconductor's OneWire interface protocol. Install the DHT library by Adafruit and the Adafruit Unified Sensor library using the Library Manager in the Arduino IDE.

Include the library at the top of your code, and make a constant for the pin number. Then initialize the sensor.

In the main loop, you just need to read the sensor and print the results. `.readTemperature()` defaults to Celsius, but if you want Fahrenheit, call `.readTemperature(true)`. The temperature values you'll see in this project will be in Celsius.

```
/*
  mqtt client to control air conditioner
  context: Arduino
 */
#include <DHT.h>            // include DHT library
const int sensorPin = 4;    // pin number for the sensor
DHT dht(sensorPin, DHT11); // make an instance of the library

void setup() {
  Serial.begin(9600);       // start serial communications
  dht.begin();              // start the sensor
}

void loop() {
  float temperature = dht.readTemperature();// read temperature
  float humidity = dht.readHumidity();      // read humidity
  Serial.print(temperature);                // print them
  Serial.print("°C ,");
  Serial.print(humidity);
  Serial.println("% rH");
}
```

Control It

Now it's time to add the relay control. You'll need variables for the air conditioner's properties, so add them now. New lines are in blue as usual.

```
/*
  mqtt client to control air conditioner
  context: Arduino
*/
#include <DHT.h>              // include DHT library
boolean deviceIsOn = false;      // device state
int temperature = 0;             // current temp.
int lastTemperature = 0;         // previous temp. read for comparison
int setPoint = 18;               // thermostat setpoint
int mode = 3;                    // off, on, or auto (1, 2, 3)
boolean deviceConnected = false;// network connection status
```

Add a constant for the relay's I/O pin, and set it to be an output in the `setup()`. Take an initial temperature read in the `setup()` as well.

```
const int sensorPin = 4;         // pin number for the sensor
const int relayPin = 5;          // relay pin
DHT dht(sensorPin, DHT11);       // instance of the library

void setup() {
  Serial.begin(9600);            // start serial communications
  dht.begin();                   // start the sensor
  temperature = dht.readTemperature();  // take an initial reading
  pinMode(relayPin, OUTPUT);            // make the relay pin an output
}
```

In the `loop()`, check to see if the temperature has changed by comparing it to the previous reading. If so, print it. Then save the current reading as the previous for next time.

The air conditioner will have the following modes: on, off, and auto. In auto mode, it will turn on and off depending on the temperature like a thermostat. Regardless of the mode, control the relay using the `deviceIsOn` variable.

The relay shouldn't be switched more than once every couple seconds, and the temperature sensor changes quite slowly, so add a delay at the end of the `loop()`.

```
void loop() {
  temperature = dht.readTemperature();          // read the temperature
  if (abs(temperature - lastTemperature) > 0) { // see if it's changed
    Serial.print("temperature changed to: ");
    Serial.println(temperature);
  }
  lastTemperature = temperature;          //save for comparison next time

  if (mode == 3) {    // auto
    checkThermostat();
  }

  digitalWrite(relayPin, deviceIsOn);    // turn on or off the relay pin
  delay(2000);
}
```

Add a new method, `checkThermostat()`, at the end of your sketch. It checks the temperature and compares it to a setpoint. If the temperature is greater than the setpoint, it turns the relay on. If the temperature is less, the relay is turned off.

```
void checkThermostat() {
  if (temperature > setPoint) {
    if (!deviceIsOn) {
      deviceIsOn = true;         // turn on
    }
  }
  if (setPoint > temperature) {
    if (deviceIsOn) {
      deviceIsOn = false;        // turn off
    }
  }
}
```

Before you run this sketch, replace the relay connection with an LED so you can make sure it's behaving properly before you try to control alternating current.

When you run the sketch at this point, you'll see the LED (in place of the relay) turn on or off depending on the `setPoint` value. Try changing the value to something less than the temperature sensor's reading and uploading your code again to see what happens. You should be able to control the output by changing the temperature sensor, either with your breath or with a cold object next to the temperature sensor. When you're satisfied that it's working properly you can connect the relay again. Now you're ready to add the MQTT client code.

First you'll need an MQTT client library for Arduino. In the Library Manager, you'll see there are several. The following code uses MQTT by Joel Gaehwiler. The source code can be found at github.com/256dpi/arduino-mqtt. Install it from the Library Manager.

Since you're adding network connections, you'll need the credentials to connect to your local network as usual. Use the Sketch menu→Add File... to add the **config.h** file to your Arduino project that you've used with every other networked microcontroller project in the book.

Credentials In addition to the usual SSID and password strings, you'll be adding MQTT credentials to it this time. It should look like this. Fill in your network credentials and make up an MQTT password. Remember it for later, since you'll need it when you configure the broker.

```
/*
  mqtt client to control air conditioner
  context: Arduino
*/
char ssid[] = "network";              //  your network SSID (name)
char password[] = "password";         // your network password
char mqttUser[] = "airConditioner";   // mqtt username
char mqttPass[] = "something";        // mqtt password
```

Connect It Start by adding the global variables for networking (new material shown in blue).

```
/*
  mqtt client to control air conditioner
  context: Arduino
*/
#include <DHT.h>              // include DHT
#include <SPI.h>              // include SPI library
#include <WiFi101.h>          // include WiFi101 library
//#include <ESP8266WiFi.h>    // use instead of WiFi101 for ESP8266 modules
#include <MQTTClient.h>       // include MQTT Client library
#include "config.h"           // include credentials

WiFiClient netSocket;         // network connection instance
MQTTClient client;            // MQTT client instance
char serverAddress[] = "192.168.0.15";  // MQTT server address
boolean deviceIsOn = false;
// ... previous global variables should go here
```

▸▸ **Change this address to the IP address of the machine where you will run the broker.**

»

In the `setup()` add the same WiFi connection routine that you've used before. Make sure to add the **config.h** file as well.

Continued from previous page.

```
void setup() {
  // connect to WiFi
  Serial.begin(9600);              // initialize serial communication
  // while you're not connected to a WiFi AP,
  while ( WiFi.status() != WL_CONNECTED) {
    Serial.print("Attempting to connect to Network named: ");
    Serial.println(ssid);          // print the network name (SSID)
    WiFi.begin(ssid, password);    // try to connect
    delay(2000);
  }
  // When you're connected, print out the device's network status:
  IPAddress ip = WiFi.localIP();
  Serial.print("IP Address: ");
  Serial.println(ip);
```

At the end of the setup, start the MQTT client using `client.begin()`, and call `mqttConnect()`, a function you'll write shortly.

```
  // start the temperature sensor and get an initial reading:
  dht.begin();
  temperature = dht.readTemperature();

  pinMode(relayPin, OUTPUT);    // make the relay pin an output
  // start the mqtt client and try to connect:
  client.begin(serverAddress, netSocket);
  mqttConnect();
}
```

At the beginning of the `loop()`, add a routine that checks whether the client is connected and attempts to reconnect if it's not connected.

```
void loop() {
  client.loop();               // check for new messages
  if (!client.connected()) {  // if the client disconnects,
    Serial.println("disconnected from server");
    deviceConnected = false;
    mqttConnect();            // reconnect
  }
```

Any time one of the device's properties changes, you want to publish to the broker. In the conditional statement that checks for temperature change, add a line to publish the temperature if it changes.

The rest of the `loop()` stays the same.

```
  temperature = dht.readTemperature();          // read the temperature
  if (abs(temperature - lastTemperature) > 0) { // see if it's changed
    Serial.print("temperature changed to: ");
    Serial.println(temperature);
    client.publish("airConditioner/temperature", String(temperature));
  }
  lastTemperature = temperature;         // save for comparison next time

  if (mode == 3) {    // auto
    checkThermostat();
  }

  digitalWrite(relayPin, deviceIsOn);    // turn on or off the relay pin
}
```

In the `checkThermostat()` function, you'll add `client.publish()` statements for when the air conditioner gets turned on or off.

The MQTT client library expects strings for both the topic and the message, so you're converting in the `.publish()` calls.

```
void checkThermostat() {
  if (temperature > setPoint) {
    if (!deviceIsOn) {
      deviceIsOn = true;        // turn on
      client.publish("airConditioner/on", String(deviceIsOn));
    }
  }
  if (setPoint > temperature) {
    if (deviceIsOn) {
      deviceIsOn = false;       // turn off
      client.publish("airConditioner/on", String(deviceIsOn));
    }
  }
}
```

Next, add the `mqttConnect()` function mentioned earlier. This will try to connect. If it fails, it will return to the function that called it. If it succeeds in connecting, it will subscribe to all the topics for properties that might be changed by a remote client. Then it will call `publishAll()`, which you'll write next.

```
// attempt to connect to the broker:
void mqttConnect() {
  if (!client.connect("airConditioner", mqttUser, mqttPass)) {
    return;   // skip the rest of the function
  }
  Serial.println("connected to server");
  deviceConnected = true;
  // when you're connected, subscribe to properties
  // that might be changed remotely:
  client.subscribe("airConditioner/on");
  client.subscribe("airConditioner/setPoint");
  client.subscribe("airConditioner/mode");
  client.subscribe("airConditioner/update");
  publishAll();
}
```

`publishAll()` gets called whenever the microcontroller connects, or when it gets a message in the /update topic, as you'll see soon. It publishes all the air conditioner's properties.

```
// publish all properties:
void publishAll() {
  client.publish("airConditioner/on", String(deviceIsOn));
  client.publish("airConditioner/temperature", String(temperature));
  client.publish("airConditioner/setPoint", String(setPoint));
  client.publish("airConditioner/mode", String(mode));
  client.publish("airConditioner/connected", String(deviceConnected));
}
```

Finally you need a function to receive incoming MQTT messages. The library requires it to be called `messageReceived()` and requires the parameters shown here, though you'll only use the topic and message parameters. You need to parse the property from the topic. Start by finding the position of the `/` in the incoming `topic`, then use that to split the topic into two strings, `deviceName` and `property`.

If the property is "update" and the message is "all," call the `publishAll()` function.

If the property is "on" or "setPoint," then convert the message to an int and save it in the appropriate variable. The `checkThermostat()` function can then use them to update the device's behavior.

If the property is "mode," then convert it to an int and use it to update the device's behavior appropriately. The `loop()` function also uses mode to determine whether or not to call `checkThermostat()`.

That's the whole function. When you upload this to the microcontroller, you should have full control of the air conditioner from the web client that you'll write shortly, and you should get updates from the air conditioner in the web client when appropriate.

Continued from previous page.

```
void messageReceived(String topic, String payload,
                     char bytes[], unsigned int length) {
  Serial.println(topic);
  Serial.println(payload);
  // parse the topic
  int divider = topic.indexOf('/');
  String deviceName = topic.substring(0, divider);
  String property = topic.substring(divider + 1);

  // if there's a request for all data, publish all:
  if (property == "update" && payload == "all") {
    publishAll();
  }
  if (property == "on") {
    deviceIsOn = payload.toInt();
  }

  // if there's a setPoint, set it
  if (property == "setPoint") {
    setPoint = payload.toInt();
  }

  // if there's an on/off, act on it
  // if there's a mode switch, change the mode (auto/on/off)
  if (property == "mode") {
    mode = payload.toInt();
    switch (mode) {    // check mode
      case 1:   // off
        digitalWrite(relayPin, LOW);  // turn off AC
        deviceIsOn = false;
        client.publish("airConditioner/on", String(deviceIsOn));
        break;
      case 2:   // on
        digitalWrite(relayPin, HIGH); // turn on AC
        deviceIsOn = true;
        client.publish("airConditioner/on", String(deviceIsOn));
        break;
      case 3:   //auto
        checkThermostat();                 // check thermostat
        break;
    }
  }
}     // end of messageReceived() function
```

Mosquitto—An MQTT Broker

Rather than write your own MQTT broker, you're going to use an existing one, Mosquitto. It's an open source program that runs on macOS, Windows, and Linux. You can learn all about it at mosquitto.org. You should run this on your public web host.

For Windows, you can download it from mosquitto.org. However, there are several dependencies you'll need to install first. See the readme file for details. You may find it easier to use your web host or a local Raspberry Pi.

For macOS, you can install it using Homebrew, a package manager for macOS. You can learn more about it at brew.sh, and you can install it like so:

```
$ /usr/bin/ruby -e "$(curl -fsSL https://raw.githubusercontent.
com/Homebrew/install/master/install)"
```

Once Homebrew's installed, you can install Mosquitto like so:

```
$ brew install mosquitto
```

You might need to add it to your `$PATH` environment variable as well. To do this, type:

```
$sudo nano /etc/paths
```

Then add `/usr/local/sbin/` to the end of the file and save it. Then log out and open a new terminal window.

For Linux, including web hosts and Raspbian, you can use the `apt` package manager like you did in the last chapter:

```
$ sudo apt-get install mosquitto
```

Once it's installed, you're ready to test it and your client. Make note of the IP address of the machine on which you're running it, and fill it in for the value of the `serverAddress` variable in the preceding microcontroller sketch.

Try running Mosquitto from the command line like so:

```
$ mosquitto
```

You should get the following response:

```
1489964660: mosquitto version 1.4.11 (build date 2017-03-14
19:27:44+0000) starting
1489964660: Using default config.
1489964660: Opening ipv6 listen socket on port 1883.
1489964660: Opening ipv4 listen socket on port 1883.
```

To close it, press control-C.

If you get an error when you start it, then Mosquitto is already running as a daemon. Kill it (see sidebar) and then try starting it again.

If the microcontroller running the previous sketch has the IP address of the computer on which you're running Mosquitto as its server, then you'll see it connect as soon as you run Mosquitto, like so:

```
New client connected from 192.168.0.14 as airConditioner (c1,
k60, usoneone).
```

That means Mosquitto is working and the client can connect. You can see the IP address of the client in the log messages, and its username.

Make a project directory for this project called **catCamDeux** on the same machine on which you're running Mosquitto. You'll be connecting to Mosquitto from both your microcontroller and a web client, as shown in Figure 10-4. The latter will connect over a webSocket, so you'll need a custom configuration file for the broker to allow this. Mosquitto allows anonymous clients, which is not good for security, so you should block anonymous clients as well. Make a file called **mq-local.conf** in your project directory. Here's the file:

```
port 1883
listener 8888
protocol websockets
password_file mq-pwds
log_type all
allow_anonymous false
```

Save the file, then add two users, `airConditioner` for the microcontroller and `webClient` for the web client you'll write next, to a password file using the `mosquitto_passwd` command. For the first user you need to create the password file as well, like so:

```
$ mosquitto_passwd -c mq-pwds username
```

When you do this, you'll be asked to enter and confirm the password. For subsequent users, you don't need the `-c` flag. The users will be listed in plaintext in the **mq-pwds** file, but the passwords will be encrypted.

Now you're ready to write the web client for this application. Make a new p5.js project in your project directory, called **public**, as usual. Make a server script as well, called **httpServer.js**. The code for both follows.

Killing Your Daemons (and Other Processes)

On Linux, Mosquitto runs by default as a *daemon*, meaning it's a background process on your machine that starts automatically on startup or when you install it. If you get an `Error: Address already in use` message when you try to run it, you'll need to kill the daemon. To do that, you need to know the *process number*. Every running process in an operating system gets a unique number. You can see all the processes you're running by typing:

```
$ ps -u username
```

Replace `username` with your username. You'll get a list like this:

```
PID TTY          TIME CMD
  744 ?        00:00:00 sshd
  746 pts/0    00:00:01 bash
 1461 pts/0    00:00:01 mosquitto
 1902 pts/0    00:00:00 ps
```

To see all processes running, use `ps -e`. It will be a long list! To kill a process, type:

```
$ kill processnubmber
```

Replace `processnumber` with the number of the process you want to kill. In the previous list, Mosquitto is 1461.

With daemons, you don't know the process number because you didn't start it; the system did. But you know something about it. The mosquitto daemon defaults to listening on port 1883, so you can use the `netstat` command to find out its process number. Type:

```
$ sudo netstat -atnp
```

This tells netstat to get all sockets currently in use (`-a`), list the TCP sockets (`-t`) and list their IP numbers (`-n`) and process numbers (`-p`). You'll get a response like this:

```
Active internet connections (servers and established)
Proto  Recv-Q  Send-Q Local Address           Foreign Address          State        PID/Program name
tcp         0       0 0.0.0.0:22              0.0.0.0:*                LISTEN       573/sshd
tcp         0       0 0.0.0.0:1883            0.0.0.0:*                LISTEN       844/mosquitto
tcp6        0       0 :::22                   :::*                     LISTEN       573/sshd
tcp6        0       0 :::1883                 :::*                     LISTEN       844/mosquitto
tcp6        0       0 fe80::7cbf:8e5:995:5:22 fe80::cb6:e1b0:ca:62727  ESTABLISHED  613/sshd: pi [priv]
```

You can see the protocol on the socket, the local address and port, the remote address if someone is connected, the state of the socket, and the process and program name that's in control of it. The useful line is the one that says:

```
tcp  0 0 *:1883  ...LISTEN  1461/mosquitto
```

That tells you that Mosquitto is listening to port 1883 and its process number 1461. To kill it, type:

```
$ sudo kill 1461
```

You're using `sudo` because you didn't start the process, so you need to be a superuser to kill it. Replace the `1461` with your own process number.

Using ps and kill, you can find and kill any process, including daemons. If you know the name of the process, for example, even if you didn't start it, you can find it's process number by combining ps with the grep search tool, like so:

```
$ sudo ps -e | grep 'mosquitto'
```

Both ps and kill have many more features, and you can use the man pages (type `man ps` or `man kill`) to learn more. Be careful, though, about killing processes that you don't know; doing so can cause your system to crash or do unstable things.

X

Serve It The server is just a basic static file server. Install express using npm as usual, then open **httpServer.js** and write the following. This is the whole server. This should run on the same public web host as the MQTT broker.

```
/*
  A static file server.
  context: node.js
*/
var express = require('express');              // include the express library
var server = express();                        // create a server using express
server.listen(8080);                           // listen for HTTP
server.use('/',express.static('public'));      // set a static file directory
```

Mark It Up The web page needs a new library, mqtt.js. You can download the library from unpkg.com/mqtt@2.5.0/dist/mqtt.min.js. Save this in the libraries directory of your p5.js sketch; then modify the **index.html** file by adding one line in the <head> as shown in blue. The rest of the HTML file stays the same.

```
<script src="libraries/p5.js"></script>
<script src="libraries/p5.dom.js"></script>
<script src="libraries/p5.sound.js"></script>
<script src="libraries/mqtt.min.js"></script>
<script src="sketch.js"></script>
```

Web Client The **sketch.js** file for the web client starts with client connection options and device definition similar to those you set in the Arduino sketch. clientOptions holds the client MQTT connection options. Use the username and password you set earlier.

```
/*
  mqtt air conditioner controller web interface
  context: p5.js
*/
var clientOptions = {             // mqtt client options
  port: 8888,
  host: self.location.hostname,   // IP address of the host machine
  username: 'webClient',          // mqtt username
  password: 'something',          // mqtt password
};
```

device holds the same device properties as the microcontroller client publishes. You also need a variable for the client instance, and a variable to track if this web client is connected.

```
var device = {                    // device properties
  on: false,
  temperature: 24,
  setPoint: 18,
  mode: 1,
  connected: false,
  name: 'airConditioner'
};
var client;                       // mqtt client
var webConnected = false;         // if this client is connected
```

Next, set some variables for the HTML page elements: an image of the cat, a timeStamp span, a device status span, a connect button, radio buttons for the state, and a slider for the setpoint.

```
var img;                          // the image from the cam
var timeStamp;                    // a timeStamp span
var deviceStatus;                 // display objects on page
var connectButton, modeControl;   // input objects on page
```

»

In the `setup()`, initialize and position all the page elements. The user input elements will get event handlers for the `touchEnded()` event, so the user can connect and disconnect and change the setpoint. The mode controller will get a `changed()` handler for which the user picks a new mode.

Continued from previous page.

```
function setup() {
  frameRate(0.033);                              // once every 30 seconds
  noCanvas();                                     // no canvas needed
  deviceStatus = createSpan(device.name);         // device info label
  deviceStatus.position(10, 10);
  connectButton = createButton('Connect');       // connect button
  connectButton.position(10, 100);
  connectButton.touchEnded(connectMe);
  setPointSlider = createSlider(10, 40, 21, 1); // setPoint slider
  setPointSlider.position(180, 100);
  setPointSlider.touchEnded(changeSetpoint);
  var sliderLabel = createSpan('Setpoint: ');
  sliderLabel.position(100,100);
  modeControl = createRadio();                    // on/off mode controller
  modeControl.option('off', 1);
  modeControl.option('on', 2);
  modeControl.option('auto', 3);
  modeControl.style('width', '65px');
  modeControl.value(device.mode);
  modeControl.position(220, 10);
  modeControl.changed(changeMode);
  timeStamp = createSpan(new Date());            // add image and timeStamp
  timeStamp.position(10, 160);
  img = createImg('./image.jpg?');
  img.position(10, 180);
}
```

In the `draw()` function, you'll update the image element. By using the date stamp as a parameter like you did in Chapter 3's CatCam project, you can ensure that the image will reload.

```
function draw() {
  var now = new Date();                      // get a current timeStamp
  img.elt.src = './image.jpg?' + now;        // update the src of the image
  timeStamp.html(now);                       // update the timeStamp span
}
```

In the event handlers for the setpoint slider and the mode buttons, call a function called `update()`, which you'll write shortly, to publish the new values to the MQTT broker.

```
// event handler for setPoint slider:
function changeSetpoint() {
  update('setPoint', setPointSlider.value());   // publish new value
}

// event handler for mode control radio buttons:
function changeMode() {
  update('mode', modeControl.value());           // publish new value
}
```

In the event handler for the connect button, `connectMe()`, check to see if this web client is connected to the MQTT broker. If not, attempt to connect using `mqtt.connect()` and the client options you set at the beginning of the script. `mqtt.connect()` has a callback handler, `announce()`, which you'll write next. On connection, set listeners for `connect` and `message` events from the broker. If you are connected, disconnect with `client.end()`.

In the callback for the `mqtt.connect()` function, `announce()`, you'll subscribe to all the air conditioner's properties, and update the web page interface with a function you'll write shortly, `updateInterface()`. You'll also call `update()` to publish to the /update topic. This will force the microcontroller to publish updates for all the topics.

```
// connect button event handler:
function connectMe() {
  if (!webConnected) {                    // if not connected
    client = mqtt.connect(clientOptions); // connect
    client.on('connect', announce);       // listener for connection
    client.on('message', readMessages);   // listener for incoming messages
  } else {
    client.end(quit);                     // disconnect from server
  }
}

// on connect, announce your properties:
function announce() {
  connectButton.html('Disconnect');       // update connect button
  webConnected = client.connected;        // update connected status
  // loop over all properties in device:
  for (property in device) {
    client.subscribe(device.name + '/' + property);  // subscribe to them
  }
  updateInterface();    // update user interface
  update('update', 'all');
}
```

The next function, `quit()`, is the callback for the `client.end()` function. It updates the connect button label when you disconnect, and updates the `webConnected` variable.

```
// on disconnect, update connected status
function quit() {
  connectButton.html('Connect');  // update connect button
  webConnected = client.connected;
}
```

Next comes the `update()` function. In this function, you'll publish to the broker whatever topic and message are passed to this function, if you're connected to the broker.

```
// event handler for incoming messages from server
function update(property, value) {
  device[property] = value;       // update the property
  var topic = String(property);   // convert both  to strings for .publish()
  var message = String(value);
  if (webConnected) {             // if connected, publish
    client.publish(device.name + '/' + topic, message);
  }
}
```

When the broker sends messages, it will generate a `message` event, which you'll handle with `readMessages()`. In this function, you'll split the incoming topic at the slash to separate the device name from the property name; then you'll convert the message strings into the appropriate variable types. temperature, mode, and setPoint will be numbers, and the others are Booleans. At the end of the `readMessages()` function, update the page elements with `updateInterface()`.

Continued from previous page.

```
// Read incoming MQTT or serial messages:
function readMessages(topic, message) {
  topic = topic.toString();             // convert topic to String
  var strings = topic.split('/');        // split at the slash
  var origin = strings[0];               // origin comes before the slash
  var property = strings[1];             // property comes after the slash

  if (property === 'temperature' ||      // these properties need to
    property ==='mode' ||                // be converted to numbers
    property === 'setPoint') {
    device[property] = Number(message);
  } else {                              // the other properties are boolean
    // tricky way of getting the boolean value:
    device[property] = (String(message) == '1');
  }
  updateInterface();    // update user interface
}
```

Finally, write the `updateInterface()` function to update the page elements using the values from the `device` object when needed. That's the end of the client script.

```
// UI update function:
function updateInterface() {
  var onState;      // whether AC motor is off or on
  if (device.on) {  // if on === true
  onState = 'on';   // convert true to 'on'
  } else {
  onState = 'off';  // conver false to 'off'
  }
  // update device label with name, on/off, and temperature
  deviceStatus.html(device.name
  + '<br>' + onState
  + '<br>temperature: ' + device.temperature
  + '<br> thermostat setpoint: ' + device.setPoint);

  setPointSlider.value(device.setPoint);     // update setPoint slider
  modeControl.value(device.mode);             // update mode radio buttons
}
```

Now you've written the microcontroller script, the HTTP server, and the web client. To test this, run Mosquitto on your server like so:

```
$ mosquitto -c mq-local.conf &
```

By using the ampersand (&), you're running the program in the background. The program will launch using the settings in the **mq-local.conf** file, and return control to the keyboard. You can press any key to return to the command prompt so you can run the http server as well:

```
$ node httpServer.js
```

You'll see messages from the server commingled with messages from Mosquitto. To kill the server, use control-C as usual. To kill Mosquitto, use the kill process described in the "Killing Your Daemons" sidebar.

With both the browser and the server running, and the microcontroller connected to the network, open a browser and go to http://server.address:8080. You should see a window like Figure 10-9, without the image. Tap the Connect button, and you should be connected to the mqtt broker. On the server command line, you should see messages like this:

```
1490449929: Config loaded from mq-local.conf.
1490449929: Opening websockets listen socket on port 8888.
1490449929: Opening ipv4 listen socket on port 1883.
1490449929: Opening ipv6 listen socket on port 1883.
1490449930: New client connected from 74.72.23.0 as mqttjs_
ce14eb0b (c1, k10000, u'webClient').
1490449930: Sending CONNACK to mqttjs_ce14eb0b (0, 0)
1490449930: Received SUBSCRIBE from mqttjs_ce14eb0b
1490449930:       airConditioner/on (QoS 0)
1490449930: mqttjs_ce14eb0b 0 airConditioner/on
1490449930:       airConditioner/temperature (QoS 0)
...
1490449948: New connection from 74.72.23.0 on port 1883.
1490449948: New client connected from 74.71.6.0 as
airConditioner (c1, k60, u'airConditioner').
1490449948: Sending CONNACK to airConditioner (0, 0)
1490449948: Received SUBSCRIBE from airConditioner
1490449948:       airConditioner/on (QoS 0)
1490449948: airConditioner 0 airConditioner/on
```

That means both clients can connect and are exchanging messages. When the temperature changes on the air conditioner, you should see it reflected in the web client, and when the web client changes the controls, you'll see the air conditioner print it out into the Serial Monitor.

Network Cameras

The last thing to add is the web cam so you can see the cat. There are two ways you can do this: use a commercial networked camera, or build your own.

Cameras that connect to the internet have been available for several years now, and the prices on them, predictably, get cheaper and cheaper. For about $40–$60, you can get a small camera that has a WiFi module onboard and that runs as its own server. The one used for this project, the D-Link DCS-930L, was purchased at an office-supply store for $70. It's no longer available, but many of the D-Link cameras or other brand cameras will work. Check the user manual for FTP or SFTP functionality. The D-Link DCS-5222L and DCS-960L have this feature.

Setting up these cameras is very straightforward and is explained in the documentation that comes with them. First, you need to connect to the camera through a wired Ethernet connection, and open its administrator page in a browser. There, you configure the WiFi network to which you want to connect, save it to the camera's memory, and restart the camera.

In the control panel of cameras like this one, there's usually an option for FTP or SFTP. Using this, you can have the camera send its image to a remote public server. Although you could configure the camera as a server itself and make it public through port forwarding, it's more secure to pass the image to your public web host. Always use SFTP when available, as it uses an encrypted connection, and is therefore more secure. You want to configure the camera to send its image into the public directory of your project, where the HTML and JavaScript for the p5.js script live. That way, it will be loaded with the page when a client connects.

You could also use a Raspberry Pi and a Pi Cam. A Raspberry Pi Zero W, a Pi Camera Module, and a case will cost you about as much as a low-end commercial network camera, but there's more you can do with it. See www.raspberrypi.org/products for the parts. The Pi Zero Camera Pack from Adafruit (www.adafruit.com/product/3414) combines them all in a kit: a Pi Zero W, a Pi Camera module, and a case. The details for expanding the server script and configuring the Pi as a camera are on the pages that follow. X

Making a Pi Web Cam

Setting up a Pi web cam takes two steps: first, program your HTTP server to take file uploads like you did for the first CatCam project, and second, write a shell script on the Pi to automate the camera and upload.

Add the following lines to the end of the **httpServer.js** script on your web host. This code supports file uploads, as you saw in Chapter 4.

Next add the Pi Cam to your Raspberry Pi, and enable it by running

```
$ sudo raspi-config
```

Choose Interfacing Options; then choose Camera and enable it. Then restart the Pi, and add a file called **piCam.sh** to the project directory. *Shell scripts* like this one are used in operating systems to automate the running of different programs.

Upload it To program the **httpServer.js** script to take uploads, add the multer library that you used in Chapter 4 using npm:

```
$ npm install multer
```

Then add the functions that you used in the **catUploader.js** script from Chapter 4. New lines are shown in blue.

You're adding options for storing the uploaded files using multer's `diskStorage()` function first. Configure it to save files in the **public** directory.

Next, initialize multer to take single file uploads, with the type "image." The client will make POST requests to send the image, with the "image" parameter being the file itself.

At the end of the script you'll add a listener for the /upload POST request, so write a callback function for it called `getUpload()`.

You also need to add `saveUpload()`, which gets called on every upload and saves the file.

Finally, add the POST listener to the end of the script.

```
/*
  A static file server in node.js that takes uploads.
  Put your static content in a directory next to this called public.
  context: node.js
*/
var express = require('express');          // include the express library
var server = express();                     // create a server using express
var multer  = require('multer');        // middleware for file uploads

// set up options for storing uploaded files:
var imgStore = multer.diskStorage({
  destination: __dirname + '/public/', // where you'll save files
  filename: saveUpload       // function to rename and save files
});

// initialize multer bodyparser using storage options from above:
var upload = multer({storage: imgStore});
// set file type: single file, with the type "image"
// (the client must have the same file type for the uploaded file):
var type = upload.single('image');

// callback function to handle route for uploads:
function getUpload(request, response) {
  // print the file info from the request:
  var fileInfo = JSON.stringify(request.file);
  console.log(fileInfo);
  response.end( fileInfo + '\n');
}

// callback function for file upload requests:
function saveUpload(request, file, save) {
  // this calls a function in the multer library that saves the file:
  save(null,file.originalname);
}

server.listen(8080);                          // listen for HTTP
server.use('/',express.static('public'));   // set a static file directory
server.post('/upload', type, getUpload);  // upload a file
```

Automate It This shell script will run the raspistill program to make an image from the camera and save it as **image.jpg**. Then it will run curl to upload the image to your web host.

Change the address for the upload to the address of your web host.

```
while [ : ]  # run forever
do
    # Take a picture, 100ms delay, 400x300, 80% quality, no preview
    raspistill -t 100 -w 400 -h 300 -q 80 -o image.jpg -n
    # upload to the server using curl
    curl -F "image=@image.jpg" 'http://example.com:8080/upload'
    # print out a confirmation
    echo "picture uploaded"
    # sleep for 30 seconds
    sleep 30
done
```

▸▸ Change this address to the IP address of the machine where you are running the HTTP server and MQTT broker.

Once you've saved this file, give it permission to be executed on the command line using the change mode command (chmod) like so:

```
$ chmod u+x piCam.sh
```

You just gave the current user (that's pi) execute permissions on the file. To run it, type:

```
$ ./piCam.sh
```

The script will now try to upload an image to the server every 30 seconds. If the server is running and you've got the web client open, you should see the image updated in the web client as shown in Figure 10-9.

The great thing about using the browser as your application interface for mobile applications is that you have to write the app only once for all platforms. Many of the popular mobile applications are developed by making a basic browser shell without any user interface, then developing the user interface in HTML and JavaScript. Because so many mobile phone applications are mainly network applications, it's a practical approach. The web approach to mobile apps means you get to take advantage of protocols that don't run in a browser but still rely on HTTP. You'll see this in practice in the next project.

X

Figure 10-9
Screenshot of the final CatCam Thermostat control, taken on an Android phone.

 **Project 30**

Phoning the Thermostat

You put a lot of work into the last project. Fortunately, you get to reuse it in this project. You're going to keep the hardware exactly the same, but change the software in order to build an interface that lets you call the thermostat on the phone, hear the temperature and status by voice, and set the thermostat with your phone keypad.

MATERIALS

- **» Completed Project 29**
- **» Twilio account**
- **» Telephone**

Luba is a bit of a luddite when it comes to new technologies, and she's not really fond of the mobile web interface for the thermostat. "But it's a phone!" she complains. "Couldn't I just call someone and have them stop by to change the temperature?" It's a fair point: if you've got a phone, you should be able to make things happen with a phone call.

IP-based telephony has taken leaps and bounds in the last several years, to the point where the line between a phone call and a web page is very blurry. Server applications such as Google Voice and Asterisk are like virtual switchboards—they connect the *public switched telephone network* (*PSTN*) and the internet. These servers use a protocol called *Session Initiation Protocol*, or *SIP*, to establish a connection between two clients and determine what services they are capable of sending and receiving. For example, a SIP client might be able to handle voice communications, text messages, route messages to other clients, and so forth. Sometimes a SIP server sets up the connection between the clients, then gets out of the way and lets them communicate directly. Other times it manages the traffic between them, translating the protocols of one into something that the other can understand. It's the 21st-century version of the old switchboard operators. When application designers have done their job right, you never need to know anything about SIP, because your phone or software just tells you what it can do, and provides you with a way to address other people. The phone does this by presenting you with a keypad and a call button.

If you've called an automated help service in the last few years, chances are you were talking to a SIP server. When you spoke, it tried to recognize your words using speech-to-text software, or it directed you to touch numbers on your keypad using text-to-speech software. When you entered numbers or words, it might have translated your input into HTTP GET or POST requests to query a remote server or local database. When it got results, it read those back to you using text-to-speech again. If it couldn't understand what you were saying (perhaps you were screaming "Operator! Operator! Give me a human being!"), it rerouted your call to a number where a human would answer.

For this project, you're going to use a commercial SIP service from Twilio to make a voice interface to the thermostat you just built. Twilio provides a variety of *Voice over IP* (*VoIP*) services like voicemail, conference calling, and more. With their commercial accounts, you can buy phone numbers to which you attach these services, so your customers can call your service directly. They also have a free trial service. With the free service, you must use the phone number they assign to you, and you have to use a passcode to access your application once you've called in. Twilio offers voice and SMS numbers in a dozen or so countries as of this writing. See support.twilio.com/hc/en-us/articles/223183068-Twilio-international-phone-number-availability-and-their-capabilities to see if yours is available. Phone number exchange is unfortunately one place the PSTN lags considerably behind the internet, mostly for commercial and political reasons. It's also why this is the one application in this book that requires a commercial service rather than running on your own server. If you really want to get deep into VoIP and SIP, check out **Asterisk: The Future of Telephony** by Jim Van Meggelen, Jared Smith, and Leif Madsen (O'Reilly).

X

What's the Standard?

Here's the bad news about SIP and VoIP applications: there's not a standard approach to them yet. They all offer slightly different services, though the basics of making calls, sending SMS messages, recording calls, and reading touchtones are available almost everywhere. Each server and each commercial provider has a different approach to providing an application programming interface (API) for its service. The markup you learn here for Twilio won't apply when you're building an application using another service like Tropo, Google Voice, or any of the dozens of other providers out there.

There are four questions that can help you choose which tool to use for any project:

- Does it offer the features I need?
- Is it available in my area?
- Is it simple to use?
- Will it work with my existing tools?

For this application, I chose Twilio because its markup language, TwiML, is very simple, and their examples make it clear how to separate the markup language from another server-based language, like node.js. Everything happens through GET or POST. Twilio has excellent node.js examples, but you don't need them to get started. Its debugger is useful, and its technical support is good as well. It lacks features that other services have—like the ability to get the audio level while recording, or speech-to-text conversion—but, on the whole, the benefits of its simplicity outweigh any of the missing features.

To complete this project, you'll need an account on www.twilio.com. You can build and run this project with a free account, or you can use a paid account. If you use the free account, you will have to listen to an introduction every time you call, and you'll only be able to support one application at a time.

You'll also need the URL of your web host from the previous project. For example, if your server is at 63.118.45.189, the address for this project will be http://63.118.45.189:8080/voice.xml.

Figure 10-10
The Twilio dashboard. Enter the URL for your web host in the "A Call Comes In" box.

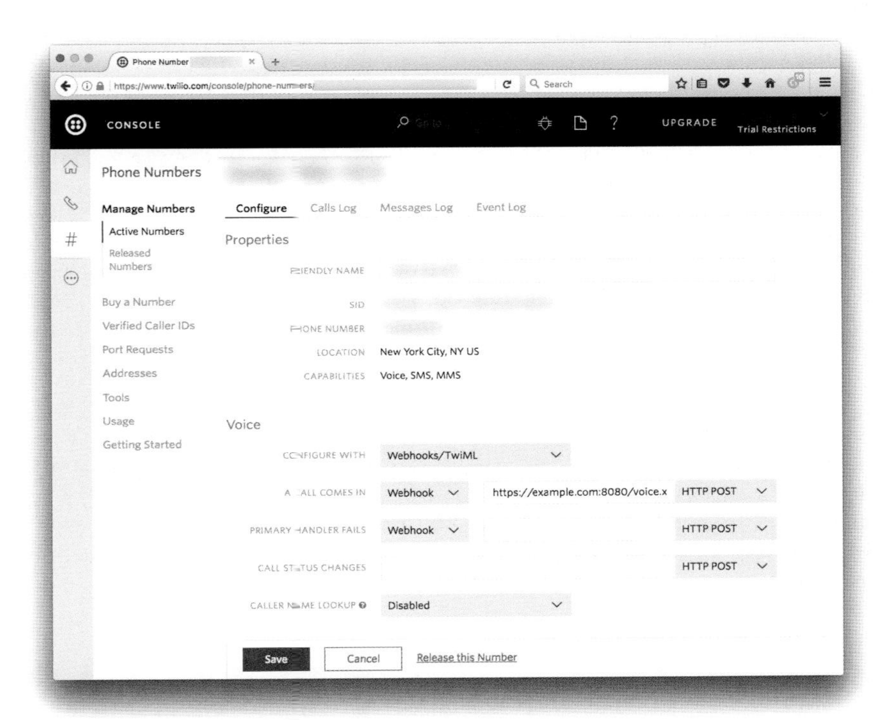

Log in to your Twilio account and go to the dashboard. Figure 10-10 shows the dashboard panel. Click All Products and Services, then Phone Numbers to get a phone number. Your first one is free, even though it says "buy." The phone number allows you to connect your application to gateway using HTTP server whose address you provide. You need to register your receiving phone number as well. Click Phone Numbers, then Verified Caller IDs and enter the number from which you'll call Twilio.

The default voice application is a Webhook/TwiML application, meaning that when a call comes in, Twilio will make a POST request to your server looking for some TwiML that tells it what to do. Put in the address of your server. Once you've done that, it's time to write an XML file for the server and to modify the server script so it will respond as needed.

A Brief Introduction to XML

XML, or *eXtensible Markup Language*, is a general markup language used by many web and database services. XML allows you to describe nearly anything in machine-readable form. XML is made up of *tags* that begin and end with `<` and `>`. Tags describe *elements*, which can be any concept you want to label. Elements can have subelements; for example, a `<body>` might have `<paragraph>` subelements, marked up like so:

```
<body>
   <p>This is the content of the paragraph</p>
</body>
```

The stuff between the tags is the *content* of an element. It's usually the stuff that humans want to read but that the machines don't care about.

Every element should have an opening and closing tag, though sometimes a tag can close itself, like this:

```
<Pause length="10" />
```

What's inside a tag other than the tag's name are its *attributes*. Here, the `length` is an attribute of `Pause`, and this specific pause's length is `10` seconds. Attributes allow you to describe elements in great detail.

If you're thinking all of this looks like HTML, you're right. XML and HTML are related markup languages, but XML is a more general markup language.

TwiML

Twilio's markup schema, TwiML, is a description of the functions Twilio offers, written in XML. It describes what you can do with Twilio. There are elements for handling a voice call, and elements for handling an SMS message. The list of elements is pretty short:

Voice elements:

```
<Dial>
   <Client>
   <Conference>
   <Number>
   <Queue>
   <Sip>
<Enqueue>
<Gather>
<Hangup>
<Leave>
<Pause>
<Play>
<Record>
<Redirect>
<Reject>
<Say>
<Sms>
```

SMS elements:

```
<Sms>
<Redirect>
```

They're all explained in depth on Twilio's documentation pages (www.twilio.com/docs/api/twiml), but you can probably guess what most of them do.

For this project, you're going to describe a `<Response>`. Within that, you'll use `<Gather>` to collect keypresses from the caller. The `<Gather>` element is TwiML's version of an HTML form, so it has an `action` attribute and a `method` attribute that you'll use to tell it where and how to send its results. You'll also use the `<Say>` element to speak to the caller using text-to-speech.

You'll serve this XML from your web host by extending the HTTP server you wrote in node.js for the previous project. You can either duplicate that project, or just continue to edit it in place. You'll need one additional node.js library, the mqtt library. This library is the same as the one for client-side scripts. Install it using npm:

```
$ npm install mqtt
```

X

Mark It Up You need two TwiML documents for this project. When a caller dials in, this document will be the initial response; when she enters keypresses, the document will call itself again to update the settings. Save it as **voice.xml** to the **public** directory from your previous project.

You can see the same variables from your previous project: `#temperature`, `#setPoint`, and `#on`. The server will replace them just like it does with the HTML documents. The hashtag (#) helps make the variables replaceable without replacing those same words that describe them. You'll see how shortly.

```
<?xml version="1.0" encoding="UTF-8"?>
<Response>
    <Gather numDigits="2" timeout="5" action="set-temp.xml" method="POST">
        <Say>
            The current temperature is
            #temperature
            degrees Celsius.
            The thermostat is set to #setPoint degrees Celsius.
            The air conditioner is set to #on.
            If you would like to change the thermostat,
            please enter a new setting.
            If you are satisfied, please hang up.
        </Say>
    </Gather>
    <Say>
          You didn't give a new setting,
          so the thermostat will remain at #setPoint degrees. Goodbye!
    </Say>
</Response>
```

The second TwiML document, **set-temp.xml**, should also be saved in the **public** directory. You may have noticed it's the action on the `<Gather>` tag in the first document. This is what gets called when a caller responds to the first document by pressing numbers.

```
<?xml version="1.0" encoding="UTF-8"?>
<Response>
        <Say>
            The thermostat is now set to #setPoint degrees.
            The temperature is #temperature degrees.
            Thank you. Goodbye.
        </Say>
        <Pause length="2"/>
        <Hangup/>
</Response>
```

Here's how the application will work: The user will call the Twilio number, and Twilio make a POST request for **voice.xml**. The server will fill in the necessary variables to the file and send the result to Twilio. Twilio will read back the text of the first TwiML document. Then it will wait for two digits, and will request the second document when you type them. If not, it will time out after 5 seconds and read the text after the `<Gather>` tag.

The numbers the caller enters will be sent to the server in the body of the second POST request, for **set-temp.xml**. The server will collect the input and publish it to the MQTT broker. Then it will send Twilio the second XML file, and Twilio will read that to the caller. The caller then hangs up, having changed the thermostat. This application doesn't allow the caller to change the air conditioner's mode, but you can add that part on your own once you see how the application works.

The changes that need to be made to the server sketch follow on the next page. It needs three additional features: to respond to POST requests for XML documents, to read those documents and modify them before serving them, and to subscribe and publish to the MQTT server. X

Modify the Server

Open the **httpServer.js** script from the previous project and make the following modifications. New lines are shown in blue as usual. Add the fs (filesystem) and mqtt libraries at the top of the program. Add the same `clientOptions` and `device` variables from your web client script. Note, though, that the host is changed to localhost. That's because this node.js server also acts as an MQTT client, and runs on the same computer as the Mosquitto server.

```
/*
  TwiML and  static File server and MQTT client
  context: node.js
*/
var express = require('express');          // include the express library
var server = express();                    // create a server using express
var multer  = require('multer');           // middleware for file uploads
var fs = require('fs');                    // local file access library
var mqtt = require('mqtt');                // mqtt client library

var clientOptions = {      // mqtt client options
  port: 1883,
  host: 'localhost',
  username: 'xmlClient',
  password: 'something',
  keepalive: 10000
};
var device = {             // device properties
  on: false,
  temperature: 24,
  setPoint: 18,
  mode: 1,
  connected: false,
  name: 'airConditioner'
};
```

You'll be adding a listener for POST requests for XML files shortly. After the `saveUpload()` function and before the server listeners at the end, add a new callback function to handle these requests called `postFile()`. This function will open the file and save it in a string called `output`. Then it will use JavaScript's search and replace functions to replace all the hashtagged properties with the same ones from the `device` variable. Finally, it will send the modified string as a response to Twilio's request.

```
function postFile(request, response) {
  var fileName = __dirname + '/public/' + request.path;
  var data = fs.readFileSync(fileName);
  output = String(data);

  for (property in device) {
    var searchTerm = new RegExp('#'+ property, 'g');
    var value = String(device[property]);
    output = output.replace(searchTerm, value);
  }
  response.writeHead(200, {'Content-Type': 'text/xml'});
  response.end(output);
}
```

Of course you'll need to be able to publish and subscribe to the MQTT server as well. You can use the same functions, `announce()` and `readMessages()`, that you wrote for the web client, with minimal modifications.

```
function announce() {
  for (property in device) {
    client.subscribe(device.name + '/' + property);  // subscribe to them
  }
}

function readMessages(topic, message) {
  topic = topic.toString();                 // convert topic to String
  var strings = topic.split('/');           // split at the slash
  var origin = strings[0];                  // origin comes before the slash
  var property = strings[1];                // property comes after the slash

  if (property === 'temperature' ||         // these properties need to
    property ==='mode' ||                   // be converted to numbers
    property === 'setPoint') {
    device[property] = Number(message);
  } else {                                  // the other properties are boolean
    // tricky way of getting the boolean value:
    device[property] = (String(message) == 'true');
  }
}
```

At the end of the script, add a listener for POST requests for any XML file, and give it the `postFile()` function as a callback. Then add the `mqtt.connect()`, call, and the event listeners for MQTT as well. That's the end of the server script.

```
server.listen(8080);                           // listen for HTTP
server.use('/',express.static('public'));   // set a static file directory
server.post('/upload', type, getUpload);    // upload a file
server.post('/*.xml', postFile);

client = mqtt.connect(clientOptions); // connect
client.on('connect', announce);       // listener for connection
client.on('message', readMessages);   // listener for incoming messages
```

When you've made these changes, restart the server and call your Twilio number. Follow the directions, and the phone will ring. If you did everything right, you'll hear the text of the document read out to you, and you'll be able to enter a new thermostat setting using your phone keypad. When you're done, hang up. You should see the changes you made through voice propagate to both the web client and the microcontroller client via the MQTT broker.

Congratulations, you've built your first voice-based telephony application! As you can see, there is a lot more you can do with this, just by structuring your TwiML pages carefully and having the HTTP server read any changes from the POST request and pass them to the MQTT broker, and vice versa. In fact, there is a full node.js API that mimics TwiML, but it doesn't add significant features to the simple search-and-replace approach taken here.

X

Text-Messaging Interfaces

Text messaging is arguably the most common use for mobile phones. Many people regard texts as less intrusive than voice calls in the flow of daily life. With a text message, you can get information across quickly and with no introduction. In addition, it's easy to send email to SMS, and vice versa; they're not platform-dependent—as long as you have a mobile phone or email account. For situations where you need an immediate, unobtrusive notification, or to give a single instruction, they work wonderfully.

SMS, or *Short Messaging Service*, began as a way of sending information over the mobile phone networks' signaling channel. The idea was to send bytes as part of the data sent to signal incoming calls, but to do so when there was no incoming call. These short messages could be used for diagnostic purposes, to notify the receiver of voicemail, or for other quick notifications. When SMS was rolled out as a commercial service option for customers, however, it became much more than that. SMS may be limited to 160 characters per message, but people have found many creative ways to pack a lot of info into those 160 characters.

Almost all mobile carriers provide an SMS-to-email gateway as part of their service, which means you can send an SMS from an email client and receive SMS messages in your inbox. To test this out, send a text message from your phone, but instead of sending it to a phone number, enter your email address as the destination. Depending on your carrier, it may be sent via *MMS*, or *Multimedia Message Service*, or it may simply go as an SMS. Check your inbox, and you'll see a message from yourself. Now you'll know the email address to use if you want to send yourself a text message via email as well. Most of the time it's simply your phone number @ your carrier's email address. Here are some common SMS-to-email servers for a few U.S., Canadian, and European carriers:

AT&T: `phonenumber@txt.att.net`
T-Mobile: `phonenumber@tmomail.net`
Virgin Mobile: `phonenumber@vmobl.com`
Sprint: `phonenumber@messaging.sprintpcs.com`
Verizon: `phonenumber@vtext.com`
Bell Canada: `phonenumber@txt.bellmobility.ca`
Telenor Norway: `phonenumber@mobilpost.no`
Telia Denmark: `phonenumber@gsm1800.telia.dk`
Swisscom: `phonenumber@bluewin.ch`
T-Mobile Austria: `phonenumber@sms.t-mobile.at`
T-Mobile Germany: `phonenumber@t-d1-sms.de`
T-Mobile UK: `phonenumber@t-mobile.uk.net`

A longer list can be found at www.emailtextmessages.com (warning: I have not verified every one of these). With U.S. carriers, many expect a simple 10-digit phone number without the leading country code (which is +1 for the U.S.), The international standard is called *E.164*. and looks like this: +nnnnnnnnnnnnnnn where n is 12–15 digits with the first 1–3 digits as the country code. Carriers in the United States and around the world tend to be idiosyncratic about this. Check with your carrier's customer support to find out how they handle it.

SMS Sending with Twilio

The node.js script on the next page creates an HTTPS client that connects to your Twilio account and sends an SMS message. It can be used as an SMS gateway script for a networked device. Create a new project folder called **TwilioSms**, with two files: **smsClient.js** and **twilioCreds.js**. The latter, **twilioCreds.js**, is a module that contains your Twilio API key, auth token, and phone number, which you can find on your Twilio dashboard. It looks like this. Fill in your own details:

```
module.exports = {
apiKey: 'Axxxxx', // Twilio API key
auth: '1xxx',     // Twilio auth token
number: '+15555555555' // Twilio phone number
}
```

> **Consider encrypting this script, or removing it after you're done testing. It could become the target of abuse.**

Send It To send an SMS with Twilio, you make an HTTPS POST call to the Twilio API. Start by including node.js's built-in https library and the querystring library, which formats JSON objects as HTTP-style query strings. Include **twilioCreds.js** as well, so you have your account details. You'll enter the recipient phone number from the command line.

```
/*
HTTPS client for Twilio SMS API
context: node.js
*/
var https = require('https');            // include https library
var querystring = require('querystring'); // include querystring library
var creds = require('./twilioCreds.js');  // your Twilio credentials
var recipient = process.argv[2];   // recipient phone number from command
line
```

The format for the SMS message is simple: To, From, Body. The body can be no more than 1600 characters. Phone numbers must be E.164 format. The `postData` variable converts it to a query string.

```
var message = {          // the SMS message you plan to send
  To: recipient,
  From: creds.number,
  Body:'Hi There!'
}
// convert message to HTTP-style query string:
var postData = querystring.stringify(message);
```

The options for the HTTPS request are like the options you've seen for a POST request before. Twilio uses HTTP Basic authorization as well, with the API key and auth token as username and password. The body has to be encoded as a form, which is why you're using the querystring library.

```
// the HTTPS request options:
var options = {
  host: 'api.twilio.com',
  auth: creds.apiKey + ':' + creds.auth,
  port: 443,
  path: '/2010-04-01/Accounts/' + creds.apiKey + '/Messages',
  method: 'POST',
  headers: {
    'Content-Type': 'application/x-www-form-urlencoded',
    'Content-Length': postData.length
  }
};
```

The actual HTTPS call is at the end of the script as usual. Write a callback for it called `confirm()`. This function launches two listeners for the response: `'data'` and `'end'`. The response may come in as several chunks. so the `gather()` function joins them together, and the `printResult()` function prints them out.

```
// request callback function:
function confirm(response) {
  var result = '';           // string to hold the response

  function gather (data) {  // response 'data' callback function
  result += data;
}

function printResult () {   // response 'end' callback function
console.log(result);
}

response.on('data', gather);     // add each chunk to the result string
response.on('end',printResult); // print result when done
}
```

Finally, make the actual HTTPS request. The body of the request is written second; then `end()` closes the request.

```
// make the actual request:
var request = https.request(options, confirm);  // start it
request.write(postData);                        // send the data
request.end();                                  // end it and send it
```

To run this script. type:

```
$ node smsClient.js +15555555555
```

Replace the phone number with the number that you'd like to text. You should get an XML reply from Twilio, and a text on the phone to which you sent, saying "Hi There!" Just as you can only call Twilio from an authorized number if you're on a free account, you can also only text from those numbers.

The XML response contains metadata about the message: the sender and recipient, the time sent, the status (queued, sent, etc.), and more. Twilio has a REST API in addition to TwiML that allows you to query for all kinds of metadata about the messages and calls that you initiate or receive through your account.

Although this script doesn't do much, by now you've seen enough node.js examples that you should be able to expand this into a full application, whether you're running this as a server on a web host or using it as a client on an embedded board to intersect with the physical world via SMS.

SMS Receiving with Twilio

Sending is only half the fun, of course. The following is a simple client for receiving messages from Twilio. Save this as **smsServer.js** in the same project folder. To run it, you'll need to install express and `body-parser`:

```
$ npm install express body-parser
```

In the Twilio Dashboard, go to Programmable SMS, click Messaging Service, and click the Create new Messaging Service button. Give it a name you like, and click Create. This will take you to the configuration screen. Check the "process inbound messages" box, then for the inbound settings, enter the URL of your web host with the server you'll write as follows: http://your.web.host.address:8080/sms. Click Save. Then click Numbers and add your existing Twilio number. Now you're ready to write and run the following server.

Receive It This is a simple express server that takes a POST request for /sms and sends a TwiML response.

```
/*
  TwiML server and text parser
  context: node.js
*/
var express = require('express');          // include the express library
var server = express();                    // create a server using express
var bodyParser = require('body-parser');   // include body parser middleware
```

You'll write the listener for the POST request at the bottom as usual, so write a callback function for it called `getText()`. Enter your own phone number in place of the one shown here. This just gets the From and Body of the text, but try logging the whole `request.body` as well. You'll see all the metadata that Twilio gives you about the text you sent it.

```
function getText(request, response) {
  console.log(request.body.Body);
  var twiml = '<?xml version="1.0" encoding="UTF-8"?> \
    <Response>  \
        <Message to="+15555555555">Thanks for the message</Message> \
    </Response>';
  response.writeHead(200, {'Content-Type': 'text/xml'});
  response.end(twiml);
}
```

At the end, write the server listener. Tell it to use the body parser, and to respond to a POST request for SMS.

```
server.listen(8080);                                   // listen for HTTP
server.use(bodyParser.urlencoded({extended: true})); // enable body parsing
server.post('/sms', getText);                         // listen for /sms
```

Run this script, then send it a text message to your Twilio number!

Microcontrollers on Mobile Phone Networks

Cellular modems are devices that allow you to connect a microcontroller to mobile phone networks directly. Using one of these, your microcontroller connects to the internet the same way that your mobile phone does. A modem has a phone number, and it can send and receive SMS messages, make HTTP requests, and do anything else you can do on the internet. They have their limitations, but when you need to connect when other networks are not available, they are excellent tools to know.

Mobile phone networks have made significant advances in data connectivity the last couple of decades. In the early 2000s, connecting to the internet from your phone was a slow and awkward process. Now it's the most common way to use a mobile phone for most people. Similarly, data rates worldwide have gotten fast enough that in many places mobile connectivity is as good or better than wired connectivity, for less media-intensive applications.

The first generation of mobile phone communications involved analog radio communications between phones and carriers, and standards differed from region to region globally. As mobile technology has advanced, connection standards have increased in speed and compatibility globally. Second-generation systems (2G) introduced digital communications and SMS messaging. The *Global System for Mobile Communications (GSM)* was the main 2G protocol. Although digital, GSM was still a circuit-switched network. Late in the transition from second- to third-generation systems, the *General Packet Radio Service (GPRS)* was introduced, and mobile networks got packet switching. (See Chapter 3 for the difference between packet and circuit switching.) The third generation of mobile networks (3G) introduced the *Universal Mobile Telecommunications System (UMTS)*, based on GSM. 3G systems brought faster speeds and new frequency ranges for mobile communication, and faster packet switching became more common using the *Enhanced Data Rates for GSM Evolution (EDGE)* protocol. The fourth generation of mobile networks boosted speeds again added services like IP telephony, high-definition video delivery, and more. With the advent of 4G systems, mobile telephony resembles the internet more than it does historical telephone networks. *Long-Term Evolution (LTE)* is the 4G upgrade on packet networking, and can theoretically achieve up to 100 megabits-per-second download and 50 megabits-per-second upload speeds.

Data modems for mobile systems have kept pace with the network development, of course, but the price and ease-of-use tend to lag behind. Many of the modems available at a hobbyist level are still 2G modems. 2G modems use GPRS to make data connections like HTTP requests and email delivery. This presents a challenge, as carriers in many countries are dropping support for 2G data connectivity. 3G modems and even LTE modems are available, if you're willing to pay more for them. Fortunately, many microcontroller-based applications can be built on SMS, which works even on 2G modems.

Cellular communications come at a cost, both in terms of power usage and connectivity costs. Besides the cost of the modem, you'll most likely need a mobile data account too. Fortunately most carriers offer pay-as-you-go data-only accounts relatively inexpensively, but the modem and the account together can add up. Finding a carrier who will offer 2G service can be a challenge these days, so the example that follows will use SMS to get around that problem so you can use less expensive modems.

Cellular modems generally connect to microcontrollers and other devices through asynchronous serial communication. Most of them use an AT-style command set like you saw with the Digi radios in Chapter 7. In fact, Digi sells a cellular module that communicates over 4G LTE networks.

GSM introduced the world to the *subscriber identity module (SIM) card*. Your SIM card is associated with your subscriber account, and contains a unique ID number called an *international mobile subscriber identity (IMSI)*. Similarly, your phone has a unique *international mobile equipment identifier (IMEI)*. Your IMSI and IMEI are how the carriers identify you, connect to you, and bill you.

Cellular Modems Compared

Different electronics retailers all have their own preferences for cellular modems. Figure 10-11 shows the range of modules discussed here.

SparkFun sells Spreadtrum Technologies' SM5100B GSM/GPRS module. It has an AT command set, and it can send and receive SMS messages and make network connections. It has GPIO pins, including two analog pins, a battery charger, and connections for a keypad and an LCD interface. They no longer sell a breakout board for it but they do sell a separate SIM card board.

Adafruit sells the Fona line of products based on SIMTech's 2G SIM80x modules, and a board based on the 3G SIM5320 module. All of their breakout boards have a SIM card holder and antenna connector on board, and the Feather Fona has an M0 processor on board as well. Seeed Studio sells the rePhone lines of GSM modules, also based on the SIMTech SIM80x modules. The RePhone core 2G-AtmelSAMD21 is similar to a Feather Fona, with the M0 on board, in a different form factor. Their Xadow GSM + BLE board, one of the least expensive modules on the market, uses mostly the same AT command set as the SIMTech modules. The Xadow board is difficult to work with if you don't have the correct connecting cables, though. Libelium (www.libelium.com) sells a number of different modules through their Cooking Hacks site (www.cooking-hacks.com). Their modules are particularly popular in the European market. Many of their products are also based on SIMTech's modules.

Digi sells a cellular modem in the XBee series that has the same footprint and AT command set that you used in the projects in Chapter 7. It's the highest-priced module I tested for this book, but also the easiest to use, and the only one that's a 4G LTE modem, and therefore capable of direct internet connectivity on modern networks. The Digi XBee Cellular kit comes with a SIM card and limited data account out of the box. Contact Digi for details on the various plans they offer.

Particle (www.particle.io) make a cellular module as well, the Electron. Particle's boards use their own IDE and API system that, though similar in appearance to Arduino, is not compatible with it. Using their boards requires learning their system from the ground up, a longer explanation than is available here. Particle's networking for the Electron centers around their cloud service and uses MQTT, which you learned about in the first project in this chapter.

Of the modems mentioned here, the Fona and the RePhone offer the best balance of features and reasonable cost. The example that follows uses the SIM800 AT command set and will work with those modules.

The two challenges to using any of these modules are power and price. The SIM800 module, for example, can draw up to 2 amps when it's making a call. The board won't draw all that current the whole time, but if it can't get it when it needs it, you won't make the connection. These modules are best powered from a battery, and many of them feature a connection and charging circuit for a Lithium-Polymer (LiPoly) battery.

For any of these modules, you'll need a mobile subscription and working SIM card from your favorite mobile carrier. Unless you have a flat-rate data and text plan, you can spend a lot of money testing and debugging GPRS projects. A second alternative is a pay-as-you-go plan for your SIM card. Neither is an ideal choice, unfortunately, so do as much troubleshooting as you can offline to save money where possible. X

Figure 10-11
Cellular modems L to R: Adafruit Fona 800 and Fona Feather, Seeed Studio Xadow GSM+BLE, XBee Cellular, Particle Electron. Note the antennas on the Fona, Xadow, and Electron. All of these have antennas, and you need them to get a good signal.

SMS Sending with a Cellular Modem

This exercise shows you how to connect a SIM800 modem to a microcontroller and send an SMS. It's not a full project, just an introduction to cellular modems. It sends a text message with the reading from a light-dependent resistor whenever you press a pushbutton. Any of the SIM80x modules mentioned will work for this.

Figure 10-12 shows the Fona800 connected to a 101. The pin numbers were chosen to work on a Feather Fona board. If you're using that board, these are the pins to which the on-board modem is connected. Except for the Fona Feather, the modems used here need to be powered separately from the microcontroller. The Fona modules from Adafruit and the RePhone modules from Seeed Studio all have LiPoly battery connectors. Get a battery that's at least 1200mA to be safe, and make sure you get a battery that matches the connector on your board.

Even though the modem is powered separately, the V_{io} and ground pins should be connected to the microcontroller's voltage output and ground, so that the modem has a voltage reference for the serial transmit and receive lines.

The antenna is not shown here, but you need one. None of the modems mentioned here work adequately without an antenna.

Insert your working SIM card (it's best to initialize it in your phone first, to know it works). Then power the modem.

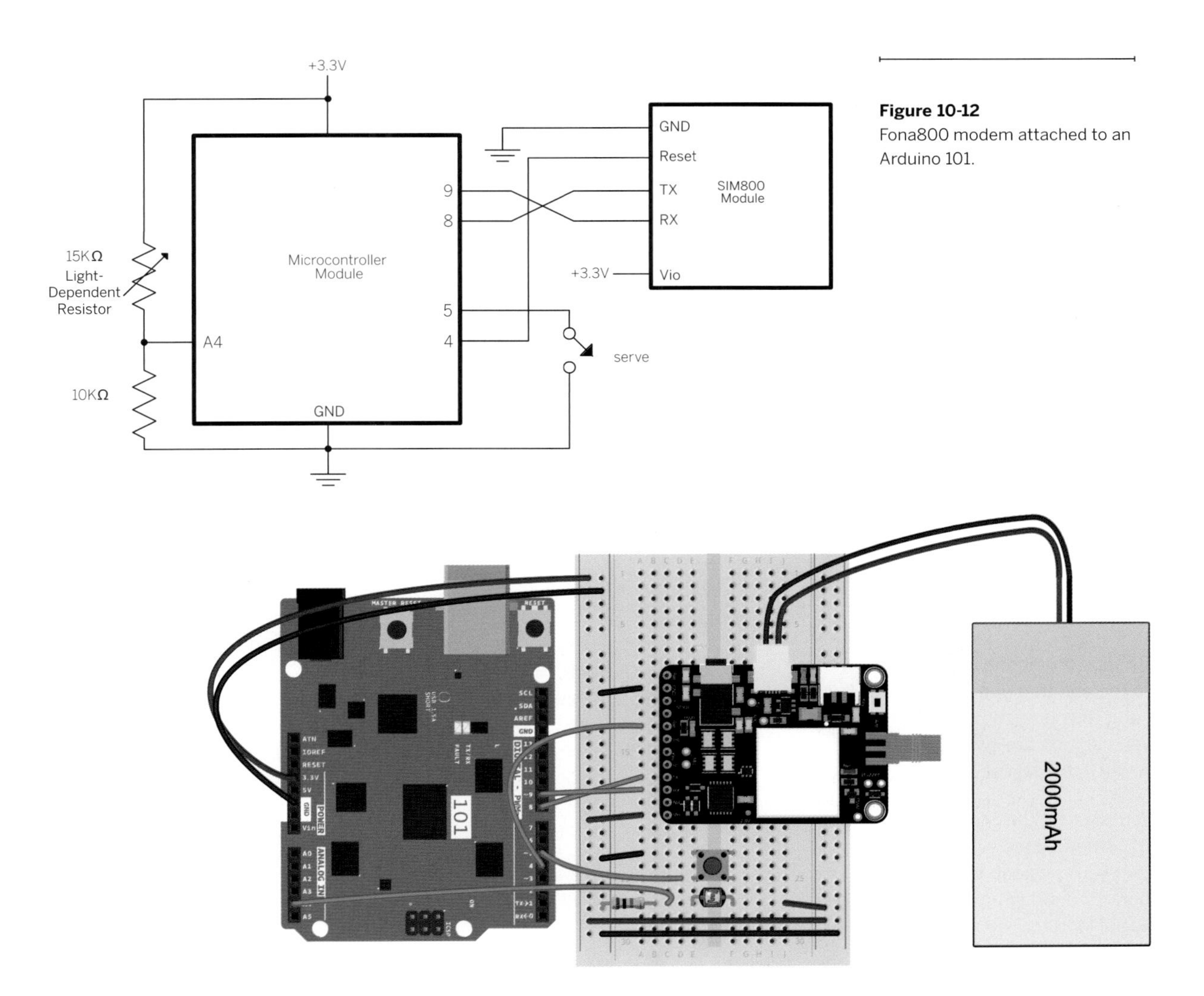

Figure 10-12
Fona800 modem attached to an Arduino 101.

Cellular modems connect to a host computer, whether it's a personal computer, embedded board, or microcontroller, through an asynchronous serial link. To get started with them, connect the TX of your modem to the RX of a USB-to-serial adapter and vice versa. Connect their grounds together as well, but power the modem from a separate source. Once you're connected, open a serial connection to the device at 9600bps and type:

```
AT
```

Hit the Enter key, and you should get this reply:

```
OK
```

AT commands generally receive an OK or ERROR response. Enable verbose error messages like so:

```
AT+CMEE=2
```

Now run a few diagnostic checks, starting with the battery level:

```
AT+CBC
```

You should get something like `+CBC: 0,100,4214`. That means the battery is at 100%, or 4.214V. Now check the IMEI:

```
AT+CIMI
```

You'll get a unique 15-digit number that's the IMEI for your modem. Next check the IMSI:

```
AT+GSN
```

You should get another 15-digit number. Next check if you're connected:

```
AT+COPS?
```

If you're connected, you'll get something like `+COPS: 0,0, "T-Mobile USA"` or another carrier name, and if not, you'll get `+COPS: 0`. Check the antenna connection and coverage in your area. You can also check the signal strength:

```
AT+CSQ
```

You should get something like `+CSQ: 18, 99`. If you didn't get a good response, check your antenna connection and signal strength in your area using your mobile phone.

If you got a positive response on all of those, you're ready to send some SMS messages. Here's how to send one:

```
AT+CMGF=1
OK
AT+CMGS="+15555556666"
> Your text goes here
```

Finish it by sending control-Z (ASCII 0x1A). You should get the text on your phone. Now you're ready to program a microcontroller to do it.

X

Text It This sketch was originally written for a Feather Fona, but it works on any model Arduino, if you change the pin numbers for the TX, RX, and reset pins, as needed.

Change the phone number shown here to the one that you want to text. If you set it to your incoming Twilio number, you can text the TwiML server that you wrote in the last section.

You're using the SoftwareSerial library to communicate with the modem so that you can keep the hardware serial port for debugging messages to the Serial Monitor.

```
/*
  SIM80 GSM modem SMS sender
  context: Arduino
*/
#include <SoftwareSerial.h>          // include softwareSerial library

const int sim800Tx = 8;              // TX and RX pins for SIM800
const int sim800Rx = 9;
const int sim800Reset = 4;           // reset for SIM800
const int pushButton = 5;            // pushbutton
int lastButtonState = HIGH;          // previous button state
String phoneNum = "+15555556666"; // number to text

// make a softwareSerial port on the TX and RX pins:
SoftwareSerial sim800 = SoftwareSerial(sim800Tx, sim800Rx);
```

In the `setup()` function, start serial communications on both hardware and software serial ports. Then initialize the input and output pins. Note that the pushbutton is wired using the internal pullup resistor; it goes low when pressed.

Take the modem's reset pin low, then high to reset the modem. Then send an OK message and listen for the `OK` reply using the Stream `.find()` command. This command will return false if it doesn't get an incoming matching string within a second.

```
void setup() {
  Serial.begin(9600);                    // start serial
  sim800.begin(9600);                    // start modem serial
  pinMode(sim800Reset, OUTPUT);          // set I/O pins
  pinMode (pushButton, INPUT_PULLUP);

  digitalWrite(sim800Reset, LOW);       // pulse modem reset pin
  delay(200);
  digitalWrite(sim800Reset, HIGH);
  delay(5000);                           // give the modem time to start

  while (!sim800.find("OK")) {          // send initial "AT"
    sim800.println("AT");
  }
  Serial.println("GSM modem ready");
}
```

In the `loop()` function you'll read a pushbutton and compare it to its previous state, as you've done before (see the IR Camera Control project in Chapter 6 for comparison). If it's changed, you'll send a text using the `sendSMS()` function, which you'll write next.

You're using the String `.replace()` function here to quickly format the analog sensor data as a JSON string. When you send this to the TwiML server from the previous section, it will be easy to parse out the sensor reading.

```
void loop() {
  // read pushbutton:
  int buttonState = digitalRead(pushButton);
  if (buttonState != lastButtonState) {            // if it's changed,
    delay(100);                                    // debounce delay
    if (buttonState == LOW) {                      // and it's pushed,
      int reading = analogRead(A4);                // read analog input
      String sensorReading = "{\"Sensor\":X}";     // make JSON format
      sensorReading.replace("X", String(reading)); // add sensor value

      Serial.println("sending...");                // send an SMS
      int result = sendSMS(phoneNum, sensorReading);
      Serial.println(result);                      // 1 = success
    }
  }
  lastButtonState = buttonState;                   // save for next loop
}
```

To send the actual SMS message, write a function called `sendSMS()`. In this function, send the commands you tested with earlier. First send the `AT+CGMF=1`, then wait for an `OK`. Then send `AT+CGMS=` with your phone number, and wait for the `>` prompt. Then send your sensor message string, followed by control-Z, which is 0x1A in ASCII. Then the modem will send the text. That's the whole program.

```
// sends an SMS message using the modem:
int sendSMS(String phoneNumber, String message) {
  sim800.println("AT+CMGF=1");        // set message format
  if (!sim800.find("OK")) {           // if you don't get OK
    return -2;                        // return error
  }
  sim800.print("AT+CMGS=\"");         // start message
  sim800.print(phoneNumber);          // add phone number
  sim800.println("\",145");           // add message format (from datasheet)
  if (!sim800.find(">")) {            // if you don't get >
    return -1;                        // return error
  }
  sim800.print(message);              // add body of message
  sim800.write(0x1A);                 // control-Z ends and sends message
  return 1;                           // return success!
}
```

Modem Programming Tips

Program this sketch with the phoneNum variable set to your mobile phone's number, initially. Assuming your phone is operational, when you press the button you should get a text with the message:

```
{sensor: 234}
```

The actual sensor value will depend on your sensor, of course.

There are a few details of this sketch worth noting, as they're useful techniques with any asynchronous serial device that uses an AT-style command set.

Since you're always waiting for an OK after most commands, the Arduino API's `Stream` class is a handy way to do this. The `.find()` command will read the whole incoming string, and return true if it finds a matching substring to the parameter you give it. There's also `.findUntil()`:

```
.findUntil("target", '\n');
```

This function will do the same as `.find()`, but will stop if it finds the character in the second parameter (in this case, the newline, `\n`). That way you can still read or parse what comes after the second parameter.

The `Stream` class also has a `.timeout()` function, used by its parsing functions. The default timeout is 1 second. For example, if `.find()` doesn't find your string within one second, it returns 0. You can change the timeout in this sketch like so:

```
sim800.timeout(5000);  // set a 5-second timeout
```

As mentioned in Chapter 4, the `Stream` class is the basis for many asynchronous serial communication libraries, including Serial, SoftwareSerial, WiFi, and others.

Your `sendSMS()` function returned one of three values: -2, -1, or 1. It's common for functions to return negative numbers as error codes. In this case, -2 means the `CGMF` command failed, -1 means the `CGMS` command failed, and 1 means success. Although you're not using the result in this sketch, it's a handy debugging technique when developing the code.

Debug It When you have to connect to an asynchronous serial device like these modems through a microcontroller, it's useful to write a serial passthrough program that just takes the input from the hardware serial and passes it to the software serial, and vice versa. This is similar to the technique you saw in Chapter 2's sidebar, "Using an Arduino As a USB-to-Serial Adapter." This way, you can use the Serial Monitor or any serial terminal application to test your modem commands. This sketch works well for testing the SIM800 on a Feather Fona board, though you might want to add in the reset pin and the reset block from the `setup()` of the previous sketch.

X

```
/*
  SoftwareSerial passthrough
  context: Arduino
*/
#include <SoftwareSerial.h>
const int sim800Tx = 8;              // TX and RX pins for SIM800
const int sim800Rx = 9;

// make a softwareSerial port on the TX and RX pins:
SoftwareSerial sim800 = SoftwareSerial(sim800Tx, sim800Rx);

void setup() {
  Serial.begin(9600);                // start hardware serial
  sim800.begin(9600);                // start software serial
 }

void loop() {
  if (Serial.available()) {          // read data from hardware serial
   sim800.write(Serial.read());      // send it to software serial
  }
    if (sim800.available()) {        // read data from software serial
   Serial.write(sim800.read());      // send it to the hardware serial
  }
}
```

Native Applications for Mobile Phones

Even though web and SMS interfaces offer many possibilities, there are projects for which a mobile app makes more sense than a browser-based interface. If you're aiming to make an application for all mobile phones, you need to become familiar with the different mobile operating systems and the programming tools available for them. However, there are simpler alternatives.

The mobile device industry has seen a lot of change, as have smartphone operating systems. Google's Android OS and Apple's iOS are the dominant players at the moment, having beaten out several competitors as the market matures. As of the third quarter of 2016, Android was the most popular of smartphone operating systems, taking 81.7 percent of the market, with iOS taking most of the remaining share, according to the Gartner Group (www.gartner.com/newsroom/id/3609817). Android's share is up from 35% in 2010. If you're going to develop native apps, Android and iOS are your likely targets.

If you're developing for iOS, you'll need at least an Apple ID and a developer account from developer.apple.com. With the free account, you can load apps to your device via USB (called *sideloading*), but can't distribute via the App Store. Android requires no pre-registration, so all you need is the toolchain and the knowledge of how to use it.

PhoneGap

If you're interested in getting access to the sensors on your phone, and you're comfortable working in JavaScript, PhoneGap (www.phonegap.com) is a good option. PhoneGap is a platform that gives you access to all the phone's hardware sensors through HTML5 and JavaScript. Basically, PhoneGap embedded the phone's built-in browser engine (which implements all of HTML5's standards), added a bunch of hooks into useful hardware functionality, and released it as a basic framework in which you can develop apps. You download the application framework, which is written in your phone's preferred programming language (Java on Android, Swift on iOS). You don't need to do anything to that code—it's just there to act as a shell for your app. You write your own HTML, CSS, and JavaScript documents, which form the core of your app. Then you compile it all and upload it to your phone.

PhoneGap is now an Adobe product, but Adobe donated the code base to the Apache Foundation as the open source project Cordova (cordova.apache.org). The two are basically the same as of this writing. PhoneGap uses Cordova as its core tools, and offers a few extensions as well.

PhoneGap also offers some tools to skip a few steps. PhoneGap Build (build.phonegap.com) is an online compilation service where you can upload your HTML files and it will compile an application for you to download to your phone. PhoneGap Build lets you skip the part where you install your own toolchain for your mobile platform. PhoneGap Desktop is a desktop app that creates new projects and runs a local HTTP server so you can test your project's interface in a web browser, not unlike you've been doing with node.js and p5.js. PhoneGap Developer is a mobile app that loads your project from the PhoneGap Desktop server. With Build, you're relying on Adobe's servers for your compilation, and with Desktop and Developer, your app is never truly standalone. To make a finished app that can run on its own, you need to install the toolchain. That's what you'll do here.

When combined with p5.js, PhoneGap is a relatively easy place to start developing your own apps for iOS and Android. You will need an Apple Developer account if you want to make iOS application, but once you do, applications you develop using this toolkit can be used on both Android and iOS devices. In the following section, you'll learn how to set up the PhoneGap environment and create a sample PhoneGap/p5.js app. After that, you'll build a full project.

> **This project is the most complex of the book, because it involves many loosely connected programming pieces: the PhoneGap command-line interface (CLI), the iOS or Android Software Development Kits (SDKs), p5.js, and in the final project, Bluetooth LE on a microcontroller and construction of wearable electronics. This section relies on concepts you've learned earlier in the book, which is why it's been saved for last. Enjoy the challenge!**

Installing the PhoneGap Toolchain

The PhoneGap toolchain consists of the PhoneGap command-line interface, the Android or iOS SDK, a text editor, and a mobile device.

You install PhoneGap using npm, like so:

```
$ npm install -g phonegap
```

Once it's installed, you'll need to install the software development kit for your mobile platform to complete the toolchain, and you'll be ready to develop.

The iOS Toolchain

iOS apps can only be developed on macOS. You'll need XCode. macOS users can install XCode from the Mac App Store. Once XCode is installed, you'll also need to install the XCode command-line tools. Do this from the command line like so:

```
$ xcode-select --install
```

To develop iOS apps, you'll also need an Apple Developer Account. You can get one at developer.apple.com. Follow Apple's instructions for connecting your device and your developer account with XCode.

You also need to

```
$ npm install -g ios-deploy
```

> **if you're not using a node version manager like nvm or n, you may have to use the following to install:**
>
> ```
> $sudo npm install -g ios-deploy
> --unsafe-perm=true --allow-root
> ```

For more on setting up Cordova/PhoneGap with iOS, see cordova.apache.org/docs/en/latest/guide/platforms/ios.

The Android Toolchain

Android apps can be developed on macOS, Windows, or Linux. You'll need to install the Android Software Development Kit (SDK) to do so. The SDK comes as part of Android Studio, which you can download from developer.android.com/studio. Download Android Studio and follow the install instructions. Also download the command-line tools, version r25.2.3 and unzip them. You can find them at:

dl-ssl.google.com/android/repository/tools_r25.2.3-windws.zip
dl-ssl.google.com/android/repository/tools_r25.2.3-linux.zip

or

dl-ssl.google.com/android/repository/tools_r25.2.3-macosx.zip

Figure 10-13
The "About phone" menu, showing your Android version.

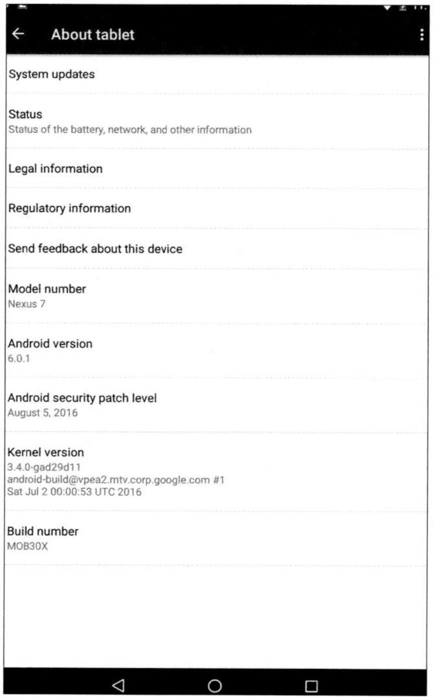

Once you've installed Android Studio, click the Configure option in the splash screen and choose the SDK manager. In the SDK manager, open System Settings and click Android SDK. A new pane will open showing you the ADK platforms. Add older Android versions back to 5.1 Lollipop to make sure you're compatible with your device. You can view your device's Android version in the Settings under About Tablet or About Phone. Figure 10-13 shows that screen.

Once you've installed the Android SDK, you need to add it to your system's `PATH` environment variable, so that other applications, including PhoneGap, know where the SDK tools are on your system.

Open a terminal window and open your **.bash_profile** file. This is your profile for every time you log into the command-line interface (also called the *bash shell*):

```
$ nano .bash_profile
```

Add the following line:

On macOS:

```
export PATH=$PATH:~/Library/Android/sdk/tools:~/Library/
Android/sdk/platform-tools:
```

On Linux:

```
export PATH=$PATH:~/Android/sdk/tools:~/Android/sdk/
platform-tools:
```

On Windows 10, using Cygwin, it is:

```
export PATH=$PATH:/cygdrive/c/Users/username/AppData/Local/
Android/Sdk/tools/
```

When you've added this line, press control-X and then y and Enter to save changes. Then log out of the command line and log back in for your changes to take effect.

Finally, move the command-line tools you downloaded into the tools directory. For macOS, that's **~/Library/Android/sdk/**, for Linux it's **~/Android/sdk/**, and for Windows, it's **C:\Users**`username`**\AppData\Local\Android\Sdk**. Replace the existing tools directory if there is one. PhoneGap and Cordova need this earlier version of the tools. Now you're ready to write a PhoneGap app.

Preparing Your Android Device

Before you can load an app to an Android device, you have to enable Developer Mode, then enable USB debugging. On any Android device, go to the home screen and choose Settings. Scroll to the bottom and choose "About phone." In that menu, you'll see a listing of the Android version. Figure 10-13 shows the menu. Make sure you have version 5.1 or later for these examples.

To enable Developer Mode, tap the Build number seven times, and you'll get a dialog informing you that you're a developer. If you go back to the Settings menu, you'll have a new option, Developer Options. Tap it, and enable USB Debugging, and your hardware is ready to go.

Create a PhoneGap Project

To understand how PhoneGap works with the mobile SDKs and p5.js, let's build a simple app. Open your command-line interface and type the following to create a new PhoneGap project:

```
$ phonegap create buttonApp com.example.buttonapp ButtonApp
```

Why three names? PhoneGap will create a new directory called **buttonApp**. Operating systems often use reverse domain name notation to keep track of applications, so the operating system name for this app is **com.example.buttonapp**. Replace `com.example` with your own domain name. Finally, **ButtonApp** is what the project name will be in XCode or Android Studio.

Change directories to the **buttonApp** directory and list files:

```
$ls
```

You'll get

```
CONTRIBUTING.md
config.xml
platforms
www
README.md
hooks
plugins
```

The **config.xml** file describes your project. It's like a package.json file for node.js projects, in that it lists author name, dependencies, target operating systems, and other details needed to build the app. You won't need to edit it by hand most of the time, as the PhoneGap CLI edits it when you make changes to your project. However, you'll need to remove some elements by hand to work with p5.js. Open the **config.xml** and find the `<platform name="ios">` or `<platform name="android">` element. Within that, remove any `<icon>` or `<splash>` tags, so PhoneGap doesn't look for these when it compiles. It's a good idea to simply remove any platforms for which you're not planning to build, like Windows or Windows 8.

The **platforms** directory is where the mobile SDKs' project files live. There's nothing in it initially, but you add your preferred mobile platform like so:

```
$ phonegap platform add android
```

or

```
$ phonegap platform add ios
```

When you do this, PhoneGap will call on whichever mobile SDK you're using to install a new project for that platform in the **platforms** directory. You can also use `phonegap platform rm` to remove a platform.

iOS users: Before you can compile for iOS, there's a little work you need to do in XCode to get ready. From the **platforms/ios/** directory, open **ButtonApp.xcodeproj** in XCode, and the project editor will open. Click the project in the left tab, and you'll get the project properties pane. Pick a developer team or individual developer (that's you). Save and close the file.

The **plugins** directory is where PhoneGap installs its extension libraries for things like Bluetooth, sensors, and so forth. You'll see this in action in the final project later in this section. The **hooks** directory is used for scripts that help with the build of your app. You won't use hooks in these exercises.

The **www** directory is where you build your app's interface in HTML, CSS, and JavaScript. PhoneGap has its own template for this, but you're going to use p5.js instead. You'll modify the HTML file slightly to make it compatible, however. Delete the **www** directory and replace it with a p5.js project called **www** instead. Copy

Try It This is the **index.html** for your project. The meta tags in the head of the document are from PhoneGap's default settings for new apps. They set details of the window format, and the content security policy, so your app can only load JavaScript and other content from the app itself, or from trusted sources.

You'll also need to add a link to the cordova.js library, which PhoneGap will add when it compiles your app. After that come your usual p5.js libraries and your sketch.

Notice that the script links come in the body of the document, not the head. PhoneGap works better if it loads scripts after the page is loaded.

```
<!DOCTYPE html>
<html>
<head>
  <meta charset="UTF-8">
  <meta name="format-detection" content="telephone=no" />
  <meta name="msapplication-tap-highlight" content="no" />
  <meta name="viewport" content="user-scalable=no, initial-scale=1, maximum-
scale=1, minimum-scale=1, width=device-width" />
<meta http-equiv="Content-Security-Policy" content="default-src 'self' data:
gap: 'unsafe-inline' https://ssl.gstatic.com; style-src 'self' 'unsafe-inline';
media-src *" />
  <title>ButtonApp</title>
</head>
<body>
  <script type="text/javascript" src="cordova.js"></script>
  <script src="libraries/p5.js"></script>
  <script src="libraries/p5.dom.js"></script>
  <script src="sketch.js"></script>
</body>
</html>
```

in the requisite files from the p5.js example project. If you've been using the p5-manager command-line tool, you can remove the **www** directory and add your p5.js project like so:

```
$ rm -rf www
$ p5 g -b www
$ cd www
$ p5 update
```

Now you're ready to write some code. The HTML page, shown on page 420, is different than the HTML you've written for normal p5.js projects. It includes settings for the mobile app view, and security settings for content that the app can load. Use it to replace the normal p5.js **index.html** file.

Build and Run the App

Add the following sketch, and you'll be ready to build the app. First, plug in your device and prepare it:

iOS: open **ButtonApp.xcodeproj** in XCode (it's in the **platforms/ios** directory). Figure 10-14 shows the app on Android. Click the ButtonApp project in the file browser to the left, and you'll see the project info in the main window. Under Signing, choose your iOS developer account. Then plug in your device and look at the top of the window, where you see `ButtonApp > Generic iOS device`. Click the app name there and you'll get a drop-down menu that includes your device. Choose it, and look in the Signing tab to register your device. Now you're ready.

Android: Plug in your device, and you'll get a dialog asking you to enable USB debugging from this computer. Tap yes.

Finally, you're ready to build and run the app. On the command line, type:

```
$ phonegap run android --device
```

If you're building for iOS, replace `android` with `iOS`. If you've installed only one platform, you can just type `phonegap build`. If you installed everything correctly, PhoneGap will build the app and pass it to the platform's toolchain for compilation. You'll see the compiler messages go by, and at the end, success. Then PhoneGap will attempt to deploy it to your device.

If you're working on iOS, you may want to use the Run button in XCode to build and deploy instead, so you see the error messages in XCode.

For the sketch.js, let's reuse the sneaky button sketch you wrote in Chapter 1. It will test the basic UI elements we need: making a UI element (a button), responding to user input (`touchEnded`), and writing to a UI element (the response div). Here it is again:

```
/*
  Sneaky button
  context: p5.js
  Moves a button when you click it.
*/
var myButton, responseDiv;     // DOM elements

function setup() {
  createCanvas(windowWidth, windowHeight);  // create the canvas
  myButton = createButton('click me');      // create the button
  myButton.touchEnded(changeButton);         // set button's listener
function
  myButton.position(10, 10);                // position the button
  responseDiv = createDiv('catch me');      // create a text div
  responseDiv.position(10, 40);             // position it
}

// runs when you release the mouse or stop touching the screen:
function changeButton() {
  var x = random(windowWidth) - myButton.width;   // a new x position
  var y = random(windowHeight) - myButton.height; // a new y position
  myButton.position(x, y);                        // move the button
  responseDiv.html(x + ',' + y);                 // update the responseDiv
}
```

Where Does the App Run?

You have two choices as to where your app will run: it can run in an emulator on your desktop, or it can run on your actual phone. The emulator is useful if you don't have a mobile device handy. Both XCode for iOS and Android Studio include emulators for the devices running their operating systems. To run the emulator on iOS, pick the device you want to emulate in XCode instead of your plugged in device. To run the Android emulator, type `android avd` on the command line to launch the Android Virtual Device creator. You'll also need to open the SDK Manager in Android Studio and install system images for the SDK versions that you're using. Emulators tend to be slow, though, and they're nowhere near as exciting as running the app on your actual phone. Though it used to be difficult, the iOS and Android toolchains now make it easy to run on a connected device as well. Given the choice, I prefer running on the actual device.

Summary

That may seem like a lot of steps, but now that you've done it once, it will get faster. Here's a checklist of the main steps, for future projects:

1. Create the project using the PhoneGap CLI
2. Edit the **config.xml** file appropriately
3. Add your platform
4. Add your credentials, if you're using iOS
5. Remove the **www** directory, replace with a p5.js project
6. Edit the **index.html** file appropriately
7. Write your code
8. Build the project
9. Run

With this checklist in hand, you're ready to build the final project.

X

Figure 10-14
The app that you just wrote, on Android.

 Project 31

Personal Mobile Datalogger

One popular reason to develop native applications on a mobile phone is to use the phone's Bluetooth radio as a connection to other devices. In this way, your phone can become a mobile conduit to send the data to a site on the internet. In this project, you'll sense your Galvanic skin response using a microcontroller, send the data via Bluetooth to your phone, and log the result to a file on the internet.

MATERIALS

- **Android or iOS device**
- **RedBear BLE Nano**
  - **Features used: analog I/O, Bluetooth LE**
- **MK20 USB board for BLE Nano**
- **Extra-long male header pins, 0.1" spacing**
- **Battery holder and battery, see chapter head**
- **1 resistor, 270-kilohm**
- **Sewable snaps**
- **Conductive thread**
- **Shieldit Super 13" fabric**
- **Velcro**
- **Hoodie**
- **Embroidery thread**

A number of personal data enthusiasts gather personal biometric data for many different purposes, from visualizing their activity patterns in order to improve exercise habits, to tracking sleep patterns in order to find solutions to insomnia. Quantified Self meetups (quantifiedself.com) have popped up around the world for people to share tips and tricks on how to do this, and devices like FitBit (www.fitbit.com) and the Apple Watch have come on the market to make biometric tracking easier. What those services don't often do, though, is give you control of your data. By building your own services, you can do that.

This project is based on the work of ITP alumnus Mustafa Bağdatlı, shown in Figure 10-15. Mustafa wanted to track his Galvanic skin response (GSR) and heart rate against his calendar, so he could see when his mood—as reflected in his heart rate and GSR—were affected by the events of the day. His project, Poker Face, tracked the two biometric characteristics on a LilyPad Arduino, transmitted them via Bluetooth serial to a mobile phone, and logged the result on the web. You can find more on Poker Face at mustafabagdatli.com.

In this project, you'll build the same, but using Bluetooth LE, and without the heart rate sensor to keep it simple. Feel free to change the sensor for whatever you want to track yourself.

Figure 10-16 shows the system for this project. The microcontroller, acting as a Bluetooth LE peripheral, reads the analog voltage from the sensor and advertises it as a service, as you did with the camera shutter in Chapter 6. An app on the phone connects to the peripheral and subscribes to updates. When there's an update, the app saves it locally. When you tap the Upload button on the app, it makes an HTTP GET request to a node.js-based server on a web host. The server saves the incoming data to a file. What you do with the data from there is up to you.

Figure 10-15
Mustafa Bağdatlı wearing Poker Face, a biometric datalogger linked to a mobile phone. *Photo courtesy of Mustafa Bağdatlı.*

The Circuit

To measure Galvanic skin response, all you need is a high-value resistor and your skin. As a test, take a multimeter and measure the resistance across your wrist. You'll find that the resistance is high, probably in the megohm range if your skin is cool and dry. Work up a sweat and measure it again. You'll see that the resistance has gone down. When you exercise, or when you're faced with stressful or arousing situations (good or bad), you perspire more, changing the conductance of your skin. That's what this project will measure. It's only a rough approximation, as it doesn't measure any chemical changes in your sweat, but it will give a decent approximation of the changes in your skin conductance.

The sensor is basically a voltage divider, with your skin as the variable resistor and a fixed resistor completing the circuit. A 270-kilohm resistor is used here (see Figure 10-18 and Figure 10-19), but feel free to change it to suit your skin. Generally, below 10 kilohm won't work.

Conductive fabric and conductive thread were used to make the sensor for this circuit, and to attach the microcontroller to the garment. Different conductive fabrics and threads have different material and electrical properties, so you might want to experiment with a few to find what works for you. Mustafa used stretchable conductive fabric sewn into wristbands for his cuffs, as you can see in Figure 10-17. For this project, I used Shieldit Super fabric with adhesive backing.

The microcontroller module for this project is a RedBear Labs BLE Nano. It's got an ARM M0 inside that you can program with the Arduino IDE, and a Nordic Labs 51822 Bluetooth LE radio inside, both in a compact package. This is the same radio that's in the Arduino 101, and this project will work with the 101 by modifying the code to use the CurieBLE library instead of the BLEPeripheral library. The Arduino 101 is less well suited to construction of wearables due to its form factor, however. Since it uses the BLEPeripheral library, this code will also work with any Arduino using a Nordic Labs nRF51822 or nRF8001 radio, like the BLE camera control project in Chapter 6. See that project for a comparison of the libraries.

The power supply is a LilyPad Coin Cell Battery Holder from SparkFun. If you prefer a power supply with more charge capacity, the LilyPad Simple Power board contains a connector for a LiPo battery and a battery charger connected to a USB connector.

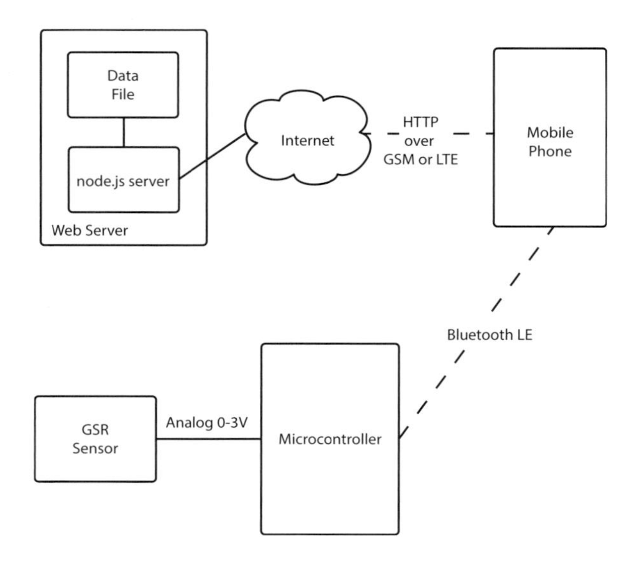

Figure 10-16
The system diagram for the mobile datalogger project.

Figure 10-17
Detail photo of the Poker Face GSR wristband. Here, the wristband is inside-out to show the conductive fabric contacts. *Photo courtesy of Mustafa Bağdatlı.*

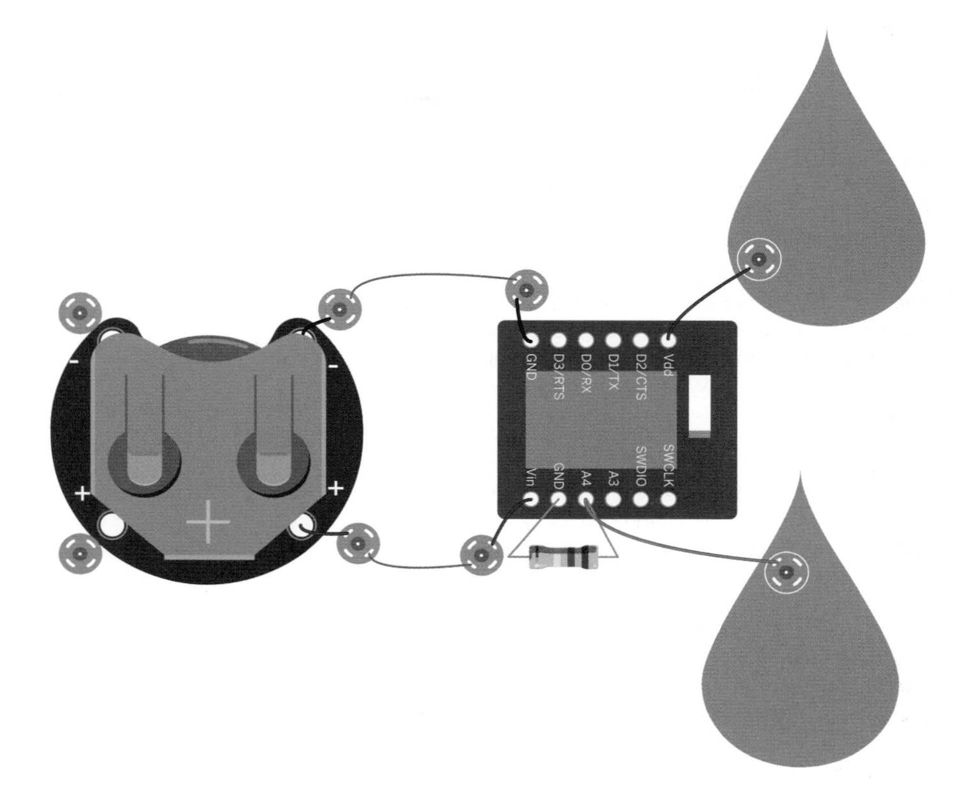

Figure 10-18
The mobile datalogger circuit. The contacts for the sensor are conductive fabric. The snaps are soldered via wires (red, blue, black) to the components and sewn via conductive thread (gray) to each other as shown. This way, the electronics are removable so you can wash the garment.

Figure 10-19
The mobile datalogger circuit schematic.

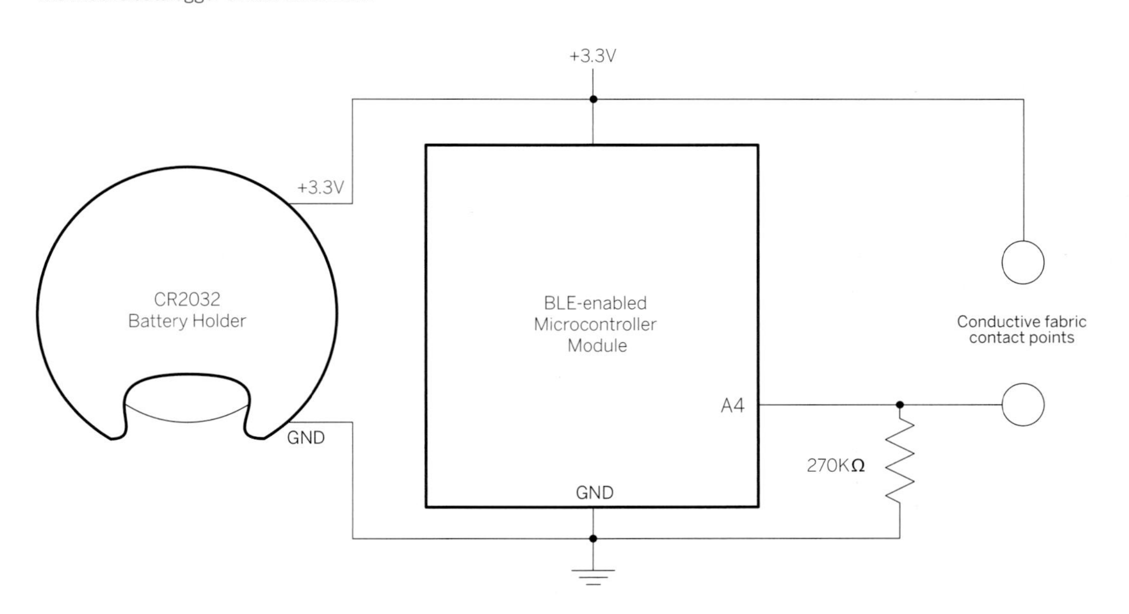

Testing the Circuit

Wearable devices are difficult to test once you've built them into the garment. Although wearable toolkits like the LilyPad and Flora lines aim to make it easier, it's important to test your circuit before you attach it to a garment. Program the board and test it with a Bluetooth LE app like LightBlue for iOS or macOS or nRF Connect for Android to ensure that it's working before you attach it to the garment. The BLE peripheral sketch for this project includes a writable characteristic that you can use to tune the sensitivity of the GSR sensor, so that you don't have to reprogram the board if conditions change once the device is in the garment.

The BLE Nano is normally programmed by soldering headers to it and plugging it into the MK20 board. You don't want pins on the board for a wearable, because they'll stick into the wearer and create discomfort. If the headers aren't soldered, though, you'll have difficulty getting a good connection to the MK20 board. Figure 10-21 shows a method for using extra-long header pins to get a good connection. You can also use IC hooks to connect the BLE Nano to the MK20.

The BLE Nano will be connected to a coin cell battery via conductive thread once you sew it into the garment, so you need to connect it temporarily in order to test it. IC hooks work well for this purpose. Figure 10-20 shows the components connected together via IC hooks for testing. They're like alligator clips, but smaller. They work better than alligator clips for small, closely spaced connections like these.

The Code

The program for this device follows. Program the board, then assemble it with the sensor and power supply using IC hooks to test that it's working properly.

If you haven't used the BLE Nano previously, you'll need to add it to the Arduino IDE using the Boards Manager. Open the Arduino Preferences and add the following URL to the Additional Boards Manager URLs field: http://redbearlab.github.io/arduino/package_redbearlab_index.json. The software library for this is the BLEPeripheral library mentioned in Chapter 6, by Sandeep Mistry. Install it using the Library Manager. The documentation is at github.com/sandeepmistry/arduino-BLEPeripheral. The code follows.

Read It At the beginning of the code, include the libraries you need and establish the pin numbers for the SPI radio connection. Make two global variables for the sensor as well.

```
/*
 Galvanic Skin Response reader
 Context:  Arduino, for nRF51822 or nRF8001
 */
#include <SPI.h>
#include <BLEPeripheral.h>

// define pins (varies per shield/board)
#define BLE_REQ   10
#define BLE_RDY   2
#define BLE_RST   9
```

Finish off the global variables by initializing the `BLEPeripheral` object and its service and characteristics as well. You're making both a sensor characteristic to store the sensor reading, and a threshold characteristic so that you can change the sensitivity without reprogramming the board.

```
int lastInput = 0;     // previous input from the sensor
int threshold = 10;    // sensor difference threshold

BLEPeripheral blePeripheral = BLEPeripheral(BLE_REQ, BLE_RDY, BLE_RST);
// create service
BLEService sensorService("0927AA6A-3588-11E7-A919-92EBCB67FE33");
```

Wearable Microcontrollers

If you're interested in making wearable electronics, there are two electronics lines of electronics modules to know about. The LilyPad line, originally designed by Leah Buechley and made by SparkFun, has a number of useful components in it, and the power modules for this project come from that line. The Flora line, made by Adafruit, is similar. The LilyPad line is focused primarily on using the GPIO of your microcontroller with other modules, while the Flora line connects components primarily with digital communications buses like I2C or the proprietary protocols of the WS2812 and other programmable LEDs. Both product lines feature a central microcontroller module, sensors and actuators, power connectors, and radio modules. Seeed Studio's approach to wearables is oriented more toward watches, with their Pebble development kit.

At the time of this writing, however, the only module with a Bluetooth LE radio was the Flora Bluefruit LE from Adafruit. In fact, that board uses the same controller and radio as the BLE Nano, but it doesn't give you access to the I/O pins, which defeats the point of using it as a standalone module. It's designed to be used as a serial connector to the main Flora processor. Although the BLE Nano's connections are more difficult to sew connections to, it's the most flexible board for this particular project. It's sold without pins soldered to the terminals, so you can sew it into a garment, and it's low-power, so it's easy to power using the available power modules from the LilyPad or Flora lines.

For more information about conductive fabrics, see Leah Buechley's book Sew Electric (HLT press); Kate Hartman's Make: Wearable Electronics (Maker Media); Syuzi Pakhchyan's book Fashioning Technology: A DIY Intro to Smart Crafting (O'Reilly); or Hannah Perner-Wilson's online resource at www.plusea.at. Becky Stern (beckystern.com) has also created dozens of great wearable tutorials for Adafruit, Make, and elsewhere.

X

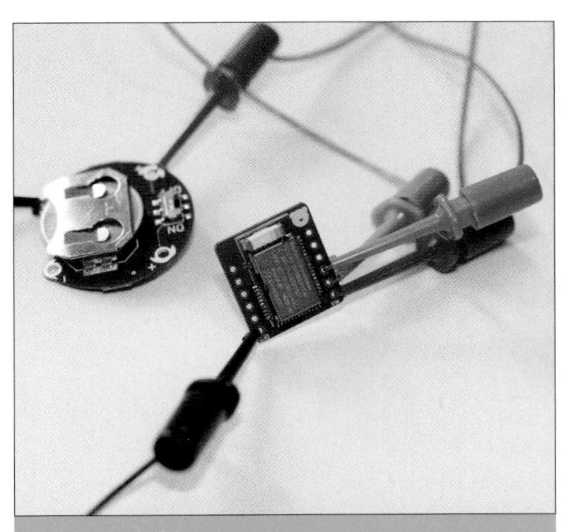

Figure 10-20. IC hooks are good for wearables testing.

Figure 10-21. Connecting the BLE Nano to the MK20 without soldering the header pins can be tricky. Tilt the board and use extra-long male header pins to establish a solid connection.

You're using long UUIDs for this application (as you should always) because Android 7 will reject short UUIDs that it doesn't already know.

In the `setup()`, configure the peripheral. Add the local name, set the advertised service, add the services and characteristics, and set an initial value for the threshold characteristic.

In the `loop()`, listen for new connections from central devices. When you get one, you'll do two things: listen for the central to change the threshold characteristic, and read the sensor and update the sensor characteristic. If the sensor reading and the previous reading differ by more than the threshold, publish a new value for the sensor characteristic.

Make sure to save the current sensor value as `lastInput` so that you can compare them next time through the `loop()`.

Finally, if the central disconnects, print it out for diagnostic purposes.

Continued from previous page.

```
// create LED characteristic and threshold:
BLEIntCharacteristic sensorCharacteristic( \
"0927ADA8-3588-11E7-A919-92EBCB67FE33", BLERead | BLENotify);
BLEIntCharacteristic thresholdCharacteristic( \
"0927AF9C-3588-11E7-A919-92EBCB67FE33", BLERead | BLEWrite);

void setup() {
  Serial.begin(9600);          // initialize serial
  // set advertised local name and service UUID:
  blePeripheral.setLocalName("BleNano");
  blePeripheral.setAdvertisedServiceUuid(sensorService.uuid());

  // add service and characteristics to device
  blePeripheral.addAttribute(sensorService);
  blePeripheral.addAttribute(sensorCharacteristic);
  blePeripheral.addAttribute(thresholdCharacteristic);

  // set initial value for threshold, and begin:
  thresholdCharacteristic.setValue(threshold);
  blePeripheral.begin();
  Serial.println("BLE LED Peripheral active");
}

void loop() {
  BLECentral central = blePeripheral.central();  // listen for connections
  // if you get a connection:
  if (central) {
    Serial.print("Connected to central: "); // print central's address
    Serial.println(central.address());
    //  while central is connected:
    while (central.connected()) {
      if (thresholdCharacteristic.written()) {
        threshold = thresholdCharacteristic.value();
      }
      int input = analogRead(A4);
      if (abs(input - lastInput) > threshold) {
        sensorCharacteristic.setValue(input);
        Serial.print(threshold);
        Serial.print(",");
        Serial.println(input);
      }
      lastInput = input;
    }

    // central disconnected
    Serial.print("Disconnected from central: ");
    Serial.println(central.address());
  }
}
```

Test the Code

Upload this sketch to your board and test it by opening your favorite Bluetooth LE testing app (LightBlue or nRF Connect) and connecting to the peripheral. Subscribe to the sensor characteristic. Touch the sensor pads and see if the value changes. If it doesn't, use LightBlue or nRF Connect to write a lower value to the threshold characteristic, and then check the sensor characteristic and see if you get a change. Once you know you're getting consistent readings, you're ready to sew the garment. You'll need to fine-tune the threshold once everything is sewn together.

The Garment Construction

For this project, you'll use iron-on conductive fabric attached inside the pocket of a hoodie as your sensor contacts. When you reach in your pocket, you can touch the fabric contacts with the palm of your hand, and take a reading on your phone with the other hand. Or, you can let the mobile app run continually to take a reading every two minutes.

You're not actually going to sew the microcontroller or battery into the garment. Instead, they'll snap on, so you can remove them to wash the garment. You can use 30AWG wire wrap (the same kind you used in Chapter 3) to solder to the snaps. When you solder to the snaps, keep the snap's backing on so that the front and back both adapt to the heat together. That way they'll be able to snap together again when they cool. You can solder the snap fronts directly to the contacts on the battery holder.

Conductive thread connects the fabric contacts on one side of the garment to sewable snap backs on the other side of the garment. The microcontroller and battery are soldered to the snaps' fronts, and snap into the inside of the hoodie. Figure 10-22 shows the layout on the inside of the garment. It's just like Figure 10-18.

If your conductive fabric has adhesive backing (the LessEMF version does), you can iron the conductive fabric contacts into the pocket. Space them so they cover about the width of the heel of your palm, and contact it comfortably when you put your hand in the pocket. Figure 10-23 shows you the inside of the pocket. If necessary, sew the edges with embroidery thread. Make sure you don't sew your power and ground terminals through the sensor contacts or you'll create a short circuit!

At this point, the circuit should be fully functional. To test it, connect to it through LightBlue or nRF Connect again, check the sensor reading, and change the threshold characteristic as needed. Once you know it's working as expected, you can write the mobile client.

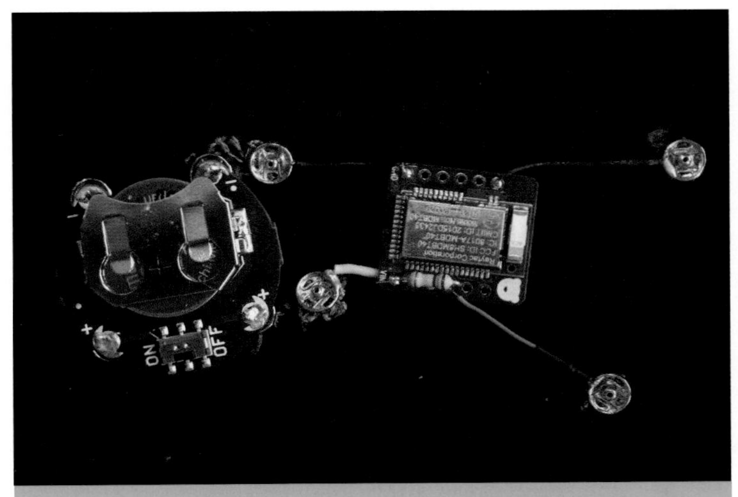

Figure 10-22. The layout of the hoodie components on the inner lining: The microcontroller and the battery are snapped in. Sewable snaps solder nicely, but keep the backing on them when you solder so the heat doesn't deform the front. Make sure your power snaps are not sewn through the sensor contacts on the front of the hoodie or you'll create a short circuit.

Figure 10-23. The conductive fabric contacts inside the hoodie pocket.

The PhoneGap Mobile Client

The central piece of this project is the PhoneGap app that connects to the wearable device, reads the sensor, and uploads the data to the server. Now it's time to make that app. Start by making a new PhoneGap project on the command line like you did before:

```
$ phonegap create BleDatalogger com.example.bledatalogger
BleDatalogger
```

Then:

```
$ cd BleDatalogger
```

Then edit the **config.xml** file, eliminating all `<icon>` or `<splash>` tags from the `<platform name="ios">` or `<platform name="android">` elements, and removing the other platforms. Save it, then add your platform. Replace `ios` with `android` if you're building for iOS:

```
$ phonegap platform add android
```

You'll need the BLE central plugin for PhoneGap and Cordova as well, so add it like so:

```
$ phonegap plugin add cordova-plugin-ble-central
```

Next, replace the **www** directory with a p5.js project:

```
$ cd BleDatalogger
$ rm -rf www
$ p5 g -b www
$ cd www
$ p5 update
```

If you're developing for iOS, open the **BleDatalogger.xcodeproj** file from **/platforms/ios/** and add your developer credentials.

Now you're ready to write the app. The **index.html** and **sketch.js** files follow. Figure 10-24 shows the user interface elements of the app, in Android.

>> **Note: for more on the BLE Central plugin, see its documentation at github.com/don/cordova-plugin-ble-central.**

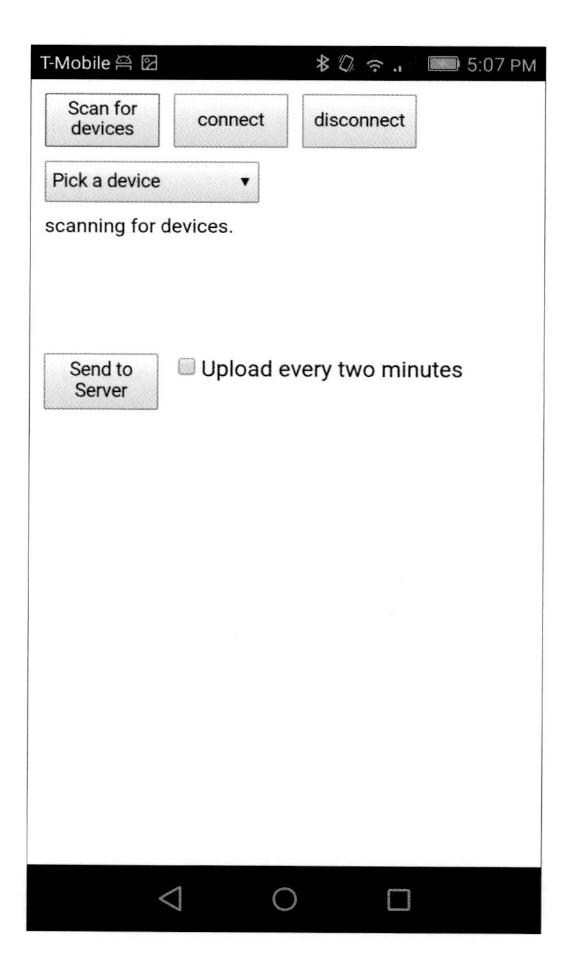

Figure 10-24
The app that you're about to write, on Android.

The Index Page

Here's the **index.html** page. The main difference between this one and the page from the previous PhoneGap app is the addition of the `connect-src` attribute in the `Content-Security-Policy` tag. This allows your app to connect to a remote server.

```
<!DOCTYPE html>
<html>
<head>
  <meta charset="UTF-8">
  <meta name="format-detection" content="telephone=no" />
  <meta name="msapplication-tap-highlight" content="no" />
  <meta name="viewport" content="user-scalable=no, initial-scale=1, maximum-
scale=1, minimum-scale=1, width=device-width" />
    <meta http-equiv="Content-Security-Policy" content="default-src
'self' data: gap: 'unsafe-inline' https://ssl.gstatic.com; connect-src
http://192.168.0.10:8080; style-src 'self' 'unsafe-inline'; media-src *" />
  <title>ButtonApp</title>
</head>
<body>
  <script type="text/javascript" src="cordova.js"></script>
  <script src="libraries/p5.js"></script>
  <script src="libraries/p5.dom.js"></script>
  <script src="sketch.js"></script>
</body>
</html>
```

Replace the IP address here with the IP address of your web host. Later, your app will connect to the web host to log the data from the sensor.

The p5.js Sketch

In the p5.js sketch, start by establishing the UI elements. You need a button to scan for new peripherals, a connect and disconnect button, and a select menu for the discovered peripherals. You also need a div for the sensor data, and one for responses from the peripheral or data server. You'll also need a checkbox for auto-upload and a button for manual upload.

You'll also need a variable to hold a timer for auto-upload, one to check the state of the peripheral connection, a string for the server URL, an array of new readings, and a JSON object to describe the peripheral.

```
/*
Bluetooth Central for PhoneGap
context: p5.js in PhoneGap
uses Don Coleman's cordova-plugin-ble-central in PhoneGap/cordova
*/
// UI elements:
var deviceList, responseDiv, dataDiv, autoUpload;
var scanButton, connectButton, disconnectButton, uploadButton;

var autoUploadTimer;          // timer for auto-upload
var connectState = false;     // whether connected to the peripheral
// replace with your server's address:
var dataServer = 'http://192.168.0.10:8080';
var readings = new Array();  // readings not yet uploaded

// parameters of the BLE peripheral you're looking for:
var myDevice = {
  serviceUUID: '0927AA6A-3588-11E7-A919-92EBCB67FE33',
  sensorCharacteristic: '0927ADA8-3588-11E7-A919-92EBCB67FE33',
  id:''
};
```

In the `setup()` you'll initialize and position all the UI elements, and give them callback functions for touch or change. You'll also do some styling, to make them big enough to operate on a small screen.

There is no `draw()` function for this sketch. All of the other functions are either callbacks for the UI elements, or helper functions to those callbacks.

Continued from previous page.

```
function setup() {
  // create scan, connect, disconnect, and upload buttons:
  scanButton = createButton('Scan for devices');
  scanButton.touchEnded(scanForDevices);
  scanButton.position(10, 60);
  scanButton.size(80,40);

  connectButton = createButton('connect');
  connectButton.touchEnded(connectToDevice);
  connectButton.position(100, 60);
  connectButton.size(80,40);

  disconnectButton = createButton('disconnect');
  disconnectButton.touchEnded(disconnectFromDevice);
  disconnectButton.position(190, 60);
  disconnectButton.size(80,40);

  uploadButton = createButton('Send to Server');
  uploadButton.touchEnded(sendToServer);
  uploadButton.position(10, 250);
  uploadButton.size(80,40);

  // create auto-upload checkbox:
  autoUpload = createCheckbox('Upload every two minutes', false);
  autoUpload.position(100, 250);
  autoUpload.changed(setAutoUpload);

  // create response and data divs:
  responseDiv = createDiv('tap the scan button to begin');
  responseDiv.position(10, 150);
  responseDiv.style("font-size", "14px");

  dataDiv = createDiv('');
  dataDiv.position(10, 200);
  dataDiv.style("font-size", "14px");
```

Having created all the UI elements, make sure Bluetooth is enabled at the end of the `setup()`.

```
  // check if BLE is enabled:
  ble.isEnabled(scanForDevices, bleError);
}
```

The first callback function is for the Scan for Devices button. It scans for new peripherals. First, remove the `deviceList` select element if it's there so you can update it. Then re-create it as a new select element, set its size and position it, and add a callback function for when it changes.

Once you've made the `deviceList` element, start a BLE scan using the `ble.scan()` plugin. The Cordova library automatically added the `ble` object for you, which is your instance of the BLE plugin. As shown here, the `scan()` will only scan for devices advertising the service you're looking for, `myDevice.serviceUUID`. Set a timeout to notify the user when the scan is finished too.

```
// initiates BLE scan for devices:
function scanForDevices() {
  responseDiv.html('scanning for devices.');
  // if device list is already populated, remove it to refresh:
  if (deviceList) deviceList.remove();

  deviceList = createSelect();          // create a new select element
  deviceList.position(10, 110);
  deviceList.size(150, 30)
  deviceList.option('Pick a device', '');  // set default option
  deviceList.changed(selectDevice);

  //scan for the serviceUUID for 5 seconds:
  ble.scan([myDevice.serviceUUID], 5, discoverDevice, bleError);
  // set a timeout to announce the end of the scan:
  setTimeout(ble.stopScan, 5000, scanFinished, bleError);
}
```

When the `ble.scan()` function discovers a new device, it calls a function called `discoverDevice()`. In that function, add the newly discovered device as a new option in the `deviceList` element. Make a string of the name and RSSI, and use the device's MAC address (aka its ID) as the value of this option.

```
// runs when a new device is discovered:
function discoverDevice(device) {
  var result = device.name + ' ' +    // make a string with the new device
  'RSSI: ' + device.rssi;
  deviceList.option(result, device.id);  // add it to the select element
}
```

`scanFinished()` is the callback for the timeout you set for the scan. It just updates the `responseDiv`.

`selectDevice()` is the callback for when the user selects from the `deviceList` element. It sets the ID for the current device using the MAC address of the device the user chose.

```
function scanFinished() {
  responseDiv.html('scan complete. Pick a device.');
}

// callback for when the user chooses from the select element:
function selectDevice() {
  myDevice.id = deviceList.value();
  responseDiv.html('Selected: ' + deviceList.value());
}
```

`connectToDevice()` is the callback for the connect button. If you're not connected, it attempts to make a connection. If it succeeds, it calls `onConnect()`, and if it fails, it calls `bleError()`.

Continued from previous page.

```
// callback for connect button:
function connectToDevice() {
  if (!connectState) {
    responseDiv.html('connecting to ' + myDevice.id);
    ble.connect(myDevice.id, onConnect, bleError);
    connectState = true;
  }
}
```

The `onConnect()` function is the success callback for when you try to connect. In it, you'll subscribe to the sensor characteristic using the `ble.startNotification()` function, and update the `responseDiv`.

```
function onConnect() {
  // subscribe to incoming data
  ble.startNotification(myDevice.id, myDevice.serviceUUID,
    myDevice.sensorCharacteristic, onData, bleError);
  responseDiv.html('Waiting for data from ' + myDevice.id);
}
```

Whenever the sensor characteristic changes on the peripheral, the `onData()` function will be called. The data arrives in an `ArrayBuffer` object, so you need to convert it to an array of 16-bit unsigned integers (`Uint16Array`) and then the result will be the first element of that array. Create a JSON object using the current timestamp and the sensor value; then push it to the `readings` array. Also display it in the `dataDiv`.

```
function onData (data) {
  var input = new Uint16Array(data);     // get the reading from the ArrayBuffer
  var reading = {                        // make a JSON object
    timestamp: new Date(),               // add timestamp
    value: input[0]                      // add reading value
  }
  readings.push(reading);                // add reading to readings array
  dataDiv.html('latest: ' + JSON.stringify(reading));
}
```

The callback for the Send To Server button is called `sendToServer()`. In this function, iterate over the readings array and make an HTTP GET call with each one, to log it to the server. The format of the GET message is http://example.com:8080/data/timestamp/sensorValue. Pop each reading off the array when you send it, so the array is empty when you are done.

```
function sendToServer() {
  responseDiv.html('uploading data...');
  while(readings.length > 0) {       // as long as there are unsent readings,
    var reading = readings.pop();    // pop the last off the readings array
    // format it for sending as a GET request:
    var path = '/data/' + reading.timestamp + '/' + reading.value;
    // send it
    httpGet(dataServer + path, serverReply);
  }
}
```

Next you need a callback for when the auto-upload checkbox changes. `setAutoUpload()` either sets or clears an interval to call `sendToServer()` every two minutes, depending on the state of the checkbox.

```
function setAutoUpload() {
  if (!autoUpload.checked()) {
    responseDiv.html('Auto-upload off');
    // cancel the timer
    clearInterval(autoUploadTimer);
  } else {
    responseDiv.html('Auto-upload on');
    // set timer to upload every 2 minutes
    autoUploadTimer = setInterval(sendToServer, 2*60000);
  }
}
```

Whenever you make an HTTP request to the server, `serverReply()` will display the server's response.

```
function serverReply(data) {
  dataDiv.html('');
  responseDiv.html('Server said: ' + data);
}
```

When the user taps the Disconnect button, the `disconnectFromDevice()` function will run. This function attempts to disconnect, and runs `onDisconnect()` to update the `responseDiv` if it succeeds.

```
function disconnectFromDevice() {
  if (connectState) {
    ble.disconnect(myDevice.id, onDisconnect, bleError);
    connectState = false;
  }
}

function onDisconnect() {
  responseDiv.html('disconnected from ' + myDevice.id);
}
```

Finally, `bleError()` is a general error function, called by many of the other functions if they fail. It prints the error to the `responseDiv`.

```
function bleError(error) {
  responseDiv.html(' there was a BLE error' + JSON.stringify(error));
}
```

That's the end of the sketch. Save it, then run PhoneGap from the command line to upload to your mobile device:

```
$ phonegap run --device
```

If everything went right, you should now be able to scan for and connect to the Bluetooth LE wearable device you made earlier. Figure 10-27 shows what it should look like. Connect to your wearable device, then wait for a few readings. Once you know that it's working, you're ready to write the server.

Save the server shown next to your web host or whatever IP address you used in the `dataServer` variable as **server.js**. You'll need to install express.js using npm as usual in order to run this, and you'll need an empty file called **data.csv** in the same directory.

Log It The server for this project is only written for program-to-program interaction, so there is no client-side interface for it. It accepts GET requests in the format that the client is sending:

/data/timestamp/sensorReading

It reads the parameters from each request, formats them as comma-separated values, and appends them to a file called **data.csv**. If the file doesn't exist, the script will create it. Then it gives the client a reply with the time and date of the upload.

The `fs.appendFile()` function, which writes to the data file. needs a callback function. Call that function `confirmSave()`, and use it to log to the server's command line whether or not the data got saved to the file.

```
/*
Datalogger
context: node.js

expects HTTP GET request with /data/timestamp/reading
*/
var express = require('express');      // include the express library
var fs = require('fs');                // include the filesystem library
var server = express();                // create a server using express

function log(request, response) {
  // format the params into a comma-separated string:
  var newData = request.params.timestamp + ','
  + request.params.reading + '\n';

  // add the data to the data file and send the client a response:
  fs.appendFile('data.csv', newData, confirmSave);
  response.end('Last upload at ' + new Date());
}

function confirmSave(error) {
  var now = new Date();                // make a date string
  if (error) {                         // if there's an error,
    console.log(now + ': ' + error);  // log it
  } else {                             // if not, log the successful save
    console.log(now + ': ' + 'Saved to file');
  }
}

server.listen(8080);                             // listen for HTTP
server.get('/data/:timestamp/:reading', log);   // wait for GET request
```

Troubleshooting

When you run this server, you should be able to upload data to it by tapping the Send to Server button, or by enabling Upload Every Two Minutes. You should see the server respond on the command line, and the client will get a response with the time of the upload. If you look at the **data.csv** file, you should see comma-separated readings like this:

```
Sat Apr 15 2017 14:59:47 GMT-0400 (EDT),20
Sat Apr 15 2017 16:48:41 GMT-0400 (EDT),20
Sat Apr 15 2017 16:48:41 GMT-0400 (EDT),22
Sat Apr 15 2017 16:48:40 GMT-0400 (EDT),18
```

If everything didn't go right, here are a few things to check:

Did the app compile? If not, go back to the PhoneGap project checklist and make sure you covered all the steps.

Did the app open with the right UI elements? If not, check for an error in your **sketch.js** file. You can check the basic UI elements by opening the **index.html** file in a browser. The Bluetooth functionality won't work, but the buttons, divs, and select items should show up.

Did the Bluetooth scan produce any results? If not, make sure your Bluetooth device is powered and in range. You can also try removing `myDevice.serviceUUID` in `scanForDevices()` and just scan for any device instead.

Did the server receive your request? If not, make sure you have the right URL in the **index.html**'s Content Security Policy and in the **sketch.js** `dataServer` variable.

Remote Debugging

If you can't figure out what's wrong but you know your app is running on your device, you can use the JavaScript console in the desktop Chrome browser or Safari on macOS to inspect the code while it's running on your device. Here's how.

Make sure your device is connected with USB to your personal computer. Open chrome://inspect in Chrome on your personal computer. Figure 10-25 shows the inspection window.

Choose the device running your app. A new window opens and you're debugging the HTML and JavaScript on your phone. You can see console log messages, inspect the page elements, and all the things you can do with the JavaScript console when debugging regular web pages.

The process for iOS and Safari is similar. On your mobile device, open Settings panel, choose Safari→Advanced→Web Inspector, and turn it on. On your Mac, open Safari and choose Preferences→Advanced and turn on Show Develop Menu in Menu Bar. After uploading your iOS PhoneGap app, choose Develop→ Phone→App→index.html.

Remote debugging is a very powerful tool. You can solve most of the problems you might encounter in deploying a PhoneGap app with these tools.

X

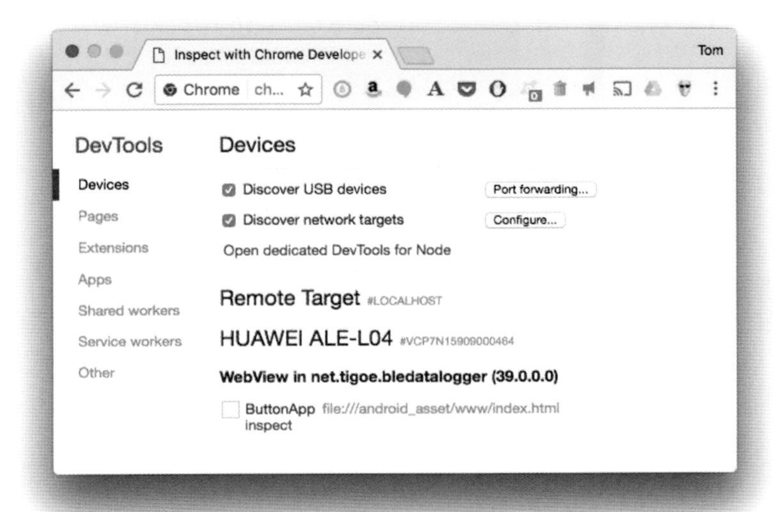

Figure 10-25
The Chrome device inspection pane.

The app you just built combines many of the systems you've been using throughout this book: microcontroller programming, HTTP requests, client- and server-side JavaScript, mobile networks, and Bluetooth LE. If you got all of that working and understood it all, congratulations! You've mastered the basics of a lot of communications technologies.

The connection between the internet and mobile networks widens your options by offering a ubiquitous connection and a variety of communication methods. Taking advantage of it requires some creative thinking about how to hop across different interfaces, systems, and protocols.

These days, mobile phone networks cover almost the entire planet, more than any other form of network connection. The technology of mobile networks changes quickly, and many of the tools disappear almost as fast as they appeared. If you're making projects using mobile network technologies, it's best to take a broad view. Look for the things that seem simpler and more stable—like SMS and HTTP—and be prepared to shift your approach when the tools change. X

Figure 10-26
The datalogger sketch in action.

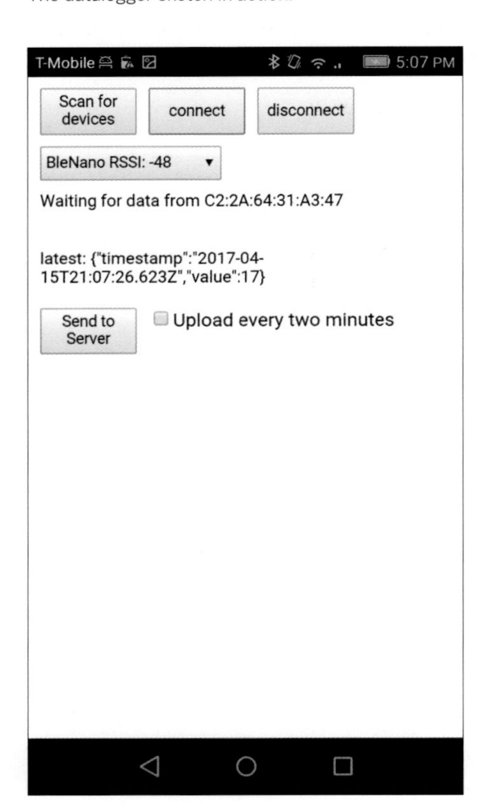

Figure 10-27
The datalogger hoodie in action. How excited is he to be wearing it? Check his GSR readings in the `data.csv` file to find out!

Conclusion

By definition, networked devices don't stand alone. If you're connecting a device to the internet, you should take advantage of the power it offers. The more tools and protocols you know, the easier and more fun that is. Consider advantages offered by servers that have public addresses—and plenty of computing horsepower to spare. Consider combinations of wired and wireless protocols to give the things you build maximum freedom to respond to their physical environments.

As you can see by now, the physical interfaces that are the most flexible are usually the ones that are the most ephemeral on the network. They get turned on and off, they connect through private networks, and they use limited processors to save size, weight, and power. They're not always capable of doing everything that a dedicated web server can do, nor are they available at all times. However, these are not limitations when such devices are used in conversation with dedicated servers.

As hubs of your projects, use public, persistent servers that have public addresses and are always on the internet. They can store messages for physical devices that aren't online and deliver them later. They can act as proxies to take care of things for which simpler devices aren't designed, like complex data management or authentication. Don't be afraid to use tools and protocols in ways not thought of by their original designers. Make the technology fit the person's needs.

Making things talk to each other always comes back to two questions. Who are you serving in making what you're making? What does that person need, and how can you best give it to them?

To design a flexible network of physical things that's really useful to humans, you have to think about the person's needs and actions first. Then, it's all about the rules of love and networking:

- Listen More than You Speak
- Never Assume
- Agree on How You Say Things
- Ask Politely for Clarification

The Internet of Things is useless if those things don't improve the quality of our lives and the ways in which we communicate with each other.

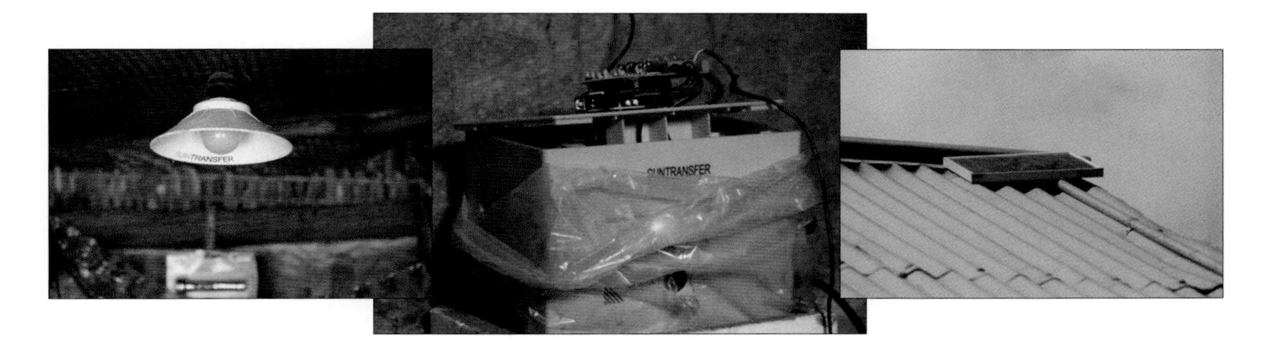

↑ SIMbaLink, by Meredith Hasson, Ariel Nevarez, and Nahana Schelling. The SIMbaLink team developed a device that remotely monitors the health of a solar home system and reports it back to the SIMbaLink client website via a GPRS modem. Working in conjunction with a solar company in Ethiopia, SIMbaLink remotely monitored solar home systems outside Awassa, Ethiopia. The systems consisted of a 10W solar panel, a battery, and four 1-watt LED lamps. This simple setup brings light to an otherwise dark and unpowered rural home. *Photos courtesy of Meredith Hasson.*

3.3V Zener Diodes
TIP120 Transistors
2N2222 Transist
green LEDs
red LEDs
yellow LEDs
Photo resistors
DC power connectors
DSUB9 female connectors
nectors
Jumpers
straight headers
right angle headers

MAKE: PROJECTS

Where to Get Stuff

Many different hardware suppliers and software sources are mentioned in this book. This appendix provides a list of those parts and a summary of the vendors, along with a brief description of each. It's organized into three sections: Supplies, Hardware, and Software.

Here's a shopping list of the parts used in this book. Each part lists the projects in which it's used. The landscape has changed somewhat, as some companies have been acquired and others have gone out of business. Fortunately, there are still lots of small, interesting electronics vendors out there for you to get to know.

If there are updates to this list, you'll find a link to them at http://oreilly.com/catalog/0636920010920.

DISTRIBUTOR KEY

A Arduino Store (store.arduino.cc)
AF Adafruit (www.adafruit.com)
AMZ (www.amazon.com)
B Belkin (www.belkin.com)
D Digi-Key (www.digikey.com)
DL D-Link (www.dlink.com)
F Farnell (www.farnell.com)
ID Identive (www.identiveusa.com)
J Jameco (jameco.com)
L LessEMF (www.lessemf.com)
MS Maker SHED (www.makershed.com)
P Pololu (www.pololu.com)
PS PowerSwitch Tail (www.powerswitchtail.com)
PX Parallax (www.parallax.com)
RS RS (www.rs-online.com)
SF SparkFun (www.sparkfun.com)
SS Seeed Studio (www.seeedstudio.com)

Infrastructure

» **Personal computer** Used in all projects.
» **WiFi connection to the internet** Used in most projects.
» **Project enclosure** Used in several projects.
» **1/16-inch Mat board** Used in projects 8 & 9, 26, 27 for project enclosures.
» **1 Bluetooth-enabled personal computer** Used in projects 3, 12, 18, 31. If your laptop doesn't have a Bluetooth radio, use a USB Bluetooth adapter. **AF** 1327, **RS** 807-7742, **SS** 113990026
» **Android or iOS device** Used in project 31.

Microcontrollers, Shields, and Prototyping Boards

» **Arduino 101 module** Used in most projects.
D 1660-1003-ND, **J** 2239331, **SF** DEV-13787, **AF** 3033, **F** 2520713, **RS** 913-9999, **SS** 114990575, **A** ABX00005, GBX00005 (Africa and EU)
» **Arduino MKR1000 module** Used in most projects.
AF 3156, **RS** 124-0657, **A** ABX00004, GBX00011 (Africa and EU), **D** 1659-1005-ND
» **Arduino Uno module** Used in most projects.
D 1050-1024-ND, **J** 2151486, **SF** DEV-11021, **A** A000099, **AF** 50, **F** 1848687, **RS** 715-4081, **SS** ARD132D2P
» **ATtiny84 microcontroller** Used in project 4.
D ATTINY84A-PU-ND, **SF** COM-11232, **F** 1455160, **RS** 738-0684
» **ESP8266** Either the Adafruit ESP8266 Huzzah! or the SparkFun EsP8266 Thing Dev board will work. **SF** WRL-13231, **AF** 2471
» **Raspberry Pi** Used in several projects. A BeagleBone Green or other embedded Linux processor will work, too. **SF** DEV-13825, **AF** 3055 or 3400, **SS** 102010048 or 114990584, **RS** 896-8660, **F** 2525225
» **Solderless breadboard** Used in most projects. **D** 438-1045-ND, **J** 20723 or 20601, **SF** PRT-12615 or PRT-12002, **F** 4692810, **AF** 64, **SS** 319030002 or 319030001
» **Prototyping shields for Arduino** An alternative to solderless breadboards in all projects, but you'll still need a tiny breadboard with them.
AF 2077, **A** TSX00083, **SF** DEV-07914
Breadboards for protoshields: SF PRT-12044, **AF** 65, **D** 923273-ND
» **Perforated printed circuit board** Used in projects 8 & 9, 27.
AF 1609, **D** V2018-ND, **J** 616673, **F** 4903213, **RS** 159-5420

Communications Modules

» **USB-to-TTL serial adapter** Used in several projects. **SF** DEV-09716 or DEV-14050, **AF** 3309 or 284, **SS** 317990026
» **Bluetooth Serial module** Used in projects 3, 18.
AF 1588 **SF** WRL-12580 or WRL-12576
» **Digi XBee or XBee Pro S2C 802.15.4 RF module** Used in project 14.
AF 128, **D** 602-1892-ND, **SF** WRL-08665, **J** 2253722, **PX** 32416
» **HopeRF RFM95W or Semtech SX1276 radio modules** Used in project 11.
AF 3072, **SS** 113060006
» **USB-XBee adapter** Used in project 14.
J 32400, **SF** WRL-11812, **AF** 247, **PX** 32400
» **Bluetooth LE radio** Used in Project 12 if you're not using an Arduino 101 or BLE Nano.
AF 1697
» **RedBear BLE Nano and MK20 USB board** Used in project 31. Alternative part for project 12. Sold together. **MS** MKRBL5, **SF** WRL-14071
» **RFID reader** Used in projects 24, 25, 27.
SCL3711 USB Smart Card Reader
ID SCL3711

- PN532 NFC reader module **AF** 364, **D** 1528-1781-ND, **SS** 113030001
- Mifare Classic RFID tags Used in projects 24, 25, 26, 27. **AF** 359 or 360, **SF** SEN-10128 or SEN-11319, **SS** 113990013
- IP-based camera Used in projects 29, 30. The text explains how to make one with a Raspberry Pi and Pi Cam. Raspberry Pi cam: **SF** DEV-14028, **AF** 3099, **SS** 113990214, **RS** 913-2664
  Pi Zero W Camera Kit A 3414
  D-link IP Cameras if you opt to not use a Raspberry Pi above **DL** DCS-930L, DCS-5222L, or DCS-960L
- **Cellular modems:**
- Adafruit Fona 800 USA **AF** 3147, Europe **AF** 2691
- Adafruit Fona Feather **AF** 3027
- SeeedStudio Xadow GSM+BLE **SS** 102040005
- XBee Cellular **D** 602-1976-ND,
- Particle Electron Kit Americas/Australia **SF** WRL-14211, Africa/Asia/Europe **SF** WRL-14212

Breakout Boards and Connectors

- Gas sensor breakout board Used in project 14. **SF** BOB-08891, **P** 1479 or 1639
- 3-wire JST connector pigtail Used in projects 13, 15. **SF** SEN-08733
- 9V battery clip Used in projects 3, 13. **D** 1568-1237-ND, **J** 2207056, **SF** PRT-09518, **A** 80, **F** 1650675
- Power plug, 2.1mm inside diameter, 5.5mm outside diameter Used in projects 3, 27. **D** CP3-1000-ND , **J** 28760, **SF** PRT-10287, **A** 369 **F** 1737256
- Wire-wrapping wire Used in projects 5, 31. **D** K386-ND, **J** 22577, **F** 09WX4670
- Wire-wrapping tool Used in project 5. **D** K445-ND or WSU-30M, **J** 2150361, **F** 441089
- 0.1-inch male header pins Used in most projects. **D** A26509-20-ND, **J** 103377, **SF** PRT-00116, **F** 1593411
- 0.1-inch female headers **D** ED7102-ND, **F** 1122344, **SF** PRT-00115
- XBee breakout board Used in project 14. **SS** 113100001 **SF** BOB-08276
- Extra-long male header pins, 0.1" spacing Used in multiple projects. **AF** 400, **SF** PRT-12693
- 2mm female header rows Used in project 14. **SF** PRT-08272, **D** 3M9406-ND

Common Components

- 100-ohm resistor Used in projects 8 & 9, 29, 30. **D** 100QBK-ND, **J** 690620, **F** 9337660, **RS** 755-0707
- 220-ohm resistor Used in several projects. **D** 220QBK-ND, **J** 690700, **F** 9339299, **RS** 707-7612
- 1-kilohm resistor Used in several projects. **D** 1.0KQBK-ND, **J** 690865, **F** 9339051, **RS** 707-7666
- 10-kilohm resistors Used in projects 2, 6, 10. **D** 10KQBK-ND, **J** 691104, **F** 9339060, **RS** 707-7745
- 47-kilohm trimmer potentiometer Used in project 14. **D** A105657-ND, **J** 254028, **RS** 186-205
- 10-kilohm potentiometers Used in several projects. **J** 29082, **SF** COM-09939, **F** 350072, **RS** 249-9294
- 270-kilohm resistor Used in project 31. **J** 691446, **D** CF14JT270KCT-ND, **RS** 845-7577
- LEDs Used in projects 7, 8 & 9, 11, 14, 26. **D** 160-1144-ND or 160-1665-ND, **J** 34761 or 94511, **F** 1855510, **RS** 228-5972 or 826-830, **SF** COM-09592 or COM-09590
- RGB LED, common cathode Used in projects 1, 4. **D** 754-1492-ND, **J** 2125181, **SF** COM-00105, **F** 2290374, **RS** 861-4290
- Infrared LED Used in projects 10, 12. **J** 106526, **A** 387, **SF** COM-09469, **F** 1716710, **RS** 577-538, **SS** MTR102A2B
- 5V regulator Used in project 14. **J** 51262, **D** LM7805CT-ND, **F** 9756078, **RS** 918-1971
- 3.3V regulator Used in project 14. **D** 497-1491-5-ND, **J** 242115, **F** 1703357, **RS** 438-4885
- 1µF capacitor Used in project 14. **D** 1189-1324-ND, **J** 94161, **F** 8126933, **RS** 475-9009
- 10µF capacitor Used in projects 14, 15. **D** P11212-ND, **J** 29891, **F** 1144605, **RS** 762-1736
- 100µF capacitor Used in project 14. **D** P10269-ND, **J** 158394, **F** 1144642, **RS** 762-1746
- TIP120 Darlington NPN transistor Used in projects 14, 27. **D** TIP120-ND, **J** 32993, **F** 9804005 **RS** 808-0502
- 9V battery Used in project 3.
- 9–12V DC power supply Used in project 14. **SF** TOL-00298, **AF** 798, **J** 170245, **F** 1176248
- 3.3V-5V Battery Supply Used in project 14. **AF** 771, **D** BC4AAW-ND, **SS** 320180002
- Lithium Polymer battery Used in projects 3, 13. **SF** PRT-13813 or PRT-08483; **AF** 258 or 2011
- CR2032 Battery Holder and Battery Used in project 31. **SF** DEV-10730 or DEV-13883 with battery PRT-00338, **AF** 1870 or 1871 with 654.

Specialty Components

- Voltmeter Used in project 7. **SF** TOL-10285, **F** 1015878, **RS** 244-890

- **Oscilloscope** Used in project 10. **SS** 109990013, **SF** TOL-11702, **AF** 468
- **IR phototransistor** Used in projects 10, 12. **D** 365-1068-ND, **RS** 654-8542
- **WS2812 programmable LEDs (NeoPixels)** Used in project 13. **AF** 2226, 2858, or 2859 **D** 1528-1610-ND, **J** 2247947, **SF** BOB-13282, **SS** 104990139
- **Belkin WeMo Switch modules** Used in project 26. **B** P-F7C027
- **Relay Control Kit** Used in projects 29, 30. **SF** KIT-13815
- **Solenoid Lock** Used in project 27. **AF** 1512 **AMZ** "uxcell DC 12V Open Frame Type Solenoid for Electric Door Lock"
- **PowerSwitch Tail** Alternative part used in projects 29, 30. **SF** COM-10747, **AF** 2935, **PS** 240vac-kits (240V AC version)
- **4N35 optoisolator** used in projects 29, 30. **J** 41056, **F** 1244500, **D** 160-1304-5-ND, **RS** 597-302 Alternatively, SparkFun makes a breakout board. **S** BOB-09118

Sensors

- **Flex Sensor resistors** Used in projects 2, 3. **D** 905-1000-ND, **J** 150551, **SF** SEN-10264, **AF** 182, **RS** 708-1277
- **Momentary switches or pushbuttons** Used in projects 2, 8, 9, 10. **D** GH1344-ND or SW400-ND, **J** 2231822 or 119011, **SF** COM-09337, **F** 1634684, **RS** 718-2213
- **Rotary encoder with pushbutton** Used in projects 8 & 9. **AF** 377, **SF** COM-10982 with BOB-11722, **SS** 311130001
- **Force-sensing resistors, Interlink 400 series** Used in project 5. **D** 1027-1001-ND, **J** 2128260, **SF** SEN-09375, **A** 166
- **USB camera** Used in projects 5, 21, 22, 23.
- **Photocells (light-dependent resistors)** Used in project 6. **D** PDV-P9200-ND, **J** 202403, **SF** SEN-09088, **F** 7482280
- **Hanwei gas sensor** Used in project 14. **SF** SEN-09405, **P** 1481, **PX** 605-00009
- **Sharp GP2Y0A21YK infrared distance ranger** Used in projects 12, 14 **J** 2150256, **D** 425-2063-ND, **AF** 164, **F** 1243869, **RS** 666-6564
- **Ultrasonic rangefinder, SRF04, HC-SR04, or equivalent** Used in project 16. **SF** SEN-13959, **D** 1568-1421-ND
- **GPS receiver** Used in project 18. **AF** 746, **SF** GPS-1275
- **Garmin GLO GPS** available from buy.garmin.com.
- **Bad Elf GPS Pro+** available from bad-elf.com
- **ST Microelectronics LSM303DLH digital compass** Used in projects 19, 20. **AF** 1120, **SF** BOB-13303, **P** 1250, **SS** 101020081
- **LED tactile button** Used in projects 8 & 9. **SF** COM-10443 and BOB-10467
- **DTH11 Temperature and Humidity** Sensor used in projects 29, 30. **AF** 386, **J** 2245415, **SS** 101020011

Miscellaneous

- **Rubber bumpers** **D** 3M156065-ND, **RS** 120-6041, **J** 2119718, **A** 550 , **SF** COM-10594, **F** 1165068
- **hex standoffs, 4–40 threads female-female** Used in projects 8, 26, 27. You may need different part numbers to match your project's particular height (standoffs, also known as spacers, come in different body heights), but these will get you started on a search. **D** 36-2204-ND, **RS** 123-6835, **F** 2301244
- **4–40 thread pan head machine screws** **D** 36-9300-ND, **RS** 274-5086, **F** 2500400
- **Ping-pong ball** Used in project 1.
- **Small pink monkey** Used in projects 2, 3.
- **Cat (Felis silvestris catus)** Used in projects 5, 29, 30.
- **Cat mat** Used in project 5.
- **Thick pieces of wood or thick cardboard, about the size of the cat mat** Used in project 5.
- **Lighting filters** Used in project 6.
- **IR camera remote** Used in projects 10, 12.
- **Frosted plexi tube, 4" diameter** Used in project 13.
- **Candy tin, 4" diameter** Used in project 13.
- **Cymbal monkey** Used in project 14.
- **IC hooks with Pigtails** Used in project 31. **SF** CAB-09741 and CAB-00501
- **Sewable Snaps, 5mm diameter** Used in project 31. **AF** 1126 **SF** DEV-11347
- **Conductive Fabric** Used in project 31. **AF** 1168 **L** 1220
- **Thick conductive thread** Used in project 31. **AF** 603, **SF** DEV-11791 **L** 304
- **Hoodie** Used in project 31.
- **Embroidery thread** Used in project 31.

Hardware

This list includes vendors for current and past editions which are still doing business.

Abacom Technologies
Abacom sells a range of RF transmitters, receivers, and transceivers, and serial-to-Ethernet modules.

www.www.abacom-tech.com
email: abacom@abacom-tech.com

Acroname Robotics
Acroname sells a wide variety of sensors and actuators for robotics and electronics projects. They've got an excellent range of esoteric sensors like UV-flame sensors, cameras, and thermal-array sensors. They've got a lot of basic distance rangers as well. They also have a number of good tutorials on their site on how to use their parts.

www.acroname.com
email: info@acroname.com

Adafruit Industries
Adafruit makes a number of useful open source DIY electronics kits, including an AVR programmer, an MP3 player, and more.

www.adafruit.com
email: sales@adafruit.com

Arduino Store
The Arduino Store sells Arduino microcontroller boards, shields, and accessories for Arduino.

store.arduino.cc

Atmel
Atmel makes the AVR microcontrollers that are at the heart of the Arduino, Wiring, and BX-24 modules and many others. Atmel is now owned by Microchip.

www.atmel.com

Charley Chimp
Come on, where else are you going to get a cymbal clanging monkey?

charleychimp.com
email: customercare@charleychimp.com

CoreRFID
CoreRFID sells a variety of RFID readers, tags, and other RFID products.

www.rfidshop.com
e-mail: info@corerfid.com

D-Link

D-link makes a number of USB, Ethernet, and WiFi products, including the D-link IP-based WiFi camera used in Chapter 10.

www.dlink.com
e-mail: sales@dlink.com

Devantech/Robot Electronics
Devantech makes ultrasonic ranger sensors, electronic compasses, LCD displays, motor drivers, relay controllers, and other useful add-ons for microcontroller projects.

www.robot-electronics.co.uk
e-mail: sales@robot-electronics.co.uk

Digi
Digi makes XBee radios, radio modems, and Ethernet bridges.

www.digi.com

Digi-Key Electronics
Digi-Key is one of the U.S.'s largest retailers of electronics components. They're a staple source for things you use all the time—resistors, capacitors, connectors, some sensors, breadboards, wire, solder, and more.

www.digikey.com

ELFA
ELFA is one of Northern Europe's largest electronics components suppliers.

www.elfa.se

Farnell/Element14
Farnell supplies electronics components for all of Europe. Their catalog part numbers are consistent with the online store Newark, in the U.S., so if you're working on both sides of the Atlantic, sourcing Farnell parts can be convenient.

uk.farnell.com
email: sales@farnell.co.uk

Figaro USA, Inc.
Figaro Sensor sells a range of gas sensors, including volatile organic-compound sensors, carbon monoxide sensors, oxygen sensors, and more.

www.figarosensor.com
email: figarousa@figarosensor.com

Future Technology Devices International, Ltd. (FTDI)
FTDI makes a range of USB-to-Serial adapter chips, including the FT232RL that's been popular with makers for many years.

www.ftdichip.com
email: sales1@ftdichip.com

Glolab

Glolab makes a range of electronic kits and modules, including several useful RF and IR transmitters, receivers, and transceivers.

www.glolab.com
email: lab@glolab.com

Gridconnect

Gridconnect distributes networking products, including those from Lantronix.

www.gridconnect.com
email: sales@gridconnect.com

Images SI, Inc.

Images SI sells robotics and electronics parts. They carry a range of RFID parts, force-sensing resistors, stretch sensors, gas sensors, electronic kits, speech-recognition kits, solar energy parts, and microcontrollers.

www.imagesco.com
email: imagesco@verizon.net

Interlink Electronics

Interlink makes force-sensing resistors, touchpads, and other input devices.

www.interlinkelectronics.com
email: specialty@interlink electronics.com

IOGear

IOGear make computer adapters. TheirUSB-to-Serial adapters are good, and they carry Powerline Ethernet products.

www.iogear.com
email: sales@iogear.com

Jameco Electronics

Jameco carries bulk and individual electronics components, cables, breadboards, tools, and other staples for the electronics hobbyist or professional.

www.jameco.com
email: domestic@jameco.com
international@jameco.com
custservice@jameco.com

Lantronix

Lantronix makes serial-to-Ethernet modules: the XPort, the MatchPort, and many others.

www.lantronix.com
email: sales@lantronix.com

Libelium

Libelium makes an XBee-based product and other wireless products.

www.libelium.com

Linx Technologies

Linx makes a number of RF receivers, transmitters, and transceivers.

www.linxtechnologies.com
email: info@linxtechnologies.com

Low Power Radio Solutions

LPRS makes a number of RF receivers, transmitters, and transceivers.

www.lprs.co.uk
email: info@lprs.co.uk

Maker SHED

Launched originally as a source for back issues of MAKE Magazine, the Maker SHED now has a variety of stuff for makers, crafters, and budding scientists.

www.makershed.com
email: help@makershed.com

Maxim Integrated Products

Maxim makes sensors, communications chips, power-management chips, and more. They also own Dallas Semiconductor. Together, they're one of the major sources for chips related to serial communication, temperature sensors, LCD control, and much more.

www.maximintegrated.com
email: info2@maxim-ic.com

Microchip

Microchip makes the PIC family of microcontrollers. They have a very wide range of microcontrollers, for just about every conceivable purpose. They own Atmel now as well.

www.microchip.com

Mouser

Mouser is a large retailer of electronic components in the U.S. They stock most of the staple parts used in the projects in this book, such as resistors, capacitors, and some sensors. They also carry the FTDI USB-to-Serial cable.

www.mouser.com
email: help@mouser.com

NetMedia

NetMedia makes the BX-24 microcontroller module and the SitePlayer Ethernet module.

www.basicx.com
siteplayer.com
email: sales@netmedia.com

Newark/Element14

Newark supplies electronics components in the U.S. Their catalog part numbers are consistent with Farnell/Element14 in Europe, so if you're working on both sides of the Atlantic, sourcing parts from Farnell and Newark can be convenient.

www.newark.com
email: order@newark.com

New Micros

New Micros sells a number of microcontroller modules. They also sell a USB-XBee dongle that allows you to connect Digi XBee radios to a computer really easily. Their dongles also have all the necessary pins connected for reflashing the XBee's firmware serially.

www.newmicros.com
email: nmisales@newmicros.com

Parallax

Parallax makes the Basic Stamp family of microcontrollers. They also make the Propeller microcontroller, and a wide range of sensors, beginners' kits, robots, and other useful tools for people interested in electronics and microcontroller projects.

www.parallax.com
email: sales@parallax.com

Phidgets

Phidgets makes input and output modules that connect desktop and laptop computers to the physical world.

www.phidgets.com
email: sales@phidgets.com

Pololu

Pololu makes a variety of electronic components and breakout boards for robotics and other projects.

www.pololu.com
email: www@pololu.com

RS Online

RS Online is one of Europe's largest electronics retailers. They sell worldwide.

www. rsonline.com
email: general@rs-components.com

Samtec

Samtec makes electronic connectors. They have a very wide range of connectors, so if you're looking for something odd, they probably make it.

www.samtec.com
email: info@samtec.com

Seeed Studio

Seeed Studio makes many useful and inventive open source electronics parts. They also offer manufacturing services, PCB production services, and more.

www.seeedstudio.com
email: order@seeed.cc

SkyeTek

SkyeTek makes RFID readers, writers, and antennas. They have been acquired by Jadaktech, which makes a number of products for healthcare OEMs.

www.skyetek.com
www.jadaktech.com

Smarthome

Smarthome makes a wide variety of home-automation devices, including cameras, appliance controllers, X10, and INSTEON.

www.smarthome.com
email: custsvc@smarthome.com

SparkFun Electronics

SparkFun makes it easier to use all kinds of electronic components. They make breakout boards for sensors, radios, and power regulators, and they sell a variety of microcontroller platforms.

www.sparkfun.com
email: customerservice@sparkfun.com

Symmetry Electronics

Symmetry sells ZigBee and Bluetooth radios, serial-to-Ethernet modules, WiFi modules, cellular modems, and other electronic communications devices.

www.semiconductorstore.com

TI-RFID

TIRIS is Texas Instruments' RFID division. They make tags and readers for RFID in many bandwidths and protocols.

www.tiris.com

Trossen Robotics

Trossen Robotics sells a range of RFID supplies and robotics. They have a number of good sensors, including Interlink force-sensing resistors, linear actuators, Phidgets kits, RFID readers, and tags for most RFID ranges.

www.trossenrobotics.com
email: trsupport@trossenrobotics.com

Software

Most of the software listed in this book is open source. Many of the software platforms here are no longer used in the current edition, but are still useful in general.

Arduino

Arduino is a programming environment for AVR microcontrollers. It's based on Processing's programming interface. It runs on macOS, Linux, and Windows operating systems.

www.arduino.cc

Asterisk

Asterisk is a software private branch exchange (PBX) manager for telephony. It runs on Linux and Unix operating systems.

www.asterisk.org

AVRlib and avr-gcc

AVRlib is a library of C functions for a variety of tasks using AVR processors. It runs on macOS, Linux, and Windows operating systems as a library for the avr-gcc compiler.

www.nongnu.org/avr-libc

The GNU avr-gcc is a C compiler and assembler for AVR microcontrollers. It runs on macOS, Linux, and Windows operating systems. The simplest way to use it on Windows is to use AVR Studio: www.atmel.com/tools/atmelstudio.aspx. On Linux, you can download the toolchain here: www.atmel.com/tools/atmelavrtoolchainforlinux.aspx, and on macOS, CrossPack for AVR does the job: www.obdev.at/products/crosspack. The avr-gcc toolchain is built into the Arduino IDE, so when you're using the IDE, you're using avr-gcc.

ble-central

ble-central is a Bluetooth LE central plugin for PhoneGap and Cordova.

github.com/don/cordova-plugin-ble-central/

CoolTerm

CoolTerm is a freeware (though not open source) serial terminal application for macOS and Windows, written by Roger Meier.

freeware.the-meiers.org

Cygwin

Cygwin is a POSIX-like command-line interface for Windows. It allows you to use many Linux command-line tools on Windows.

www.cygwin.com

Dave's Telnet

Dave's Telnet is a telnet application for Windows.

dtelnet.sourceforge.net

Eclipse

Eclipse is an integrated development environment (IDE) for programming in many different languages. It's extensible through a plug-in architecture, and there are compiler links to most major programming languages. It runs on macOS, Linux, and Windows.

www.eclipse.org

Express.js

Express.js is a JavaScript web server framework for node.js.

expressjs.com

GitHub

GitHub is a host for git, a version-control tool for programming source code of any language. Git and GitHub are great tools to share your code.

git-scm.com
github.com

Java

Java is a programming language. It runs on macOS, Linux, and Windows operating systems, and many embedded systems as well. Originally developed by Sun Microsystems, it has been acquired by Oracle.

http://www.oracle.com/technetwork/java

JavaScript

JavaScript is a fast, lightweight scripting language. It is one of the three core technologies of web development, alongside HTML and CSS. It's also a great language for command-line and embedded programming, thanks to node.js.

developer.mozilla.org/en-US/docs/Web/JavaScript

libnfc

libnfc is an API for Near Field Communication. It's available on macOS, Windows, and POSIX systems.

nfc-tools.org

Max/MSP

Max is a commercial graphic data-flow authoring tool. It allows you to program by connecting graphic objects rather than writing text. Connected with Max are MSP, a real-time audio-signal processing library, and Jitter, a real-time video-signal processing library. It runs on macOS and Windows operating systems.

www.cycling74.com

Mosquitto

Mosquitto is a cross-platform MQTT broker application. It's available for macOS, Windows, and POSIX systems,

mosquitto.org

noble

noble is a Bluetooth LE central API for node.js. Its complement, bleno, is a Bluetooth LE peripheral API.

github.com/sandeepmistry/noble

Node.js

Node.js is a JavaScript development for server, desktop, and embedded environments.

nodejs.org

Nodemailer

Nodemailer is an email client library for node.js.

nodemailer.com

node-serialport

Node-serialport is an asynchronous serial port library for node.js.

github.com/EmergingTechnologyAdvisors/node-serialport

npmjs.org

npmjs.org is the Node Package Manager repository. It hosts or links to a wide variety of libraries for node.js, and other platforms.

npmjs.org

onoff

onoff is a library for node.js that enables GPIO control on embedded processors.

www.npmjs.com/package/onoff

p5.js

p5.js is a JavaScript framework based on Processing. It is browser-based, and designed as an introductory programming environment. It's a good way to make graphic interfaces in a browser.

p5js.org

PEAR

PEAR is the PHP Extension and Application Repository. It hosts extension libraries for PHP scripting.

pear.php.net

PHP

PHP is a scripting language that is especially suited for web development and can be embedded into HTML. It runs on macOS, Linux, and Windows operating systems.

www.php.net

PicBasic Pro

PicBasic Pro is a commercial BASIC compiler for PIC microcontrollers. It runs on Windows.

melabs.com

PhoneGap/Cordova

PhoneGap, owned by Adobe, is a framework for programming native apps on mobile phone operating systems using HTML, CSS, and JavaScript. Cordova is the open source reference implementation of PhoneGap.

phonegap.com
cordova.apache.org

Processing

Processing is a programming framework and environment designed for the nontechnical user who wants to program images, animation, and interaction. Based on Java, it runs on macOS, Linux, and Windows.

www.processing.org

Puredata (PD)

Puredata (PD) is a graphic data-flow authoring tool. It allows you to program by connecting graphic objects rather than writing text. It's developed by one of the original developers of Max, Miller Puckette. It runs on macOS, Linux, and Windows operating systems.

puredata.info

PuTTY SSH

PuTTY is a telnet/SSH/serial port client for Windows.

www.puttyssh.org

QRCode.js

QRCode.js is a QR code reading library for node.js.

github.com/IagoLast/qrcodejs

Dan Shiffman's Libraries

Dan Shiffman has written a number of useful libraries for Processing and p5.js. He's also made countless tutorials for beginning programmers.

github.com/shiffman
shiffman.net

Tor

Tor is a browser and network that enables anonymous browsing through a network of proxy nodes.

www.torproject.org

Tracking.js

Tracking.js is a face tracking library for node.js.

trackingjs.com

Twilio

Twilio is a commercial IP telephony provider. They provide application programming interfaces that allow you to connect telephone calls to web applications.

www.twilio.com

Visual Studio

Visual Studio is Microsoft's main IDE for Windows-based development. The community edition is free, and useful to keep on your Windows machine, since it installs many programming frameworks and tools on which other programming tools depend.

www.visualstudio.com

Wiring

Wiring is a programming environment for AVR microcontrollers. It's based on Processing's programming interface. It runs on macOS, Linux, and Windows operating systems.

www.wiring.org.co

ws

ws is a WebSocket library for node.js

github.com/websockets/ws

XCode

XCode is Microsoft's main IDE for macOS- and iOS-based development. It's free, and useful to keep on your macOS machine, since it installs many programming frameworks and tools on which other programming tools depend.

developer.apple.com/xcode

Index

J

K

L

M

Q

R

X

Y

Z